U0632026

现代医学分子生物学双语精编
Medical Molecular Biology

伍欣星 李 晖 赵 旻 主编

国家自然科学基金资助（30772308）

科学出版社

北 京

内 容 简 介

本书的编写具有鲜明特色，采用了中、英文双语形式，共精编了 11 章内容，分别用中、英文阐述了当前医学分子生物学的基础知识和前沿进展，包括基因组、转录组、蛋白质组学的基本概念，RNA 及非编码 RNA、基因表达的调控、细胞周期与细胞凋亡、细胞信号转导、肿瘤分子生物学及生物信息学等基本原理和应用。

本书是为帮助我国医学院校高年级学生（七年制、八年制）和硕士、博士研究生学习医学分子生物学的基础知识，熟练运用专业英语查阅国外文献，从事科学研究而专门编写的教材，旨在帮助那些具备一定英语基础并初步掌握医学分子生物学知识的人士提高生物医学专业水平及专业英语水平。

图书在版编目（CIP）数据

现代医学分子生物学双语精编/伍欣星等主编. —北京：科学出版社，2009

ISBN 978-7-03-025153-4

Ⅰ. 现… Ⅱ. 伍… Ⅲ. 医药学：分子生物学-汉、英 Ⅳ. Q7

中国版本图书馆 CIP 数据核字（2009）第 134101 号

责任编辑：李 悦 陈珊珊/责任校对：钟 洋
责任印制：钱玉芬/封面设计：陈 敬

科学出版社 出版

北京东黄城根北街 16 号
邮政编码：100717
http://www.sciencep.com

新蕾印刷厂 印刷
科学出版社发行 各地新华书店经销

*

2009 年 8 月第 一 版　　开本：787×1092 1/16
2009 年 8 月第一次印刷　　印张：28
印数：1—2 500　　　　　字数：627 000

定价：55.00 元

（如有印装质量问题，我社负责调换〈长虹〉）

编者名单

主　编　伍欣星　李　晖　赵　旻

编　者（按姓氏笔画排序）

王又又　武汉大学医学部

方　勤　中国科学院武汉病毒研究所

左泽华　武汉大学医学部

朱　涛　中国科学技术大学

伍欣星　武汉大学医学部

孙　军　华中科技大学同济医学院

李　晖　武汉大学医学部

邱小萍　武汉大学医学部

汪亚平　中国科学院水生生物研究所

陈　庆　武汉大学医学部

张　翼　武汉大学生命科学学院

武军驻　武汉大学医学部

赵　旻　武汉大学医学部

姚　军　台州学院医学院

袁　萍　华中科技大学同济医学院

前　言

为帮助我国医学院高年级学生（七年制、八年制）、硕士研究生和博士研究生学习医学分子生物学的基础知识，熟练运用专业英语查阅国外文献，从事科学研究，编者专门编写了本教材。

分子生物学是一门新兴的、发展迅速的学科。从诞生到取得巨大成就的数十年里，其为生命科学的研究带来了诸多令人振奋的科学技术和成果。分子生物学的理论和技术手段，已经深入到生命研究领域的各个层面。因此对每一位相关领域的研究人员来说，掌握足够的分子生物学知识是开展研究工作的基础。人类对待科学技术不应仅仅停留于使用的层面，而应懂得必要的原理，只有这样才能真正掌握并运用自如。这就是编写本书的初衷，即通过对基础知识、原理和技术方法的介绍，为医学分子生物学这门课程的学习、应用提供简洁、高效的平台。

20 世纪 50 年代 DNA 双螺旋结构的发现到 90 年代人类基因组计划的实施，以及目前后基因组时代的深入研究，都是人类在生命科学研究领域的划时代成就。分子生物学在这些研究中起着重要作用，其自身也得到了巨大的发展。从分子水平探求生命现象的奥秘，曾是每位研究者梦寐以求的方式；而现在以分子方式探寻生命活动规律、分析繁杂的生物信息，则已经成为相关科学研究的基本模式。因此，如何透彻、高效地学习和掌握分子生物学的理论和技术已经成为分子生物学教学工作的重点，如同万丈高楼建设的关键在于扎实的根基一样。为人所熟悉的分子杂交、PCR、DNA 测序等分子生物学技术，是许多新兴的、更为复杂技术的基础。它们是如何创立、改进，如何使用并衍生出新的技术等问题，都将是本书描述的主要内容。

本书的编写有着鲜明的特色，采用了中、英文双语形式。众所周知，欧美国家的课程和教学体系与我国有很大差异，特别是在基础教育部分，其教材中的部分内容和我们的教学需求对接不上。我们试图以国外优秀教材的内容为基础，根据国内的教学情况进行改编、自编、翻译，使其尽量接近我国教学的需求。这种本土化程度更高的双语教材的出版，反映了我国专业教材从全中文—全英文（原版引进）—原版加少量中文注释—原版编辑加中文翻译需求的变化，方便读者在学习分子生物学基础知识的同时，提高专业英语水平。这对于相关文献的阅读和科技论文的书写具有重要意义，这也是编者在多年教学工作中的深刻体会和收获。

本书精编了 11 个单元（章），分别用中、英文阐述了当前医学分子生物学的基础知识和前沿进展，包括基因组、转录组、蛋白质组学的基本概念，RNA 及非编码 RNA、基因表达的调控、细胞周期与细胞凋亡、细胞信号转导、肿瘤分子生物学、生物信息学等基本原理和应用。本书除介绍基因工程、分子杂交、基因转移等经典实验技术外还涉及分子生物学前沿技术，如染色质免疫共沉淀、噬菌体展示、质谱分析等。限于篇幅，本书无法对许多重要理论和实验进行详述，而往往倾向于陈述事实和阐明原理。

本书的主要编者均是从事生物化学与分子生物学教学工作多年的教师，或是承担着国家"863"计划、"973"计划、国家自然科学基金课题的科研工作者，绝大多数有着国外留学、工作和参加国际学术交流的经历。编者对如何学习并运用双语从事医学分子生物学研究的重要性有着深刻的体会和认识，他们也更加了解中国学生和读者的需求。

在本书全部完稿之际，编者要特别感谢美国密歇根州立大学（Michigan State University）魏赛男助理教授，在百忙之中审阅本书的英文部分；感谢美国乔治城大学（Georgetown University）病理学系刘学锋助理教授，在教学和科研方面，给予的最直接帮助；感谢华中科技大学冯作化教授，华中农业大学郑用琏教授，许多疑惑都是从他们那里得到了启发；感谢中国科技大学朱涛教授，在百忙中翻译了第五章英文书稿；编者还要感谢苏建国博士后，吴启家、周宇博士，博士研究生谭胜，硕士研究生佘茜、钱鹏旭；他们年轻有为，充满活力，毫无怨言地协助他们导师承担了资料查询、文字和图表的修改及校对等工作。本书的出版得到了国家自然科学基金及病毒学国家重点实验室经费资助，在此一并感谢！

由于分子生物学是一门发展十分迅速的学科，大量新的理论和技术不断涌现，而以往的知识也在不断地更新，因此要编写好这本医学分子生物学双语教材是十分困难的，书中的不妥之处敬请广大读者和专家批评、指正。

伍欣星　李　晖　赵　旻

2009 年 4 月

Preface

This book is used as a textbook specially for senior medical students (seven-year or eight-year), Master and PhD graduate students, as well as specialized talents who study basic knowledge of medical molecular biology. It can be also used by skillfully professionals for their English literature search and academic research. Our aim is to help those who possess the necessary professional English and the preliminary molecular biological knowledge to improve their bio-medicine knowledge and professional English.

Molecular biology is a new but fast-growing subject. The great achievements have been made since its birth. It has brought a lot of exciting science, technology and results for the life science research. The theory and technology of molecular biology have penetrated into various aspects and levels of our life from the unattainable Hall of Science research. Therefore, for every researcher, possessing enough knowledge of molecular biology is the basis of studies. Individuals dealing with science and technology should not only stop at the level of application, but also need to understand the principles and explore its theoretical principal; this is the only way to really grasp and tools the molecular biological knowledge. We are here providing simple and efficient platform for studying and application of this course via introduction of the basic knowledge, principles and technical methods of molecular biology. Indeed, this is the original intention to write this molecular biology book.

From the DNA double helix structure in 1950s to the beginning of the Human Genome Project in 1990s, as well as in-depth research of post-genomic era in current, all of them are the epoch-making achievements in the field of human life science research. With the paces of such achievements, molecular biology has played an important role and obtained great achievement. To explore the phenomenon of the mysteries of life from the molecular level is the dream of all researchers; but now it has become a basic study model to explore the molecular law of life activities and analyze complex biological information. Therefore, how thoroughly and efficiently to study molecular biology has become the focus of the teaching work, as a solid foundation is the key to the high-rise building. As we all know, the hybridization, PCR, DNA sequencing and other molecular biology techniques which are known to people, are the basis of some more complex technologies. Their invention, advancement and usage would all be described in this book.

Bilingual version in both Chinese and English is the distinct characteristics of this book. As we all known, European and American countries' curriculum and teaching system were quite different from Chinese, particularly in the part of elementary education, portion of the contents of its teaching material is not matched with our teaching requirement. According to the domestic teaching situation, we attempt to take the contents of overseas outstanding teaching material as the foundation, carry on the reorganization, arrangement and translation, in order to match up with our national teaching demand as much as possible. The publication of such a highly localized bilingual teaching materials, has reflected the required movement of our national specialized teaching material from the entire Chinese, to entire English (first edition introduction) to the first edition, by adding a few Chinese annotations to the second edition, then adding the Chinese translation. At the same time for learning basic knowledge of molecular biology, readers can study the professional English. It is very important for reading literatures and writing scientific papers, this is also a deep harvest and teaching experiences of all editors for many years.

This book is composed of eleven chapters written in Chinese and English separately, both of them covered medical molecular biology fundamental knowledge and the current progression including basic concepts of genome, transcriptomics and proteomics; RNA and expressional regulation of uncoding RNA; cell cycle and apoptosis; signal transmit network, epigenetics decoration, tumor and application of bioinformatics. Besides gene engineering, molecule hybridization, transgenesis and other classic techniques, this book also refers to the front edge of molecular biology, such as chromatin coimmunoprecipitation, phage display, mass spectral analysis, and so on. Due to space limitations, this book can not detail the important theories and experiments at most and often tend to clarify the principles and a statement of fact.

The main authors of the book were engaged in teaching work on biochemistry and molecular biology for many years, and some of them have undertaken the researches supported by the 863, 973, National Natural Science Foundation. Most of the authors possess experiences of overseas study, work and international academic network, this grant them with clear understanding of the needs of Chinese students and readers more than do foreign experts.

The authors would like to thank the following professionals for their contributions on the book's publication in the past few years. They are Sainan Wei, assistant professor in the department of Pediatrics and Human Development at Michigan State University and Director of Prenatal Screening, DNA Diagnostics and Cytogenetics Laboratories, who spent their time on the revision of the English version. Dr. Xuefeng Liu, assistant professor of Department of Pathology, Georgetown University. Either teaching or research, he gave the most direct help to the editors; Zuohua Feng professor of the Huazhong University of Science and Technology; Yonglian Zheng, professor of Huazhong Agricultural University. They kindly gave us lots of inspiration and help; We gratitude to Dr. Tao Zhu, professor of University of Science and Technology of China, who in charge of the manuscript translation of the fifth chapter; At the same time, we must give our cordial thanks to post doctoral fellow Dr. Jian Guo Su, Dr. Qijia Wu, Dr. Yu Zhou, doctorate candidates Sheng Tan, and master graduate students Qian She, Pengxu Qian ; they sincerely offer assistance to the authors for querying data, writing, editing charts and proof-reading . The publication of this book was supported by National Natural Science Foundation and State Key Laboratory of Virology of China . We would also like to thank them for all these supports.

Finally, we must mention that the molecular biology is really a fast growing science, a large number of new theories and technologies are constantly emerging and, in turn, the previous knowledge has been updated and revised. We have tried our best to prepare this text book in both Chinese and English, the errors and inappropriateness in the book are still unavoidable. We would greatly appreciate if readers and experts bring us the criticism and correction.

Xinxing Wu, Hui Li, Min Zhao
April, 2009

目　录

Contents

第一章 概 论

一、分子生物学基本概念

分子生物学（molecular biology）是从分子水平研究生物大分子（核酸、蛋白质）的结构与功能，从而阐明生命现象本质的学科。生物大分子均具有较大的相对分子质量，由简单的小分子，如核苷酸或氨基酸等排列组合而成并蕴藏各种生物信息，还具有复杂的空间结构以形成精确的相互作用系统，由此构成生物的多样性和生物个体生长发育和代谢调节控制系统。在分子水平上研究生命的本质则是指对遗传、生殖、生长和发育等生命基本特征的分子机制进行阐明。阐明这些复杂的结构及功能的关系是分子生物学的主要任务。作为 20 世纪发展最快的学科之一，分子生物学已经成为生命科学研究领域必不可少的强有力工具；同时生命科学及其相关学科的发展，也为分子生物学的发展不断融入新的理论、技术和方法。

二、分子生物学理论和技术发展史

20 世纪是分子生物学诞生和发展的重要时期，在过去的几十年里分子生物学的理论和技术均取得了许多重大的突破和进展，由此可将分子生物学的发展大致划分为三个阶段。

第一阶段 主要是指 19 世纪后期到 20 世纪 50 年代初这段时间，是现代分子生物学诞生的准备和酝酿阶段。在这个时期，通过两类学派的科学家们的协同努力，确定了蛋白质是生命的主要物质基础，而生物遗传的物质基础是 DNA。一派是以英国的 Astbury 等为代表的所谓结构学派（structurist），他们主要用 X 射线衍射技术研究蛋白质和核酸的空间结构，认为只有搞清生物大分子的三维结构，才能阐明生命活动的本质，分子生物学一词正是 Astbury 在 1950 年根据他的这一思想首先提出来的。另一派称为信息学派，他们着眼于遗传信息的研究。它的创始人之一，德国的 Delbruck 以噬菌体为研究对象，把噬菌体看作最小的遗传单位，研究其遗传信息的表达和调控，这一派也称为噬菌体学派。在这个时期，分子生物学研究的最重要成果是证明了遗传的物质基础是 DNA 而不是蛋白质，Avery 等（1944）证明了使肺炎双球菌由粗糙型转为光滑型的转化因子是 DNA。随后，噬菌体学派的 Hershey 和 Chase 进一步发现只有噬菌体的 DNA 被"注射"到细菌体内并在其中繁殖，而蛋白质则留在细胞之外。但在当时，对 DNA 的结构研究尚少，还未能确认 DNA 成为遗传物质的基础。

第二阶段 这一阶段是从 20 世纪 50 年代初到 70 年代初。以 1953 年 Watson 和 Crick 提出的 DNA 双螺旋结构模型作为现代分子生物学诞生的里程碑。DNA 双螺旋结构的发现正是结构学派和信息学派汇合所结出的硕果。双螺旋学说第一次用分子结构的特征解释生命现象的最基本问题之———基因复制的机制，从而使人类真正进入到分子生物学时代。在这以后 20 年里，许多重大的发现，如 mRNA 分子、DNA 聚合酶、RNA 聚合酶、限制性内切核酸酶、连接酶、质粒的利用以及遗传密码的破译等得以实现。在此基础上，确立了遗传信息传递的中心法则（central dogma），并建立了以中心法则为基础的分子遗传学（molecular genetics）基本理论体系。

第三阶段　此阶段的标志就是 20 世纪 70 年代初重组 DNA 技术的建立。这项技术使分子生物学家能够在体外按照主观意愿切割和拼接 DNA 分子，借助细菌制造大量所需的 DNA 片段，极大地促进了 DNA 本身结构和功能的研究。这项技术标志着人类改造生命、创建物种成为可能。

三、分子生物学研究内容

1. 核酸的结构与功能

研究内容包括核酸/基因组的结构，遗传信息的复制、转录与翻译，遗传信息的突变与修复，基因的表达调控以及基因工程技术的发展和应用等。其中基因组、基因的表达调控和重组 DNA 技术的应用是当前研究的重点。

基因组　测定一个生物基因组核酸的全序列对理解这一生物的生命信息及其功能具有重大意义，在此思想的指导下，基因组核酸序列的测定成为早期基因组研究的重点。1977 年 Sanger 测定了 ΦX174 基因组全部 DNA 序列；随后一些病毒包括乙型肝炎病毒、艾滋病毒等基因组的全序列也陆续被测定；1996 年底测出了大肠杆菌（*Escheridua coli*，*E.coli*）基因组 DNA 的全序列长约 4×10^6 个核苷酸。1990 年人类基因组计划（human genome project，HGP）开始实施，这是生命科学领域有史以来最庞大的研究计划，目前已经测定出人类基因组全部 DNA 序列，并确定人类约 30 000 个基因的一级结构，这些结果使人类进入到了后基因组时代。

基因表达调控　分子遗传学基本理论建立者 Jacob 和 Monod 提出的操纵子学说建立了基因表达调控模式研究的基础。从开始认识到原核生物基因表达调控的一些基本规律，到逐渐认识了真核基因组结构和调控的一些现象和复杂性，基因表达调控的基本规律正在被人类所逐渐认识。1977 年最先发现猴 SV40 病毒和腺病毒中编码蛋白质的基因序列的不连续性，内含子在真核基因表达调控中的作用功能得以发现。1981 年 Cech 等发现四膜虫 rRNA 的自我剪接，从而发现核酶（ribozyme）。另外，真核基因的顺式作用元件与反式作用因子、核酸与蛋白质间的分子识别与相互作用等，均是基因表达调控领域的重大发现。

重组 DNA 技术　限制性内切核酸酶和连接酶的发现和利用为基因工程提供了有力的工具；1972 年 Berg 等将 SV40 病毒 DNA 与噬菌体 P22DNA 在体外重组并转化大肠杆菌获得成功，使蛋白质的合成打破了种属界限。随后几十年内，大量真核蛋白通过重组 DNA 技术得以表达，包括重组人生长激素、人胰岛素、人干扰素、人白细胞介素-2、人集落刺激因子、重组人乙型肝炎疫苗等多种基因工程药物和疫苗进入了生产或临床应用，相关领域已成为医药业发展的重要方向，将对医学和工农业发展作出新贡献。转基因动植物和基因敲除动植物的成功也是基因工程技术发展的结果。基因诊断与基因治疗是基因工程在医学领域发展的另一个重要方面。伴随着基因工程的迅速发展，许多分子生物学新技术不断涌现，包括从手工到全自动化学合成核酸，DNA 序列的快速测定法，全自动核酸序列测定仪的问世，1985 年 Cetus 公司 Mullis 等发明的聚合酶链反应（polymerase chain reaction，PCR）等，它们对分子生物学的发展起到了重大的推动作用。

2. 蛋白质的分子生物学

　　蛋白质的分子生物学主要研究执行各种生命功能的主要大分子——蛋白质的结构与功能。尽管人类对蛋白质的研究比对核酸研究的历史要长得多，但由于其研究难度较大，与核酸分子生物学相比其发展较慢。近年来在认识蛋白质的结构及其功能关系方面取得了较大进展。

3. 细胞信号转导的分子生物学

　　细胞信号转导的分子生物学研究细胞内、细胞间信息传递的分子机制。在外源信号的刺激下，细胞可以将这些信号转变为一系列的生物化学变化，信号转导研究的目标是阐明这些变化的分子机制，明确每种信号转导与传递途径的作用和调节方式，认识各种途径间的网络控制系统。在理论和技术方面研究信号转导机制，是当前分子生物学发展最迅速的领域之一。细胞信号转导机制的研究可以追溯至 20 世纪 50 年代。1957 年 Sutherland 发现 cAMP，1965 年提出第二信使学说；1977 年 Ross 证实 G 蛋白的存在和功能，并将 G 蛋白与腺苷酸环化酶的作用联系起来，深化了对 G 蛋白偶联信号转导途径的认识。70 年代中期以后，癌基因和抑癌基因的发现，蛋白酪氨酸激酶的发现及其结构与功能的深入研究，各种受体蛋白基因的克隆和结构功能的探索，免疫活性细胞对抗原的识别及其活化信号的传递研究等，使细胞信号转导研究有了长足的进步。

四、分子生物学在医学中的应用及未来

　　医学分子生物学（medical molecular biology）已发展成为分子生物学最重要的分支和应用领域，是从分子水平研究人体在正常和疾病状态下生命活动规律的一门学科。它主要研究人体生物大分子——蛋白质和核酸的结构、功能及其相互作用，同疾病发生、发展的关系，探讨分子生物学在医学领域中形成的各专门研究领域或交叉学科的相关知识。在医学相关的各个学科中，包括微生物学、免疫学、病理学、药理学以及临床各学科，分子生物学都正在广泛地形成交叉与渗透，这些已成为现代医学重要的特征。所形成的分子检验医学、分子肿瘤学、分子病毒学、分子病理学、分子药理学等，大大促进了医学的发展。目前分子生物学技术和方法在临床医学中的应用包括遗传病诊断与治疗，产前筛查及干预治疗，肿瘤诊断，治疗及预后判断，微生物耐药机制的研究等方面，已经取得较突出的成绩。

1. 分子生物学在检验医学中的应用

　　20 世纪 90 年代起，分子生物学技术应用于临床检测。短短的几年中，分子生物学技术已深入到检验医学的多个方面。例如，在细菌鉴定及分型中常应用核酸扩增技术，核酸探针杂交技术以及新近开发的 16S rRNA 相关度分析等。这些技术具有敏感、特异、安全和快速等特点，在微生物检验中发挥着日益重要的作用。另外，在检测耐药基因方面分子生物学技术日益受到重视。临床上可用 PCR 方法检测耐药基因，来判断待检菌对某种抗菌药物是否具有耐药性，以及可能存在的沉默耐药基因。

2. 分子生物学在遗传学研究中的应用

　　分子生物学技术在遗传学研究中的应用已取得了许多进展，包括基因表达和基因突变的检测，真核基因表达调控机制的研究等许多方面。例如，流式细胞仪（flow cytometry）和图像细胞仪（image cytometry）等可以对细胞群体和特定蛋白或核酸的

含量进行分析；PCR 及其衍生技术、显微解剖法（microdissection）、DNA 芯片技术（DNA chip）可以分析基因的表达和结构；酶促切割错配法（enzyme mismatch cleavage，EMC）、切割片段长度多态性（cleavage fragment length polymorphism，CE-FLP）、双脱氧指纹图谱法（dideoxy fingerprinting，ddF）、错配接合蛋白质截短测试法（protein truncation test，PTT）等可对未知突变基因进行分析；引物延伸法（primer extension，PEX）、寡核苷酸链接检测法（oligonucleotide ligation assay，OLA）、毛细管电泳法（capillary electrophoresis，CE）等方法则可对已知突变基因进行分析。值得一提的是，目前几乎所有检测基因突变的分子诊断技术都是建立于 PCR 的基础之上，目前已达 20 余种，自动化程度、分析结果的准确性和精确性均有很大提高。

3. 分子生物学在肿瘤研究中的应用

肿瘤的发生、发展是医学分子生物学研究的重点。特别是在肿瘤的分子机制、易感性预测、早期诊断和预后监测等方面，分子生物学发挥着巨大作用。由于部分肿瘤的发生具有遗传学基础，因此肿瘤遗传易感性的检测对于肿瘤高危人群的筛检及确定具有重要价值；同时某些肿瘤的发生和病毒感染有关。例如，人乳头状瘤病毒（HPV）各亚型与良性、恶性子宫颈病变之间存在密切联系，因而相关病毒的检测对于监控肿瘤易患人群和预防治疗具有重要意义。目前，核酸杂交技术与 PCR 技术仍是有效的病毒检测手段。在肿瘤疗效监测上，分子生物学技术在判断肿瘤的微小病灶及某些白血病的疗效方面发挥重要作用，有助于采取或调整治疗措施及方案。例如，肿瘤治疗过程中会产生多药耐药性（multidrug resistance，MDR），因此用 PCR 法检测多药耐药性基因对疗效判断有帮助。而且分子生物学技术在肿瘤基因治疗中也起重要作用，基因修饰和基因替换等方法已经广泛应用于肿瘤的治疗。另外，基因的突变与扩增决定着患者的预后。例如，有研究表明 *ras*、*c-myc*、*p53* 等基因突变与大肠肿瘤、卵巢癌等的临床分期和预后有密切联系，因此分子生物学技术也是肿瘤预后判断的重要依据。

（赵　旻）

参 考 文 献

朱玉贤，李毅. 2002. 现代分子生物学. 第二版. 北京：高等教育出版社

Burkhard Ziebolz，Marcus Droege. 2007. Toward a new era in sequencing. Biotechnology Annual Review. 13：1～26

Cavalli-Sforza L L. 2005. The human genome diversity project：past，present and future. Nat. Rev. Genet. 6（4）：333～340

Malacinski G M. 2002. 分子生物学精要. 第三版. 北京：科学出版社

William C S Cho. 2007. Proteomics technologies and challenges. Genomics，Proteomics & Bioinformatics. 5（2）：77～85

Chapter 1　Introduction

1.1　The Basic Concepts of Molecular Biology

Molecular biology is the study of the structure and function of biological macromolecules (nucleic acid, protein). Biological macromolecules, arranged and assembled from smaller simple molecules such as amino acids and nucleotides, have a greater molecular weight, and contain a variety of biological information. They have complex structures, which account for biological diversity, regulate metabolism as well as control individual growth and development. To clarify the complex relationship between structure and function is the main task of molecular biology. As one of the fastest growing disciplines in the twentieth century, molecular biology has become a powerful tool in the field of science research and medicine; at the same time, life sciences and related disciplines have provided the new theories, techniques and methods for the continued development of molecular biology.

1.2　History of the Theory and Technology of Molecular Biology

The 20th century was an important period in the birth and development of molecular biology. The past few decades have seen the emergence of molecular biology as a major player in significant break-throughs and progress in the life sciences. The development of molecular biology can be broadly divided into three phases.

1.2.1　The First Stage

Mainly refers to the period from the late 19th century to the early 1950s. This period of time was the birth and preparation stage of modern molecular biology. During this period, two schools of scientists together determined that protein is the main material basis of life, and that genetic material is represented by DNA. One school is the so-called structurist, represented by William Astbury. Based on the notion that an understanding of the three-dimensional structure of biological macromolecules is necessary to clarify the nature of life, they used X-ray diffraction technique to study the space structure of protein and nucleic acid. The word "molecular biology" was first put forward according to Astbury's idea in 1950. Another school was called the school of information; they focused on genetic information. Germany's Delbruck, one of the founders of the school of information, used phage as the research material to study the expression and regulation of genetic information-the phage was regarded as the smallest genetic unit. This school is also known as the phage school. During this period, the most important mission of molecular biology was to prove that DNA rather than protein was the genetic material. Although Avery et al. (1944) proved that the rough-smooth conversion factor of pneumococcus is DNA, at that time, the structure of DNA was still unclear.

1.2.2　The Second Stage

This stage is from the early 1950s to early 1970s. In 1953, Watson and Crick made the DNA double helix model as a milestone of modern molecular biological. DNA double helix structure is the fruit of the merger of data from both the schools of structure and information. The double helix theory explained for the first time one of the most fundamental issues about the characteristic of life—the mechanism of gene replication at the molecular level. Following this, many important discoveries such as mRNA molecules, DNA polymerase, RNA polymerase, restriction endonuclease, ligase, as well as the use of the plasmid and deciphering of the genetic code could be achieved, Based on this, genetic information of central dogma was established, as well as the elementary theory system of molecular genetics, which based on central dogma. And a basic system of molecular genetics was established.

1. 2. 3　The Third Stage

The paradigm of this stage is recombinant DNA technology, which was established in the early 1970s. This technique allows molecular biologists to cut and splice DNA molecules accordingly, in vitro, and to manufacture large amounts of DNA fragments using bacteria. It greatly promotes research into the structure and function of DNA. This technique marked the transformation of life and made possible the creation of species.

1. 3　The Content of Molecular Biology

1. 3. 1　Structure and Function of Nucleic Acid

The study of the structure and function of nucleic acid includes reproduction of genetic information, transcription and translation, mutation and repair of genetic information, gene expression and regulation, development and applications of genetic engineering techniques. The genome, gene expression, regulation and genetic recombination technique are the focus of this study.

1. 3. 1. 1　Genome

Determination of nucleic acid sequences of a genome is of great significance in understanding the functions and regulation of a living organism's genetic information. Therefore analysis of nucleic acid sequences became the early focus of genomic research. In 1977, Sanger sequenced the 5375 bp of ΦX174 DNA; the whole genome sequences of some viruses have also been determined, including hepatitis B virus and HIV. By the end of 1996, the 4×10^6 bp of the *Escherichia coli* (*E. coli*) genomic sequence was determined. In 1990 the Human Genome Project started; it was the largest research project in the history of the field of life sciences. The genomic DNA of human being (3×10^9 bp) was sequenced and the primary structure of 30 000 human genes was identified The human race moved into the post-genome era.

1. 3. 1. 2　Regulation of gene expression

Jacob and Monod put forward the operon theory, paving the way for gene expression research. The knowledge of regulation of gene expression in prokaryotes gradually cumulated into an understanding of the structure and complexity of eukaryotic genome regulation. In 1977, the discontinuity of protein encoding sequence was first recognized in SV40 monkey virus and adenovirus genome, followed later by an understanding of the function of intron in eukaryotic gene expression. In 1981, Cech et al. observed self-editing in tetrahymena rRNA, leading to the discovery of ribozyme. Other important discoveries in the field of gene expression include *cis*-regulatory elements and *trans*-transcription factors of eukaryotic gene, as well as the interaction and molecular recognition between proteins and nucleic acids.

1. 3. 1. 3　Recombinant DNA technique

The discovery of restriction endonuclease and ligase provided a powerful tool for genetic engineering; in 1972, Berg reorganized SV 40 virus DNA and P22 phage DNA *in vitro* and successfully transformed *E. coli*, During subsequent decades, recombinant DNA technique was used to express a large number of eukaryotic proteins, including recombinant human growth hormone, insulin, human interferon, human interleukin 2, human colony-stimulating factor, recombinant hepatitis B vaccine, and others genetically engineered medicines and vaccines were produced and used in clinical trials. This field has become an important direction for the development of medical industry and has made new contributions to the development of industry and agriculture. Furthermore, transgenic gene or gene knockout animals and plants are developed using genetic engineering technology. Gene therapy and genetic diagnoses are other important aspects of medical science that have benefited from genetic engineering. Along with the rapid development of genetic engineering, many new technologies of molecular biology are emerging, including automated synthesis of nucleic acid. Also, the Cetus company Mullis invented polymerase chain reaction (PCR) in 1985. These have played a major role in promoting the development of molecular biology.

1. 3. 2　Molecular Biology of Protein

Protein molecular biology is concerned with understanding the structure and function of major molecules—protein's functions in life. Although the study of proteins started long before the study of nucleic acid, protein research has proven to be more complex hence protein molecular biology has progressed rather slowly compared with nucleic acid research. Nonetheless, in recent years, substantial progress has been made in understanding the relations between structure and function of proteins.

1. 3. 3 Molecular Biology of Cell Signal Transduction

Cell signal transduction is the molecular basis of intracellular transmission of information. Cells can transform outside stimulation signal into a series of biochemical reactions. The goal of signal transduction research is to clarify the molecular mechanism of these changes, to elucidate the role and regulation of each signal transduction, and to understand the network control system among signal pathways. The theory and technology of signal transduction is one of the fastest-growing areas in molecular biology. The study of cell signal transduction mechanism can be traced back to the 1950s. Sutherland found cAMP in 1957, and postulated the second messenger theory in 1965; Ross confirmed the existence and function of G-protein in 1977, and linked G-protein with the role of adenylylcyclase, shedding more light on G-protein-coupled signal transduction pathway. Since the mid-1970's, discovery of oncogenes and tumor suppressor genes, progress has been made in various aspects of cell signal transduction research. Areas of progress include in-depth study of the structure and function of protein tyrosine kinase, cloning and exploration of a variety of receptor gene structure and function, as well as study of the reorganization and activation of immune cells in response to antigens.

1. 4 Molecular Biology in Medicine

Molecular biology has been widely cross-penetrated into various medical-related disciplines, including microbiology, immunology, pathology, pharmacology, and other clinical disciplines; it has become an important component of modern medicine. The formation of a number of cross disciplines, such as molecular laboratory medicine, molecular oncology, molecular virology, molecular pathology, and molecular pharmacology and so on, greatly promoted the advancement of human medicine. There is a wide application of molecular biology in various medicine-related disciplines, including microbiology, immunology, pathology, pharmacology as well as clinical disciplines, making it an important component of modern medicine.

1. 4. 1 Application of Molecular Biology in Laboratory Medicine

Molecular biology has entered into the clinical diagnostic medicine since early 1990s. Just a few years later, techniques of molecular biology have important impact into medical detection in many aspects. For example, PCR and nucleic acid probes have been applied for the identification of bacterial type, diagnosis of genetic diseases, as well as newly developed 16S rRNA correlation analysis, and so on. These technologies are sensitive, specific, safe and fast, have been well recognized and playing an increasingly important role in disease diagnosis. In addition, molecular biology techniques have also received increasing attention on the detection of drug resistance in pharmacogenetics recently.

1. 4. 2 Application of Molecular Biology in Genetics Research

Techniques of molecular biology have been used in genetic research. For example, flow cytometry and image cytometry is used for analysis of cellular groups, and the content of particular protein and nucleic acid; PCR and its derivative technologies, microdissection, DNA microarray is used for analyses of gene expression and gene structure; enzyme mismatc cleavage (EMC), cleavage fragment length polymorphism (CFLP), dideoxy fingerprinting (ddF) and protein truncation test (PTT) have been used for mutation screening of given genes. Primer extension (PEX), oligonucleotide ligation assay (OLA) and capillary electrophoresis (CE) are used to analyze known gene mutations.

1. 4. 3 Application of Molecular in Cancer Research

The occurrence and development of tumor has been the focus of the molecular biology study. Molecular biology plays an important role, in particular, in studies of molecular mechanism of tumorigenesis, early diagnosis and prognostic predication, determination of drug susceptibility and detection of minima residual diseases. Occurrences of some tumors are close related to viral infection, such as the human papilloma virus (HPV) subtypes with benign and malignant cervical lesions. So detection of the relevant virus is significant for prevention, diagnosis and treatment of cancers in susceptible population. Molecular biology techniques also play an important role in cancer gene therapy. In addition, the mutant and amplification of gene help determine the prognosis of patients, for example, studies have shown that mutations of *ras*, *c-myc*, *p53* genes closely correlated with the clinical stage and the prognosis of colon cancer and ovarian cancer, since the molecular biology techniques has became an important means for judging the prognosis.

(Min Zhao)

第二章 基因、基因组与基因组学

第一节 基 因

基因是生物遗传的基本单位。所有的生命体都依靠基因保存信息从而构建和维系子代细胞，传递遗传性状。

一、基因的概念

1. 历史回顾

基因的存在首先由 Gregor Mendel 在研究豌豆遗传杂交时发现，假定其为传递性状的因子。然而他并没有使用"基因"这个名词，而是用"遗传因子"解释他的发现。1889 年 Mendel 的观点被 Hugo de Vries 在他的《细胞内泛生论》一书中命名为"泛子"（pangen）——遗传性状的最小粒子。20 年后 Wilhelm Johannsen 将其简化为"基因"（gene）。1910 年 Thomas Hunt Morgan 发现基因位于染色体的特殊位置。1928 年，Frederick Griffith 发现基因可以移动。1941 年，George Wells Beadle 和 Edward Lawrie Tatum 发现特定的基因编码特定的蛋白质，提出"一个基因，一个酶"假说。1944 年 Oswald Avery、Colin Munro MacLeod 和 Maclyn McCarty 通过"转化"实验证实基因的本质是 DNA。1953 年 James D. Watson 和 Francis Crick 证明了 DNA 的双螺旋分子结构。1958 年 Crick 证明了生物遗传的中心法则，即 DNA 转录成 RNA，RNA 翻译成蛋白质。这个法则也有例外，如逆转录病毒的逆转录。

1972 年，Walter Fiers 和他的团队第一次测定了噬菌体 MS2 衣壳蛋白的基因序列。1977 年 Richard J. Roberts 和 Phillip Sharp 发现基因能断裂成片段，从而使一个基因能产生几种蛋白质。近年来，生物学进展使基因的概念不断延伸和扩展。

2. 基因简介

在细胞中，基因是含有编码序列和非编码序列的 DNA 序列。当基因表达时，编码序列和非编码序列被转录，产生基因信息的 RNA 拷贝。RNA 能通过遗传密码指导蛋白质合成。某些情况下，RNA 能直接使用，如核酶有酶的功能，miRNA 有调节作用。转录这些 RNA 的 DNA 序列被称为 RNA 基因。基因表达的分子，无论是 RNA 或蛋白质，都是基因产物，对所有生命体的发育和功能有重要作用。

一些病毒以 RNA 的形式保存整个基因组，根本不含 DNA。因为利用 RNA 保存基因，感染细胞宿主后能立即开始合成蛋白质，不需要等待转录过程。另外，RNA 逆转录病毒，如 HIV，在蛋白质合成前需要将它们的基因从 RNA 逆转录为 DNA。2006 年，法国研究者发现了小鼠中 RNA 介导遗传的现象。RNA 作为遗传物质在病毒中很普遍，在哺乳动物中 RNA 遗传非常罕见。

基因是基因组序列上的遗传单位，有调节区、转录区和（或）其他功能序列区。机体发育和生物个体表型可以看作是基因之间和基因与环境相互作用的产物。

3. 基因的功能结构

许多原核基因有操纵子（operon）结构，或一组产物功能相近并作为一个整体转录的基因。真核基因一次只转录一个，但可能包括转录但不翻译成蛋白质的内含子（翻译前被剪接）。剪接同样能发生在原核基因，但非常少见。所有基因除了编码蛋白质或RNA产物的区域外都有调节区域。

启动子（promoter） 启动子是 DNA 依赖的 RNA 聚合酶（简称 RNA 聚合酶）识别、结合的 DNA 序列。启动子位于结构基因的上游（5′端）。DNA 序列的长度为20～200 bp。比较了许多细菌启动子发现，它们都有相同或几乎相同的序列，一是 TATA-AT 序列，又称 TATA 框或 Pribnow 盒，是 RNA 聚合酶的结合部位；二是 TTGACA 序列，是 RNA 聚合酶的识别序列。真核生物 RNA 聚合酶有三类（Ⅰ、Ⅱ、Ⅲ），RNA 聚合酶Ⅱ负责转录Ⅱ类基因，即所有蛋白质编码基因和部分 snRNA 基因，有着类似于 Pribnow 盒的 TATAAA 序列，称为 Hogness 盒。而Ⅰ类和Ⅲ类基因的启动子差别很大。除 RNA 聚合酶外，启动子还可以结合其他蛋白质。与这些蛋白质结合后，RNA 聚合酶才能结合到启动子区。然后这些蛋白质激活，控制 RNA 聚合酶的活性从而影响基因解码和转录。

衰减子（attenuator） 衰减子序列发现于编码氨基酸生物合成的细菌基因簇。衰减子位于前导序列（leader sequence）内。随着氨基酸水平在细胞内的升降，衰减子调节转录水平以适应不断改变的氨基酸水平。高浓度的氨基酸导致结构基因低水平的转录，而低浓度的氨基酸导致基因高水平的转录。衰减作用不依赖于阻遏，这两种现象是彼此独立的。衰减作用引起结构基因转录的早期终止。

增强子（enhancer） 增强子首先发现于动物 DNA 病毒 SV40。这个控制元件靠近病毒复制起点。同样在真核细胞和 RNA 病毒中也发现了增强子。增强子序列的功能表现为增加转录结构基因的 RNA 聚合酶分子的数量。增强子似乎不像启动子一样有着和结构基因的起始位点相关的特殊位点。实际上，它们常被发现位于所控制基因起始位点的上游或下游的某些区域。增强子和序列特异性蛋白结合而被激活。增强子与组织特异性调节和发育时期的时序调节有关。

二、DNA 损伤、修复和基因突变

1. DNA 损伤

DNA 损伤是指 DNA 正常的化学或物理结构的改变。碱基杂环上的一些氮原子和碳原子，即一些环外功能团具有相当的化学活性，许多外源物质能引起这些位点的改变。碱基的化学特性的变化会导致不能配对或错配。如果这种损伤在 DNA 中保留下来，这一突变就会通过诱变而被固定下来。这些化学改变也可以造成 DNA 的结构改变从而阻断复制和（或）转录，导致细胞死亡。因此 DNA 损伤可能引起突变或致死。

（1）DNA 的自发损伤

有些 DNA 损伤是自发性的，是由于 DNA 内在化学活性以及细胞中存在的正常活性分子所致。DNA 的自发损伤（DNA lesion of spontaneous）有以下几种。

互变异构移位（tautomeric shift） 每个核苷酸碱基可表现为两种互变异构体中的

任意一种，这些互变异构体处于动态平衡中。例如，胸腺嘧啶（T）存在两种互变异构体：酮式和烯醇式，平衡更倾向于酮式，但偶尔在复制叉移动时模板 DNA 中出现烯醇式胸腺嘧啶。这将导致一次"错误"，因为烯醇式胸腺嘧啶与鸟嘌呤（G）而不是腺嘌呤（A）配对。复制完成后，罕见互变异构体将逆转为其常见形式，造成子代双链中的一个错配。

脱氨基作用（deamination） 它是指移除分子上的一个氨基。例如，胞嘧啶（C）会自发水解脱氨变成尿嘧啶（U），如果未被修复，产生的尿嘧啶会在接下来的复制中与腺嘌呤配对，发生点突变。

脱嘌呤作用（depurination） 脱嘌呤作用是腺嘌呤（A）和鸟嘌呤（G）的 N9 及脱氧核糖 C1′之间 N-糖苷键发生断裂的自发水解反应，DNA 失去了嘌呤碱基。DNA 的糖-磷酸骨架依然是完整的，产生的脱嘌呤位点呈现非编码损伤，即该嘌呤碱基所编码的遗传信息丢失了。脱嘌呤作用在 37℃时以每小时丢失 10 000 个嘌呤的速率在人体细胞中发生。脱嘧啶也可能发生，但频率很低。

复制滑移（replication slippage） 重复序列可诱发复制滑移，即模板链及其拷贝发生相对移动，使部分模板被重复复制或被遗漏。其结果是新的多聚核苷酸拥有或多或少的重复单位。复制滑移也可能与近年发现的人类三核苷酸重复序列扩增疾病（trinucleotide repeat expansion disease）有关。

氧化性损伤（oxidative damage） 由于超氧化物、过氧化物以及羟自由基等活性氧（ROS）的存在，所有需氧细胞会在正常条件下发生氧化损伤。这些自由基可在许多位点上攻击 DNA，产生一系列特性变化了的氧化产物。

（2）DNA 的诱发损伤

诱变剂（mutagen） 它导致突变的化学或物理因素。很多环境中自然存在的化学物质具有致突变性，近年来人类的工业活动产生了更多的化学诱变剂；物理因素，如辐射也有致突变性；大部分生物都或多或少的暴露在这些诱变剂中，使其基因组受到损害。诱变剂以三种不同的方式导致突变：一是某些诱变剂作为碱基类似物，在 DNA 合成时被错误地当作底物。二是某些诱变剂直接与 DNA 反应，造成结构改变，导致 DNA 复制时模板链的错误复制。三是某些诱变剂间接作用于 DNA，它们自身不影响 DNA 结构，而是使细胞合成，如过氧化物等可直接导致突变的化学物质。诱变剂的范围极广，常见的化学诱变剂有以下几类。

碱基类似物（base analog） 它指那些与标准碱基非常类似的嘌呤或嘧啶碱基，在细胞合成核苷酸时作为 DNA 的成分掺入到 DNA 分子中去，使 DNA 复制时发生配对错误，从而引起有机体变异。

脱氨剂（deaminating agent） 它是能除去氨基的化学物。在基因组 DNA 分子中自发发生一定数量的碱基脱氨基作用，某些化学物质可提高其发生率，如亚硝酸等能使嘌呤或嘧啶脱氨，改变核酸结构和性质，造成 DNA 复制紊乱。

烷化剂（alkylating agent） 烷化剂带有一个或多个活泼的烷基。通过烷基置换，取代其他分子的氢原子（称为"烷化作用"）。可将烷基加入到核酸各位点，但其加入位点有别于正常甲基化酶的甲基化位点。烷化剂可导致 DNA 断裂、缺失或修补。烷基化的影响取决于核苷酸被修饰的位点及所添加烷基基团的类型。

嵌入剂（intercalating agent） 它是扁平分子，可沿着 DNA 走向滑入双螺旋碱基对之间，轻微解开螺旋而使相邻碱基对间距扩大，可影响 DNA 的结构，阻止聚合酶和其他 DNA 结合蛋白发挥正常功能。其结果是阻止了 DNA 合成，抑制了转录（transcription）和突变的发生。

物理诱变剂最重要的类型有以下两类。

紫外辐射（UV radiation） 它诱发相邻嘧啶碱基的二聚化，尤其当两个碱基都是胸腺嘧啶时，形成环丁基二聚体。另外一类 UV 诱导的光产物（photoproduct）是 6-4 位损伤，即相邻嘧啶的 4-位碳原子和 6-位碳原子发生共价交联。

电离辐射（ionizing radiation） 电离辐射对 DNA 有多种影响。某些类型的电离辐射直接作用于 DNA，其他则通过在细胞内激发形成过氧化物类的反应分子而间接起作用。

2. DNA 修复

由于基因组每天受到数以千计的损伤以及复制时出现的错误，细胞必须具备有效的修复系统。没有这些修复系统，基因组维持细胞的基本功能不超过几个小时，随后关键基因就会因 DNA 损伤而失活。与此类似，细胞系将高速积累复制错误，以致细胞分裂之后，基因组将失去功能。多数细胞具有 5 类不同的 DNA 修复系统。

（1）直接修复

直接修复（direct repair），顾名思义，直接作用于受损核苷酸，使之恢复为原来的结构。直接修复系统通过填补切口并纠正某些类型的核苷酸来进行修复，但仅有少数几种 DNA 损伤能够直接被修复。

如果切口两端核苷酸的 5′-磷酸和 3′-羟基没有损伤，只是磷酸二酯键的断裂，则切口（nick）可被 DNA 连接酶修复。电离辐射所造成的切口常是这种情况。

某些类型的烷基化（alkylation）损伤可以被一些酶直接逆转，这些酶可以将烷基从核苷酸转移到自身肽链上。已知多种不同生物中存在这种功能的酶。

环丁基二聚体（cyclobutyl dimer）由被称为光复活（photoreactivation）的光依赖性直接修复系统所修复。在大肠杆菌中该过程需要 DNA 光解酶（DNA photolyase，即脱氧核糖二嘧啶光解酶）。受到波长为 300～500 nm 的光激发时，酶结合环丁基二聚体并将其转化为原来的单体核苷酸。光复活是一种广泛但不普遍存在的修复类型，存在于很多但并非所有的细菌和极少的真核生物中。上述直接的损伤修复类型很重要，但它们仅是大多数生物 DNA 修复机制的次要组成部分。

（2）切除修复

切除修复（excision repair）指先切除一段含有损伤部位的多核苷酸，然后利用 DNA 聚合酶重新合成正确的核苷酸序列。切除修复分为两类。第一类是碱基切除修复（base excision repair），指除去受损的核苷酸的碱基，在产生的 AP 位点周围切除一小段多聚核苷酸，并利用 DNA 聚合酶重新合成。碱基切除是最基本的一种，它用来修复众多受损相对较轻的被修饰的核苷酸碱基，如暴露于烷化剂或电离辐射的碱基。所有细胞中都带有不同类型、能识别受损核苷酸位点的糖苷水解酶，它能够特异性切除受损核苷酸上的 N-β-糖苷键，在 DNA 链上形成去嘌呤或去嘧啶位点，统称为 AP 位点。DNA 聚合酶利用 DNA 分子另一条链中未受损核苷酸为模板填补单核苷酸缺口，以保证正确核苷酸的插入。缺口填补后，最后的磷酸二酯键由 DNA 连接酶形成，见图 2-1。

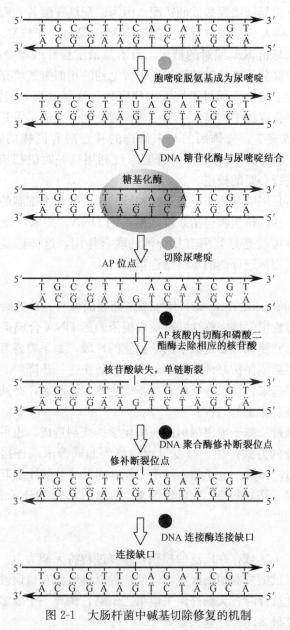

图 2-1　大肠杆菌中碱基切除修复的机制

第二类是核苷酸切除修复（nucleotide excision repair），与碱基切除修复类似，但并不去除受损碱基，而是作用于更严重的 DNA 受损区。核苷酸切除修复比碱基切除系统适用性更广，能处理更严重的损伤形式，如链内交联和巨大化学基团修饰的碱基。它也能通过暗修复（dark repair）过程校正环丁基二聚体，为不具有光复活系统的生物提供了修复途径。在核苷酸切除修复中，含有受损核苷酸的一段单链 DNA 被切除并被新的 DNA 取代。因此该过程与碱基切除修复类似，只是没有选择性去除碱基的过程而且被切除的聚核苷酸片段更长。

（3）错配修复

直接修复和切除修复都是识别并作用于诱变剂所致的损伤。这些修复系统寻找异常的化学结构而不能校正复制错误造成的错配。因为错配核苷酸各方面没有异常，只是正常的 A、C、G 或 T 插入到错误位置上。修正复制错误的错配修复（mismatch repair）系统必须检测子链和母链碱基配对的消失，而不是错配核苷酸本身。一旦发现一个错配，修复系统就切除部分子链并以切除修复中所见到的相似方式填充缺口。

错配修复发生在子链上，而母链有正确的序列。修复过程如何能正确区别二条链呢？大肠杆菌中，新合成的子链是未甲基化的，因此能与完成了甲基化的母链区分。母链的甲基化不会导致突变，修饰后的与未修饰的核苷酸有同样的碱基配对特性。在DNA 复制与子链甲基化间有一段延迟，修复系统利用这一时机扫描 DNA 寻找错配并对未甲基化的子链进行必要的校正。

真核细胞可能不是用甲基化来区别子链和母链。在哺乳类细胞的错配修复中曾提示过存在甲基化，但一些真核生物包括果蝇或酵母的 DNA，并未广泛甲基化。这些生物采用其他方法，一种可能是修复酶与复合体的联合作用，这样修复与 DNA 合成相偶联；另一种可能是利用标记母链的单链结合蛋白。

（4）重组修复

重组修复（recombination repair）用于双链断裂的修补。单链断裂并不会给细胞带来严重的后果，双螺旋仍保持完整性并可被模板依赖的 DNA 合成系统修复。而双链的断裂更为严重，因为它使双螺旋变成了两个独立的片段，如果要修复断裂，必须将它们再放在一起，并且还必须保护好两个断裂的末端，防止进一步降解。修复过程还必须保证将正确的末端连接起来：如果在核中有两条染色体发生断裂，必须将断裂片段正确地匹配并连接，才能恢复原来的结构。

双链断裂可以通过暴露于电离辐射或一些化学诱变剂造成，也可以由细胞在一些重组事件中以一种可控的方式产生。修复过程被称为非同源末端连接（non-homologous end joining，NHEJ），即两个分子的末端不需要有同源性就能连接起来。NHEJ 被看作是一种重组，因为除了修复断裂，它能将原来并未连在一起的分子或片段连接在一起，产生新的组合，见图 2-2。

（5）SOS 修复

如果基因组的一个区域存在广泛的损伤，修复过程将无能为力。细胞将面对一个严酷的选择：死亡还是试图复制损伤区域，即使这种复制可能有错误倾向并导致突变的子代分子。在面对如此选择时，大肠杆菌采取第二条路，诱导一个或多个应急途径，绕过主要损伤位点。研究最多的绕过途径是 SOS 应答。

"SOS"是国际上通用的紧急呼救信号。SOS 修复（SOS repair）是指 DNA 受到严重损伤、细胞处于危急状态时所诱导的一种 DNA 修复方式。它是一种旁路系统，允许新生的 DNA 链越过胸腺嘧啶二聚体而延长。SOS 应答最初被认为是最后的最佳选择，使细菌能复制 DNA 并在不利环境下生存。然而生存的代价是突变率的增加，因为"突变体"不能修复损伤，它仅仅是使损伤区的多核苷酸能被复制。有时尽管合成了一条和亲本一样长的 DNA 链，但常常是没有功能的，其原则：丧失某些信息而存活总比死亡好一些。增加突变率被认为是 SOS 应答的一个目的，突变在某种程度上是对损伤的有

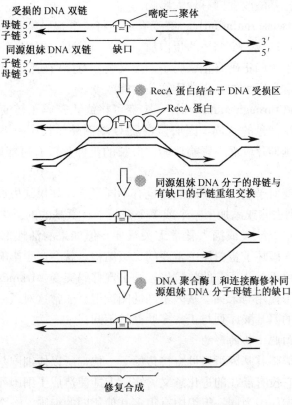

图 2-2　大肠杆菌中重组修复的机制

利应答，但这种观点仍存在争议。

3. 基因突变

突变（mutation）是基因组小范围的核苷酸序列的改变。许多突变是单个核苷酸替换的点突变（point mutation），其他形式包括一个或多个核苷酸的插入（insertion）或缺失（deletion）。突变来源于 DNA 复制的错误或是诱变剂的破坏作用。这些化学或物理诱变剂能与 DNA 作用并改变个别核苷酸的结构。所有的细胞都具有 DNA 修复酶，以尽可能减少突变的发生。

（1）突变对基因组的影响

很多由突变引起的核苷酸序列改变对基因组功能没有影响。这些沉默突变（silent mutation）实际也包括所有出现在基因以外的 DNA 及基因组非编码区和基因相关序列的突变。约 98.5% 的人类基因组基因可以突变而无显著影响。基因编码区的突变要重要得多。改变三联体密码子的点突变有以下 4 种类型。

同义突变（synonymous mutation）　突变的密码子与未突变的密码子编码同样的氨基酸。同义突变是沉默突变，因为它对于基因组的编码功能没有影响：突变基因与未突变基因编码完全相同的蛋白质。

错义突变（missense mutation）　突变改变了密码子，从而编码一个不同的氨基酸。突变基因编码的蛋白质有一个氨基酸的改变常常对蛋白质的生物学功能没有太大影响：多数蛋白质允许少量的氨基酸改变，这对其在细胞内的功能无明显影响，但有些，如处于酶的活性部位的氨基酸改变则有较大影响。

无义突变（nonsense mutation） 突变有可能将一个编码氨基酸的密码子转变为一个终止密码子。它可造成一个缩短的蛋白质，因为 mRNA 的翻译停止在新的终止密码子而不能到达位于下游的正确终止密码子。无义突变对蛋白质活性的影响取决于失去了多少多肽：影响通常是巨大的，蛋白质往往功能改变。

通读突变（read through mutation） 突变可将终止密码子转变为一个编码氨基酸的密码子，造成终止信号的通读，因而蛋白质 C 端额外延长了一系列氨基酸。多数蛋白质可允许有短的延伸片段而不影响功能，但长的延伸片段有可能影响蛋白质的折叠，造成活性下降。

缺失与插入突变对于基因的编码能力也有不同影响。如果缺失或插入的核苷酸数是 3 或 3 的整数倍，则去除或添加了一个或多个密码子，氨基酸的丢失或增加对所编码的蛋白质影响不同。这类缺失或插入常常无关紧要，但如果酶活性部位相关的氨基酸丢失，或氨基酸的插入破坏了蛋白质中重要的二级结构，就会产生影响。另外，如果缺失或插入的核苷酸数目不是 3 或 3 的整数倍，则出现移码突变（frameshift mutation），所有突变下游的密码子使用与未突变基因不同的阅读框。通常这对蛋白质功能产生重大影响，因为突变多肽的氨基酸序列与正常多肽截然不同。

（2）突变对多细胞生物的影响

由于体细胞不能将其基因组拷贝直接传给下一代，所以体细胞基因突变仅对于所在机体是重要的，但它没有潜在的进化意义。实际上即使造成了细胞死亡，大部分体细胞突变也没有大的影响，因为同一组织中有很多其他相同的细胞，一个细胞的损失没有实质性的影响。例外情况是当突变造成了细胞某种有损机体的功能异常时，如诱发肿瘤形成或其他致癌性体细胞突变，突变就会产生巨大影响。

生殖细胞中的基因突变更为重要，因为它可以传至下一代，并存在继承突变个体的所有细胞中。对于机体的表型有影响的可分为两类。

功能丢失（loss of function） 通常是一个减弱或消除蛋白质活性的突变造成的结果。多数功能丢失突变是隐性性状，因为杂合子中，第二条染色体带有未突变的同一基因编码的功能完全的蛋白质，弥补了突变的影响。但有一些功能丢失突变是显性的，如单倍体失活（haploin sufficiency），杂合子不能承受约 50% 蛋白质活性的下降。

功能获得（gain of function） 非常少见。这种突变赋予了蛋白质异常活性。很多功能获得性突变发生在调节序列而不是编码区，可产生多种后果。另外，突变可造成一个或多个与控制细胞周期有关的基因过表达，从而使细胞分裂失控而致癌。功能获得性突变通常是显性的。

第二节 基 因 组

每种生物都包含一套遗传信息，这些遗传信息是每个有生命的个体建立和维持其生物学特征所必需的。绝大多数基因组（genome），包括人类基因组和其他细胞生命形式，都由 DNA 组成，但是也有一些病毒基因组由 RNA 组成。基因组是一套完整单倍体遗传物质的总和。

一、原核生物基因组

传统观点认为，典型的原核生物基因组是一个环状 DNA 分子，位于类核（nucleoid）中。虽然大多数细菌和古细菌染色体的确是环形的，但越来越多的线性基因组被发现。

1. 原核生物基因组特点

从整体上讲，原核生物基因组要小得多。大多数原核生物基因组小于 5 Mb，只有少数比 5 Mb 大，如巨大芽孢杆菌有一个 30 Mb 的巨大基因组。通常情况下，原核生物基因比其对应的真核生物基因要短，细菌基因的平均长度大约是真核基因的 2/3，即使把后者的内含子去除也是如此。许多原核基因是由不被非编码序列中断的核苷酸序列组成，即原核 DNA 大部分为编码序列。但是，在某些真细菌和古细菌中也发现了非编码序列。

原核生物的结构基因通常成簇活化，通常被称为多顺反子转录单位（polycistronic transcription unit），即转录单位含有超过一个多肽的信息。所谓转录单位，是 RNA 聚合酶识别起始位点（启动子）到终止位点之间的 DNA 序列。从基因的 5′端开始有着以下的排列：前导序列、起始序列、编码多肽的序列、终止序列和 5～20 个核苷酸的间隔序列。细菌基因组最典型的是操纵子结构。操纵子是一组在基因组中彼此相邻的基因，两基因头尾之间可能仅隔一两个核苷酸。操纵子中所有基因都是作为一个单位表达。这种排列在原核生物基因组中很常见。大多数原核生物基因组没有高拷贝数的全基因组范围的重复序列。然而，可能在基因组的某处含有一些重复序列。

2. 质粒

质粒（plasmid）是一小段 DNA，常以环状形式在细菌中与宿主染色体共存。一些质粒可以整合到基因组中，但有一些质粒则永远是独立的。质粒所含的基因通常在宿主染色体中不存在，许多情况下，这些基因对细菌是非必需的，但它们赋予细菌的某些表型，如抗生素抗性的表型等，使细菌能在外界环境不适宜的情况下也能存活。除此之外，许多质粒能从一个细胞转移到另一个细胞，有时可在不同种属的细菌中找到相同的质粒。质粒的这些特征提示质粒是独立存在的。

二、真核生物基因组

真核生物基因组包含两个部分：染色体基因组和染色体外基因组（线粒体 DNA）。植物和其他可进行光合作用的真核生物具备定位于叶绿体的第三个基因组的普遍特征。

1. 染色体基因组

尽管所有真核生物的细胞核基因组的基本物理结构都相似，但在不同生物中基因组大小有很大差异。最小的真核生物基因组长度不到 10 Mb，最大的超过了 100 000 Mb。真核生物基因组远大于原核生物基因组，并且复杂多样。

与原核生物相比，真核生物通常采用单顺反子（monocistron）转录单位。这些生物的基因不是成簇活化而是作为单个实体。mRNA 携带的信息只编码一个多肽。真核生物基因和原核生物基因在序列排列上通常一致（前导、起始等）。但是真核生物基因有编码序列和非编码序列。因此，真核生物基因有时被称为断裂基因（split gene）。

（1）外显子和内含子

真核生物基因的显著特征是在编码（氨基酸）序列中含有非编码序列，编码序列叫作外显子（exon），非编码序列叫作内含子（intron）。内含子的大小、数目、定位和核苷酸序列对于不同基因变化很大。通常一个基因内含子的总长超过该基因外显子的 2～10 倍或更多。内含子的数量随着基因大小的增加而增加。有趣的是，内含子并不随机地分布在基因中，它们倾向于出现在特殊位置。在真核生物蛋白合成时，基因解码和 mRNA 产生后，内含子很快从 mRNA 分子上除去。

（2）基因家族

在 DNA 测序的早期，就发现具有相似或相同序列的多基因家族（multigene family）是许多基因组的共同特征。例如，所有真核生物都有多拷贝非编码的核糖体 RNA（ribosomal RNA，rRNA）基因。rRNA 基因是"简单"或"经典"的多基因家族的例子，其中各成员有相同或几乎相同的序列。这些家族是由基因倍增产生的，在某些进化作用下，家族成员保持了序列的一致性。其他多基因家族的成员虽然序列相似，但基因产物特征却明显不同，称为"复合物"，这种情况在高等真核生物中比低等真核生物中更常见。这一类多基因家族最好的例子是哺乳动物的珠蛋白基因。

一些多基因家族的成员簇集在一起，而另一些家族的成员则散布于整个基因组中。家族成员虽然分散，但仍具有序列相似性，说明它们有相同的进化起源。有时序列比较不仅可以看出家族成员之间的关系，而且可以比较不同家族之间的关系。例如，α 珠蛋白和 β-珠蛋白家族中的所有基因都有序列相似性，被认为是从一个原始珠蛋白进化而来的，因此这两个多基因家族组成一个珠蛋白基因超家族（gene superfamily）。

（3）假基因

假基因（pseudogene）与有功能的基因结构类似，但是没有编码蛋白质的能力或不能在细胞中表达。主要有两种类型的假基因。

保守假基因（conventional pseudogene）是由核苷酸序列突变而来。许多突变对基因活性影响很小，但有些突变则影响很大，单个核苷酸突变导致整个基因丧失功能是完全有可能的。一旦基因丧失功能，它就会随着突变的积累而逐渐变化，最终难以作为一个基因的残迹被辨别出来。

加工假基因（processed pseudogene）并非由进化衰退而来，它是基因表达的异常产物。加工假基因来自基因的 RNA 转录物，合成的 DNA 拷贝随后重新插入基因组中。因为加工假基因是从 mRNA 分子拷贝而来，所以它不具有其母基因的内含子。同时它没有母基因 5′-UTR 上游近端的核苷酸序列，母基因表达启动信号即定位于这个区域，加工假基因由于缺乏这些信号，因此不具有活性。

2. 细胞器基因组

几乎所有真核生物都有线粒体基因组，所有能进行光合作用的真核生物都有叶绿体基因组。大多数线粒体基因组和叶绿体基因组是环状的，但在不同生物中有许多可变性。在许多真核生物中，环状基因组与线性基因组共存于细胞器中，在叶绿体中，环状基因组与含有基因组亚单位的环状小分子作为一个整体共存。

线粒体基因组（mitochondrial genome，mtDNA）变化很大，与生物的复杂程度无关。大多数多细胞动物的线粒体基因组较小，结构紧密，基因间几乎没有间隔，人线粒体基因组就是这种类型的典范。

三、重复序列

所有生物中都有重复 DNA 序列（repetitive DNA sequence），在一些生物（包括人）中，重复 DNA 序列是整个基因组的重要成分。重复 DNA 序列有许多类型，并有几种分类方法，本书将其分成串联重复 DNA 序列和散在重复 DNA 序列。

1. 串联重复 DNA 序列

串联重复序列就是重复单元首尾相连，串接在一起的重复序列。串联重复 DNA（tandemly repeated DNA）是真核生物基因组的普遍特征，真核生物 DNA 密度梯度中的卫星带由很长的串联重复序列组成，长度可有几十万个碱基对。而在原核生物中极少见。这种重复序列也叫卫星 DNA（satellite DNA）。因为用密度梯度离心法分离基因组片段 DNA 时，含有串联重复序列的 DNA 片段形成卫星带。一个基因组可以含有几种不同类型的卫星 DNA，各含有一个不同的重复单位，大小为 5～200 bp。例如，将人类基因组截成 50～100 kb 的片段，离心后就会形成一个主带和三个卫星带。主带的 DNA 片段多由单拷贝序列组成，GC 含量接近人类基因组的平均值 40.3%，卫星带含有重复 DNA 片段。在染色体着丝粒区发现的类 α-DNA 重复就是一种人类卫星 DNA。虽然一些卫星 DNA 散布于整个基因组，但大多位于着丝粒。

串联重复 DNA 序列中有两种不出现在密度梯度的卫星带中，但也可以归为卫星 DNA，它们是小卫星（minisatellite）和微卫星（microsatellite）。小卫星形成 20 kb 长的聚集区，每个重复单位最多 25 bp。人类端粒就是一种小卫星 DNA，它含有几百个 5′-TTAGGG-3′ 的基序。除了端粒小卫星，一些真核生物基因组还含有其他小卫星 DNA，它们中有许多但并非全部靠近染色体末端。微卫星较短，通常小于 150 bp，其重复单位只有 2～6 bp 或更少。虽然每个微卫星很短，但在基因组中数量却很多。例如，人类基因组中 CA 重复的微卫星，如 $\begin{array}{l}5′\text{-CACACACACACACAC-}3′\\3′\text{-GTGTGTGTGTGTGTG-}5′\end{array}$，占基因组的 0.25%，共 8Mb；单碱基对重复，如 $\begin{array}{l}5′\text{-AAAAAAAAAAAAAAA-}3′\\3′\text{-TTTTTTTTTTTTTTT-}5′\end{array}$，占基因组的 0.15%。

许多微卫星是可变的，即重复单位的数目（n）在相同物种的不同个体中也不一样，这是因为当 DNA 复制时微卫星拷贝有时发生滑移，导致插入或缺失（这种情况较少）一个或多个重复单位。这种变化使得任何两人都不会有完全相同的微卫星长度变异组合。如果检查的微卫星足够多，就能为每个人建立一个遗传图谱（genetic profile）。遗传图谱是法医鉴定中常用的一个工具，但只是微卫星多态性应用中很小的一部分。微卫星可以用于建立同源关系和人群的姻亲关系，不仅可用于人类，也可用于其他动物和植物。

2. 散在重复 DNA 序列

上述串联重复 DNA 序列被认为是由原始序列通过复制中的滑移或 DNA 重组产生的。这两类事件都可能导致一系列相连的重复，而不是散布于整个基因组的单个重复单位。因此，散在重复（interspersed repeated）的发生一定有不同的机制，可以在基因组中离原序列较远的位置产生一个重复的拷贝。最常见的机制是转座（transposition），最分散的重复序列往往有天然的转座活性。转座有两种方式，一种需要 RNA 介导，另一种不需要。

有 RNA 参与的称为反转录转座（retrotransposition），包括以下三个步骤。正常转录合成转座子（transposon）的 RNA 拷贝；RNA 再拷贝成 DNA，最初以独立于基因组外的分子存在；转座子的 DNA 拷贝整合进入基因组，有可能返回到原来的染色体中，或进入不同的染色体。最终结果是产生了转座子的两个拷贝，并分布在基因组的不同位置。RNA 转座子或反转录元件（retroelement）是真核生物基因组的特点，在原核生物中尚未发现。反转录病毒（retrovirus）、内源性反转录病毒（endogenous retrovirus，ERV）、反转录转座子（retrotransposon）都是 LTR 元件，因为其两端都有长末端重复序列，在转座过程中发挥作用。其他反转录元件没有 LTR，称为反转录子（retroposon），在哺乳动物中包括长散布核元件（long interspersed nuclear element，LINE），含有一个反转录酶样基因，可能参与反转录转座过程。例如，人的 LINE-1 长 6.1 kb，在人类基因组中有 516 000 个拷贝。短散布核元件（short interspersed nuclear element，SINE），没有反转座酶但仍能转座，可能是用其他反转录元件合成的反转座酶。人类基因组中最常见的 SINE 是 Alu，其拷贝数接近 100 万。Alu 可能是由参与细胞内蛋白质运转的 *7SLRNA* 基因衍生而来。第一个 Alu 元件可能来自 7SLRNA 分子意外的反转录，将 DNA 拷贝整合入人基因组中。因为 Alu 元件有转录活性，这就为元件的复制提供了可能。

转座的第二种方式为 DNA 转座。植物 DNA 转座子家族——玉米的 Ac/Ds 转座元件（DNA）是第一个被发现的可转座元件。许多转座子可以由一种更为直接的 DNA 至 DNA 的方式转座。有两种不同的转座机制，一种是通过供体转座子和靶位点之间的直接相互作用，产生供体转座元件的拷贝（复制型转座）；另一种是将元件切下来再重新整合到新的位点（保守型转座）。

真核生物中，DNA 转座子比反转座子少见。但 DNA 转座子在遗传学上却有特殊的地位。而在原核生物基因组结构中 DNA 转座子比 RNA 转座子更为重要。一个大肠杆菌基因组可含有多达 20 种不同种类的 DNA 转座子。原核生物中典型的 DNA 转座子有插入序列、复合型转座子、Tn3 转座子和转座噬菌体。

第三节　基因组学

一、人类基因组

人类基因组是一个十分稳定的体系，不同的民族、群体和个体染色体数目都是 46 条，并有相同数量的基因和基因分布，也有基本相同的核苷酸序列。正是基因组结构的这种稳定性保证了人类作为一个物种的共同性和稳定性，也决定了人类基因组计划的代表性和重要意义。然而人类基因组又十分复杂，在长期的进化过程中，基因组 DNA 序列也会发生变异，这些变异有的是有益的，有的是中性的，有的是有害的。其中一些被保存下来，导致了不同民族、群体和个体之间基因组的差异或多态性。

人类基因组保存在 23 条染色体上。人类基因组共有超过 30 亿个碱基对，包含 20 000～25 000 个编码蛋白质的基因，远远小于预期。事实上，只有基因组的 1.5% 编码蛋白质，剩下的由 RNA 基因、调节序列、内含子和人们认为有争议的"无用"DNA 组成。

1. 人类（核）基因组

如上所述，人类基因组有 20 000～25 000 个编码蛋白质的基因。这个数字小于许多更简单的生物，如线虫和果蝇。然而，人类细胞广泛利用选择性剪接由一个基因产生许多不同的蛋白质，人类蛋白质组比上述生物大得多。另外，大多数人类基因有多重外显子，人类内含子通常比两侧外显子长。

人类基因在染色体上不是均匀分布的。每个染色体都有不同的基因丰富区和基因较少区，可能与染色体带和 GC 含量有关。基因密度非随机模式的意义还不十分清楚。除了编码蛋白质的基因，人类基因组含有许多 RNA 基因，包括 tRNA、rRNA、micro-RNA 和其他非编码 RNA 基因。

人类基因组有许多不同的控制基因表达的调节序列。典型的短序列出现在基因附近或在基因中。

除了已知调节序列，人类基因组含有大量未知功能的 DNA 区域。估计这些区域在人类基因组中约占 97%，其中许多属于重复序列、转座子和假基因。然而，仍然有大量的序列不能被归类。这些序列大多数可能是演化的产物，现在已经没有作用，也因此有时被称作"垃圾"DNA（"junk"DNA）。不过有一些迹象显示，这些序列可能会经由某些仍然未知的方式产生作用。最近一些利用微点阵技术的实验发现，大量非基因 DNA 事实上会被转录成 RNA，这显示转录作用背后可能还存在一些未知的机制。此外，不同种类的哺乳动物在演化的过程中共同保留了这些序列，也显示基因组中还有很多作用未知的部分。

2. 人类线粒体基因组

早在 20 多年前人类线粒体基因组的完整序列就已经测定。线粒体基因组全长 16 569 bp，其中仅有 37 个基因，其规模比核基因组小得多。其中 13 个基因编码的蛋白质参与呼吸链复合体，这是线粒体产生能量的主要生化成分；剩下的 24 个基因则对应非编码 RNA 分子，这些 RNA 分子是线粒体基因组表达所必需的。与核基因组相比，线粒体基因组中的基因组装得更为紧密，不具有内含子且部分区域存在重叠，因此任何突变都可能会累及 DNA 的重要功能区；同时，线粒体 DNA（mtDNA）缺乏有效的 DNA 修复系统及组蛋白保护，其突变频率较高，为染色体 DNA 的 10～20 倍。因此近年来由于 mtDNA 突变引起的人类各种疾病日益引起重视。

3. 人类基因组多态性

1）限制性片段长度多态性（restriction fragment length polymorphism，RFLP）限制性内切核酸酶在特异的识别序列切割 DNA 分子，这种序列特异性意味着一种 DNA 分子用限制性内切核酸酶处理后总是可以产生相同的片段。但对于基因组 DNA 来说，并不总是这样。因为一些限制性位点是多态性的，以两种等位形式存在。一种等位形式有正确的限制位点序列，因此可以被识别该位点的酶切开；另一种等位形式的序列有所改变，因此该限制性位点不能被识别。序列改变的结果使得用内切核酸酶处理后的两个相邻的限制性片段仍然连接在一起，从而导致了长度多态性。在人类基因组中大约有 10^5 个 RFLP，但是每个 RFLP 只能有两种等位形式（有或没有这个位点），这就限制了 RFLP 在人类基因作图上的应用价值，因为一个家庭的所有成员很可能都是某一个 RFLP 的纯合子。

为了量化 RFLP，有必要在很多不相关的片段背景中确定一个或两个限制性片段的长度。用覆盖多态性限制位点的探针进行 Southern 印迹，是检测 RFLP 的方法，但是目前 PCR 更常用。PCR 的引物与多态性位点的两侧经退火复性，用限制酶处理扩增的片段，然后进行琼脂糖凝胶电泳，从而进行 RFLP 分型。

2）简单序列长度多态性（simple sequence length polymorphism，SSLP） SSLP 用于分析一系列不同长度、不同数目重复单位等位基因的重复序列。与 RFLP 不同，由于每个 SSLP 可以有很多不同长度的变异体，所以它可以是多等位形式的。SSLP 有两种类型，一种是小卫星，又称可变数目串联重复序列（variable number of tandem repeat，VNTR），重复单位可以长达 25 个核苷酸；另一种是微卫星，又称简单串联重复序列（simple tandem repeat，STR），重复单位长度更短，通常为 2～4 个核苷酸。

微卫星比小卫星更常用于 DNA 标记，因为小卫星在基因组中并不均匀分布，而是更常见于染色体末端的端粒区。因此微卫星在基因组中的分布更便于用来定位。另外，长度多态性分型最快的方法是通过 PCR，但是 PCR 分型用于小于 300 bp 的序列更快而且更准确。大多数的小卫星等位基因比 300 bp 长，而典型的微卫星含有 10～30 个重复的拷贝，每个拷贝通常不大于 4 bp，因此更适合用于 PCR 分型。人类基因组中有 6.5×10^5 个微卫星。

3）单核苷酸多态性（single nucleotide polymorphism，SNP） 基因组中的某些位置上，有些个体具有一个与其他个体不同的核苷酸。在每一个基因组中都有大量的 SNP，有些也可形成 RFLP，但许多都不能，因为它们所处的序列不能被限制酶识别。人类基因组中至少有 1 420 000 个 SNP，其中只有 100 000 个可以形成 RFLP。

4 种寡核苷酸的任何一种都可以位于基因组的任意位置，因此可以设想每种单核苷酸多态性应该有 4 种等位基因。这在理论上是可能的，但是实际上大多数的 SNP 只能以两种变异体的形式存在。这是由产生 SNP 的方式和其在种群中的分布决定的。当基因组中发生点突变时产生 SNP，使一个核苷酸变成另一个。如果该突变是发生在一个个体的生殖细胞中，那么一个或多个后代就可能遗传了这个突变，经多代之后，这个 SNP 可能在种群中建立，但是仅有两个等位基因——原始序列和突变序列。如果要产生新的突变，必须在另一个个体基因组的同一位置产生一个新的突变，该个体及其后代必须遗传该突变才能产生新的等位基因。这种情况不是绝对不可能，但是很罕见；因此绝大多数 SNP 都是双等位的。

大多数 SNP 仅存在两个等位基因，所以这些标记在人类遗传学作图方面与 RFLP 存在同样的缺陷：在所研究的家族中，SNP 极有可能不能显示任何变异。SNP 的优点是它数目庞大，而且分型不需要用凝胶电泳的方法。由于 SNP 以寡核苷酸杂交分析为基础，其检测更快速。已经发明了多种筛选策略，包括 DNA 芯片（DNA chip）技术和液相杂交技术。

4. 人类基因组计划

人类基因组计划是为了测定人类基因组序列和鉴定其含有的基因而进行的国际性研究。这项计划由美国国立卫生研究院和能源部组织协调。美国、英国、法国、德国、日本和中国的国际合作机构也参与其中。人类基因组计划正式开始于 1990 年，完成于 2003 年，比原计划提前了 2 年。

人类基因组计划的主要目标是提供构成人类基因组的完整准确的 30 亿个碱基对和发现所有预测的 20 000～25 000 个人类基因。该计划同样也对在医学研究中重要的其他生物，如小鼠和果蝇进行了测序。

除了 DNA 测序，人类基因组计划试图发现新的工具获得和分析数据，使信息广泛可用。同样，因为遗传学的发展对个人和社会产生影响，人类基因组计划承诺通过伦理、法律和社会问题计划探讨基因组研究的影响。

2003 年 4 月，研究者宣布人类基因组计划完成了高质量的整个人类基因组的测序。这个序列填补了 2001 年完成的基因组工作草图的空缺，确定了许多人类基因的定位，提供有关结构和组成的信息。人类基因组的序列和分析数据的工具可以从网上免费获得。

除了人类基因组，人类基因组计划对其他生物的基因组进行了测序，包括啤酒酵母、线虫和果蝇。2002 年，研究者宣布已经完成了小鼠基因组的草图。通过研究人类基因组和其他生物基因组的相似和差异，研究者可以发现特殊基因的功能并鉴定哪些基因对生命是重要的。研究者对基因和蛋白质的功能了解得越来越多，将对药物、生物工程和生命科学领域产生重大影响。

二、基因组学

基因组学（genomics）研究生物的基因组，研究领域包括测定生物的全部 DNA 序列和精细的遗传作图，还包括基因组内现象，如杂种优势、基因多效性和其他基因组内基因位点和等位基因的相互作用。相反，现代医学和生物学研究中的分子生物学主要关注的是单个基因及其功能和作用的研究，不属于基因组学的范畴。

1. 比较基因组学

比较基因组学（comparative genomics）——并列分析人和模式生物的 DNA 序列，已经成为鉴定基因和解释其功能的最强大的策略。比较基因组学的基础是相关生物基因组的相似性。两种具有较近共同祖先的生物，其具有种属差别的基因组是由祖先基因组进化而来的。两种生物在进化阶段上越相近，它们的基因组相关性就越高。如果两生物非常相近，那么它们的基因组会表现出同线性（synteny），即基因序列的部分或全部保守。

进行人类基因组测序的主要原因之一是为了获得与人类疾病相关的基因序列，希望疾病基因的序列能有助于认识疾病，并提供预防和治疗疾病的方法。比较基因组学在疾病基因的研究中有重要的作用，因为在其他生物中发现一个人类疾病基因的同源基因常常是阐明此基因生化功能的关键。如果同源基因的功能已被鉴定，那么对应的人类基因的功能也就清楚了，反之，就有必要直接研究这一同源基因。

2. 药物基因组学

药物基因组学（pharmacogenomics）研究基因如何影响人对药物的反应。这个新领域结合了药理学和基因组学，研究根据人的遗传组成定制安全有效的治疗药物和剂量。现阶段许多药物还不是针对每个个体进行治疗，而实际上药物对每个个体的作用是不一样的。预测谁会对药物治疗有效，谁对药物没反应，谁会发生副作用（不良药物反应）是很困难的。在美国不良药物反应是导致患者住院治疗和死亡的重要原因。根据从人类基因组计划中得到的知识，研究者逐渐发现基因的遗传差异是如何影响个体对药物治疗的反应。这些遗传差异会用来预测药物治疗是否对特殊的人有益，并帮助预防不良药物反应。药物基因组学仍然在研究初期。现在它的运用十分受限，但是新的方法正在临床试验研究应用中。相信不远的将来，药物基因组学将用于研究药物定制，治疗更广泛的健康问题，包括心血管疾病、老年痴呆、癌症、艾滋病和哮喘。

第四节　基因与疾病

掌握了人类基因组的完整序列，科学家可以将单基因病的表型和相对应的基因配对。通过分析复杂的谱系，遗传学家可以发现特殊疾病过程和基因序列改变的联系。一旦基因与疾病相关的 DNA 序列改变被鉴定，确定对应基因产物（蛋白质）的结构如何影响生物功能的改变就比较容易。了解蛋白质结构和功能的哪些改变导致疾病，有利于我们设计更有效的特异性针对靶向突变蛋白的药物。

尽管 1822 个人类蛋白编码基因确定与单基因病有关，但是其中仍有 1500 多个基因是未知的，这在很大程度上是因为许多单基因病很罕见，只发生在家庭的少数成员。然而，大多数常见疾病表现出更复杂的遗传模式，与多重基因的突变有关。所以，研究工作的重点已经从单基因病转移到多基因病。

一、单基因病

最简单的遗传性状取决于一个基因位点的基因型，有时在正常的遗传和环境背景下，一个基因位点的特殊基因型就是性状表达的重要条件，这样的性状称为孟德尔性状。单基因病，又称孟德尔遗传病，是因为个体所有细胞的单个基因改变导致的。已知的人类单基因病超过 10 000 种。

1. 单基因病的遗传方式

性状在杂合子出现为显性，不出现则为隐性。显性和隐性是性状而并非基因的性质。决定孟德尔性状的基因可位于常染色体、X 染色体和 Y 染色体上。常染色体性状和女性的 X 连锁性状可为显性或隐性。男性是 X 染色体、Y 染色体的半合子，只有性染色体基因的单个拷贝，所以 X、Y 连锁的性状对于男性不存在显性和隐性。因此有 5 种基本的遗传方式。

常染色体显性遗传（autosomal dominant inheritance，AD）　通常患者的父/母至少一位是患病；后代男女均可患病；男女均可传给后代；通常患病父母是杂合子，那么父母一方正常、一方患病所产生的后代有 50% 的机会患病。

常染色体隐性遗传（autosomal recessive inheritance）　患者的父母通常未患病，通

常是无症状的携带者；后代男女均可患病；如果出生的是一个患儿，那么再出生的子代有25％的机会患病。

X连锁隐性遗传（X-linked recessive inheritance，XR） 大多数患者为男性，患者的父母通常不发病；患者的母亲通常是无症状的携带者，可能有患病的男性亲属；如果父亲患病，母亲是携带者，女性可能患病，或者是偶然的由于非随机X染色体失活；家系中不存在父子遗传（然而患病男性和携带者女性的结合看起来像父子遗传），典型情况是祖父通过女儿遗传给孙子（隔代发病）。

X连锁显性遗传（X-linked dominant inheritance） 不存在父子遗传；男女均可患病，但女性多于男性；相对于男性，女性通常症状较轻且症状多样；女性患者和正常男性的后代，无论男女，有50％的机会患病；男性患者和正常女性结合后所有女儿患病而儿子不患病。

Y连锁遗传（Y-linked inheritance） 只有男性患病；男性患者的父亲通常患病（除非产生新的突变）；男性患者的所有儿子均患病。

2. 基本遗传方式的复杂性

不外显（nonpenetrance） 在显性条件下，不外显经常出现。根据定义，显性性状在杂合子个体显现，并且表现出100％的外显率。然而，许多通常为显性遗传的人类性状，偶尔跳过一代。通常性状的出现或缺失在正常情况下取决于基因型，但是特殊的遗传背景、生活方式或纯粹意外会使性状不显现。

延迟显性（delayed dominance） 尽管基因型在受孕时就已经固定，但是表型可能直到成年才显现。这种情况下外显率和年龄有关。如果活的时间足够长，外显率可能变成100％，或者不管携带这种基因的人活的多久也不会出现症状。

变异表达（variable expression） 显性条件下，变异表达（相同基因型的人群有着程度和强度上变异的表型特征）也频繁出现。这种情况与不外显一样：其他基因、环境因素或纯粹意外影响了性状的发生。不外显和变异表达是显性性状的典型问题，但是偶尔也出现在隐性条件下。

早现遗传（anticipation） 早现遗传是指一些遗传病（通常为显性遗传病）在连续几代的遗传中，发病年龄提前并且病情严重程度增加。随着三核苷酸重复序列扩增疾病的发现，早现遗传立刻被人们接受。这些疾病的严重程度和发病年龄与重复长度有关，而重复长度在基因传给下一代时增长。

3. 遗传印迹

子女继承两套基因，一套来自母亲，一套来自父亲。通常这两套基因都是有活性的。然而在某些情况下，只有一套基因有活性。哪套基因有活性取决于起源，有些基因只在遗传自父亲的情况下有活性，另一些只在遗传自母亲的情况下有活性。这种现象叫做遗传印迹（genomic imprinting）。它在胎儿生长和发育中起重要的作用，由DNA甲基化和染色质结构调节。

4. 线粒体遗传

线粒体基因组相对于核DNA小，但是不稳定，可能是因为线粒体的DNA复制更易出错而且复制次数更高。线粒体编码的遗传病有两个罕见的特点：母系遗传和异质性。遗传是母系的，因为精子不向受精卵提供线粒体。因此线粒体遗传能影响男性和女性，但只由母亲遗传，表现出可识别的谱系模式。

细胞包含很多线粒体基因组。某些线粒体病患者的每一个线粒体基因组带有突变（同型异源性），但另一些患者正常基因组和突变基因组混合存在于每个细胞（异质性）。线粒体异质性能从母亲遗传给孩子。

二、多基因病

遗传疾病也可能是多基因的，意味着疾病可能与多个基因、生活方式和环境因子的作用有关。虽然多基因病通常有家族聚集性，但是没有明确的遗传方式，使得判断个人继承或遗传这些疾病的风险非常困难。多基因病的研究和治疗也很困难，因为导致大多数疾病的特殊因子还不清楚。数学方法已经用于研究多基因病，其中利用广泛基因组联合研究（genome-wide association study，GWAS）确定大多数多基因病的遗传因素是较有希望的方法。

1. 多因子模型

有限数量的基因位点与性状的表达有关，没有显性和隐性的区别，每个基因位点作用微小，通过累加作用形成一个明显的表型性状，环境和基因型的相互作用产生最终的表型。多因子导致两个后果，一是基因型和表现型之间简单的一对一关系消失了，除了极端表型，不可能从基因型推论出表现型。二是随着位点数目的增加，分布越来越接近正态分布曲线。

2. 阈值模型

多因子性状是连续分布的数量性状，它们是如何控制疾病的呢？图 2-3 通过阈值模型给出了最好的解释。

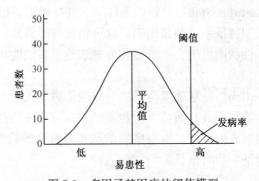

图 2-3　多因子基因病的阈值模型

人可能有或没有多基因病，但每个人都有对多基因病的易感性。易感性可能有高有低，是多基因的，并在群体中遵循正态分布。随着性状的多因子基因增加，疾病的易感性增加，当达到阈值时，疾病就产生了。超过阈值，产生疾病；低于阈值，表现为正常状态。

3. 遗传度

遗传度（heritability）是亲代传递其遗传特性的能力。它可用一定的数值表示遗传因素在性状表现中所起作用的大小。如果性状变异完全是由环境因素造成的，则遗传度等于 0。如果性状变异完全取决于遗传因素，则其遗传度为 1（100%）。遗传度和遗传方式完全不同。遗传方式是性状的固定特性，而遗传度不是。遗传度是一个群体概念。

遗传度 0.40 是指平均大约 40% 的个体差异可能是由于遗传个体差异。

4. 多基因病的特点

疾病能孤立发生，患者的父母可以不发病。虽然家族聚集性也很普遍（一个家庭中有多个患者），但是没有明显的孟德尔遗传模式。环境的影响能增高或降低发病风险。疾病的发生有性别倾向，但不是限性遗传。另外，较少发病的性别患者的一级亲属发病风险较高。同卵双生和异卵双生的发病率与孟德尔比例不相符。疾病更频繁地出现于特定的种族。

三、基因治疗

利用 DNA 治疗遗传性疾病的概念被称作基因治疗，基因治疗基于遗传学和分子生物学基础和原理。

1. 基因治疗策略

基因治疗最直接的应用是纠正导致遗传性疾病的单基因的遗传缺陷。既然病因清楚——单基因的突变，那么可以用正常基因取代错误基因。然而，机体的大多数细胞生命周期有限，不能复制，因此，导入的基因只有短暂的作用。为了完成永久的治愈，干细胞必须携带新的基因，不断遗传给子代细胞正确的基因型。导入设计的基因到人的生殖细胞（甚至胚胎）会产生遗传学人为改变的个体，使人类物种产生永久的改变。考虑到生殖细胞遗传修饰传递的伦理和安全的问题，目前所有的研究都集中在体细胞的基因转移。该方法的基因转移不是永久的，需要重复治疗。

最初基因转移研究集中在基因突变导致的疾病，如囊性纤维病、家族性高胆固醇血症和腺苷酸脱氨酶缺陷。基因治疗同样也研究多基因改变引起的基因病，包括癌症和心肌缺血。这些应用的策略是导入新的治疗性的基因而不是修正遗传异常的基因。

另外一个主要研究领域是用基因让肿瘤有更多的免疫原性或者增强对肿瘤的免疫应答。基因同样被用来诱导对潜在病原体的免疫应答。基因治疗的其他方面包括制造遗传上修饰的病毒在肿瘤细胞中特异性复制。

2. 基因转移系统

导入的基因必须进入靶细胞的细胞核，这个过程包括几个阶段：基因递送到细胞表面，结合到细胞膜，穿过细胞内部，通过细胞内运输系统到达核膜，入核。当基因进入细胞核，DNA 必须是未包装的而且释放后可以立即进行转录。当然，每一步都会有障碍，包括机体的保护性机制阻止外源物质，特别是遗传物质。

基因转移系统（gene delivery system）可分为病毒和非病毒方法。主要问题是如何导入足够的 DNA 到足够的靶组织细胞中以达到治疗所需的基因表达水平。另外一个问题是如何平衡基因表达水平和载体的潜在毒性。

1）非病毒转移系统　其包括裸露 DNA、粒子轰击和脂质体载体。

2）病毒转移系统　经过几百万年，病毒进化出克服阻碍将遗传物质导入细胞核的机制。因此病毒成为受关注的基因治疗载体。虽然所有的病毒转移遗传物质的方法比较相似，但是一些特殊病毒有着独特性质，是合适的基因治疗应用载体。病毒通常通过表面受体感染细胞。受体促进病毒内摄进入细胞的内涵体。病毒也有从内涵体逃逸和避免溶酶体降解的机制。逆转录病毒、腺病毒、腺相关病毒、疱疹病毒、痘苗病毒、慢病毒等经改造，都可作为基因治疗的载体，它们各有其优缺点。

另外细菌，特别是弗氏志贺菌，还有病毒，如昆虫杆状病毒、猫细小病毒、麻疹病毒等，人们都在研究其用于基因转移的可能。

3. 基因靶向转移

靶向转移基因增强了特异性，限制转移的基因仅在靶组织表达更为安全。另外，避免转移基因到非靶组织，也可以提高转移效率。靶向性转移有助于克服许多转移 DNA 入核的障碍。靶向转移最简单的方法是直接给药或载体到靶组织。初期的实验大多数采用这个方法。病毒或脂质体直接注射到肿瘤块进行癌症基因治疗或直接给药于呼吸道表面治疗囊性纤维病。这个方法的主要优点是简单。

受体介导的基因转移涉及细胞表面受体，将基因递送到细胞表面，通过内吞（endocytosis）作用进入细胞，然后载体及 DNA 内化作用（internalization）入核。

配体（ligand） 为了靶向导入特定的细胞类型，细胞必须表达独特的受体或过表达普通的受体，受体的配体必须清楚。当复合物递送到细胞表面，受体介导的内摄用作用能帮助基因送递。但基因表达的低效力和 DNA 在细胞中的降解妨碍了基因治疗的应用。

抗体 许多细胞靶位特异的单克隆抗体已经研制出来。直接将抗特异靶位的抗体与脂质复合体相连，再经载体靶向导入需要的细胞。逆转录病毒和腺病毒都能通过构建抗病毒外壳蛋白和细胞靶位的双特异性抗体而具有靶向性。

病毒蛋白修饰 病毒能有效转移遗传物质到靶细胞的细胞核。一个主要特征是能与细胞表面的受体结合。不同的病毒利用不同的细胞受体。病毒靶向受体的选择能影响其可感染的组织范围。例如，腺病毒和柯萨奇病毒识别的 CAR 受体在许多上皮细胞表面表达，但在造血细胞低表达。因此腺病毒是靶向上皮细胞的良好载体但却不是靶向造血细胞的适宜载体。

4. 位点特异性复制

癌症基因治疗要在肿瘤细胞中达到足够水平的基因表达量，从而确保治疗效力（细胞毒性）；同时，在非靶细胞中保持基因表达的低水平来确保安全。利用病毒能在肿瘤细胞中特异性复制，几千倍的增加局部剂量是个令人兴奋的好方法。一些利用腺病毒的创新性策略，已经获得了在肿瘤细胞中特异性复制的目标。例如，改变腺病毒复制必需的 $E1a$ 基因的启动子。天然的 $E1a$ 基因启动子在所有细胞中均具有活性，用在肿瘤中特异性表达的启动子替代该启动子，能使病毒仅在靶向细胞中复制。靶细胞可以因病毒感染后细胞裂解死亡或治疗性病毒基因的导入而死亡。另外一种方法是利用病毒和肿瘤细胞生物学的相似性。本质上，肿瘤细胞和腺病毒感染的细胞有两个共性，即进行细胞分裂和克服凋亡信号，尽管细胞分裂不受控制。这个复制功能用于病毒在肿瘤细胞靶向复制。腺病毒需要阻断宿主肿瘤抑制基因 $p53$ 的功能才能使宿主细胞分裂和使病毒有效复制。然而，大多数肿瘤的 $p53$ 已经突变，p53 蛋白无功能。如果病毒的 p53 阻断基因（$E1b$）被去除，就不能在正常细胞中复制。然而，E1b 缺陷病毒能在 p53 失活的肿瘤细胞复制。类似的方法还有利用腺病毒 $E1a$ 基因的突变来靶向 Rb 突变细胞和利用呼肠病毒靶向 Ras 过表达细胞。

小　结

　　基因是生物遗传的基本单位。在真核细胞中，基因是含有编码序列和非编码序列的DNA区域。所有基因除了编码蛋白质或RNA产物的区域外都有调节区域。细胞中的DNA会受到许多不同程度的损伤。多数细胞具有4类不同的DNA修复系统：直接修复、切除修复、错配修复和重组修复。突变是基因组小范围的核苷酸序列的改变。基因编码区的突变要重要得多。基因组是一套完整单倍体遗传物质的总和。绝大多数基因组，包括人类基因组和其他细胞生命形式，都由DNA组成，但是也有一些病毒基因组由RNA组成。真核生物和原核生物的基因组都有各自的特点。人类基因组计划是为了测定人类基因组序列和鉴定其含有的基因而进行的国际性研究。人类基因组只有20 000～25 000个功能基因。许多基因在人群中都存在许多不同的形式（等位基因），每个人是这些基因形式的独特的复合体。基因组学研究生物的基因组，比较基因组学和药物基因组学在医学研究中很重要。单基因病是由于单个基因的改变，其在家族中的分布符合孟德尔遗传规律。大多数人类疾病都是多因素引起的，是许多基因和环境因素累积的相互作用。利用DNA治疗遗传性疾病的概念被称作基因治疗。

（王又又　伍欣星）

参 考 文 献

冯作化. 2005. 医学分子生物学. 北京：人民卫生出版社

谷志远，赵亚力. 2004. 现代医学分子生物学. 北京：人民军医出版社

特纳等著；刘进元等译. 2001. 分子生物学. 北京：科学出版社

Brown S M. 2003. Essentials of Medical Genomics. Hoboken. NJ：Wiley-Liss, Inc

Brown T A. 2002. Genomes. New York and London：Garland Science

http：//www. nature. com/scitable/topic/Genes-and -Disease-1

http：//en. wikipedia. org/wiki/Gene

http：//ghr. nlm. nih. gov/handbook

Peter P. 1998. Introduction to Molecular Biology. Boston，Mass.：McGraw-Hill

Strachan T，Andrew P. 1999. Human Molecular Genetics. New York and London：Garland Science

Chapter 2   Gene, Genome and Genomics

Section 1   Gene

A gene is the basic unit of heredity in a living organism. All living things depend on genes. Genes hold the information to build and maintain their cells and pass genetic traits to offspring.

1.1   History

The existence of genes was first suggested by Gregor Mendel, who, in the 1860s, studied inheritance in peaplants and hypothesized a factor that conveys traits from parent to offspring. Although he did not use the term *gene*, he explained his results in terms of inherited characteristics. Mendel's concept was given a name by Hugo de Vries in 1889, who in his book *Intracellular Pangenesis* coined the term "pangen" for "the smallest particle (representing) one hereditary characteristic". Wilhelm Johannsen abbreviated this term to "gene" ("gen" in Danish and German) two decades later. In 1910, Thomas Hunt Morgan showed that genes occupy specific locations on the chromosome. In 1928, Frederick Griffith showed that genes could be transferred. In 1941, George Wells Beadle and Edward Lawrie Tatum showed that specific genes code for specific proteins, leading to the "one gene, one enzyme" hypothesis. Oswald Avery, Colin Munro MacLeod, and Maclyn McCarty showed in 1944 that DNA holds the gene's information. In 1953, James D. Watson and Francis Crick demonstrated the molecular structure of DNA. Together, these discoveries established the central dogma of molecular biology, which states that proteins are translated from RNA that is transcribed from DNA. This dogma has since been shown to have exceptions, such as reverse transcription in retroviruses.

In 1972, Walter Fiers and his team were the first to determine the sequence of a gene: the gene for Bacteriophage MS2 coat protein. Richard J. Roberts and Phillip Sharp discovered in 1977 that genes can be split into segments. This leads to the idea that one gene can make several proteins. Recently, biological results let the notion of gene appear more slippery.

1.2   Brief Introduction

In cells, a gene is a portion of DNA that contains both "coding" sequences that determine the sequence of corresponding protein, and "non-coding" sequences that are components but does not determine protein sequences. When a gene is active, the coding and non-coding sequences are copied in a process called transcription, producing an RNA copy of the gene's information. This piece of RNA can then direct the synthesis of proteins via the genetic code. In other cases, the RNA is used directly, for example, RNAs known as ribozymes are capable of enzymatic function, and miRNAs have a regulatory role. The DNA sequences from which such RNAs are transcribed are known as RNA genes. The molecules resulting from gene expression, whether RNA or protein, are known as gene products, and are responsible for the development and functioning of all living things.

Some viruses store their entire genomes in the form of RNA, and contain no DNA at all. Because they use RNA to store genes, their cellular hosts may synthesize their proteins as soon as they are infected and without the delay in waiting for transcription. On the other hand, RNA retroviruses, such as HIV, require the reverse transcription of their genome from RNA into DNA before their proteins can be synthesized. In 2006, French researchers came across a puzzling example of RNA-mediated inheritance in mouse. While RNA is common as genetic storage material in viruses, in mammals in particular RNA inheritance has been observed very rarely.

A gene is a locatable region of genomic sequence, corresponding to a unit of inheritance, and is associated with regulatory regions, transcribed regions and/or other functional sequence regions. The physical development and phenotype of organisms can be thought of as a product of genes interacting with each other and with the environment.

1.3　Functional Structure of a Gene

Many prokaryotic genes are organized into operons, or groups of genes whose products have related functions and which are transcribed as a unit. By contrast, eukaryotic genes are transcribed only one at a time, but may include long stretches of DNA called introns which are transcribed but never translated into protein (they are spliced out before translation). Splicing can also occur in prokaryotic genes, but is less common than in eukaryotes.

All genes have regulatory regions in addition to regions that explicitly code for a protein or RNA product.

1.3.1　Promoters

Promoters are DNA nucleotide sequences recognized by the DNA-directed RNA polymerases as their attachment sites. Promoters are located upstream, or to the $5'$ end of the structural gene. The DNA sequences range from $20\sim200$ bp.

Common to many bacterial promoters is that they have two identical or nearly identical sequences. The first is a TATAAT sequence (Pribnow box). The second is a TTGACA suquence. In eukaryotes, a sequence analogous to the Pribnow box, called the Hogness box is a TATAAA sequence.

Promoters may bind proteins other than the RNA polymerases. These other proteins bind before the polymerase and are required to allow binding of the RNA polymerase itself. These proteins, then, exert a positive, or stimulatory, control on RNA polymerase activity and consequently on gene decoding or transcription.

1.3.2　Attenuators

The attenuator sequences are found in bacterial gene clusters that code for enzymes involved in amino acid biosynthesis. Attenuators are located within so-called leader sequences, a unit of about 162 nucleotide pairs situated between the promoter-operator region and the first structural gene start site of the cluster. As the level of an amino acid in the cell rises and falls, attenuation adjusts the level of transcription to accommodate the changing levels of the amino acid. High concentrations of the amino acid result in low levels of transcription of the structural genes, and low concentrations of the amino acid result in high levels of transcription. Attenuation proceeds independently of repression; the two phenomena are not dependent on each other. Attenuation results in the premature termination of transcription of the structural genes.

1.3.3　Enhancers

Enhancers were first discovered in the animal DNA virus SV40. These control elements are located near the virus's replication origin. Enhancers have now been found in eukaryotic cells and in RNA viruses as well. The function of enhancer sequences appears to increase the number of RNA polymerase molecules transcribing a structural gene. The enhancers seem not to have particular positions relative to the initiation site of the structural gene as the promoters do. In fact, they have been found some distance upstream and downstream of the start site of the genes they control.

Enhancers are activated by the binding of sequence-specific proteins. Enhancers have been implicated in tissue-specific regulation and temporal regulation during the development phases of growth.

1.4　DNA Damage

A lesion is an alteration to the nodal chemical or physical structure of the DNA. Some of the nitrogen and carbon atoms in the heterocyclic ring systems of the bases and some of the exocyclic functional groups (i. e., the keto and amino groups of the bases) are chemically quite reactive. Many exogenous agents, such as chemicals and radiation, can cause changes to these positions. The altered chemistry of the bases may lead to loss of base pairing or altered base pairing (e. g. an altered A may base-pair with C instead of T). If such a lesion was allowed to remain in the DNA, a mutation could become fixed in the DNA by direct or indirect mutagenesis. Alternatively, the chemical change may produce a physical distortion in the DNA which blocks replication and/or transcription, causing cell death. Thus, DNA lesions may be mutagenic and/or lethal. Some lesions are spontaneous and occur because of the inherent chemical reactivity of the DNA and the presence of nodal, reactive chemical species within the cell.

1.4.1 DNA Lesions

1.4.1.1 Tautomeric Shift

Each nucleotide base can occur as either of two alternative tautomers, structural isomers that are in dynamic equilibrium. For example, thymine exists as two tautomers, the *keto* and *enol* forms, with individual molecules occasionally undergoing a shift from one tautomer to the other. The equilibrium is biased very much towards the *keto* form but every now and then the *enol* version of thymine occurs in the template DNA at the precise time that the replication fork is moving past. This will lead to an "error", because *enol*-thymine base-pairs with G rather than A. After replication, the rare tautomer will inevitably revert to its more common form, leading to a mismatch in the daughter double helix.

1.4.1.2 Deamination

For example, the base cytosine undergoes spontaneous hydrolytic deamination to give uracil. If left unrepaired, the resulting uracil would form a base pair with adenine during subsequent replication, giving rise to a point mutation.

1.4.1.3 Depurination

Depurination is a spontaneous hydrolytic reaction that involves cleavage of the N-glycosylic bond between N9 of the purine bases A and G and C1′ of the deoxyribose sugar and hence loss of purine bases from the DNA. The sugar-phosphate backbone of the DNA remains intact. The resulting apurinic site is a noncoding lesion, as information encoded in the purine bases is lost. Depurination occurs at the rate of 10 000 purines lost per human cell per hour at 37℃. Though less frequent, depyrimidination can also occur.

1.4.1.4 Replication slippager

Repeated sequences can induce replication slippage, in which the template strand and its copy shift their relative positions so that part of the template is either copied twice or missed out. The result is that the new polynucleotide has a larger or smaller number, respectively, of the repeat units. Replication slippage is probably also responsible for the trinucleotide repeat expansion diseases that have been discovered in humans in recent years.

1.4.1.5 Oxidative damage

This occurs under normal conditions due to the presence of reactive oxygen species (ROS) in all aerobic cells, for example superoxide, hydrogen peroxide and, most importantly, the hydroxyl radical (—OH). This radical can attack DNA at a number of points, producing a range of oxidation products with altered properties.

1.4.2 DNA Lesions of Mutagenesis

The definition of the term "*mutagen*" is a chemical or physical agent that causes mutations. Many chemicals that occur naturally in the environment have mutagenic properties and these have been supplemented in recent years with other chemical mutagens that result from human industrial activity. Physical agents such as radiation are also mutagenic. Most organisms are exposed to greater or lesser amounts of these various mutagens, their genomes suffering damage as a result.

Mutagens cause mutations in three different ways: Some act as base analogs and are mistakenly used as substrates when new DNA is synthesized at the replication fork; Some react directly with DNA, causing structural changes that lead to miscopying of the template strand when the DNA is replicated; Some mutagens act indirectly on DNA. They do not themselves affect DNA structure, but instead cause the cell to synthesize chemicals such as peroxides that have a direct mutagenic effect.

The range of mutagens is so vast that it is difficult to devise an all-embracing classification. We will therefore restrict our study to the most common types.

The chemical mutagens are as follows:

Base analogs are purine and pyrimidine bases that are similar enough to the standard bases to be incorporated into nucleotides when these are synthesized by the cell. The resulting unusual nucleotides can then be used as substrates for DNA synthesis during genome replication.

Deaminating agents are chemicals that have capacity of removing amino group. a certain amount of base deamination occurs spontaneously in genomic DNA molecules, with the rate being increased by chemicals.

Alkylating agents are electrophilic chemicals which readily add alkyl (e. g. methyl) groups to various positions on nucleic acids distinct from those methylated by normal methylating enzymes. The effect of alkylation depends on the position at which the nucleotide is modified and the type of alkyl group that is added.

Intercalating agents are flat molecules that can slip between base pairs in the double helix, slightly unwinding the helix and hence increasing the distance between adjacent base pairs.

The most important types of physical mutagen are as follows:

UV radiation of 260 nm induces dimerization of adjacent pyrimidine bases, especially if these are both thymines, resulting in a cyclobutyl dimer. Another type of UV-induced photoproduct is the (6-4) lesion in which carbons number 4 and 6 of adjacent pyrimidines become covalently linked.

Ionizing radiation has various effects on DNA depending on the type of radiation and its intensity. Some types of ionizing radiation act directly on DNA, others act indirectly by stimulating the formation of reactive molecules such as peroxides in the cell.

1.5 DNA Repair

In view of the thousands of damage events that genomes suffer every day, coupled with the errors that occur when the genome replicates, it is essential that cells possess efficient repair systems. Without these repair systems a genome would not be able to maintain its essential cellular functions for more than a few hours before key genes became inactivated by DNA damage. Similarly, cell lineages would accumulate replication errors at such a rate that their genomes would become dysfunctional after a few cell divisions. Most cells possess 5 different categories of DNA repair system.

1.5.1 Direct Repair

Direct repair, as the name suggests, act directly on damaged nucleotides, converting each one back to its original structure. Direct repair systems fill in nicks and correct some types of nucleotide modification. Only a few types of damaged nucleotide can be repaired directly.

Nicks can be repaired by a DNA ligase if all that has happened is that a phosphodiester bond has been broken, without damage to the 5'-phosphate and 3'-hydroxyl groups of the nucleotides either side of the nick. This is often the case with nicks resulting from the effects of ionizing radiation.

Some forms of alkylation damage are directly reversible by enzymes that transfer the alkyl group from the nucleotide to their own polypeptide chains. Enzymes capable of doing this are known in many different organisms.

Cyclobutyl dimers are repaired by a light-dependent direct system called photoreactivation. In *E. coli*, the process involves the enzyme called DNA photolyase (more correctly named deoxyribodipyrimidine photolyase). When stimulated by light with a wavelength between 300 and 500 nm the enzyme binds to cyclobutyl dimers and converts them back to the original monomeric nucleotides. Photoreactivation is a widespread but not universal type of repair: it is known in many but not all bacteria and also in quite a few eukaryotes.

The direct types of damage reversal described above are important, but they form a very minor component of the DNA repair mechanisms of most organisms.

1.5.2 Excision Repair

Excision repair involves excision of a segment of the polynucleotide containing a damaged site, followed by resynthesis of the correct nucleotide sequence by a DNA polymerase. Excision repair fall into two categories: *Base excision repair* involves removal of a damaged nucleotide base, excision of a short piece of the polynucleotide around the AP site thus created, and resynthesis with a DNA polymerase. Base excision is the least complex of the various repair systems that involve removal of one or more damaged nucleotides followed by resynthesis of DNA to span the resulting gap. It is used to repair many modified nucleotides whose bases have suffered relatively minor damage resulting from, for example, exposure to alkylating agents or ionizing radiation. The process is initiated by a DNA glycosylase that cleaves the β-N-glycosidic bond between a damaged base and the sugar component of the nucleotide. A DNA glycosylase removes a damaged base by "flipping" the structure to a position outside of the helix and then detaching it from the polynucleotide. This creates an AP or baseless site that is converted into a single nucleotide gap in the second step of the repair pathway. This step can be carried out in a variety of ways. The standard method makes use of an AP endonuclease. The single nucleotide gap is filled by a DNA polymerase, using base-paring with the undamaged base in the other strand of the DNA molecule to ensure that the correct nucleotide is inserted. After gap filling, the final phosphodiester bond is put in place by a DNA ligase. *Nucleotide excision repair* is similar to base excision repair but is not preceded by removal of a damaged base and can act on more substantially damaged areas of DNA. Nucleotide excision repair has a much broader specificity than the base excision system and is able to deal with more extreme forms of damage such as intrastrand crosslinks and bases that have become modified by attachment of large chemical groups. It is also able to correct cyclobutyl dimers by a dark repair process, providing those organisms that do not have the photoreactivation system (such as humans) with a means of repairing this type of damage. In nucleotide excision repair, a segment of single-stranded DNA containing the damaged nucleotide (s) is excised and replaced with new DNA. The process is therefore similar to base excision repair except that it is not preceded by selective base removal, and a longer stretch of polynucleotide is excised (Figure 2-1).

Figure 2-1　Mechanism of excision repair in *E. coli* (see page 13)

1.5.3 Mismatch Repair

Direct and excision repair recognize and act upon DNA damage caused by mutagens. This means that they search for abnormal chemical structures. They cannot, however, correct mismatches resulting from errors in replication because the mismatched nucleotide is not abnormal in any way. It is simply an A, C, G or T that has been inserted at the wrong position. The mismatch repair system that corrects replication errors has to detect not the mismatched nucleotide itself but the absence of base-pairing between the parent and daughter strands. Once it has found a mismatch, the repair system excises part of the daughter polynucleotide and fills in the gap, in a manner similar to excision repair.

The repair must be made in the daughter polynucleotide because it is in this newly synthesized strand that the error has occurred; the parent polynucleotide has the correct sequence. How does the repair process know which strand is which? In *E. coli* the answer is that the daughter strand is, at this stage, undermethylated and can therefore be distinguished from the parent polynucleotide, which has a full complement of methyl groups. These methylations are not mutagenic, the modified nucleotides having the same base-pairing properties as the unmodified versions. There is a delay between DNA replication and methylation of the daughter strand, and it is during this window of opportunity that the repair system scans the DNA for mismatches and makes the required corrections in the undermethylated, daughter strand.

The one difference in eukaryotes is that methylation might not be the method used to distinguish between the parent and daughter polynucleotides. Methylation has been implicated in mismatch repair in mammalian cells, but the DNA of some eukaryotes, including fruit flies and yeast, is not extensively methylated; it is thought that these organisms must therefore use a different method. Possibilities include an association between the repair enzymes and the replication complex, so that repair is coupled with DNA synthesis, or use of single-strand binding proteins that mark the parent strand.

1.5.4　Recombination Repair

Recombination repair is used to mend double-strand breaks. A single-stranded break in a double-stranded DNA molecule does not present the cell with a critical problem. The double helix retains its overall intactness and the break can be repaired by template-dependent DNA synthesis. A double-stranded break is more serious because this converts the original double helix into two separate fragments which have to be brought back together again in order for the break to be repaired. The two broken ends must be protected from further degradation. The repair processes must also ensure that the correct ends are joined: if there are two broken chromosomes in the nucleus, then the correct pairs must be brought together so that the original structures are restored.

Double-strand breaks are generated by exposure to ionizing radiation and some chemical mutagens, and are also made by the cell, in a controlled fashion, during recombination events the repair process is called non-homologous end joining (NHEJ), the name indicating that there is no need for homology between the two molecules whose ends are being joined. NHEJ is looked on as a type of recombination, as well as repairing breaks, because it can be used to join molecules or fragments that were not previously joined, producing new combinations (Figure 2-2).

Figure 2-2　Mechanism of recombination repain in *E. coli* (see page 15)

1.5.5　SOS Repair

If a region of the genome has suffered extensive damage then it is conceivable that the repair processes will be overwhelmed. The cell then faces a stark choice between dying or attempting to replicate the damaged region even though this replication may be error-prone and result in mutated daughter molecules. When faced with this choice *E. coli* cells invariably take the second option, by inducing one of several emergency procedures for bypassing sites of major damage. The best studied of these bypass processes is the SOS response.

The SOS response is primarily looked on as the last best chance that the bacterium has to replicate its DNA and hence survive under adverse conditions. However, the price of survival is an increased mutation rate because the mutasome does not repair damage. It simply allows a damaged region of a polynucleotide to be replicated. It has been suggested that this increased mutation rate is the purpose of the SOS response, mutation being in some way an advantageous response to DNA damage, but this idea remains controversial.

1.6　Mutation

A mutation is a change in the nucleotide sequence of a short region of a genome. Many mutations are point mutations that replace one nucleotide with another; others involve insertion or deletion of one or a few nucleotides. Mutations result either from errors in DNA replication or from the damaging effects of mutagens, such as chemicals and radiation, which reacts with DNA and changes the structures of individual nucleotides. All cells possess DNA-repair enzymes that attempt to minimize the number of mutations that occur.

1.6.1　The Effects of Mutations on Genomes

Many mutations result in nucleotide sequence changes that have no effect on the functioning of the genome. These silent mutations include virtually all of those that occur in intergenic DNA and in the noncoding components of genes and gene-related sequences. In other words, some 98.5% of the human genome can be mutated without significant effect.

Mutations in the coding regions of genes are much more important. First, we will look at point mutations that change the sequence of a triplet codon. A mutation of this type will have one of four effects.

It may result in a *synonymous change*, the new codon specifying the same amino acid as the unmutated codon. A synonymous change is therefore a silent mutation because it has no effect on the coding function of the genome: the mutated gene codes for exactly the same protein as the unmutated gene.

It may result in a *missense change*, the mutation altering the codon so that it specifies a different amino acid. The protein coded by the mutated gene therefore has a single amino acid change. This often has no significant effect on the biological activity of the protein because most proteins can tolerate at least a few amino acid changes without noticeable effect on their ability to function in the cell, but changes to some amino acids, such as those at the active site of an enzyme, have a greater impact.

The mutation may convert a codon that specifies an amino acid into a termination codon. This is a *nonsense mutation* and it results in a shortened protein because translation of the mRNA stops at this new termination codon rather than proceeding to the correct termination codon further downstream. The effect of this on protein activity depends on how much of the polypeptide is lost: usually the effect is drastic and the protein is non-functional.

The mutation could convert a termination codon into one specifying an amino acid, resulting in readthrough of the stop signal so the protein is extended by an additional series of amino acids at its C terminus. Most proteins can tolerate short extensions without an effect on function, but longer extensions might interfere with folding of the protein and so result in reduced activity.

Deletion and insertion mutations also have distinct effects on the coding capabilities of genes. If the number of deleted or inserted nucleotides is three or a multiple of three then one or more codons are removed or added, the resulting loss or gain of amino acids having varying effects on the function of the encoded protein. Deletions or insertions of this type are often inconsequential but will have an impact if, for example, amino acids involved in an enzyme's active site are lost, or if an insertion disrupts an important secondary structure in the protein. On the other hand, if the number of deleted or inserted nucleotides is not three or a multiple of three then a frameshift results, all of the codons downstream of the mutation being taken from a different reading frame from that used in the unmutated gene. This usually has a significant effect on the protein function, because a greater or lesser part of the mutated polypeptide has a completely different sequence to the normal polypeptide.

It is less easy to make generalizations about the effects of mutations that occur outside of the coding regions of the genome.

1.6.2 The Effects of Mutations on Multicellular Organisms

Because somatic cells do not pass copies of their genomes to the next generation, a somatic cell mutation is important only for the organism in which it occurs: it has no potential evolutionary impact. In fact, most somatic cell mutations have no significant effect, even if they result in cell death, because there are many other identical cells in the same tissue and the loss of one cell is immaterial. An exception is when a mutation causes a somatic cell to malfunction in a way that is harmful to the organism, for instance by inducing tumor formation or other cancerous activity.

Mutations in germ cells are more important because they can be transmitted to members of the next generation and will then be present in all the cells of any individual who inherits the mutation. Most mutations, including all silent ones and many in coding regions, will still not change the phenotype of the organism in any significant way. Those that do have an effect can be divided into two categories:

Loss-of-function is the normal result of a mutation that reduces or abolishes a protein activity. Most loss-of-function mutations are recessive, because in a heterozygote the second chromosome copy carries an unmutated version of the gene coding for a fully functional protein whose presence compensates for the effect of the mutation. There are some exceptions where a loss-of-function mutation is dominant, one example being haploinsufficiency, where the organism is unable to tolerate the approximately 50% reduction in protein activity suffered by the heterozygote.

Gain-of-function mutations are much less common. The mutation must be one that confers an abnormal activity on a protein. Many gain-of-function mutations are in regulatory sequences rather than in coding regions, and can therefore have a number of consequences. Alternatively the mutation could lead to over-expression of one or more genes involved in control of the cell cycle, thus leading to uncontrolled cell division and hence to cancer. Because of their nature, gain-of-function mutations are usually dominant.

Section 2　Genomes

Every organism possesses a genome that contains the biological information needed to construct and maintain a living example of that organism. Most genomes, including the human genome and those of all other cellular life forms, are made of DNA but a few viruses have RNA genomes.

2.1　Genomes of Prokaryotes

The traditional view has been that in a typical prokaryote the genome is contained in a single circular DNA molecule, localized within the nucleoid. Although the majority of bacterial and archaeal chromosomes are indeed circular, an increasing number of linear ones are being found.

2.1.1　Characteristics

On the whole prokaryotic genomes are much smaller. Most prokaryotic genomes are less than 5 Mb in size, although a few are substantially larger than this: *B. megaterium*, for example, has a huge genome of 30 Mb. In general, prokaryotic genes are shorter than their eukaryotic counterparts, the average length of a bacterial gene being about two-thirds that of a eukaryotic gene, even after the introns have been removed from the latter. Most prokaryotic genes are composed of nucletide sequences that are not interrupted by noncoding sequences; though, in some Eubacteria and Archaebacteria, noncoding sequences have been found.

The structural genes of prokaryotes are most often activated in clusters, which are often referred to as polycistronic transcription units. That is, the transcription unit contains information for the construction of more than one polypeptide. The transcription units, which are mRNA molecules, have the following nucleotide arrangements starting at the gene's 5'end: a leader sequence, a start sequence, a nucleotide sequence for a polypeptide, a stop sequence, and a spacer sequence of from 5~20 nucleotides to separate one gene from the next. *An operon* is a group of genes that are located adjacent to one another in the genome, with perhaps just one or two nucleotides between the end of one gene and the start of the next. All the genes in an operon are expressed as a single unit. This type of arrangement is common in prokaryotic genomes. Most prokaryotic genomes do not have anything equivalent to the high-copy-number genome-wide repeat families found in eukaryotic genomes. They do, however, possess certain sequences that might be repeated elsewhere in the genome.

2.1.2　Plasmid

A *plasmid* is a small piece of DNA, often, but not always circular, that coexists with the main chromosome in a bacterial cell. Some types of plasmid are able to integrate into the main genome, but others are thought to be permanently independent. Plasmids carry genes that are not usually present in the main chromosome, but in many cases these genes are non-essential to the bacterium, coding for characteristics such as antibiotic resistance, which the bacterium does not need if the environmental conditions are amenable. as well as this apparent dispensability, many plasmids are able to transfer from one cell to another, and the same plasmids are sometimes found in bacteria that belong to different species. These various features of plasmids suggest that they are independent entities.

2.2　Genomes of Eukaryotes

All of the eukaryotic nuclear genomes that have been studied are divided into two or more linear DNA molecules, each contained in a different chromosome; all eukaryotes also possess smaller, usually circular, mitochondrial genomes. The only general eukaryotic feature not illustrated by the human genome is the presence in plants and other photosynthetic organisms of a third genome, located in the chloroplasts.

2.2.1　Nuclear Genomes

Although the basic physical structures of all eukaryotic nuclear genomes are similar, one important feature is very different in different organisms. This is genome size, the smallest eukaryotic genomes being less than 10 Mb in length, and the largest over 100 000 Mb.

In contrast to prokaryotes, eukaryotes generally use monocistronic transcription units. The genes of these organisms are not activated in clusters but rather as single entities. Messenger RNA carries information for only a single polypeptide.

The eukaryotic gene is most often an arrangement of sequences similar to that of the prokaryotic gene (leader, start, etc.), but these genes also have coding and noncoding sequences of nucleotides. Consequently, the eukaryotic gene is sometimes called a split gene.

Exons and introns

In eukaryotic genes, the nucleotide sequences that specify amino acids (the coding sequences) are called exons. The noncoding nucleotide sequences are called introns. Introns vary in size, number, location, and nucleotide sequence from one gene to another. Often the total length of the introns of a gene exceeds the total length of the exons of that gene by anywhere from 2~10 times and more. The number of introns appears to increase as the size of the gene increases. Interestingly, introns do not occur randomly along a gene's length. Rather, they occur at particular locations. During the process of protein synthesis in eukaryotes, introns arc removed from the mRNA molecule soon after the gene has been decoded and mRNA has been produced.

Families of genes

Since the earliest days of DNA sequencing it has been known that multigene families—groups of genes of identical or similar sequence—are common features of many genomes. For example, every eukaryote that has been studied has multiple copies of the genes for the non-coding ribosomal RNAs (rRNAs). The rRNA genes are examples of "simple" or "classical" multigene families, in which all the members have identical or nearly identical sequences. These families are believed to have arisen by gene duplication, with the sequences of the individual members kept identical by an evolutionary process. Other multigene families, more common in higher eukaryotes than in lower eukaryotes, are called "complex" because the individual members, although similar in sequence, are sufficiently different for the gene products to have distinctive properties. One of the best examples of this type of multigene family is the mammalian globin genes.

In some multigene families, the individual members are clustered, but in others the genes are dispersed around the genome. The important point is that, even though dispersed, the members of the multigene family have sequence similarities that point to a common evolutionary origin. When these sequence comparisons are made it is sometimes possible to see relationships not only within a single gene family but also between different families. All of the genes in the α-globin and β-globin families, for example, have some sequence similarity and are thought to have evolved from a single ancestral globin gene. We therefore refer to these two multigene families as comprising a single globin gene superfamily.

Pseudogenes

Pseudogenes are defect relatives of known genes that have lost their protein-coding ability or are otherwise no longer expressed in the cell. There are two main types of pseudogene:

A *conventional pseudogene* is a gene that has been inactivated because its nucleotide sequence has changed by mutation. Many mutations have only minor effects on the activity of a gene but some are more important and it quite possible for a single nucleotide change to result in a gene becoming completely non-functional. Once a pseudogene has become non-functional it will degrade through accumulation of more mutations and eventually will no longer be recognizable as a gene relic.

A *processed pseudogene* arises not by evolutionary decay but by an abnormal adjunct to gene expression. A processed pseudogene is derived from the mRNA copy of a gene by synthesis of a DNA copy which subsequently re-inserts into the genome. Because a processed pseudogene is a copy of an mRNA molecule, it does not contain any introns that were present in its parent gene. It also lacks the nucleotide sequences immediately upstream of the 5′UTR of the parent gene, which is the region in which the signals used to switch on expression of the parent gene are located. The absence of these signals means that a processed pseudogene is inactive.

2.2.2 Eukaryotic Organelle Genomes

Almost all eukaryotes have mitochondrial genomes, and all photosynthetic eukaryotes have chloroplast genomes. Initially, it was thought that virtually all organelle genomes were circular DNA molecules. We still believe that most mitochondrial and chloroplast genomes are circular, but we now recognize that there is a great deal of variability in different organisms. In many eukaryotes the circular genomes coexist in the organelles with linear versions and, in the case of chloroplasts, with smaller circles that contain subcomponents of the genome as a whole.

Mitochondrial genome sizes are variable and are unrelated to the complexity of the organism. Most multicellular animals have small mitochondrial genomes with a compact genetic organization, the genes being close together with little space between them. The human mitochondrial genome is typical of this type.

2.3   The Repetitive DNA Content of Genomes

Repetitive DNA is found in all organisms and that in some, including humans, it makes up a substantial fraction of the entire genome. There are various types of repetitive DNA, and several classification systems have been devised. The scheme that we will use begins by dividing the repeats into those that are clustered into tandem arrays and those that are dispersed around the genome.

2.3.1   Tandemly Repeated DNA

Tandemly repeated DNA is a common feature of eukaryotic genomes but is found much less frequently in prokaryotes. This type of repeat is also called satellite DNA because DNA fragments containing tandemly repeated sequences form "satellite" bands when genomic DNA is fractionated by density gradient centrifugation. For example, when broken into fragments $50\sim100$ kb in length, human DNA forms a main band (buoyant density $1.701\text{g}\cdot\text{cm}^{-3}$) and three satellite bands ($1.687\text{g}\cdot\text{cm}^{-3}$, $1.693\text{ g}\cdot\text{cm}^{-3}$ and $1.697\text{g}\cdot\text{cm}^{-3}$). The main band contains DNA fragments made up mostly of single-copy sequences with GC compositions close to 40.3%, the average value for the human genome. The satellite bands contain fragments of repetitive DNA.

The satellite bands in density gradients of eukaryotic DNA are made up of fragments composed of long series of tandem repeats, possibly hundreds of kb in length. A single genome can contain several different types of satellite DNA, each with a different repeat unit, these units being anything from $5\sim200$ bp. The three satellite bands in human DNA include at least four different repeat types.

The alphoid DNA repeats found in the centromere regions of chromosomes is one type of human satellite DNA. Although some satellite DNA is scattered around the genome, most is located in the centromeres. Although not appearing in satellite bands on density gradients, two other types of tandemly repeated DNA are also classed as "satellite" DNA. These are minisatellites and microsatellites. *Minisatellites* form clusters up to 20 kb in length, with repeat units up to 25 bp; *microsatellite clusters* are shorter, usually less than 150 bp, and the repeat unit is usually 13 bp or less.

Telomeric DNA, which in humans comprises hundreds of copies of the motif 5'-TTAGGG-3', is an example of a minisatellite. In addition to telomeric minisatellites, some eukaryotic genomes contain various other clusters of minisatellite DNA, many, although not all, near the ends of chromosomes.

The typical microsatellite consists of a 1 bp, 2 bp, 3bp or 4 bp unit repeated $10\sim20$ times. Although each microsatellite is relatively short, there are many of them in the genome. In humans, for example, microsatellites with a CA repeat, such as

5'-CACACACACACACAC-3'
3'-GTGTGTGTGTGTGTG-5'

make up 0.25% of the genome, 8 Mb in all. Single base-pair repeats such as

5'-AAAAAAAAAAAAAAA-3'
3'-TTTTTTTTTTTTTTT-5'

make up another 0.15%.

Many microsatellites are variable, meaning that the number of repeat units in the array is different in different members of a species. This is because "slippage" sometimes occurs when a microsatellite is copied during DNA replication, leading to insertion or, less frequently, deletion of one or more of the repeat units. No two humans alive today have exactly the same combination of microsatellite length variants: if enough microsatellites are examined then a unique genetic profile can be established for every person. The only exceptions are genetically identical twins. Genetic profiling is well known as a tool in forensic science, but identification of criminals is a fairly trivial application of microsatellite variability. Microsatellites can be used to establish kinship relationships and population affinities, not only for humans but also for other animals, and for plants.

2. 3. 2　Interspersed Genome-Wide Repeats

Tandemly repeated DNA sequences are thought to have arisen by expansion of a progenitor sequence, either by replication slippage or by DNA recombination processes. Both of these events are likely to result in a series of linked repeats, rather than individual repeat units scattered around the genome. *Interspersed repeats* must therefore have arisen by a different mechanism, one that can result in a copy of a repeat unit appearing in the genome at a position distant from the location of the original sequence. The most frequent way in which this occurs is by transposition, and most interspersed repeats have inherent transpositional activity. There are two alternative modes of transposition, one that involves RNA intermediate and one that does not.

The version that involves an RNA intermediate is called retrotransposition. The basic mechanism involves three steps. An RNA copy of the transposon is synthesized by the normal process of transcription. The RNA transcript is copied into DNA, which initially exists as an independent molecule outside of the genome. The DNA copy of the transposon integrates into the genome, possibly back into the same chromosome occupied by the original unit, or possibly into a different chromosome. The end result is that there are now two copies of the transposon, at different points in the genome.

RNA transposons or retroelements are features of eukaryotic genomes but have not so far been discovered in prokaryotes. Retroviruses endogenous retroviruses and retrotransposons described so far are LTR elements, as they have long terminal repeats at either end they play a role in the transposition process. Other retroelements do not have LTRs. These are called retroposons and in mammals include the following: LINEs (long interspersed nuclear elements) contain a reverse-transcriptase-like gene probably involved in the retrotransposition process. An example is the human element LINE-1, which is 6. 1 kb and has a copy number of 516 000 in the human genome. SINEs (short interspersed nuclear elements) do not have a reverse transcriptase gene but can still transpose, probably by "borrowing" reverse transcriptase enzymes that have been synthesized by other retroelements. The commonest SINE in the human genome is Alu, which has a copy number of over 1 million. Alu seems to be derived from the gene for the 7SL RNA, a non-coding RNA involved in movement of proteins around the cell. The first Alu element may have arisen by the accidental reverse transcription of a 7SL RNA molecule and integration of the DNA copy into the human genome. Some Alu elements are actively copied into RNA, providing the opportunity for proliferation of the element.

Many transposons are able to transpose in a more direct DNA to DNA manner. With these elements we are aware of two distinct transposition mechanisms, one involving direct interaction between the donor transposon and the target site, resulting in copying of the donor element (replicative transposition), and the second involving excision of the element and re-integration at a new site (conservative transposition).

In eukaryotes, DNA transposons are less common than retrotransposons, but they have a special place in genetics because a family of plant DNA transposons, the *Ac/Ds* elements of maize, is the first transposable elements to be discovered.

DNA transposons are a much more important component of prokaryotic genome anatomies than the RNA transposons. The insertion sequences, IS1 and IS186, a single E. coli genome may contain as many as 20 of these of various types. Most of the sequence of an IS is taken up by one or two genes that specify the transposase enzyme that catalyzes its transposition. IS elements can transpose either replicatively or conservatively. Other kinds of DNA transposon known in E. coli, and fairly typical of prokaryotes in general, are as follows: insertion sequences, composite transposons, Tn3-type transposons and transposable phages.

Section 3　Genomics

3. 1　The Human Genome

Much of what makes us unique individuals is that the DNA sequence in each of us is different from that of other people. The data from the Human Genome Project indicate that on average any two people have 99. 9% identical DNA sequences. Yet that 0. 1% difference is spread over 3. 2 billion bases of DNA and thus amounts to a significant number of distinct genetic traits that uniquely distinguish the genome of every person. In fact, the HGP now estimates that there are just 32 000 functional genes in the human genome. For each of these genes, there exist many different variant forms (known as alleles) in the human population, and each person has a unique combination of these forms.

The human genome is the genome of *Homo sapiens*, which is stored on 23 chromosome pairs. Twenty-two of these are autosomal chromosome pairs, while the remaining pair is sex-determining chromosomes. The haploid human genome occupies a total of just over 3 billion DNA base pairs. The haploid human genome contains an estimated 20 000~25 000 protein-coding genes, far fewer than had been expected before its sequencing. In fact, only about 1. 5% of the genome codes for proteins, while the rest consists of RNA genes, regulatory sequences, introns and (controversially) "junk" DNA.

3. 1. 1　The Human Nuclear Genome

There are estimated 20 000~25 000 human protein-coding genes. Surprisingly, the number of human genes seems to be less than a factor of two greater than that of many much simpler organisms, such as the roundworm and the fruit fly. However, human cells make extensive use of alternative splicing to produce several different proteins from a single gene, and the human proteome is thought to be much larger than those of the aforementioned organisms. Besides, most human genes have multiple exons, and human introns are frequently much longer than the flanking exons.

Human genes are distributed unevenly across the chromosomes. Each chromosome contains various gene-rich and gene-poor regions, which seem to be correlated with chromosome bands and GC-content. The significance of these nonrandom patterns of gene density is not well understood. In addition to protein coding genes, the human genome contains thousands of RNA genes, including tRNA, ribosomal RNA, microRNA, and other non-coding RNA genes.

The human genome has many different regulatory sequences which are crucial to control gene expression. These are typically short sequences that appear near or within genes.

Aside from genes and known regulatory sequences, the human genome contains vast regions of DNA, the function of which, if any, remains unknown. These regions in fact comprise the vast majority, by estimates, 97% of the human genome size. Much of this is composed of repeat elements transposons and pseudogenes.

However, there is also a large amount of sequence that does not fall under any known classification. Much of this sequence may be an evolutionary artifact that serves no present-day purpose, and these regions are sometimes collectively referred to as "junk" DNA. There are, however, a variety of emerging indications that many sequences within are likely to function in ways that are not fully understood. Recent experiments using microarrays have revealed that a substantial fraction of non-genic DNA is in fact transcribed into RNA, which leads to the possibility that the resulting transcripts may have some unknown function. Also, the evolutionary conservation across the mammalian genomes of much more sequence than can be explained by protein-coding regions indicates that many, and perhaps most, functional elements in the genome remain unknown.

3. 1. 2　The Human Mitochondrial Genome

The complete sequence of the human mitochondrial genome has been known for over 20 years. At just 16 569 bp, it is much smaller than the nuclear genome, and it contains just 37 genes. Thirteen of these genes code for proteins involved in the respiratory complex, the main biochemical component of the energy-generating mitochondria; the other 24 specify non-coding RNA molecules that are required for expression of the mitochondrial genome. The genes in this genome are much more closely packed than in the nuclear genome and they do not contain introns. Due to the lack of a system for checking for copying errors, Mitochondrial DNA (mtDNA) has a more rapid rate of variation than nuclear DNA.

3.1.3　DNA Polymorphism

3.1.3.1　Restriction fragment length polymorphisms (RFLPs)

Recall that restriction enzymes cut DNA molecules at specific recognition sequences. This sequence specificity means that treatment of a DNA molecule with a restriction enzyme should always produce the same set of fragments. This is not always the case with genomic DNA molecules because some restriction sites are polymorphic, existing as two alleles, one allele displaying the correct sequence for the restriction site and therefore being cut when the DNA is treated with the enzyme, and the second allele having a sequence alteration so the restriction site is no longer recognized. The result of the sequence alteration is that the two adjacent restriction fragments remain linked together after treatment with the enzyme, leading to a length polymorphism. There are thought to be about 10^5 RFLPs in the human genome, but of course for each RFLP there can only be two alleles (with and without the site). The value of RFLPs in human gene mapping is therefore limited by the high possibility that the RFLP being studied shows no variability among the members of an interesting family.

In order to score an RFLP, it is necessary to determine the size of just one or two individual restriction fragments against a background of many irrelevant fragments. Southern hybridization, using a probe that spans the polymorphic restriction site, provides one way of visualizing the RFLP, but nowadays PCR is more frequently used. The primers for the PCR are designed so that they anneal either side of the polymorphic site, and the RFLP is typed by treating the amplified fragment with the restriction enzyme and then running a sample in an agarose gel.

3.1.3.2　Simple sequence length polymorphisms (SSLPs)

SSLPs are arrays of repeat sequences that display length variations, different alleles containing different numbers of repeat units. Unlike RFLPs, SSLPs can be multi-allelic as each SSLP can have a number of different length variants. There are two types of SSLP.

Minisatellites, also known as variable number of tandem repeats (VNTRs), in which the repeat unit is up to 25 bp in length; Microsatellites or simple tandem repeats (STRs), whose repeats are shorter, usually dinucleotide or tetranucleotide units.

Microsatellites are more popular than minisatellites as DNA markers, for two reasons. First, minisatellites are not spread evenly around the genome but tend to be found more frequently in the telomeric regions at the ends of chromosomes. Microsatellites are more conveniently spaced throughout the genome. Second, the quickest way to type a length polymorphism is by PCR, but PCR typing is much quicker and more accurate with sequences less than 300 bp in length. Most minisatellite alleles are longer than 300 bp. Typical microsatellites consist of $10\sim30$ copies of a repeat that is usually no longer than 4 bp in length, and so are much more amenable to analysis by PCR. There are 6.5×10^5 microsatellites in the human genome.

3.1.3.3　Single nucleotide polymorphisms (SNPs)

These are positions in a genome where some individuals have one nucleotide and others have a different nucleotide. There are vast numbers of SNPs in every genome, some of which also give rise to RFLPs, but many of which do not because the sequence in which they lie is not recognized by any restriction enzyme. In the human genome there are at least 1.42 million SNPs, only 100 000 of which result in an RFLP.

Any of the four nucleotides could be present at any position in the genome, so it might be imagined that each single nucleotide polymorphism (SNP) should have four alleles. Theoretically this is possible but in practice most SNPs exist as just two variants. This is because of the way in which SNPs arise and spread in a population. An SNP originates when a point mutation occurs in a genome, converting one nucleotide into another. If the mutation is in the reproductive cells of an individual, then one or more of the offspring might inherit the mutation and, after many generations, the SNP may eventually become established in the population. But there are just two alleles—the original sequence and the mutated version. For a third allele to arise, a new mutation must occur at the same position in the genome in another individual, and this individual and his or her offspring must reproduce in such a way that the new allele becomes established. This scenario is not impossible but it is unlikely; consequently the vast majority of SNPs are biallelic.

Most SNPs exist in just two forms, so these markers suffer from the same drawback as RFLPs with regard to human genetic mapping: there is a high possibility that a SNP does not display any variability in the family that is being studied. The advantages of SNPs are their abundant numbers and the fact that they can be typed by methods that do not involve gel electrophoresis. SNP detection is more rapid because it is based on oligonucleotide hybridization analysis. Various screening strategies have been devised, including DNA chip technology and solution hybridization techniques.

3.1.4 The Human Genome Project

The Human Genome Project was an international research effort to determine the sequence of the human genome and identify the genes that it contains. The project was coordinated by the National Institutes of Health and the U.S. Department of Energy. Additional contributors included universities across the United States and international partners in the United Kingdom, France, Germany, Japan, and China. The Human Genome Project formally began in 1990 and was completed in 2003, 2 years ahead of its original schedule.

As researchers learn more about the functions of genes and proteins, this knowledge will have a major impact in the fields of medicine, biotechnology, and the life sciences.

The main goals of the Human Genome Project were to provide a complete and accurate sequence of the 3 billion DNA base pairs that make up the human genome and to find all of the estimated 20 000~25 000 human genes. The Project also aimed to sequence the genomes of several other organisms that are important to medical research, such as the mouse and the fruit fly.

In addition to sequencing DNA, the Human Genome Project sought to develop new tools to obtain and analyze the data and to make this information widely available. Also, because advances in genetics have consequences for individuals and society, the Human Genome Project committed to exploring the consequences of genomic research through its Ethical, Legal and Social Implications (ELSI) program.

In April 2003, researchers announced that the Human Genome Project had completed a high-quality sequence of essentially the entire human genome. This sequence closed the gaps from a working draft of the genome, which was published in 2001. It also identified the locations of many human genes and provided information about their structure and organization. The Project made the sequence of the human genome and tools to analyze the data freely available via the Internet.

In addition to the human genome, the Human Genome Project sequenced the genomes of several other organisms, including brewers' yeast, the roundworm, and the fruit fly. In 2002, researchers announced that they had also completed a working draft of the mouse genome. By studying the similarities and differences between human genes and those of other organisms, researchers can discover the functions of particular genes and identify which genes are critical for life.

3.2 Genomics

Genomics is the study of the genomes of organisms. The field includes intensive efforts to determine the entire DNA sequence of organisms and fine-scale genetic mapping efforts. The field also includes studies of intragenomic phenomena such as heterosis, epistasis, pleiotropy and other interactions between loci and alleles within the genome. In contrast, the investigation of single genes, their functions and roles, something very common in today's medical and biological research, and a primary focus of molecular biology, does not fall into the definition of genomics.

3.2.1 Comparative Genomics

Comparative genomics, analyzing DNA sequence patterns of humans and well-studied model organisms side-by-side, has become one of the most powerful strategies for identifying human genes and interpreting their function. The basis of comparative genomics is that the genomes of related organisms are similar. Two organisms with a relatively recent common ancestor will have genomes that display species-specific differences built onto the common plan possessed by the ancestral genome. The closer two organisms are on the evolutionary scale, the more related their genomes will be. If the two organisms are sufficiently closely related then their genomes might display synteny, the partial or complete conservation of gene order.

One of the main reasons for sequencing the human genome is to gain access to the sequences of genes involved in human disease. The hope is that the sequence of a disease gene will provide an insight into the biochemical basis of the disease and hence indicate a way of preventing or treating the disease. Comparative genomics has an important role to play in the study of disease genes because the discovery of a homolog of a human disease gene in a second organism is often the key to understanding the biochemical function of the human gene. If the homolog has already been characterized then the information needed to understand the biochemical role of the human gene may already be in place; if it has not been characterized then the necessary research can be directed at the homolog.

3. 2. 2　Pharmacogenomics

Pharmacogenomics is the study of how genes affect a person's response to drugs. This relatively new field combines pharmacology (the science of drugs) and genomics (the study of genes and their functions) to develop effective, safe medications and doses that will be tailored to a person's genetic makeup.

Many drugs that are currently available are "one size fits all", but they don't work the same way for everyone. It can be difficult to predict who will benefit from a medication, who will not respond at all, and who will experience negative side effects (called adverse drug reactions). Adverse drug reactions are a significant cause of hospitalizations and deaths in the United States. With the knowledge gained from the Human Genome Project, researchers are learning how inherited differences in genes affect the body's response to medications. These genetic differences will be used to predict whether a medication will be effective for a particular person and to help prevent adverse drug reactions.

The field of pharmacogenomics is still in its infancy. Its use is currently quite limited, but new approaches are under study in clinical trials. In the future, pharmacogenomics will allow the development of tailored drugs to treat a wide range of health problems, including cardiovascular disease, Alzheimer disease, cancer, HIV/AIDS, and asthma.

Section 4　Genes and Disease

With the complete sequence of the human genome in hand, scientists are now poised to match monogenic disease phenotypes to their corresponding genes. By analyzing complex pedigrees, geneticists can correlate changes in gene sequence with particular disease states. After all, once a disease-associated change in the DNA sequence of a gene is identified, it is much easier to determine how the structure of the corresponding gene product (protein) might be changed in a manner that alters its biological function. The nature of disease-associated changes in protein structure and function can in turn enhance our ability to design drugs that effectively and specifically target mutant proteins.

Although 1822 of the protein-encoding genes in humans are estimated to be associated with monogenic disease, the identities of more than 1500 of these genes remain unknown, largely because many of these single-gene diseases are rare and occur in small numbers of families. However, many of the common diseases exhibit a more complex inheritance pattern and are associated with mutations in multiple genes (in other words, these conditions are polygenic). As a result, research efforts have begun to shift from a focus on monogenic disease to a focus on polygenic disease, which can involve complex interactions between genes and the environment that are not easily interpreted.

4. 1　Monogenic Diseases

The simplest genetic characters are those whose presence or absence depends on the *genotype* at a single *locus*. Sometimes a particular genotype at one locus is both necessary and sufficient for the character to be expressed, given the normal genetic and environmental background of the organism. Such characters are called mendelian characters. Monogenic diseases result from modifications in a single gene occurring in all cells of the body. Scientists currently estimate that over 10 000 of human diseases are known to be monogenic.

4.1.1　Mendelian Pedigree Patterns

A character is *dominant* if it is manifest in the **heterozygote** and *recessive* if not. Note that dominance and recessiveness are properties of characters, not genes. Mendelian characters may be determined by loci on an autosome or on the X sex chromosomes or Y sex chromosomes. Autosomal characters in both sexes and X-linked characters in females can be dominant or recessive. Males are hemizygous for loci on the X chromosomes and Y chromosomes, where they have only a single copy of each gene, so the question of dominance or recessiveness does not arise in males for X-linked or Y-linked characters. Thus there are five archetypal mendelian pedigree patterns.

4.1.1.1　Autosomal dominant inheritance

- An affected person usually has at least one affected parent.
- Affects either sex.
- Transmitted by either sex.
- A child of an affected×unaffected mating has a 50% chance of being affected (this assumes the affected parent is heterozygous, which is usually true for rare conditions).

4.1.1.2　Autosomal recessive inheritance

- Affected people are usually born to unaffected parents.
- Parents of affected people are usually asymptomatic carriers.
- Affects either sex.
- After the birth of an affected child, each subsequent child has a 25% chance of being affected.

4.1.1.3　X-linked recessive inheritance

- Affects mainly males.
- Affected males are usually born to unaffected parents; the mother is normally an asymptomatic carrier and may have affected male relatives.
- Females may be affected if the father is affected and the mother is a carrier, or occasionally as a result of non-random X-inactivation.
- There is no male-to-male transmission in the pedigree (but matings of an affected male and carrier female can give the appearance of male to male transmission).
- Trait or disease is typically passed from an affected grandfather, through his carrier daughters, to half of his grandsons.

4.1.1.4　X-linked dominant inheritance

- The trait is never passed from father to son.
- Affects either sex, but more females than males.
- Females are often more mildly and more variably affected than males.
- Matings of affected females and normal males produce 1/2 the sons affected and 1/2 the daughters affected.
- All daughters of an affected male and a normal female are affected. All sons of an affected male and a normal female are normal.

4.1.1.5　Y-linked inheritance

- Affects only males.
- Affected males always have an affected father (unless there is a new mutation).
- All sons of an affected man are affected.

4.1.2　Complications to the Basic Pedigree Patterns

4.1.2.1　Nonpenetrance

With dominant conditions, nonpenetrance is a frequent complication. By definition, a dominant character is manifest in a heterozygous person, and so should show 100% penetrance. Nevertheless, many human characters, while generally showing dominant inheritance, occasionally skip a generation. Very often the presence or absence of a character depends, in the main and in normal circumstances, on the genotype at one locus, but an unusual genetic background, a particular lifestyle or maybe just chance means that the occasional person may fail to manifest the character.

4.1.2.2　Delayed dominance

The genotype is fixed at conception, but the phenotype may not manifest until adult life. In such cases the penetrance is age-related. The penetrance may become 100% if the person lives long enough, or there may be people who carry the gene but who will never develop symptoms no matter how long they live.

4.1.2.3　Variable expression

Related to nonpenetrance is the variable expression, variable extent and intensity of phenotypic signs among people with a given genotype frequently seen in dominant conditions. The cause is the same as with nonpenetrance: other genes, environmental factors or pure chance have some influence on development of the symptoms. Nonpenetrance and variable expression are typically problems with dominant, rather than recessive, characters. However, both nonpenetrance and variable expression are occasionally seen in recessive conditions.

4.1.2.4　Anticipation

Anticipation describes the tendency of some variable dominant conditions to become more severe in successive generations. Anticipation suddenly became respectable, even fashionable, with the discovery of trinucleotide repeat expansion disease. Severity or age of onset of these diseases correlates with the repeat length, and the repeat length tends to grow as the gene is transmitted down the generations.

4.1.3　Genomic Imprinting

People inherit two copies of their genes, one from their mother and one from their father. Usually both copies of each gene are active, or "turned on" in cells. In some cases, however, only one of the two copies is normally turned on. Which copy is active depends on the parent of origin: some genes are normally active only when they are inherited from a person's father; others are active only when inherited from a person's mother. This phenomenon is known as genomic imprinting. Genomic imprinting plays a critical role in fetal growth and development. Imprinting is regulated by DNA methylation and chromatin structure.

4.1.4　Mitochondrial Inheritance

The mitochondrial genome is small but highly mutable compared to nuclear DNA, probably because mitochondrial DNA replication is more error-prone and the number of replications is much higher. Mitochondrially-encoded diseases have two unusual features, matrilineal inheritance and frequent heteroplasmy.

Inheritance is matrilineal, because sperm do not contribute mitochondria to the zygote. Thus a mitochondrially inherited condition can affect both sexes, but is passed on only by affected mothers, giving a recognizable pedigree pattern.

Cells contain many mitochondrial genomes. In some patients with a mitochondrial disease, every mitochondrial genome carries the causative mutation (*homoplasmy*), but in other cases a mixed population of normal and mutant genomes is seen within each cell (*heteroplasmy*). mitochondrial heteroplasmy can be transmitted from heteroplasmic mother to heteroplasmic child.

4.2　Polygenic Diseases

Genetic disorders may also be complex, multifactorial or polygenic, this means that they are likely associated with the effects of multiple genes in combination with lifestyle and environmental factors. Although complex disorders often cluster in families, they do not have a clear-cut pattern of inheritance. This makes it difficult to determine a person's risk of inheriting or passing on these disorders. Complex disorders are also difficult to study and treat because the specific factors that cause most of these disorders have not yet been identified. Numerous methods have been developed to study complex disorders. One of the more promising methods is the use of genome-wide association studies (GWAS) that identify the common genetic factors that underlie major complex disorders.

4. 2. 1   The Multifactorial Model

The multifactorial model is then: Several, but not an unlimited number, loci are involved in the expression of the trait; There is no dominance or recessivity at each of these loci; The loci act in concert in an additive fashion, each adding or detracting a small amount from the phenotype; The environment interacts with the genotype to produce the final phenotype.

As more loci are included we see two consequences: The simple one-to-one relationship between genotype and phenotype disappears: except for the extreme phenotypes, it is not possible to infer the genotype from the phenotype As the number of loci increases, the distribution looks increasingly like a Gaussian curve.

4. 2. 2   The Threshold Model

If multifactorial traits are quantitative traits with continuous distribution, how can they control diseases? Multifactorial diseases are best explained by the threshold model.

You may or may not have a polygenic disease, but everyone has a certain susceptibility to polygenic disease. The susceptibility may be low or high; it is polygenic and follows a Gaussian distribution in the population. As the number of multifactorial genes for the trait increases, the liability for the disease increases. When it reaches a threshold, the liability is so great that abnormality, what we call disease, results. Beyond the threshold, disease results. Below the threshold, normal development is observed. (Figure 2-3)

Figure 2-3   Threshold model of polygenic disease (see page 26)

4. 2. 3   Heritability

Heritability is the extent to which genetic individual differences contribute to individual differences in phenotypic individual differences. Because it is a proportion, its numerical value will range from 0. 0 (genes do not contribute at all to phenotypic individual differences) to 1. 0 (genes are the only reason for individual differences). Heritability is quite different from the mode of inheritance. The mode of inheritance is a fixed property of a trait, but heritability is not. Heritability is a population concept. A heritability of 40% informs us that, on average, about 40% of the individual differences may in some way be attributable to genetic individual difference.

4. 2. 4   Characteristics

A multifactorial disease has a combination of distinctive characteristics that can be differentiated from clear-cut Mendelian or sex-limited conditions. These traits include the following:

The disease can occur in isolation, with affected children born to unaffected parents. Although familial aggregation is also common (i. e., there may be multiple cases in the same family), there is no clear Mendelian pattern of inheritance.

Environmental influences can increase or decrease the risk of the disease.

The disease occurs more frequently in one gender than in the other, but it is not a sex-limited trait. In addition, first-degree relatives of individuals belonging to the more rarely affected gender have a higher risk of bearing the disease.

The concordance rates in monozygotic and dizygotic twins contradict Mendelian proportions. A concordance rate is a measure of the rate at which both twins bear a specific disease.

The disease occurs more frequently in a specific ethnic group.

4.3 Gene Therapy

The concept of using DNA as a treatment for genetic disease, called gene therapy, has its roots in the history of genetics and molecular biology.

4.3.1 Strategies of Genes Therapy

The most intuitive application for gene therapy is to correct an inherited defect within a single gene that causes a genetic disease. Since the cause of the disease is clear, a mutation within a single gene, the potential therapy is equally apparent, replace the faulty gene with a normal copy. However, most cells in the body have a limited life span and do not replicate, so an introduced gene would have only a temporary effect. For a gene therapy strategy to achieve a permanent cure, stem cells must incorporate the new gene and continually supply progeny cells with the corrected genotype. Introducing an engineered gene into human germ cells (or even into embryos) would lead to the production of genetically engineered humans, creating a permanent change in the human species. In view of the ethical and safety concerns of germ-line transmission of genetic alterations, all studies performed so far have focused on somatic-cell gene transfer. The consequence of this approach is that gene transfer is not permanent and the therapy needs to be repeated.

Initial gene-transfer studies focused on diseases caused by single-gene mutations. Some examples are cystic fibrosis, familial hypercholesterolemia, and adenosine deaminase deficiency. Gene therapy research has also examined diseases that result from alterations in many genes, including cancer and cardiac and limb ischemia. The strategy in these applications is to deliver a new therapeutic gene rather than to correct an inherited abnormal gene.

Another major area of interest is to use genes to make tumors more immunogenic, or to enhance the immune response against a tumor. Genes are also being used to induce an immune response against potential pathogens, in essence using genes as vaccines. Another area of gene therapy involves creating genetically modified viruses to replicate specifically within tumor cells.

4.3.2 Gene Delivery Systems

The introduced gene must enter the nucleus of the target cell, which is accomplished in several steps: The gene is delivered to the cell surface, binds to the cell membrane, crosses into the cell's interior, negotiates the intracellular trafficking pathways to reach the nuclear membrane, and enters the nucleus. Once the gene is in the nucleus, its DNA must be unpacked and released in a transcription-ready form. Of course, each step presents its own obstacles, including the body's protective mechanisms that prevent foreign material, particularly genetic material, from gaining access to a cell's nucleus.

The many delivery systems that have been studied can generally be divided into viral and nonviral methods. In either case, the predominant problem has been introducing enough DNA into enough cells in the target tissue to achieve therapeutic levels of gene expression. An additional problem has been balancing the level of gene expression against any potential toxicity of the delivery vehicle.

Nonviral Delivery Systems mainly include naked DNA, particle Bombardment and liposome Vectors.

Viral Delivery Systems Over many millions of years, viruses have evolved mechanisms to overcome the obstacles blocking delivery of genetic material into the cell's nucleus. Thus viruses make attractive vectors for gene therapy. Although all viruses share some similarities in the methods used to transfer their genetic material, it is the distinct properties of specific viruses (retroviruses, adenovirus, adeno-associated viruses, herpes simplex virus, vaccinia, alphaviruses and lentivirus) that dictate the choice of vector suitable for a particular gene therapy application.

Viruses usually infect cells by targeting a surface receptor that is used by the cell for other functions. This receptor targeting improves internalization of the virus into endosomes within the cell. Viruses have also developed mechanisms for escaping from the endosome in the cytoplasm and avoiding lysosomal degradation. Individual viruses have both highly desirable properties for use as gene-transfer vectors but also some problems that limit their full use.

Bacteria, in particular Shigella flexneri, and viruses, such as baculovirus, feline parvovirus, and measles virus, are being studied for their possible use in gene delivery.

4. 3. 3　Targeting Gene Delivery

Gene delivery is targeted to enhance specificity, because limiting transgene expression to the target tissue improves safety. Furthermore, by avoiding delivery of the vector to nontarget tissues, efficacy is improved. Targeting can help overcome the many obstacles and impediments to delivery of DNA to the nucleus.

The simplest way to target a tissue is direct administration of the vector to that tissue, and most initial clinical trials took this approach. Viruses or liposomes are directly injected into tumor masses for cancer gene therapy or applied directly to the surface of the respiratory tract for cystic fibrosis. The main advantage of this method is its simplicity.

Receptor-mediated gene delivery involves cell surface receptors that provide an avenue for both targeting gene delivery to the cell surface, entering the cell by endocytosis, and vector internalization into the nucleus.

4. 3. 3. 1　Ligands

To target a specific cell type, the cell in question must either express a unique receptor or overexpress a common receptor. In addition, a ligand for that receptor needs to be identified. These ligands have usually been associated with therapeutic DNA by using a polylysine bridge. When the complexes are delivered to the cell surface, receptor binding and receptor mediated internalization of the complex can aid gene delivery. Despite the elegance of this approach, this field of gene therapy has been hampered by poor efficacy of gene expression and DNA degradation within the cell.

4. 3. 3. 2　Antibodies

Monoclonal antibodies specific for a variety of cellular targets have been explored in several settings. Antibodies directed against a specific target can be added to lipoplex or polyplex complexes to target the vector to the desired cell. Both retroviruses and adenoviruses have been retargeted by constructing bispecific antibodies that bind to both a viral coat protein and to the desired cellular target.

4. 3. 3. 3　Viral Protein Modification

As noted earlier, viruses have developed many strategies for efficient delivery of their genetic material into the nucleus of the target cell. A primary characteristic is the ability to bind to receptors on the cell surface. Different viruses usurp different cellular receptors. The choice of receptor targeted by the virus influences the repertoire of tissues that can be infected. For instance, the CAR receptor for the adenovirus, which is shared by the Coxsackie virus, is expressed on many epithelial surfaces but at low levels on hematopoietic cells (blood forming stem cells). Thus the adenovirus is good vector for targeting the epithelium but a poor vehicle for hematopoietic cells.

4. 3. 4　Site-Specific Replication

A major limitation in cancer gene therapy is achieving high enough levels of the therapeutic agent within the tumor cell to ensure efficacy (cell toxicity) while maintaining low enough levels in nontarget tissues to ensure safety. The concept of using a virus that could specifically replicate in tumor cells, increasing the local dose several thousands folds, is an exciting proposition. Several innovative strategies, mainly using adenovirus, have been pursed to achieve this aim of tumor-specific viral replication. One method involves changing the promoter of the adenoviral E1a gene, which is essential for viral replication. The native E1a promoter is active in all cell types. Replacement of the viral promoter with the promoter of a gene that is preferentially expressed in a tumor can limit viral replication to the targeted cell type. The target cell can then be killed, either as a result of the lytic viral infections or by transfer of a therapeutic viral gene. Another approach is to exploit the similarities between viral and tumor cell biology. In essence, both tumor cells and adenoviral-infected cells share two common needs: to undergo cell division and to overcome signals for apoptosis (protective cell death) so they can stay alive, despite uncontrolled cell division. This duplication of function can be exploited to target viral replication to tumor cells. The adenovirus needs to block the function of the host tumor-suppressor gene *p53*, so it can push the host cell into cell division and replicate itself efficiently. However, the majority of tumors already have mutations in the *p53* gene that render the p53 protein nonfunctional. If the virus is modified so that its *p53*-blocking gene (E1b) is removed, it will not be able to replicate in normal cells. However, the E1b-deficient virus would be able to replicate in tumor cells that have inactive *p53*. Similar methods focus on mutations in the adenoviral E1a gene to target Rb-mutated cells and use of a reovirus to target Ras overexpressing cells.

Summary

A gene is the basic unit of heredity in a living organism. In eukaryotic cells, a gene is a portion of DNA that contains both "coding" sequences and "non-coding" sequences. All genes have regulatory regions in addition to regions that explicitly code for a protein or RNA product. DNA in cells suffers a wide range of damage. Most cells possess four different categories of DNA repair system: direct repair, excision repair, mismatch repair and recombination repair. A mutation is a change in the nucleotide sequence of a short region of a genome. Mutations in the coding regions of genes are much more important. A genome is an organism's complete set of DNA, including all of its genes. Most genomes, including the human genome and those of all other cellular life forms, are made of DNA but a few viruses have RNA genomes. eukaryotes and prokaryotes have quite different types of genome. The Human Genome Project was an international research effort to determine the sequence of the human genome and identify the genes that it contains. The Human Genome Project now estimates that there are just 20 000~25 000 functional genes in the human genome. For many of these genes, there exist many different variant forms (known as alleles) in the human population, and each person has a unique combination of these forms. Genomics is the study of the genomes of organisms. Monogenic diseases are caused by alterations in a single gene, and they segregate in families according to the traditional Mendelian principles of inheritance. The vast majority of human diseases can be categorized as multifactorial which result from the cumulative interaction of a number of genes with environmental factors. The concept of using DNA as a treatment for genetic disease, called gene therapy.

(Youyou Wang, Xinxing Wu)

第三章 RNA 和非编码 RNA

RNA 是生物体内一类主要的大分子。DNA 基因组中编码的遗传信息首先被转录成大量的 RNA 分子，一个基因组的所有转录物被称为转录组（transcriptome）。对某些基因来说，RNA 是它们最终的产物，如蛋白质翻译中的 tRNA 和 rRNA，以及小核 RNA（snRNA）、小核仁 RNA（snoRNA）和最近发现的微小 RNA（miRNA）等。而对于信使 RNA（mRNA）来说，它们需要被进一步翻译成多肽，经过修饰后折叠成有功能的蛋白质。需要指出的是许多 RNA 分子也需要被修饰才能具有功能，如 rRNA 和 tRNA 就需要在特定位点被修饰，这项工作是由 snoRNA 来完成的。

从是否编码遗传信息的角度来讲，RNA 可以分成 mRNA 和非编码 RNA（non-coding RNA）。需要指出的是高等真核生物中大部分编码蛋白质的基因都被内含子打断，转录出的前体 mRNA 需要被正确地剪接，移除内含子来形成成熟的 mRNA。这项剪接的任务是由一个被称为剪接体的蛋白质-RNA 复合体完成的。一些非编码 RNA 和 snRNA 是剪接体的必要组成部分。大部分前体 mRNA 具有不止一种剪接方式，从而产生多种成熟 mRNA 形式，并最终产生多种蛋白质，这种剪接方式称为可变剪接（alternative splicing）。可变剪接是一个被高度调控的分子事件，它增加了蛋白质组的复杂度，并导致蛋白质组的组织特异性。有趣的是，这其中的某些内含子还可以编码反式作用的非编码 RNA，包括 snoRNA 和 miRNA，说明内含子除了在基因调控层面上起作用外，还有另外的重要作用。

本章将首先对 RNA 结构特征进行一些介绍，然后是 RNA 的作用方式以及核酶的结构和催化。非编码 RNA 在 RNA 加工、真核以及原核生物的基因调控中的作用组成本章的第三个主题。

第一节 RNA 的结构

一、RNA 的基本结构单位和一级结构

1. 碱基、核苷与核苷酸

RNA 的一级结构是指连接在戊糖-磷酸链上的碱基构成的线性序列。参与 RNA 组成的有 4 种核苷：腺嘌呤（A）、鸟嘌呤（G）、胞嘧啶（C）和尿嘧啶（U）。一个核苷是由一个戊糖分子和一个碱基缩合而成，二者通过糖苷键相连接。不同的核苷通过 $3'$，$5'$-磷酸二酯键相连形成线性多聚体。在 RNA 分子中，核苷和相应的磷酸基团组成一个核苷酸，如图 3-1a 所示。

2. 碱基配对

在 DNA 分子中，两种最重要的碱基配对是腺嘌呤与胸腺嘧啶、鸟嘌呤与胞嘧啶之间的配对，即沃森-克里克碱基配对。除了这两种配对，在 RNA 中还存在另外一种经典的碱基配对，即鸟嘌呤与尿嘧啶之间的配对，这种配对具有类似于沃森-克里克碱基

配对的热稳定性（图 3-1b）。RNA 分子通过部分碱基的互补配对，形成局部的小双链区。

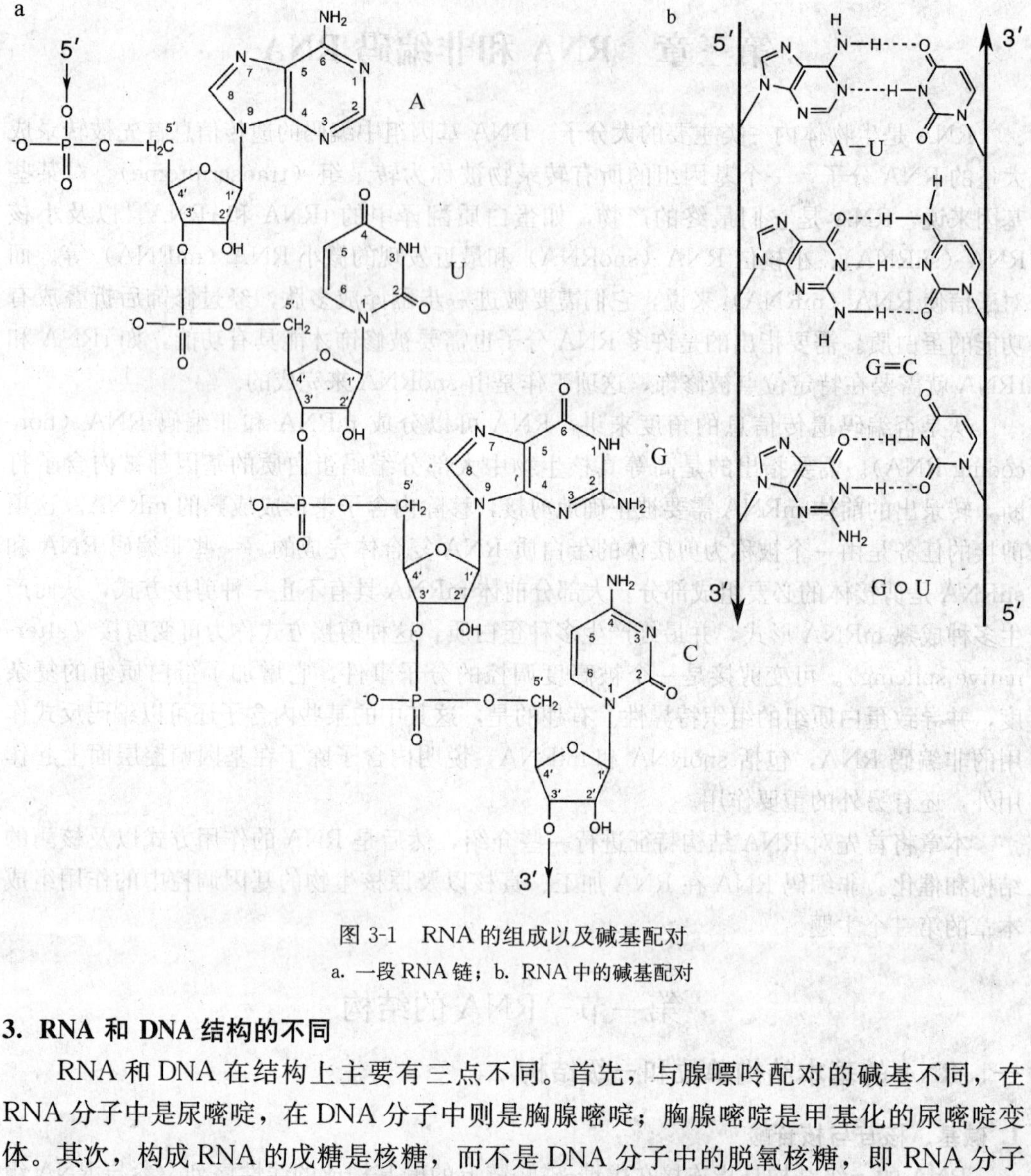

图 3-1　RNA 的组成以及碱基配对

a. 一段 RNA 链；b. RNA 中的碱基配对

3. RNA 和 DNA 结构的不同

RNA 和 DNA 在结构上主要有三点不同。首先，与腺嘌呤配对的碱基不同，在 RNA 分子中是尿嘧啶，在 DNA 分子中则是胸腺嘧啶；胸腺嘧啶是甲基化的尿嘧啶变体。其次，构成 RNA 的戊糖是核糖，而不是 DNA 分子中的脱氧核糖，即 RNA 分子中核苷糖环的第二位是一个羟基基团。羟基基团的存在可以使 RNA 分子形成 A 型双螺旋，而 DNA 分子一般形成 B 型双螺旋。因此，RNA 链的螺旋构型有深而窄的大沟和浅而宽的小沟，这和 DNA 的螺旋构型不同。并且，第 2 位的羟基基团可以攻击磷酸二酯键，切割 RNA 分子。和 DNA 分子相比，RNA 分子更不稳定，更容易发生水解。最后，RNA 链的长度通常比较短，并且没有反义链与之形成互补配对，所以 RNA 分子一般不能形成长而稳定的双链结构。

二、RNA 的二级结构

和 DNA 一样，RNA 通过碱基配对形成二级结构。但二者又有所不同，DNA 的二级结构是由两条反向平行的核苷酸链构成的右手双螺旋，脱氧核糖和磷酸分子在双螺旋外侧，可以对内部的碱基配对加以空间上的限制，因而只能形成沃森-克里克碱基配对。RNA 通常是以单链形式存在，而且非常柔韧，更容易形成复杂的碱基配对或相互作用。

RNA 的二级结构是由配对区和非配对区组成的二维结构。沃森-克里克碱基配对和 Wobble 配对是 RNA 二级结构的基本构件。在大亚基核糖体 RNA 中，超过 60％的核苷酸参与了碱基配对。在 RNA 的二级结构里，存在着不同的结构元件，如茎区、茎环、错配、内环、突起环以及多分支的环等。茎环结构是指一条 RNA 链的两端形成配对而中间部分却保持单链。内环则是配对链两侧同时被多个未配对的碱基打断。错配是内环的特殊形式，是指两条配对链各被一个碱基打断。突起环指的是配对链的一条有未配对的碱基插入，而另一条对应的位置没有碱基插入。多分支环则是指三个或多个配对区被未配对的碱基连接的结构。正如上文提到的一样，RNA 双链形成 A 型双螺旋，具有浅而宽的小沟和深而窄的大沟。这样的特征使得其他分子难以与其大沟靠近，因此 RNA 的小沟对于它与蛋白质的相互作用具有重要意义。

三、RNA 的结构花式和三级结构

几乎所有的结构 RNA 分子都需要形成特定的三级结构才能行使其功能。RNA 的三级结构由二级结构元件紧密组装而成。图 3-2 所示的是 tRNA 的二级结构和三级结构。tRNA 的二级结构是由三个茎环结构（D 臂、T 臂和反密码子臂）和一个接受臂茎区组成；其三级结构为倒"L"形。接受臂与 T 臂通过共轴堆积形成竖立的茎干，而水平的茎干则是由 D 臂和反密码子臂共轴堆积而成。D 环和 TΨC 环之间的相互作用稳定了"L"的转角部分。三级结构的形成包含复杂多样的三级相互作用，这些作用可以发生在两个茎区之间，两个单链区之间或者一个茎区和一个单链区之间。

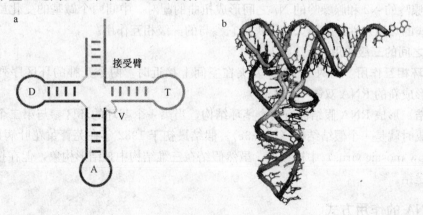

图 3-2　tRNA 的结构
a. tRNA 的二级结构；b. tRNA 的三级结构

1. RNA 双链间的相互作用

在 RNA 双链中，大多数参与形成氢键的功能基团都被包埋在核糖骨架的内部，所以两个 RNA 双链并没有很多介导它们直接相互作用的功能基团。但是它们仍然可以通过水分子介导的骨架-骨架相互作用进行双链间接触。而且，它们还可以通过 RNA 骨架的 2′羟基介导，将其中一条双链的骨架与另一条双链的小沟组装到一起。

碱基堆积是一种 RNA 的碱基排列方式，其中一个碱基位于另一个的上方。碱基堆积是稳定 RNA 结构的最基本也是最重要的方式。例如，tRNA^{Phe} 的 76 个核苷酸中的 72 个都参与了碱基堆积。同样，当两个 RNA 双螺旋末端配对的碱基堆积到一起时，这两个 RNA 双螺旋也会被上下堆积起来。这种堆积被称为共轴堆积，是大分子 RNA 形成高度有序的结构而采用的最广泛的方式。共轴堆积最先是在 tRNA^{Phe} 的晶体结构中被发现的，在该晶体中，D 臂和反密码子臂的碱基相互堆积在一起，T 臂和接受臂采用了同样的作用方式，如图 3-2b 所示。

2. RNA 链和 RNA 单链之间的相互作用

A-minor 基序　A-minor 基序是指一条 RNA 链的腺嘌呤插入到相邻双链的小沟中 CG 碱基对的位置，并与 C 或者 G 的 2 位羟基形成氢键（图 3-3a）。A-minor 基序既能稳定茎区之间的相互作用，又能稳定茎区与单链区之间的相互作用，是大分子结构 RNA 中最广泛也是最重要的远距相互作用。

碱基三联配　碱基三联配是指单链中的碱基和双链的碱基对形成附加的非经典配对（图 3-3b）。单链中的碱基能够插入双链的大沟或者小沟中，因此被分成两种形式：大沟碱基三联配和小沟碱基三联配。通过碱基三联配作用，单链 RNA 可以将两条双链拉到一起。

四元环　四元环是指由 4 个核苷酸构成的末端环（图 3-3c）。在几乎所有的结构 RNA 中，四元环都是最普遍的结构花式之一。四元环有两种常见形式："UNCG" 环和 "GNRA" 环（N 代表任何碱基，R 代表嘌呤）。所有的 "GNRA" 四元环在第一位都是一个鸟嘌呤，第四位是一个腺嘌呤。这两个碱基可以分别在鸟嘌呤的 N3、腺嘌呤的 N6、鸟嘌呤的 N2 和腺嘌呤的 N7 之间形成扭曲的配对。中间两个碱基的变化虽然对该结构花式的热稳定性影响不大，但决定它参与的三级相互作用。

3. 单链之间的三级作用

环-环相互作用　当两个 RNA 发夹在空间上接近时，两个互补的环区序列可以相互配对形成新的 RNA 双链（图 3-3d）。

假结　形成 RNA 假结需有两个茎环结构，当第一个茎环的环区参与第二个茎环的茎区形成时就是一个假结结构（图 3-3e）。假结最初于 1982 年在芜菁黄花叶病毒（turnip yellow mosaic virus）中被报道。虽然假结在三维结构中是结形构象，但在拓扑结构上并不是结。

四、RNA 的作用方式

RNA 除了基本的储存遗传信息（病毒）和传递遗传信息（大多数）的功能之外，还具有多种多样的功能，如引导修饰，调控基因表达甚至催化。总体来讲，RNA 的功能主要通过以下几种方式实现。

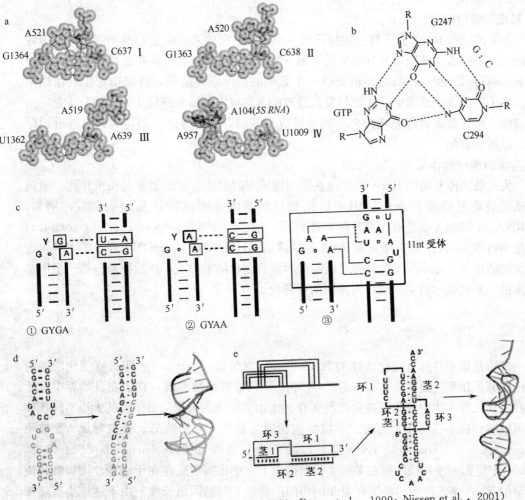

图 3-3 RNA 三级相互作用图示（Auffinger，2000；Batey et al.，1999；Nissen et al.，2001）

a. 4 种 A-minor 基序；b. 碱基三联配；c. GNRA 四元环和它们的受体；d. 环-环相互作用；e. 假结

1. 碱基配对

很大一部分功能非编码 RNA 是反式调控因子，这类 RNA 对目标的作用主要是通过碱基配对来实现的。这种作用方式可以丰富基因调控的复杂度，也就是说，一个 RNA 可以调控多个基因的表达。另外，功能 RNA 与目的基因之间的配对可以是完美的碱基配对，也可以是不完美的。例如，miRNA 一方面可以与目的基因 3'-UTR 区域非完美配对，通过翻译抑制的机制调控基因表达；另一方面可以与目的基因完美配对，通过对目的基因 RNA 的降解来调控基因表达。

2. 通过二级结构起作用

有些 RNA 是顺式作用元件，这类 RNA 通常只能通过形成二级结构来调控自身所在基因的表达，如核糖开关。核糖开关可以分成两部分，一个可以结合目标小分子的适体和一个调控平台。核糖开关有几种调控平台，首先，它可以形成一个茎环结构，通过 P 依赖转录终止来控制基因的 RNA 丰度；其次，它可以形成特定的结构来螯合核糖体结合位点，以此控制基因的蛋白质表达；最后，它可以形成特定的结构来破坏前体

mRNA 的剪接。

3. 形成三级结构

所有的结构 RNA 分子都必须折叠成一个紧密的结构来行使它们的功能，其中最好的例子就是核酶。核酶是具有催化活性的 RNA 分子，它们的底物和活性中心往往在序列上相隔很远，所以核酶必须折叠成一个紧密的三维空间结构，将底物定位到活性中心。在Ⅰ型内含子核酶中，底物与发起进攻的亲核攻击基团分别位于不同的分子上，所以核酶必须折叠成紧密的结构来容纳亲核攻击基团，并将其定位到与底物足够近的距离，以起始催化。

4. 与蛋白质一起作用

大多数功能非编码 RNA 不能独立地调控基因的表达。它们通常起桥梁作用，通过碱基配对将其招募的蛋白质因子定位到目标基因来完成对该基因的调控。例如，miRNA 和 siRNA 就是通过碱基配对将 RISC 复合体（RNA-induced silencing complex）定位到目的 RNA 上，来抑制该 RNA 翻译或切割目的 RNA。snoRNA 介导的甲基化也是类似的机制。snoRNA 首先通过碱基配对定位到目的基因上，然后招募包括一个甲基化酶的一系列蛋白质完成对目的基因的甲基化。

第二节　核　　酶

核酶是指那些具有催化活性的 RNA 分子。它与酶（enzyme），即具有催化功能的蛋白质分子相呼应而生。在过去近 30 年中有十几种核酶被发现，它们在自然界中广泛存在，在三界生物中都有发现并参与多样的生物学功能。根据一级序列的大小不同，核酶可以被分成小核酶和大核酶。小核酶包括锤头核酶、发卡核酶、链孢霉属 VS 核酶（neurospora vs ribozyme，VS 核酶）以及丁型肝炎病毒核酶（hepatitis delta virus，HDV），它们的序列长度都在 200 个核苷酸以下。小核酶主要存在于病毒、类病毒和微卫星的 RNA 基因组中，将滚环复制的中间产物加工到基因组长度。由于这类小核酶催化的都是由核酶自身的核苷发动的位点特异性的切割反应，因此又称为核苷水解核酶（nucleolytic ribozyme）。核糖核酸酶（RNase P）、Ⅰ型内含子、Ⅱ型内含子以及核糖体属于大核酶，它们的序列长度从几百到几千核苷酸不等。Ⅰ型内含子和Ⅱ型内含子在低等真核生物和植物的细胞器基因组中富集存在，催化两步连续的转酯反应，它们的准确剪接对其宿主基因所编码的 RNA 的成熟至关重要。RNase P 核酶是一类在三界生物中普遍存在、起持家功能的核酶，它由蛋白质和 RNA 组分共同组成，负责 tRNA 的加工成熟。其催化功能由 RNA 组分完成，因此称为 RNase P 核酶。另外，生物体中普通也是最重要的生物学过程——多肽合成，其中最重要的肽酰转移酶活性主要是由核糖体的rRNA 贡献的，意味着核糖体也是一个核酶。

一、核苷水解核酶

1. 核苷水解核酶的催化

核苷水解核酶催化位点特异性的切割反应将滚环复制的中间产物加工成基因组长度。该切割反应是一个转酯反应，切割位点处的 $2'$-O 攻击其相邻的 $3'$-磷酸，在 $3'$ 端形成一个 $2',3'$-环化磷酸，并在其 $5'$ 端形成一个自由的羟基。该催化反应的化学本质是一个S_N2反应，如图3-4a所示。切割位点处的$2'$-羟基被去质子化，形成的$2'$-O与临近的$3'$-磷原子形成一个较弱的键，切割位点处形成一个三棱锥式的中间态，其中$2'$-O亲核试剂和$5'$-O离去基团位于三棱锥的两个顶端，与$3'$-磷酸处在一条直线上（in-line）。该反应结束后，切割位点处的$3',5'$-磷酸二酯键被打断，$2'$-O与$3'$-磷酸之间则形成共价键连接。总体而言，这些核酶也可以催化这个切割反应的逆反应——连接反应，利用$5'$-O作为亲核攻击试剂对环化的磷酸进行攻击。

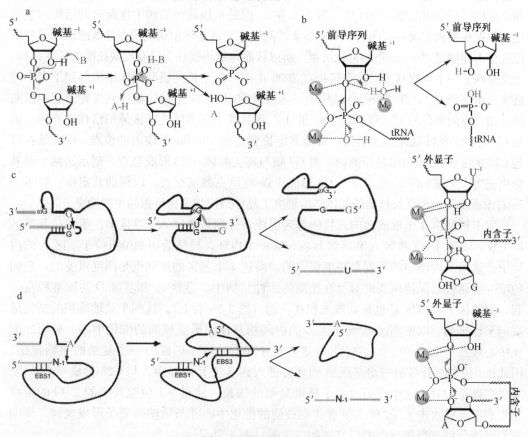

图3-4　核酶的反应机制

a. 核苷水解核酶的催化机制，广义酸（A）和广义碱（B）双面三棱锥的中间态中共平面的4个原子用灰色填充；b. RNase P核酶的催化机制；c. Ⅰ型内含子自剪接的过程及第一步反应的中间态；d. Ⅱ型内含子的自剪接过程及第一步反应的中间态

　　该切割反应是一个广义酸碱反应，其中需要一个广义碱对$2'$-羟基进行去质子化以激活$2'$-O，以及一个广义酸对离去基团$5'$-O进行质子化。核苷水解核酶之所以这样命名是因为在它们的反应中，广义酸和广义碱都是核苷，但HDV核酶例外，它可能使用了一个金属离子作为广义酸或者广义碱。

　　广义的酸碱反应是核酶的一个普遍机制。与其他核酶相比，小核酶的广义酸碱反应有两个独特之处。其中最大的不同在于它们利用与被攻击的敏感磷原子处在同一个核苷酸中的基团作为亲核攻击试剂，而不是与切割位点很远的碱基（Ⅱ型内含子）或者外源的分子（Ⅰ型内含子、RNase P和核糖体）。另外一个不同在于这类核酶使用自身的基团作为广义酸和广义碱，而其他核酶则使用金属离子作为广义酸和广义碱。

2. 核苷水解核酶的结构特征以及活性中心的组成

虽然核苷水解核酶具有不相关的一级序列和三级结构，它们都需要形成紧密的折叠以将起作用的核苷酸带到切割位点附近充当广义碱和广义酸，分别激活亲核攻击试剂和稳定离去基团。

锤头核酶由一个保守的核心和三个组装成"锤头"形状的茎区组成。锤头核酶是第一个得到原子水平分辨率结构的核酶，该晶体结构显示它的茎区II和茎区III共轴堆积在一起，而茎区I则与茎区II形成一个锐角（图3-5a左）。但是从该晶体结构中推断出的结构信息与生化数据在很多方面不相吻合，其中最大的问题在于该晶体中切割位点处敏感磷原子的定位完全不能满足"in-line"的催化机制。所以该晶体结构被认为没有反映核酶的活性结构，而是反映了一个本底状态。这个猜想在2006年由一个全长的核酶晶体结构得到了完全的证实（图3-5a中）。在这个晶体结构中，茎区 II 和茎区 III 连接区域序列戊糖骨架的扭曲使切割位点的核苷酸C17堆积到茎区 III 上，将其置于三茎区连接区域的活性中心内。碱基G5与C17的呋喃糖形成氢键，帮助其定位到适合"in-line"攻击的位点。G12处在可与C17的 2'-O 形成氢键的距离内，而G8则与离去基团 5'-O 形成氢键，暗示这两个碱基分别充当广义碱和广义酸。C3与G8形成沃森-克里克碱基配对，以帮助其定位，如图3-5a右图所示。这个全长核酶的晶体结构解决了晶体结构与生化数据的主要冲突。

发卡核酶由4个形成四向连接的茎区组成，分别是茎区A、茎区B、茎区C和茎区D（图3-5b左上）。茎区A和茎区B各包含一个内环，核酶的切割位点位于茎区A的内环中。这两个内环中的序列是高度保守的，而这4个茎区的序列则是高度可变的，它们的唯一限制就是保持碱基配对。在核酶的活性结构中，茎区A和茎区D共轴堆积在一起，而茎区B和茎区C也是共轴堆积在一起（图3-5b右上）。这两个共轴堆积的结构域反向平行排列，以使茎区A和茎区B的内环相互作用形成核酶的催化中心。核酶的催化中心涉及茎区A中A10、G11与茎区B中A24、C25形成的一个复杂的核糖拉链。因此+1位的鸟苷酸被挤出茎区A内环并插入到茎区B内环中，与C25形成一个沃森-克里克碱基配对。这个作用改变了核糖骨架的构象，使得-1位腺苷酸的 2'-O 的定位非常利于亲核攻击。G8和A38并排在核酶的催化中心并与敏感磷原子形成氢键，说明它们可能分别充当催化中的广义碱和广义酸（图3-5b下）。

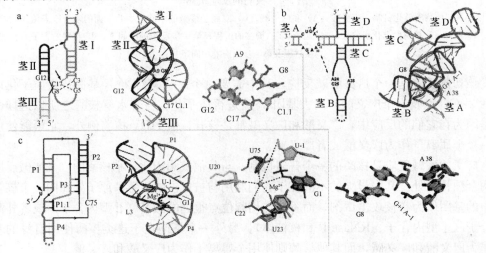

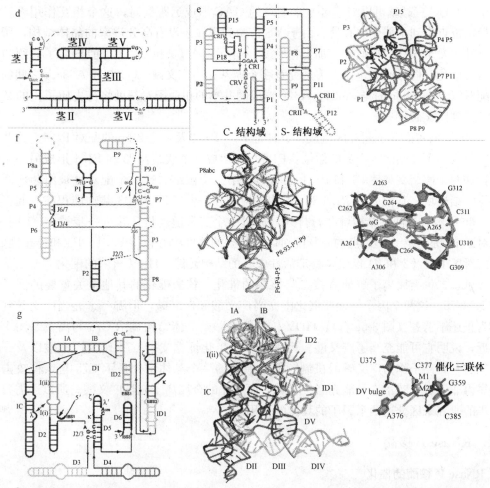

图 3-5 核酶的二级结构、三级结构以及活性中心的组成。二级结构示意图中的颜色与三级结构中的相一致。核酶的切割位点用箭头标注。如果存在晶体结构，核酶的三级结构基于这些晶体结构，用 pyMol V. 0. 99 重建（http://pymol. sourceforge. net/）。核酶切割位点处的碱基，以及组成活性中心的结构元件见标注。如果有晶体结构，核酶活性中心原子水平的组成即被从这些晶体结构中摘录出来

a. 锤头核酶，其三级结构根据 PDB 文件 2GOZ 重建；b. 发卡核酶，双向的虚线箭头表示两端的碱基形成核糖拉链，其三级结构根据 PDB 文件 1M5K 重建；c. HDV 核酶，其三级结构根据一个 C75U 的核酶突变体的晶体，PDB 文件 1SJ3 重建；d. VS 核酶的二级结构，虚线表示环 I 和环 V 之间形成远距配对；e. RNase P 核酶，二级结构中用虚线表示的结构域并未包括在晶体结构中，三级结构根据 PDB 文件 2A64 重建；f. I 型内含子，虚线的区域表示这些区域可能有周边结构插入，P1 和 P2 在晶体结构中并未被包括，一个四膜虫内含子的三级结构根据 PDB 文件 1X8W 重建；g. II 型内含子，其三级结构根据 PDB 文件 3BWP 重建

　　VS 核酶长度约为 150 nt，由 6 个茎区组成，这 6 个茎区形成两个三向连接：2-3-6 连接和 3-4-5 连接，如图 3-5d 所示。茎区 I 的内环中包含了核酶的切割位点，茎区 IV、茎区 III 和茎区 VI 共轴堆积在一起，形成一个长的柱状结构；茎区 II 和茎区 V 从这个柱中发散出来。核酶连接区域的序列对核酶茎区的共轴堆积至关重要。核酶的催化中心位于茎区 VI 的 A730 内环中，而核酶的切割位点位于茎区 I 内环中 G620 之后。茎区 I 末端

环中的 GUC 三碱基可以与茎区 V 末端环的 GAC 形成远距配对，这个相互作用将核酶的切割位点带到了催化中心处。由于 VS 核酶到现在还没有高分辨率的晶体结构，所以对比其与发卡核酶的相似性，加上生化数据显示，A765 和 G638 可能参与了核酶的广义酸碱催化，但目前为止还不清楚究竟哪个碱基充当广义碱。G620 和 A638A 之间的扭曲配对以及 G638 同时与 A621、A622 形成的共享的扭曲配对帮助将待去质子化的 $2'$ 羟基定位到广义碱附近。

HDV 核酶由 5 个配对区组成，分别是 P1～P4 以及一个 2 bp 的配对 P1.1。这些配对区形成一个复杂的双交叉的假结：P1 和 P2 形成一个假结，P3 和 P1.1 形成另一个假结，这两个假结交叉使 P1 和 P1.1 之间形成第三个假结。这 5 个配对区被组装成两个平行的结构域，其中 P1、P1.1 和 P4 共轴堆积在一起，P2 和 P3 共轴堆积在一起，如图 3-5c 所示。这两个结构域并行排列，通过 5 个交叉链连接，并进一步被 P1.1 固定。配对区 P1 是包含切割位点的底物配对区。核酶的活性中心包括 P1.1、P3 和连接区域，而配对区 P1、P2 和 P4 仅仅是稳定活性中心的结构元件。HDV 核酶的催化需要二价金属离子，二价金属离子可能直接参与核酶的催化。核酶催化的其他至关重要的因子是 C75、U75。这两个因子可能在催化中充当广义酸和广义碱，但哪一个充当广义碱到目前为止还不清楚（图 3-5c 右）。HDV 核酶的第一个晶体结构显示 C75 与离去基团十分靠近，说明它可能充当了广义酸。后来的多个生化证据支持 C75 作为广义酸以及一个水化的金属离子作为广义碱的机制。但是最近的一个晶体结构显示 C75 与亲核攻击试剂靠得非常近，暗示它可能充当广义碱，而水化的金属离子充当广义酸。所以还需要进一步的研究来区分这两个因子的具体的作用。

二、RNase P 核酶

1. RNase P 核酶的催化

RNase P 核酶是一个在三界生物中普遍存在的核酶。它利用一个水解反应将前体 tRNA 加工成成熟的 tRNA。该反应的化学本质也是一个广义酸碱反应。虽然该核酶已经有几个晶体结构，但功能基团在原子水平上的排列还没有得到，所以该反应的机制大部分是从生化数据中推断得来的。反应中的亲核攻击试剂被推断是一个被广义碱激活的水分子。该反应需要 3 个二价金属离子参与，如图 3-4b 所示。离子 A 作为广义碱对水分子进行去质子化，形成一个水化的金属离子作为亲核攻击试剂起始反应。离子 B 在稳定三棱锥中间态中起作用。前体 tRNA 切割位点（N1）处的 $2'$-羟基通过一个水分子结合第三个金属离子 C 完成对 $3'$ 离去基团的质子化。

2. RNase P 核酶的结构以及活性中心

RNase P 核酶的一级序列和二级结构是非常多样的，但所有的核酶都具有一系列高度保守的结构元件组成的核心结构。核酶的核心结构包括 5 个包含高度保守序列的保守区，分别为 CRI-CRV（图 3-5e 左）。除了这些共同的核心结构外，每一个 RNase P 核酶都具有独特的周边结构，这些周边结构在每界生物中在进化上是保守的。细菌的 RNase P 长 330～400 nt，根据其具有的周边结构可以分成两个亚类：A 型（Ancestral）和 B 型（*Bacillus*）。A 型 RNase P 核酶与 B 型 RNase P 核酸长度相当，但具有不同的周边结构。真核生物的 RNase P 核酶长度约为细菌的 2/3，并缺少很多周边结构元件。

RNase P 核酶可以被分成两个独立折叠的结构域（图 3-5e 左）。特异性结构域（S domain）与前体 tRNA 的 TψC 环相互作用，在底物识别中起作用；催化结构域（C domain）则负责前体 tRNA 接受臂和 3′-CCA 的识别，并催化水解反应。核酶的整体结构是一个由远程入邬相互作用组装到一起的、由共轴堆积的茎区组成的紧密折叠。这些在空间上相关分布的茎区形成一个特征性的平面来结合前体 tRNA。核酶的 5 个保守区和 P15 以及三个共轴堆积的结构域 P1-P4-P5、P2-P3 和 P8-P9 一起，形成活性中心（图 3-5e 右）。P4 是一个在 CRI 和 CRV 之间形成的远距配对，它对核酶的催化至关重要。最近的晶体结构显示，P4 的大沟可以结合三个二价金属离子，但这些金属离子距离活性中心太远，可能没有直接参与催化。而在连接区域结合的两个二价金属离子则与活性中心非常近，可能参与核酶的催化。核酶的周边结构分布于核酶的表面并远离活性中心，它们可以通过远程入邬作用稳定核酶的整体结构。

虽然不同的核酶具有独特的周边结构，RNase P 的整体结构却保持一致，可能是通过衍生出相应的远程相互作用得到维持的。例如，A 型核酶中 P12 和 P13 的相互作用在 B 型核酶中被 P10.1 和 P12 的相互作用所取代。另外 A 型核酶中的 P16-P17-P6 结构域在 B 型核酶中并不存在，这个结构域参与形成的三级相互作用在 B 型核酶中被 P5.1 和 P15.1 的相互作用维持。

3. RNase P 核酶的底物识别

虽然已经得到 RNase P 核酶的几个晶体结构，但尚缺少核酶-底物复合物的晶体结构，所以对底物结合的了解主要来源于生化数据以及对 tRNA 和 RNase P 晶体结构的模拟。前体 tRNA 中影响结合的决定因素包括其 5′前导序列、3′-CCA 以及 TψC 茎环。与前面提到的一样，核酶的折叠包含一个平面对 tRNA 进行结合。核酶 J5/15 连接区中一个绝对保守的腺苷酸可以与前体 tRNA 5′前导序列的 −1 位核苷酸相互作用，来识别 5′前导序列。对 TψC 茎环的识别是由核酶的特异性结构域完成的，其 P10/11 区域两个高度保守的腺苷酸参与 TψC 环的入邬。核酶 L15 与前体 tRNA 3′-CCA 的相互作用很早就被发现对水解位点的选择至关重要。最近的生化数据表明这个相互作用远非碱基配对那么简单。例如，该相互作用需要二价金属离子的参与，而且对可能的配对进行互补突变仍然导致催化活性的下降，说明该相互作用并非简单的碱基配对。

4. 蛋白质在 RNase P 核酶中的作用

RNase P 核酶是一个核酸-蛋白质复合体，其蛋白质组分的存在可以稳定核酶的结构，或帮助核酶核心结构的形成。RNase P 核酶的蛋白质组分在三界生物中差异非常大，看起来蛋白质的含量与核酶结构的复杂度有一定的联系：细菌 RNase P 核酶具有更为复杂的结构，这些由 RNA 形成的结构就可以满足催化的大部分需求，所以细菌 RNase P 核酶仅含有约 10％的蛋白质；真核生物 RNase P 核酶的结构相对简单，但其蛋白质含量达到了 70％。所以真核 RNase P 核酶中的蛋白质组分可能发挥结构作用，替代缺失掉的 RNA 结构元件的功能。

三、Ⅰ型内含子

1. Ⅰ型内含子的催化

Ⅰ型内含子催化的自我剪接反应是一个连续的两步转酯反应，包括在 5′ 和 3′ 剪接位点的顺序切割以及外显子的连接（图 3-4c 左）。在第一步切割反应中，包含 5′ 底物的茎区进入核酶的催化核心，将 5′ 剪切位点正确定位以接受攻击。一个外源的鸟苷酸也被结合到核酶的活性中心，用其 3′-羟基作为亲核试剂对 5′ 剪切位点的磷原子进行攻击。该反应导致 5′ 剪切位点磷酸二酯键的断裂，在 5′ 外显子上形成一个自由的 3′-羟基，并使外源鸟苷酸连接到内含子的 5′ 端。第一步反应完成之后，核酶经过结构的转换，将第一步中进行攻击的鸟苷酸释放，然后结合内含子末端高度保守的鸟苷酸。之后第二步反应起始，用 5′ 外显子上的 3′ 自由羟基作为亲核攻击试剂对 3′ 剪切位点进行攻击。该反应导致外显子的连接和内含子的释放。

Ⅰ型内含子催化的化学本质也是一个广义酸碱反应，至少 2 个 Mg^{2+} 直接参与了催化过程。如图 3-4c 右图所示，在第一步反应中 Mg^{2+}（M_B）结合到外源鸟苷酸的 3′-羟基，作为一个广义碱激活亲核攻击试剂。另一个 Mg^{2+}（M_A）作为广义酸稳定 3′ 离去基团。在第二步反应中这两个离子交换角色，M_A 作为广义碱而 M_B 作为广义酸。催化的中间态在第一步反应中被整合了 Mg^{2+} 的外源鸟苷酸的 2′-羟基稳定，在第二步反应中被一个氢键的网络所稳定。

2. Ⅰ型内含子的结构

Ⅰ型内含子在序列上差异很大，从几百个核苷酸到几千个核苷酸都有分布。但是所有的Ⅰ型内含子都具有一个由 9 个配对区（P1～P9）组成的高度保守的二级结构（图 3-5f 左）。这 9 个配对区组成 3 个结构域：P1-P2 结构域、P4-P6 结构域和 P3-P9 结构域，每个结构域中的配对区都是共轴堆积在一起的。P1 是底物配对区，包含了核酶的 5′ 剪切位点。P3 和 P7 形成一个对核酶催化至关重要的假结结构。核酶的三级结构是一个紧密的折叠，其中 P4-P6 结构域充当结构的骨架，而 P3-P9 结构域紧密包裹在 P4-P6 结构域的周围（图 3-5f 中）。核酶的 3 个结构域由保守的连接区连接，这些连接区在核酶整体结构的组装以及活性中心的形成中起到重要的作用。双链连接区可能通过小沟-小沟相互作用介导结构域之间的组装。单链连接区排在结构域的周边，提供一个互补的面，供结构域之间的组装。例如，J3/4 和 J6/7 可以同 P4-P6 结构域形成碱基三联配，促进 P4-P6 与 P3-P9 之间的组装（图 3-5f 左）。

除了这 9 个保守的配对区外，大多数Ⅰ型内含子还具有至少一个额外的结构域，被称为周边结构，如 *Tetrahymena* 内含子中的 P5abc、P2.1 和 P9.1-P9.2 等。这些周边结构主要是通过形成各种三级相互作用稳定核酶的核心结构，它们的缺失总体来讲会降低但不会消除核酶的剪接活性。但是某些周边结构对核酶的活性缺失是必需的。例如，*Candida* 内含子中的 P2.1，其根部配对可以与 P3 和 P6 形成一个三螺旋连配的结构，该结构对核酶催化核心紧密结构的形成起至关重要的作用。

3. 底物识别

在Ⅰ型内含子核酶的催化中，5′ 和 3′ 外显子必须在催化核心正确定位，使它们的敏感磷原子与亲核试剂足够靠近。5′ 外显子的定位首先是由内含子中的内部引导序列

（IGS）决定的，它可以与 5′ 外显子进行碱基配对，形成底物茎区 P1（图 3-5f 左）。核酶的 5′ 剪切位点是 P1 中的一个 GU 配对，它在催化核心的定位是由多个因素共同决定的。首先，P1 与 P2 是共轴堆积的，而在活性结构中，P2 与 P8 通过小沟-小沟相互作用结合到一起，所以 P1 的定位也就因此而决定。其次，J2/3 和 J8/7 作用于 P1-P2 结构域的小沟。最后，剪切位点的 GU 配对通过与 J4/5 的 A-minor 作用插入到 P4-P6 结构域。这些作用一起将 5′ 剪切位点固定到核酶的活性中心。

内含子 3′ 端是一个保守的鸟苷酸，在第二步反应中，这个残基必须进入到核酶的活性中心接受攻击。所以 3′ 剪切位点就是由这个鸟苷酸决定的。两个元件参与 3′ 剪切位点的选择，首先是一个非常短的远距配对 P9.0。它是由 P7 下游序列和内含子末端鸟苷酸上游的序列配对形成的。这个远距配对将内含子的末端鸟苷酸带到活性中心所在的 P7 附近。其次是 P10，这是一个由 3′ 外显子和部分 IGS 序列形成的配对，这个作用进一步将 3′ 剪切位点定位到活性中心。

4. Ⅰ型内含子的活性中心

Ⅰ型内含子的活性中心位于其 P7 茎区，它可以在第一步反应中结合一个外源的鸟苷酸作为亲核试剂发动攻击，也可以结合内含子的末端鸟苷酸，在第二步反应中接受攻击。该活性中心是由 4 层碱基三联配构成的。在四膜虫内含子中，鸟苷酸（外源鸟苷酸或末端鸟苷酸）与 P7 配对区的 G264-C311 配对形成一个共平面的碱基三联配（图 3-5f 右）。这个碱基三联配被夹在其他三层碱基三联配中。在其上方，C262 与 G312 配对，而 A263 与 G312 的小沟形成氢键，形成第一层碱基三联配。碱基 C262 正好位于碱基三联配的鸟嘌呤环正上方，可能通过碱基堆积稳定鸟苷酸的结合。在碱基三联配下方，A261 作用于 A256-U310 配对的大沟，形成第三层碱基三联配。最后 A306 碱基的 Hoogsteen 边作用于 C266-G309 的大沟，形成第四层碱基三联配。该活性中心可以结合两个二价金属离子充当广义酸和广义碱，分别对亲核攻击试剂进行去质子化，对 3′ 离去基团进行质子化。

四、Ⅱ型内含子

1. Ⅱ型内含子的催化

Ⅱ型内含子的催化机制与剪接体内含子完全一致，所以被认为是剪接体内含子的祖先。Ⅱ型内含子的剪接也是一个连续的两步转酯反应（图 3-4d 左）。第一步剪切中，内含子分支位点的内源腺苷酸的 2′-羟基对 5′ 剪切位点进行亲核攻击。该反应导致 5′ 剪切位点的切割，形成一个套马索结构的内含子，并在断裂的 5′ 外显子上形成一个自由的 3′-羟基。反应结束后，5′ 外显子被固定在原处，并在第二步剪切反应中，用其自由的 3′-羟基对 3′ 剪切位点进行亲核攻击，得到连接的外显子并释放套马索结构的内含子。

Ⅱ型内含子催化的化学机制与Ⅰ型内含子几乎相同（图 3-4d 右），但以下两个特征除外。首先，Ⅱ型内含子利用一个内含子内部的腺苷酸攻击 5′ 剪切位点，用其 2′-羟基作为亲核攻击试剂。其次，与Ⅰ型内含子中剪切位点处的 2′-羟基参与稳定中间态不同，该基团在Ⅱ型内含子中是自由的。

2. Ⅱ型内含子的结构

除了结构域Ⅴ（D5）中的保守序列外，Ⅱ型内含子的一级序列是高度变化的，但它们都可以形成一个高度保守的二级结构。该结构由 D1～D6 共 6 个结构域组成（图 3-5g），D1 包含了识别两端外显子以及分支位点亲核攻击试剂的必要信息。D2 和 D3 是核酶完全活性的必要组成部分，它们的连接区域（J2/3）是活性中心的一部分。D4 通常编码一个大的可读框，对核酶的折叠和催化没有作用。D5 是 Ⅱ 型内含子中最为保守的区域，它直接参与活性中心的形成，并提供核酶催化的功能基团。D6 包含分支位点腺苷酸，它的 2′ 羟基在第一步剪切反应中充当亲核攻击试剂。

根据其不同的结构特征，Ⅱ 型内含子可以分为 ⅡA、ⅡB 和 ⅡC 三个亚类。一个 ⅡC 核酶的晶体结构最近被解析。如图 3-5g 中所示，Ⅱ 型内含子的活性结构是 6 个结构域的紧密组装，其中 D1 充当整个结构的骨架，为外显子的结合和活性中心的形成提供一个预先形成的外部沟。在结构的上方，D1 的 ⅠA 茎区和 ⅠB 茎区共轴堆积在一起；处在左方的 Ⅰ(i)、Ⅰ(ii) 和 D2 也是共轴堆积在一起。结构域 Ⅰ 的 ⅠC 茎区与共轴堆积的 Ⅰ(i)-Ⅰ(ii)-D2 靠得很近并平行排列。在右方，结合底物的结构域 ⅠD1 和 ⅠD2 紧密的靠在一起。D3 和 D4 在底部共轴堆积在一起。这些结构域将保守的结构域 Ⅴ 包裹在中间。D6 虽然包含在此晶体中，但并不可见。但在该晶体中有合适的空间，D6 被认为位于 D5 旁边的缝隙中。

3. 底物识别

Ⅱ 型内含子底物的识别是由内含子和外显子的相互作用，主要是碱基配对，决定的。内含子 ⅡA 和 ⅡB 在底物的识别方面具有更多的共性。它们的 5′ 底物是由两个 6 bp 的沃森-克里克碱基配对（IBS1-EBS1 和 IBS2-EBS2）的相互作用进行识别的。而 5′ 剪接位点的选择则是由内含子的核心决定的，特异性的识别 EBS1 末端单链和双链交界处的磷酸二酯键（图 3-5g 左）。内含子 ⅡA 和 ⅡB 3′ 底物的识别则是不同的。ⅡA 内含子通过一个称为 δ-δ′ 的沃森-克里克碱基配对对 3′ 底物进行识别，而 ⅡB 内含子则是通过内含子 EBS3 与外显子 IBS3 的配对进行 3′ 底物的识别。ⅡC 内含子的 5′ 底物的识别与 ⅡA、ⅡB 内含子非常不同，跟前面介绍的一样，ⅡC 内含子的 EBS1-IBS1 配对非常短，并且缺少 EBS2-IBS2 配对。ⅡC 内含子 3′ 底物的识别与 ⅡB 内含子非常相似，都是使用 EBS3-IBS3 配对完成的（图 3-5g 左）。

4. 活性中心

核酶的活性中心涉及三个主要的结构域 D1、D5、D6 和一个重要的 J2/3 连接。其中 D1 结合底物，D6 提供第一步反应中的亲核攻击试剂，D5 和 J2/3 是催化的区域。活性中心的形成是一个对 6 个结构域进行重组的过程。除了作为核酶折叠的骨架的功能外，结构域 Ⅰ 通过 ε-ε′ 以及与 D5 相互作用 κ-κ′、ζ-ζ′ 和 λ-λ′ 将两端的剪切位点固定到活性中心。分支位点腺苷酸的定位是通过它周边的多嘧啶序列与 D1 环区序列的非标准碱基配对决定的。

结构域 Ⅴ 是内含子中最保守的结构域，对催化非常重要。D5 的大部分序列都参与形成保证 D5 正确定位的三级相互作用，如 κ-κ′、ζ-ζ′ 以及 λ-λ′。D5 中参与催化的功能基团是所谓的"催化三联体"以及 D5 中的突起。在折叠的核酶中，这两个基团靠得很近，可能是一同发挥作用的。D5 同时还具有直接参与催化过程的金属离子的结合位点（图 3-5g 右）。

连接 D2 与 D3 的 J2/3 是活性中心的重要组成部分，但是其具体的作用直到最近

Oceanobacillus iheyensis 的晶体结构被解析出来之后才被了解。它与 D5 的突起以及"催化三联体"形成一个三螺旋连配的结构（图 3-5g 中），将催化所必需的内含子残基带到一起。

五、核糖体是一个核酶

核糖体是在三界生物中广泛存在的最重要的生物大分子，它催化肽基转移反应，将 mRNA 翻译成蛋白质。核糖体由 2/3 的 RNA 和 1/3 的蛋白质组成。核糖体大亚基与底物及其产物的复合物的晶体结构显示，只有 RNA 处在肽键合成的位点。共有三个 RNA 基团处于可以与进行亲核攻击的 α-氨基基团形成氢键的距离内：P 位点 tRNA 上 76 位腺苷酸的 $2'$-羟基，23S rRNA 中 A2486 的 N3 基团，以及 A2486 的 $2'$-羟基。这些氢键可能与 α-氨基基团进行亲核攻击的定位有关，或者更为重要的是，直接参与催化。在过去的几年中多个高分辨率的核糖体-mRNA-tRNA 复合体的晶体结构被解析出来，揭示了核糖体催化的结构基础。大亚基核糖体的晶体结构显示，没有蛋白质组分位于距离肽键合成位点 18Å 以内的距离，因此排除掉了蛋白质因子在该反应中起作用的可能性。所以蛋白质对催化没有任何作用，而 RNA 是完成肽键转移的唯一组分，也就是说核糖体是一个基于 RNA 的酶，也就是核酶。

第三节　小 RNA 在转录后加工中的作用

一、snoRNA 和核糖体修饰

snoRNA 是真核生物中一大类非编码 RNA，它主要包含两个家族，按序列和结构特征划分为 C/D 盒 snoRNA 和 H/ACA 盒 snoRNA。它们的主要功能是分别指导核糖体 rRNA 的 $2'$-O-核糖甲基化和假尿嘧啶核苷化（尿嘧啶转变为假尿嘧啶）。这些修饰被严格限定在功能最保守的重要的 rRNA 结构域，而且对核糖体的正常功能是必需的。snoRNA 中的反义元件通过与 rRNA 上的相应位点反向互补形成联体结构精确地确定了被修饰的位置。修饰本身是由一个普通的蛋白酶（甲基化酶或假尿嘧啶合成酶）催化的。这些酶与一个指导 snoRNA（guide snoRNA）及其他特异蛋白组装成小分子核仁核糖核蛋白体（small nucleolar ribonucleoprotein，snoRNP）。

C/D 盒 snoRNA 与 H/ACA 盒 snoRNA 的保守序列和结构如图 3-6 所示。反义 C/D 盒 snoRNA 含有两个保守序列（图 3-6a）：C 盒（$5'$-RUGAUGA-$3'$）和 D 盒（$5'$-CUGA-$3'$）。它们分别处于距 $5'$ 端和 $3'$ 端几个核苷酸的位置，但通过 $5'$ 端和 $3'$ 端形成配对使得它们挨在一起。很多 snoRNA 在其中央部位还有另一组 C$'$/D$'$ 盒，其序列不完全符合 C/D 盒的序列特征。在 D 盒和（或）D$'$ 盒 $5'$ 处，紧接着 1 个或 2 个反义元件，它的长度为 10～20 个核苷酸，与 rRNA 上的特定序列反向互补形成指导双链区。典型的甲基化位点是与 snoRNA 的 D 盒或 D$'$ 盒（CUGA motif）上游第五个核苷酸相配对的 rRNA 链上的核苷酸。

H/ACA 盒 snoRNA 的典型二级结构特征：有两个由线性区连接的带内环的发夹结构，发夹结构的下游有一短尾，如图 3-6b 所示。在线性连接区内有一保守的 H 盒（ANANNA），短尾距 $3'$ 端 3 个核苷酸处有保守的 ACA 三核苷酸，也称为 ACA 盒。每个 H/ACA 盒 snoRNA 有 1 个或 2 个双向的反义元件，它们位于发夹结构的内环区。反义元件可与 rRNA 假尿嘧啶修饰位点两侧的序列形成长为 9～13 个核苷酸的双向指导双链区，修饰位点通常位于 H 盒或 ACA 盒上游 14～16 个核苷酸的位置。

两类 snoRNA 均与一套高度保守的核心蛋白组成小核糖核蛋白体，snoRNA 相当于一个适配器将具有催化作用的蛋白质和靶标 rRNA 连接在一起。C/D 盒 snoRNP 中的核心蛋白是原纤维 N（fibrillarin）、L7Ae 和 Nop56/Nop58（按古细菌中的命名）。真核生物中具有它们的同源物。原纤维素是 2′-O-核糖甲基化酶，它与 snoRNA 的结合需要另外两个核心蛋白 L7Ae 和 Nop56/58 形成的桥。L7Ae 集结 C/D RNP 复合体的组装。它先结合 C 盒和 D 盒形成的 K 转折基序（K-turn motif），结合后的结构经调整后能和 Nop56/Nop58 结合。然后 Nop56/Nop58 招募原纤维素从而形成完整的、有功能的复合体。

古细菌 H/ACA 盒 snoRNP 中的核心蛋白是 Cbf5、L7Ae、Gar1 和 Nop10。其中 Cbf5 是假尿嘧啶合成酶，催化 snoRNA 指导的假尿嘧啶合成反应。它具有两个结构域：催化结构域和 PUA 结构域。Cbf5 的 PUA 结构域能与 ACA 盒和下端茎区结合，从而将反义指导序列锚定在 Cbf5 的催化位点附近。Gar1 和 Nop10 能分别结合到 Cbf5 催化结构域的不同位点。现有的证据表明，Gar1 在 snoRNP 与靶标 RNA 的结合和释放中起作用。与 C/D 盒 snoRNP 一样，L7Ae 与位于 snoRNA 上端茎环环区的 K 转折基序直接结合。

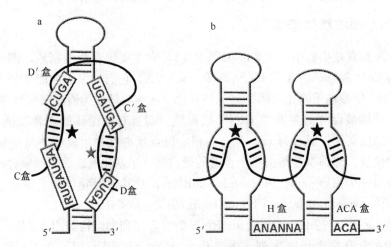

图 3-6　指导 snoRNA 及它们的靶标 RNA

C/D 盒 a. snoRNA 和 H/ACA 盒 b. snoRNA 包含保守的和特异性的序列基序（浅色）和独特的指导区域（深色）。星号标记被修饰的核苷酸的位置

除了 rRNA 的修饰，C/D 盒 snoRNA 和 H/ACA 盒 snoRNA 近来被发现也具有其他功能和靶标，在非核仁中起作用。snoRNA 可指导真核生物中的 snRNA 转录后修饰，古细菌中的 tRNA 修饰，锥形虫中的 SL RNA 修饰。还有一个 H/ACA 盒 snoRNA 能在端粒酶的合成起作用。HBII-52 snoRNA 甚至可以指导 mRNA 前体的可变剪接。一些 snoRNA 是组织特异性表达，并且是印记 RNA。值得注意的是，越来越多的孤儿 snoRNA 通过实验和计算机的方法从不同的物种里鉴定出来，它们没有明显的

rRNA 或 snRNA 靶标。有报道表明，人类孤儿 snoRNA 潜在靶标有位于可变剪接位点附近的倾向。所以还有很多 snoRNA 的靶标和功能有待进一步发现。

二、snRNA 和前体 mRNA 的剪接

前体 mRNA 剪接（precursor mRNA/pre-mRNA splicing）在基因表达中是一个关键的可调控步骤，是指内含子在剪接体（spliceosome）的催化下被准确地识别和切除。剪接体是由 5 个小核 RNA 复合体（U1、U2、U4、U5 和 U6），以及许多辅助蛋白组成。这些 snRNP 由它们相应的富含 U 的 snRNA 组分而得名。U4 snRNA 和 U6 snRNA 之间有大量的碱基配对（图 3-7a）。

U1 snRNA 由 163 个核苷酸组成，在它的 5′端有三甲基鸟嘌呤帽（m^3G cap）结构。不同物种之间的协同变异分析（covariation analysis）发现 U1 snRNA 具有 4 个茎环（stem-loop）的二级结构。其 5′端的 ACUUACCU 序列与前体 RNA 5′剪接位点的保守序列 AGGURAGU（R 代表嘌呤）配对，单链区的 Sm 位点（AUUUGUGG）是 Sm 蛋白的结合位点。人类 U1 snRNP 中有三个特异的蛋白质，它们是 U1 70K、U1A 和 U1C，其余的组成蛋白是各种 U snRNP 共有的 Sm 蛋白。U1 70K 和 U1A 分别结合到 U1 snRNA 的茎环结构 1 和 2。snRNP 有次序地参与剪接体的组装（图 3-7），U1 snRNP 通过 U1 snRNA 与 5′剪接位点结合，但是在 Hela 细胞的体外剪接体系中，经典的碱基配对并不能决定对 5′剪接位点的选择与剪接，说明可能有其他因子，如 U1 snRNP 的蛋白质，帮助了 5′剪接位点的识别与剪接体的形成。研究发现酵母的 U1C 蛋白识别 5′剪接位点的序列，该过程在 U1 snRNA 与 5′剪接位点的碱基配对形成之前发生。其他的 snRNA 与 5′剪接位点结合也可能帮助对该位点的识别。研究表明 U6 snRNA 与非常接近内含子 5′端的一段序列结合，该结合对于剪接是必要的（图 3-7）。

U2 snRNA 由 187 个核苷酸组成 5 个茎环结构：Ⅰ、Ⅱa、Ⅱb、Ⅲ 和 Ⅳ。U2 snRNP 包含很多种蛋白质组分，其中两个较大的蛋白质复合物是剪接因子 SF3a 和 SF3b。SF3b 结合到 U2 snRNA 的 5′端，SF3a 随后结合。U2 snRNA 的 5′端被大量修饰（2′-O-甲基化或假尿嘧啶化），这些修饰富集在 5′端 27 个核苷酸范围内，它们对于 SF3b 和 SF3a 与 U2 snRNA 的结合非常必要。在剪接体的复合体 A 中，分支位点（branch point）的保守序列与 U2 snRNA 通过碱基配对结合，这对剪接是必要的，因为这个配对可以把分支位点的 A 确立为第一步转酯反应的亲核攻击基团（nucleophile）。除此之外，U2 还与 U6 碱基配对（图 3-7a），这个作用在酵母中不必要，但在哺乳动物中对于高效的剪接却是十分必要的。

U5 snRNA 与其他 snRNA 和前体 mRNA 之间没有明显的碱基配对，但 U5 snRNA 中一个高度保守的环状结构可以与上游外显子的 3′端和下游外显子的 5′端结合，从而将 5′剪接位点和 3′剪接位点拉近，进行第二步的转酯反应（图 3-7a）。

U4 与 U6 之间的碱基配对形成了两个茎区（stem），称为茎区Ⅰ和茎区Ⅱ。随着剪接体的组装，U4 snRNA 和 U6 snRNA 之间的碱基配对被解开，U6 snRNA 随后与 U2 snRNA 和 5′剪接位点配对，从而替代 U1 snRNA（图 3-7a）。与此同时，U4 snRNA 从剪接体中解离，表明它的功能可能是在 U6 遇到 U2 和 5′剪接位点之前屏蔽 U6 上的结

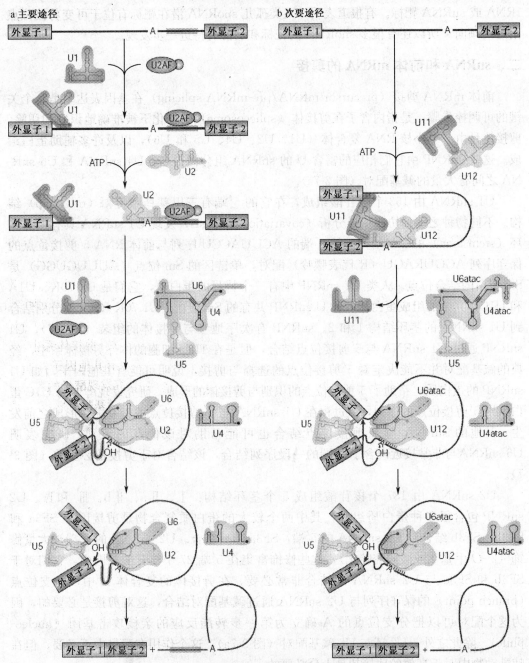

图 3-7　剪接体和稀有剪接体的组装途径以及催化机制（Patel and Steitz，2003）

合位点。U6 中一些与 U4 形成茎区 I 的碱基要与 U2 进行配对（在前面已提及该配对的重要性），由此可知 U4 的解离对 U6 随后结合 U2 形成具有活性剪接体很重要。

　　近年来，在多细胞生物中还发现一类稀有内含子，它们的保守序列与普通的内含子不同，并被另外一种低丰度的剪接体识别和切除。这种剪接体包括 4 种与普通的高丰度剪接体不同的 snRNP：U11、U12、U4atac 和 U6atac，它们分别与 U1 snRNP、U2 snRNP、U4 snRNP 和 U6 snRNP 同源。稀有内含子和普通内含子的剪接体都含有 U5 snRNP。由于稀有内含子（minor-class intron）的切除是依赖于 U12 snRNP 的，故被称为 U12 型内含子（U12-type intron）；而普通型内含子（major-class intron）则依赖于与 U12 snRNP 同源的 U2 snRNP，故称为 U2 型内含子（U2-type intron）。

综上所述，前体 mRNA 的剪接反应是由一系列有序的 RNA-RNA、RNA-蛋白质和蛋白质-蛋白质相互作用完成的。snRNP 是其中最基本的因子，而 snRNP 的 RNA 组分（snRNA）则作为"锚"与"脚手架"在剪接位点识别与两步转酯反应中发挥着作用。

三、可变剪接

随着大规模测序技术的迅猛发展，人们发现一个物种中编码蛋白质的基因数目与其细胞的复杂度之间存在巨大的落差。例如，哺乳动物中编码蛋白质的基因数目与拟南芥（20 000～25 000 个）的大致相当，而且也仅仅比酿酒酵母多三倍（约 6000 个）。这些发现暗示着在生物体内有某些机制，能够调控和多样化如此少的蛋白质编码基因功能。可变剪接（alternative splicing）从相同的转录物通过不同外显子的组合方式产生不同的成熟 mRNA，继而翻译出结构和功能不同的蛋白质。在人类基因组中，据估计有 73％的基因可以进行可变剪接，表明可变剪接可能是高等真核生物增加细胞复杂度的一种普遍机制。

在高等生物中，基因的编码序列通常被内含子打断成多个外显子，如在人类基因组中，一个基因平均含有 7 个内含子。在一个典型的含有多个外显子的 mRNA 中，剪接可以采取许多不同的方式。大多数的外显子是组成型的，它们总是被包含在最终的成熟 mRNA 中。而有些外显子在某些情况下被包含，在其他情况下又被去除，这样的外显子被称为盒式外显子（cassette exon）或者可变外显子（alternative exon）。有些外显子的剪接是互不相容的（mutually exclusive），也就是说在最终的产物中出现这些外显子中的一个时，另外一个就被排除在外。有些外显子可以改变剪接位点的位置来延长或者缩短自身的长度，这样的情况被称为可变的 5′或者 3′剪接位点（alternative 5′ or 3′ splice site）。其他的一些剪接方式请参见图 3-8。

与其他的基因调控方式类似的是，剪接的调控同样也包括顺式调控元件（*cis*-regulatory element）和反式调控因子（*trans*-regulatory factor），顺式调控元件是指前体 mRNA 上的一些特定序列，反式调控因子则是一些细胞因子，可以是蛋白质或者 RNA。剪接的核心顺式调控元件包括 5′剪接位点（5′ splice site，5′ss）、3′剪接位点（3′ splice site，3′ss）和分枝位点序列（branch point sequence，BPS），这些元件存在于每一个内含子上。但是这些核心元件提供的信息，并不足以供剪接体分辨正确和错误的剪切位点，而后者远比前者要多，这说明在剪接位点选择中还有其他的顺式调控元件在发挥着作用，其中最重要的有外显子型剪接增强子或抑制子（exonic splicing enhancer or silencer，ESE or ESS），它们存在于外显子中并促进或者抑制相应外显子的包含；内含子型剪接增强子或抑制子（intronic splicing enhancer or silencer，ISE or ISS），它们存在于内含子中促进或者抑制临近的外显子的包含或者临近的剪接位点的使用。总而言之，这些剪接调控元件（splicing regulatory element，SRE）通过不同的机制招募反式调控因子与之结合，从而增强或者抑制剪接位点的选择或者剪接体在前体 mRNA 上的组装。

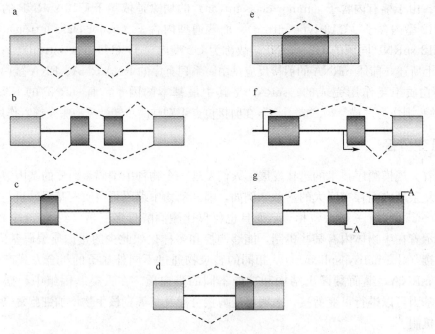

图 3-8 各种不同类型的可变剪接 （Matlin et al. ，2005）

a. 可被剪切掉的外显子；b. 相互排斥的外显子；c. 竞争性 5′剪切位点；d. 竞争性 3′剪切位点；e. 可保留的内含子；f. 多启动子的可变剪接；g. 多 poly（A）位点

ESE 具有非常多样化的序列特征，许多内含子中都存在 ESE。大多数的 ESE 通过招募 SR 蛋白（serine and arginine-rich protein）来发挥作用。SR 蛋白利用其 N 端的 RNA 结合结构域（RNA recognition motif，RRM）与 ESE 结合，C 端的 RS 结构域 （serine and arginine-rich domain）通过介导蛋白质与蛋白质的相互作用促进剪接体的组装，从而促进剪接的发生。ESS 通常被 hnRNP 家族的蛋白质结合，hnRNP 蛋白包含一个或者多个 RNA 结合结构域，它们发挥作用的机制是非常多样的。例如，PTB （hnRNP Ⅰ）可以阻断 U1 snRNP 和 U2 snRNP 之间的作用，而 hnRNP A1 可以跟外显子上下游的内含子结合，将外显子"弯出"（looping out），或者替代 snRNP 的结合。近年来，已经发展出许多用来大规模鉴定 ESE 和 ESS 的计算机方法和实验方法。ESE 已经通过指数富集配基的系统进化技术（systematic evolution of ligands by exponcntial enrichment，SELEX）在体内和体外进行了验证。通过基于细胞的荧光激发筛选（cell-based fluorescence-activated screen），在人类细胞系中具有活性的 133 个十聚核苷酸 （decanucleotide）ESS 从随机序列中被筛选出来。

同样，人们也发现了一些存在于内含子中的剪接调控元件，但是目前并没有对它们进行大规模的筛选鉴定，许多的内含子元件等待着去被发现。一个研究得比较透彻的 ISE 是 G 三联体（GGG）或者叫 G 串，它们经常成串的分布，可以促进临近的 5′剪接位点或者 3′剪接位点的使用。另外，已被鉴定出来的 ISS 包括剪接抑制子 PTB 和 hnRNP A1 的结合序列，hnRNP L 结合的富含 CA 的序列，在 FGFR2 的外显子Ⅲb 中特异的八聚核苷酸（octamer）序列，以及人类 SMN2（脊髓性肌萎缩治疗的靶点）外显子 7 中的两个元件。

可变剪接的调控蛋白大致可以分为两类。一类是表达相对比较普遍的蛋白质，它们在 mRNA 的生物学发生中具有多样化的功能。这类蛋白质可以进一步细分为两小类：SR 蛋白和 hnRNP 蛋白，SR 蛋白通常结合到外显子，促进外显子的包含，而 hnRNP 蛋白往往具有相反的作用。在不同组织中的表达量差异以及不同的翻译后修饰，可以改变上述蛋白质的活性或细胞定位，从而可以对可变剪接进行调控。另一类是功能比较单一的蛋白质，往往具有组织特异性的表达特征，调控组织特异性的剪接事件，如 Nova-1/2、TIA1/TIAR、Fox-1/2、nPTB 等。这些因子都具有 KH 型或者 RRM 型的 RNA 结合结构域。目前已经鉴定出来的剪接调控因子的数量是很有限的，在哺乳动物中只有 20 来个。近年来发展出许多基因组水平上的筛选方法，用来增加已知的剪接调控因子的数量。例如，在果蝇中进行的对 250 个 RNA 结合蛋白的 RNA 干扰筛选提供了大规模鉴定剪接调控蛋白的有效方法。

在基因组水平上，许多新技术被用来对可变剪接进行研究。生物芯片平台的发展使在基因组水平上检测可变剪接的变化成为可能，目前在人类、小鼠和黑猩猩的基因组中得到了成功的应用。另外，剪接芯片可以检测某个调控蛋白或者环境刺激对剪接的影响。此外，在系统鉴定 RNA 靶标方面，也有一些新的方法得到应用，如 CLIP（cross-linking/immunoprecipitation）、RNP 免疫沉淀（RNP immunoprecipitation，RIP）以及基因组 SELEX。这些方法与芯片或者大规模测序联用可以在基因组水平上鉴定剪接调控蛋白的 RNA 靶标，从而可以得到该蛋白质 RNA 结合序列的特征，甚至可以了解各个调控因子之间的协作或者拮抗作用。

深入了解组成型剪接以及可变剪接的调控网络是后基因组时代的一大挑战。目前非常有发展前景的思路是所谓的"从下而上"，即系统鉴定顺式调控元件，了解它们发挥作用的规律和与其他元件的相互作用。另外一个大的挑战是如何将一个系统（如一种细胞类型或者发育的一个阶段）里面得到的法则推而广之，这就需要发展新的具有普遍应用性的实验方法。希望有一天，剪接密码（splicing code），也就是预测初始转录物的剪接方式所需要的一系列法则，能够最终得到破译。

第四节　小 RNA 介导的基因表达调控

一、细菌中小 RNA 介导的基因表达调控

1. 小 RNA

细菌中的起调节作用的 RNA 俗称小 RNA（sRNA）。它们的大小为 50～500 个核苷酸，平均长度约为 100 个核苷酸。它们比真核生物中长 21～30 个核苷酸的调节 RNA 要长。它们一般不是由长的双链 RNA 前体加工而来，而是由小 RNA 基因直接编码。用计算机和实验的方法已在大肠杆菌和其他不同细菌种中鉴定了许多的 sRNA。

按照其作用方式，sRNA 被分为两种类型：反式作用的小 RNA 和顺式作用的小 RNA。对于后者，其基因调节作用主要是通过小 RNA 形成不同的结构而行使的，它们调节自身所在基因的表达。核糖开关和衰减作用就属于这种类型，它们将在后面的两节中进一步阐述。这里我们着重于前者，反式作用的小 RNA。它们主要通过和被调节的 RNA 形成配对而起作用；有少数通过调节蛋白质的活性来起作用。

那些通过配对行使功能的小 RNA 也可称为反义 RNA。在与靶标配对后，它们主要促进核酸酶对 mRNA 的降解，通过阻止核糖体与 mRNA 的结合而抑制翻译，或者通过加强核糖体与 mRNA 结合而促进翻译。一些反义 RNA 和它们的靶标能形成很长的、完全的配对。一些质粒编码的反义小 RNA 已被研究的比较清楚，它们中的一些影响质粒的复制，也有一些抑制毒性蛋白的表达。然而大多数情况下，小 RNA 和其靶标的配对是不完全的，它们的结合是比较弱的，需要 Hfq 蛋白的帮助。Hfq 是一个分子伴侣，它可以结合小 RNA 从而促进它们的稳定性。

2. 核糖开关

核糖开关是一类 RNA 调节元件，它们作为小分子代谢物的直接传感器，控制基因的转录或翻译。这些元件主要是位于它们调控基因的 5′UTR 区域。

典型的核糖开关由两个结构域组成：适体和表达平台。适体作为嵌在核糖开关里的分子传感器，高选择性地识别对应的靶标分子，如特定的代谢物。表达平台通常紧位于适体的下游。但是在很多情况下，这两个结构域可以相互重叠。表达平台的功能是对代谢物的结合作出相应的反应而调控基因的表达，主要通过 5′UTR 结构的异构化实现。即使在很相近的物种之间，表达平台的差别往往也是很大的，但作用机制除外。它们基本上是通过在配体结合适体之后改变构象从而影响基因表达的某些过程。在很多情况下，这些构象的改变是通过解除或形成沃森-克里克碱基配对从而形成相斥的不同结构。

核糖开关对基因的控制主要有两种机制。第一种，它们使用抗终止子在转录水平上进行调控。第二种，它们通过形成特定的结构以掩盖 mRNA 上核糖体结合位点在翻译水平上起作用。这两种机制是通过改变 RNA 二级结构实现的，如图 3-9 所示。

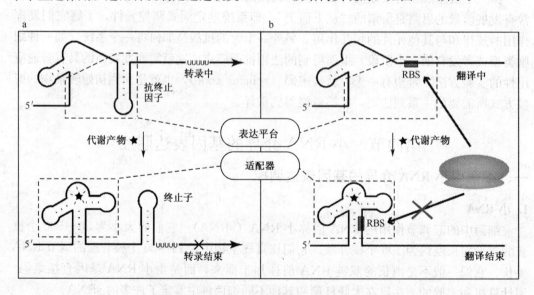

图 3-9 核糖开关控制基因的共同机制（Tucker and Breaker，2005）
a. b. 分别显示的是转录和翻译水平上的调控

核糖开关控制的基因通常编码与代谢物合成或者转运相关的蛋白质。因而，核糖开关使用的是一种反馈抑制的形式，也就是说当代谢物足够多时，制造这些代谢物的基因表达受到抑制。目前，有很多的核糖开关被鉴定出来，它们对应于很多不同的代谢物，如辅酶 TPP、辅酶 FMN、SAM、赖氨酸及其他一些氨基酸、鸟嘌呤、腺嘌呤和 Mg^{2+}。尽管核糖开关主要存在于细菌中，它们也在植物和真菌中被发现。有的在剪接连接处附近，可能调节 mRNA 的剪接。近来发现，一个基因可以被串联在一起的两个核糖开关调控。核糖开关对基因表达有效的控制可能取决于 RNA 折叠速率、代谢物与核糖开关结合的速率及 RNA 聚合酶转录速率这些因素之间的精密平衡。

3. 转录衰减作用

转录衰减作用（衰减作用）可以被定义为任何利用转录暂停或者终止来调控下游基因表达的机制。一些基因，特别是氨酰 tRNA 合成酶的基因，是由衰减子控制的。它是通过 200～300 个核苷酸长的前导 RNA 与非荷载 tRNA 直接、特异地相互作用来调节的。可以说它是一类核糖开关，但是仅对非荷载 tRNA 有反应，而对小分子代谢物没有反应，并且该调控依赖于细菌中转录和翻译的偶联。

衰减作用机制可以用大肠杆菌中的色氨酸合成操纵子的调控来解释。色氨酸 161 个核苷酸的前导 RNA 可以形成三个重叠的 RNA 二级结构：滞留结构、抗终止子和转录终止子（图 3-10）。

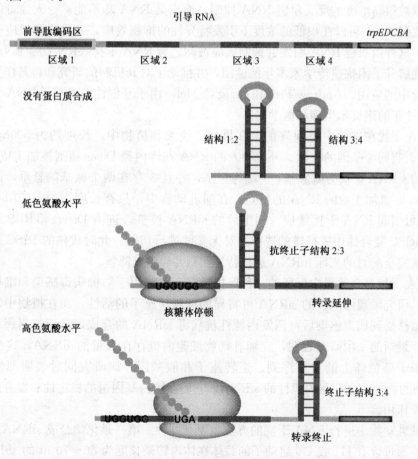

图 3-10 色氨酸操纵子衰减作用模型

衰减作用是一个典型的负反馈调节作用。其他的一些氨基酸合成操纵子，如 his、phe、leu、ilv 和 pheA 等操纵子，都是通过转录衰减作用来调节的。单独的衰减作用就可以起到强有力的调节作用，如 his 和 leu 操纵子，它们不需要转录阻抑物，而仅依靠衰减作用。

转录衰减作用也是细菌中一种高度保守的调节策略。它主要是通过，如色氨酸操纵子中前导肽的翻译来控制的。而且，RNA 结合蛋白、tRNA 和反义 RNA 也被发现可以介导转录衰减作用。

二、真核生物中小 RNA 介导的基因表达调控

1. siRNA

20 世纪 90 年代，遗传学家在构建转基因动植物的过程中发现了一个有趣的现象，在为了高表达某一基因而构建的转基因动植物中，内源基因和外源转入的基因同时被抑制。这种现象被称之为共抑制，有两种类型，转录水平的抑制（TGS）和转录后水平的抑制（PTGS）。早期的研究发现，TGS 往往伴随着特定基因的同源区域甲基化水平的升高和转录效率的降低。而 PTGS 则伴随着 mRNA 稳定性的降低。研究发现，基因表达可以被很高浓度的反义单链 RNA 抑制，而正义 RNA 却不能。令人惊奇的是，退火后的双链 RNA 可以在很低的浓度下引发特异性的抑制效应，并且这种效应可遗传到下一代。这种由双链 RNA 引发的抑制效应被称之为 RNA 干扰（RNAi）。RNA 干扰现象的发现解开了困扰遗传学家多年的谜团，并推动了对其机制的研究和将其作为药物在疾病治疗中的应用。Andrew Fire 和 Craig C. Mello 由于开创性地发现了 RNA 干扰而获得了 2006 年的诺贝尔生理和医学奖。

RNA 干扰在进化中是保守的。在果蝇、线虫和植物中，长度约为 500bp 的双链 RNA 可引起同源基因的沉默。外源导入的 RNA 在体内经 Dicer 切割被加工成 siRNA。加工后的 siRNA 5′端为磷酸基，3′端为羟基，并且 3′存在两个碱基的悬垂，这也是所有被 RNase Ⅲ 加工后的 RNA 的特点。在哺乳动物中，只有长度为 20～30 bp 的双链 RNA 才可引起 RNA 干扰效应。加工后的 siRNA 被整合进由 Dicer、TRBP 和 AGO2 组成的 RISC 复合体中进行链的选择并对无义链进行切割。此时成熟的 RISC 复合体便可对与其完全配对的靶标 mRNA 进行特异切割并导致其降解。

RNA 干扰是自然界中保留的一种保护机制，主要用于防御病毒感染和维持基因组的稳定。研究发现内源性的 siRNA 可特异的抑制转座子的活性，而在植物中的研究表明，当植株受到病毒感染后可诱发内源性抗病毒 siRNA 的合成，从而抑制病毒的增殖和侵染。最新的一项研究表明，在哺乳动物细胞内也存在大量的 siRNA，这些 siRNA 主要来源于染色体上的重复序列、逆转座子和假基因。该研究同时表明即使在没有 RdRP 酶的物种中也存在内源性的 siRNA，它们在维持基因组的稳定性和发育过程中发挥着重要作用。

在哺乳动物中进行 RNA 干扰的方法主要有三种。第一是化学合成 siRNA。第二是载体法，通过含有 H1 或 U6 启动子的载体在体内转录长度为 60～70 nt 的 shRNA，之后在 Dicer 的加工下产生特定的 siRNA。第三是 esiRNA 法，即将双链 RNA 经 Rnase Ⅲ 切割后，得到长度为 20～30 bp 的 siRNA 文库，来介导对特定靶基因的沉默。

RNA 干扰作为一种简单有效的方法被广泛应用于基因功能的研究。siRNA 作为药物已经用于治疗肝炎暴发、抗病毒以及抗肿瘤研究。siRNA 的高效性和高特异性决定了其广阔的应用前景。但目前仍然存在靶向性导入和脱靶效应问题。

2. miRNA

微小 RNA（microRNA，miRNA）是一类保守的、大小为 20～25 nt 的非编码 RNA，它们通过结合到 mRNA 3′UTR 的互补序列，介导该 mRNA 切割或是抑制其翻译来调节基因的表达。1993 年，首次在线虫中发现了 miRNA *lin-4*，2000 年 miRNA *let-7* 在果蝇中被发现。保守性分析显示，*let-7* 在果蝇、线虫和人类中存在高度的保守性。这暗示 miRNA 可能是一类在进化中保守的非编码 RNA，可能在基因调控中发挥重要作用。该突破性的发现极大地推动了 miRNA 的不断发现和鉴定，2008 年 6 月，miRbase 数据库已公布的 miRNA 数目达到 6396 个，其中有 678 个是在人类基因组中发现的。现已证实 miRNA 广泛存在于真核生物中，是真核生物中最大的基因家族之一。

除了在真核生物中发挥重要的调节作用外，miRNA 还广泛存在于单细胞生物中。近来的研究也表明一些病毒也编码某些 miRNA，在病毒的复制和感染过程中发挥着至关重要的作用。

miRNA 通常是由 RNA 聚合酶Ⅱ（Pol Ⅱ）转录的，最初产物是被称为 pri-miRNA 的前体大分子。pri-miRNA 在细胞核内被由 Drosha 和双链 RNA 结合蛋白 DGCR8 组成的复合体加工成长约 70 个核苷酸的 pre-miRNA，该分子形成一个不完全的茎环结构。pre-miRNA 被由 RAN 蛋白和 GTP 依赖的 exportin-5 蛋白组成的复合体转运到细胞质中，随后被 Dicer 加工成长约 22 个核苷酸的 miRNA：miRNA* 双链。该双链很快被整合到 RISC 复合体中，miRNA* 链在此被剪切掉，而 miRNA 被保留，形成有功能的 RISC 复合物（miRISC），如图 3-11 所示。

最近发现的一类被称为 MiRtron 的 miRNA 加工途径略有不同。MiRtron 的初步加工无需 Drosha 的切割，而是借助于细胞核内的剪接机器。miRNA 作为整个内含子被剪接出来后，首先在去环化酶的作用下解环，重折叠成 pre-miRNA，出核后在 Dicer 的加工下成熟。

动物和植物中 miRNA 抑制基因表达的作用机制存在明显的不同。植物中的 miRNA 和靶标的配对为完全配对，主要通过对靶标 mRNA 进行切割，从而介导mRNA 的降解。而动物中的 miRNA 与靶标不完全配对，主要是通过抑制翻译起作用。此外，miRNA 还可以通过引起翻译中核糖体的解离来抑制翻译。除了翻译水平的抑制，miRNA 还可促进靶标 mRNA 或新生肽的降解，但具体机制尚不清楚。

miRNA 的表达具有显著的时空和组织表达特异性。这种表达的特异性造成了其靶基因在发育的特定阶段或特定的组织被抑制，而这种调节模式可能决定了组织和器官的形态建成，如 *lin-4* 在线虫的 L1～L4 期表达，而 *let-7* 在 L4 的成虫期表达，这种表达模式决定了它们的靶基因在何时被抑制。*miRNA-124*、*miRNA-122* 和 *miRNA-1* 分别在脑、肝、骨骼肌中特异表达，通过这种表达模式它们可以调节特定组织的转录组，从而决定组织和器官的形态建成。

正常组织和癌变组织的 miRNA 表达谱有非常明显的不同，在癌变组织中大部分

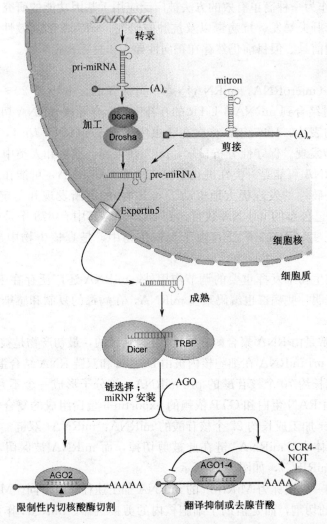

图 3-11　miRNA 的生物合成及组装（Filipowicz et al.，2008）

miRNA 的表达被明显下调，而小部分 miRNA 却被异常上调，这暗示 miRNA 在癌症的发生过程可能发挥着重要作用。对 miRNA 在染色体上定位的分析发现，50%得到注解的 miRNA 定位于染色体的脆性位点，这些位置在癌变组织中往往被缺失掉。近年的研究发现 miRNA 可以作为致癌基因和抑癌基因发挥作用，miRNA-17-92 基因簇是第一个被发现的致癌性 miRNA。而 *miRNA-16*、*let-7*、*miRNA-34a* 和 *miRNA-335* 等主要作为抑癌基因发挥作用。miRNA 在癌症发生过程中的作用为我们提供了一类富有前景的基因治疗靶标。除此之外，miRNA 还在细胞增殖和死亡、脂肪代谢、神经元定位、造血细胞系分化和发育等方面起着至关重要的调控作用。

3. rasiRNA

目前果蝇中发现有 5 种 Argonaute 蛋白：Ago1、Ago2、Ago3、Piwi 和 Aubergine（Aub）。Ago1 和 Ago2 属于 Argonaute 蛋白的 AGO 亚家族，在各种组织中普遍表达，而 Piwi、Aub 和 Ago3 属于 Piwi 亚家族，只在生殖细胞中表达。Ago1 和 Ago2 分别是

miRNA 和 siRNA 介导的基因沉默过程中的关键因子。Aubergine、Piwi 和 Ago3 与数以千计的非编码小 RNA 结合，这类小 RNA 分子被称为重复相关的小干扰 RNA（ra-siRNA），产生于基因组中的重复序列和异染色质区域。rasiRNA 是单链 RNA，与 Piwi 蛋白直接结合。它们比 Argonaute 蛋白结合的 siRNA/miRNA 要略长（在果蝇中为 24～29 个核苷酸），具有 5′端磷酸化和 3′端的 2′-O-甲基化修饰。rasiRNA 与异染色质的建立和维持，以及转座子的抑制过程相关联。

4. piRNA

Piwi 蛋白结合的 RNA 最先是在小鼠睾丸中被发现的，后来在其他哺乳动物和斑马鱼的睾丸中也有发现。这些 RNA 被称为 piRNA，它们有许多有趣的特点。piRNA 主要在动物的生殖细胞中被发现，它们（26～31 nt）比 miRNA 和 siRNA（21～24 nt）要长。piRNA 与 rasiRNA 有很多相似点，也有很多不同之处。piRNA 中与重复序列匹配的比例小于 rasiRNA。尽管如此，基于它们的共同点，rasiRNA 被归为 piRNA 的一个特殊亚类。

piRNA 的序列分析显示，它们在 5′端有很强的尿嘧啶偏向性，这与 siRNA 和 miRNA 类似。这种偏好性是双链 RNA 前体被 RNase Ⅲ 加工后产物的特征。但是，对 piRNA 两端区域序列的计算机分析没有发现类似于 miRNA 前体的二级结构，表明 piRNA 的加工成熟过程不同于 miRNA。另外，piRNA 的 3′端有核糖 2′-O-甲基化修饰，类似于植物中的 miRNA。

piRNA 在染色体上的分布是不均一的。piRNA 基因簇丛集于小鼠第 17 号、5 号、4 号和 2 号染色体上，而性染色体上几乎没有分布。它们主要存在于基因间隔区，很少存在于基因区或重复序列区。

在果蝇中，在 miRNA 和 siRNA 通路中作用的 Dicer 蛋白的缺失对 piRNA 水平没有影响，意味着 piRNA 有不同的生成途径。对果蝇中与三种 Piwi 家族蛋白结合的 piRNA 的序列分析揭示了 piRNA 的生成途径。

一个新的 RNA 介导的基因沉默途径开始展现它的奥秘。尽管我们有了这些有趣的发现，许多重要的问题仍然存在。piRNA 介导的调控途径是否保守？它们的靶标是什么？我们也不知道 piRNA 介导的基因沉默是怎样开始的，为什么基因组中有些区域集中表达 piRNA。进一步的研究也许能带来下一个突破。

小　结

RNA 是一条由磷酸二酯键连接的、由 4 种不同的核苷酸多聚形成的单链。与双链 DNA 比较起来，RNA 更容易通过碱基配对和三级相互作用形成多样的结构，这使得 RNA 具有很多生物学功能。但最近的研究发现，单纯的 RNA 在折叠时容易形成非活性的配对，使其在形成活性结构时陷入动力学或热力学陷阱。而且，RNA 在细胞内都是与蛋白质结合在一起的，这种相互作用可以阻止 RNA 这种柔性的大分子形成错误的结构。

在本章中，我们对研究的比较清楚的非编码 RNA 进行了介绍：核酶是一大类具有多样生物学功能的结构 RNA，可以在体外没有蛋白质帮助的情况下形成复杂的三级结构，并催化不同的生化反应；snoRNA 负责核糖体 RNA 的修饰；snRNA 对 mRNA 的成熟是一个必需的组分；小 RNA 在真核和原核生物中都有重要的调节作用，如 miRNA 和 RyhB RNA。上述 RNA 仅占整个基因组的很小一部分，基因组未注释的大部分转录物的生物学功能到现在为止还是一个谜。

人类基因组计划完成之后发现，人的二倍体基因组共编码约 25 000 个基因，这个数目远比以前估计的 100 000 个要小，这些编码蛋白质的基因仅占整个基因组的 1.5%。但是，整个人类基因组约有 10% 被转录成具有 poly（A）尾的 RNA，这个比例远比蛋白质编码基因占的比例高，说明很大一部分具有 poly（A）尾的 RNA 实际上是非编码 RNA。在对人的 10 条染色体的转录情况进行研究时发现，除了具有 poly（A）尾的转录物外，从基因间转录出的无 poly（A）的转录物占了整个转录组的大部分。而且，15%～20% 的蛋白质编码基因的反义链也是被转录的，这个比例在某些组织中甚至可以达到整个转录组的 50%。现在越来越清楚的是，人类基因组的大部分是被转录的，其他高等真核生物甚至原核生物也是一样，例如，至少 74.5% 的酵母基因组和 62% 的小鼠基因组是被转录了的，大肠杆菌基因组也有相当数量的反义链和基因间区域被转录。所以转录组比想像中要复杂得多。SAGE、全长 cDNA 克隆、CAGE 以及嵌合芯片等新技术的出现和发展使得转录组的复杂度被越来越清楚地认识到。这些被转录的非蛋白质编码的 RNA 通常被认为肯定编码某些未知的功能，或者调控基因表达，或者增加基因组的复杂度。

本章中介绍的非编码 RNA，除核酶外都是小 RNA。很多源自基因间区域的无 poly（A）的转录物长度甚至可以达到几万个碱基对，这些转录物是否编码了长度非常大的非编码 RNA？这些 RNA 的作用方式是什么？在完成了对基因组的测序工作后，我们又被推到了一个新的起跑线上。

（张　翼）

参 考 文 献

Abramovitz D L, Friedman R A. 1996. Catalytic role of 2′-hydroxyl groups within a group II intron active site. Science, 271 (5254)：1410～1413

Adams P L, Stahley M R. 2004. Crystal structure of a self-splicing group I intron with both exons. Nature, 430 (6995)：45～50

Andersen A A, Collins R A. 2000. Rearrangement of a stable RNA secondary structure during VS ribozyme catalysis. Mol Cell, 5 (3)：469～478

Aravin A, Gaidatzis D, Pfeffer S, et al. 2006. A novel class of small RNAs bind to Mili protein in mouse testes. Nature, 442 (7099)：203～207

Aravin A A, Hannon G J, Brennecke J. 2007. The Piwi-piRNA pathway provides an adaptive defense in the transposon arms race. Science, 318 (5851)：761～764

Ban N, Nissen P, et al. 2000. The complete atomic structure of the large ribosomal subunit at 2. 4 A resolution. Science, 289 (5481)：905～920

Barrick J E, Breaker R R. 2007. The distributions, mechanisms, and structures of metabolite-binding riboswitches. Genome Biol, 8 (11)：R239

Bartel D P. 2004. MicroRNAs：genomics, biogenesis, mechanism, and function. , Cell, 116 (2)：281～297

Batey R T, Rambo R R, Doudna J A. 1999. Tertiary Motifs in RNA Structure and Folding. Angewandte Chemie (International ed 38), 2326~2343

Black D L. 2003. Mechanisms of alternative pre-messenger RNA splicing. Annu Rev Biochem, 72: 291~336

Boudvillain M, Pyle A M. 1998. Defining functional groups, core structural features and inter-domain tertiary contacts essential for group II intron self-splicing: a NAIM analysis. Embo J, 17 (23): 7091~7104

Brannvall M, Kikovska E. 2004. Cross talk between the +73/294 interaction and the cleavage site in RNase P RNA mediated cleavage. Nucleic Acids Res, 32: 5418~5429

Breaker R R. 2008. Complex riboswitches. Science, 319 (5871): 1795~1797

Brennecke J, Aravin A A, Stark A, et al. 2007. Discrete small RNA-generating loci as master regulators of transposon activity in Drosophila. Cell, 128 (6): 1089~1103

Burke J M, Esherick J S. 1990. A 3' splice site-binding sequence in the catalytic core of a group I intron. Nature, 344 (6261): 80~82

Busch S, Kirsebom L. 2000. Differential role of the intermolecular base-pairs G291-C (75) and G293-C (74) in the reaction catalyzed by Escherichia coli RNase P RNA. J Mol Biol, 299: 941~951

Butcher, S E 2001. Structure and function of the small ribozymes. Curr Opin Struct Biol, 11 (3): 315~320

Cate J H, Gooding A R. 1996. Crystal structure of a group I ribozyme domain: principles of RNA packing. Science, 273 (5282): 1678~1685

Cech T R. 1990. Self-splicing of group I introns. Annu Rev Biochem, 59, 543~568

Cech T R. 2000. Structural biology. The ribosome is a ribozyme. Science, 289 (5481): 878~879

Chowrira B M, Berzal-Herranz A. 1995. Novel system for analysis of group I 3' splice site reactions based on functional trans-interaction of the P1/P10 reaction helix with the ribozyme's catalytic core. Nucleic Acids Res, 23 (5): 849~855

Coppins R L, Hall K B, Groisman E A. 2007. The intricate world of riboswitches. Curr Opin Microbiol, 10 (2): 176~181

Costa M, Michel F. 1995. Frequent use of the same tertiary motif by self-folding RNAs. Embo J, 14 (6): 1276~1285

Costa M, Michel F. 2000. A three-dimensional perspective on exon binding by a group II self-splicing intron. EMBO J, 19: 12

Dai L, Zimmerly S. 2003. ORF-less and reverse-transcriptase-encoding group II introns in archaebacteria, with a pattern of homing into related group II intron ORFs. RNA, 9 (1): 14~19

Das S R, Piccirilli J A. 2005. General acid catalysis by the hepatitis delta virus ribozyme. Nat Chem Biol, 1 (1): 45~52

de Lencastre A, Pyle A M. 2008. Three essential and conserved regions of the group II intron are proximal to the 5'-splice site. RNA, 14 (1): 11~24

Doherty E A, Doudna J A. 2000. Ribozyme structures and mechanisms. Annu Rev Biochem, 69: 597~615

Dorsett. Y, Tuschl T. 2004. siRNAs: applications in functional genomics and potential as therapeutics. Nat Rev Drug Discov, 3 (4): 318~329

Duarte E A, Leavitt M C. 1997. Hairpin ribozyme gene therapy for AIDS. Methods Mol Biol, 74: 459~468

Du H, Rosbash M. 2002. The U1 snRNP protein U1C recognizes the 5' splice site in the absence of base pairing. Nature, 419 (6902): 86~90

Dykxhoorn D M, Lieberman J. 2005. The silent revolution: RNA interference as basic biology, research tool, and therapeutic. Annu Rev Med, 56: 401~423

Engelhardt M A, Doherty E A. 2000. The P5abc peripheral element facilitates preorganization of the tetrahymena. group I ribozyme for catalysis. Biochemistry, 39 (10): 2639~2651

Eulalio A, Huntzinger E, Izaurralde E. 2008. Getting to the root of miRNA-mediated gene silencing. Cell, 132 (1): 9~14

Evans D, Marquez S M. 2006. RNase P: interface of the RNA and protein worlds. Trends Biochem Sci, 31 (6): 333~341

Fedorova O, Zingler N. 2007. Group II introns: structure, folding and splicing mechanism. Biol Chem, 388 (7): 665~678

Ferré-D'Amaré A R, Zhou K. 1998. Crystal structure of a hepatitis delta virus ribozyme. Nature, 395 (6702): 567~574

Filipowicz W, Bhattacharyya S N, Sonenberg N. 2008. Mechanisms of post-transcriptional regulation by microR-NAs: are the answers in sight? Nat Rev Genet, 9 (2): 102~114

Fire A, Xu S, Montgomery M K. 1998. Potent and specific genetic interference by double-stranded RNA in Cae-norhabditis elegans. Nature, 391 (6669): 806~811

Gallivan J P. 2007. Toward reprogramming bacteria with small molecules and RNA. Curr Opin Chem Biol, 11 (6): 612~619

Gesteland R F, Cech T R, Atkins J F. 2005. The RNA Wold (3rd edition), New York: Cold Spring Harbor Laboratory Press

Golden B L, Gooding A R. 1998. A preorganized active site in the crystal structure of the Tetrahymena ribozyme. Science, 282 (5387): 259~64

Golden B L, Kim H. 2005. Crystal structure of a phage Twort group I ribozyme-product complex. Nat Struct Mol Biol, 12 (1): 82~89

Gollnick P, Babitzke P. 2002. Transcription attenuation. Biochim Biophys Acta, 1577 (2): 240~250

Gottesman S. 2005. Micros for microbes: non-coding regulatory RNAs in bacteria. Trends Genet, 21 (7): 399~404

Griffin E A Jr, Qin Z. 1995. Group II intron ribozymes that cleave DNA and RNA linkages with similar efficiency, and lack contacts with substrate 2'-hydroxyl groups. Chem Biol, 2 (11): 761~770

Guo F, Gooding A R. 2004. Structure of the Tetrahymena ribozyme: base triple sandwich and metal ion at the active site. Mol Cell, 16 (3): 351~362

Hamill S, Pyle A M. 2006. The receptor for branch-site docking within a group II intron active site. Mol Cell, 23 (6): 831~840

Harris M E, Christian E L. 2003. Recent insights into the structure and function of the ribonucleoprotein enzyme ribonuclease P. Curr Opin Struct Biol, 13 (3): 325~333

Haugen P, Simon D M. 2005. The natural history of group I introns. Trends Genet, 21 (2): 111~119

He L, Hannon G J. 2004. MicroRNAs: small RNAs with a big role in gene regulation. Nat Rev Genet, 5 (7): 522~531

Hiley S L, Sood V D. 2002. 4-thio-U cross-linking identifies the active site of the VS ribozyme. Embo J, 21 (17): 4691~4698

Hoffmann B, Mitchell G T. 2003. NMR structure of the active conformation of the Varkud satellite ribozyme cleavage site. Proc Natl Acad Sci USA, 100 (12): 7003~7008

Jacquier A, Breuleux N J. 1991. Splice site selection and role of the lariat in a group II intron J. Mol Biol, 219: 14

Johnson T H, Tijerina P. 2005. Structural specificity conferred by a group I RNA peripheral element. Proc Natl Acad Sci USA, 102 (29): 10176~10181

Jones-Rhoades M W, Bartel D P, Bartel B. 2006. MicroRNAs and their regulatory roles in plants. Annu Rev Plant Biol, 57: 19~53

Kazantsev A V, Krivenko A A. 2005. Crystal structure of a bacterial ribonuclease P RNA. Proc Natl Acad Sci USA. i02 (38): 13392~13397

Kazantsev A V, Pace N R. 2006. Bacterial RNase P: a new view of an ancient enzyme. Nat Rev Microbiol, 4 (10): 729~740

Ke A, Zhou K. 2004. A conformational switch controls hepatitis delta virus ribozyme catalysis. Nature, 429

(6988): 201~215

Kishore S, Stamm S. 2006. The snoRNA HBII-52 regulates alternative splicing of the serotonin receptor 2C. Science, 311 (5758): 230~232

Konforti B B, Abramovitz D L. 1998. Ribozyme catalysis from the major groove of group II intron domain 5. Mol Cell, 1 (3): 433~441

Korostelev A, Trakhanov S. 2006. Crystal structure of a 70S ribosome-tRNA complex reveals functional interactions and rearrangements. Cell, 126 (6): 1065~1077

Kruger K, Grabowski P J. 1982. Self-splicing RNA: autoexcision and autocyclization of the ribosomal RNA intervening sequence of Tetrahymena. Cell, 31 (1): 147~157

Lafontaine D A, Norman D G. 2001. Structure, folding and activity of the VS ribozyme: importance of the 2-3-6 helical junction. Embo J, 20 (6): 1415~1424

Leontis N B, Stombaugh J, Westhof E. 2002. The non-Watson-Crick base pairs and their associated isostericity matrices. Nucleic acids research, 30: 3497~3531

Lewin A S, Hauswirth W W. 2001. Ribozyme gene therapy: applications for molecular medicine. Trends Mol Med, 7 (5): 221~228

Li F, Ding S. 2006. Virus counterdefense: diverse strategies for evading the RNA-silencing immunity. Annu Rev Microbiol, 60: 503~531

Li H. 2008. Unveiling substrate RNA binding to H/ACA RNPs: one side fits all. Curr Opin Struct Biol, 18 (1): 78~85

Lilley D M. 2003. The origins of RNA catalysis in ribozymes. Trends Biochem Sci, 28 (9): 495~501

Lilley D M. 2005. Structure, folding and mechanisms of ribozymes. Curr Opin Struct Biol, 15 (3): 313~323

Madhani H D, Guthrie C. 1994. Dynamic RNA-RNA interactions in the spliceosome. Annu Rev Genet, 28, 1~26

Martick M, Scott W G. 2006. Tertiary contacts distant from the active site prime a ribozyme for catalysis. Cell, 126 (2): 309~320

Matera A G, Terns R M, Terns M P. 2007. Non-coding RNAs: lessons from the small nuclear and small nucleolar RNAs. Nat Rev Mol Cell Biol, 8 (3): 209~220

Matlin A J, Clark F, Smith C W J. 2005. Understanding alternative splicing: towards a cellular code. Nat Rev Mol Cell Biol, 6 (5): 386~398

McKay D B. 1996. Structure and function of the hammerhead ribozyme: an unfinished story. RNA, 2 (5): 395~403

Merino E, Yanofsky C. 2005. Transcription attenuation: a highly conserved regulatory strategy used by bacteria. Trends Genet, 21 (5): 260~264

Michel F, Westhof E. 1990. Modelling of the three-dimensional architecture of group I catalytic introns based on comparative sequence analysis. J Mol Biol, 216 (3): 585~610

Nakano S, Chadalavada D M. 2000. General acid-base catalysis in the mechanism of a hepatitis delta virus ribozyme. Science, 287 (5457): 1493~1497

Nelson J A, Uhlenbeck O C. 2006. When to believe what you see. Mol Cell, 23 (4): 447~450

Nissen P, Hansen J. 2000. The structural basis of ribosome activity in peptide bond synthesis. Science, 289 (5481): 920~930

Nissen P, Ippolito J A, Ban N. et al. 2001. RNA tertiary interactions in the large ribosomal subunit: the A-minor motif. Proceedings of the National Academy of Sciences of the United States of America 98, 4899~4903

Noller H F. 2005. RNA structure: reading the ribosome. Science, 309 (5740): 1508~1514

Okamura K, Chung W, Ruby J G, et al. 2008. The Drosophila hairpin RNA pathway generates endogenous short interfering RNAs. Nature, 453 (7196): 803~806

Patel A A, Steitz J A. 2003. Splicing double: insights from the second spliceosome. Nat Rev Mol Cell Biol, 4 (12): 960~970

Perrotta A T, Been M D. 1991. A pseudoknot-like structure required for efficient self-cleavage of hepatitis delta virus RNA. Nature, 350 (6317): 434～436

Peters L, Meister G. 2007. Argonaute proteins: mediators of RNA silencing. Mol Cell, 26 (5): 611～623

Pichler A, Schroeder R. 2002. Folding problems of the 5′ splice site containing the P1 stem of the group I thymidylate synthase intron: substrate binding inhibition in vitro and mis-splicing in vivo. J Biol Chem, 277 (20): 17987～17993

Pley H W, Flaherty K M. 1994. Three-dimensional structure of a hammerhead ribozyme. Nature, 372 (6501): 68～74

Qin P Z, Pyle A M. 1998. The architectural organization and mechanistic function of group II intron structural elements. Curr Opin Struct Biol, 8 (3): 301～308

Reichow S L, Hamma T, Ferré-D'Amaré A R, et al. 2007. The structure and function of small nucleolar ribonucleoproteins. Nucleic Acids Res, 35 (5): 1452～1464

Reichow S L, Hamma T, Ferré-D'Amaré A R, et al. 2007. The structure and function of small nucleolar ribonucleoproteins. Nucleic Acids Res, 35 (5): 1452～1464

Rupert P B, Ferré-D'Amaré A R. 2001. Crystal structure of a hairpin ribozyme-inhibitor complex with implications for catalysis. Nature, 410 (6830): 780～786

Schmidt U, Podar M. 1996. Mutations of the two-nucleotide bulge of D5 of a group II intron block splicing in vitro and in vivo: phenotypes and suppressor mutations. RNA, 2 (11): 1161～1172

Schuwirth B S, Borovinskaya M A. 2005. Structures of the bacterial ribosome at 3. 5 A resolution. Science, 310 (5749): 827～834

Scott W G. 2007. Ribozymes. Curr Opin Struct Biol, 17 (3): 280～286

Selmer M, Dunham C M. 2006. Structure of the 70S ribosome complexed with mRNA and tRNA. Science, 313 (5795): 1935～1942

Shan S, Yoshida A. 1999. Three metal ions at the active site of the *Tetrahymena* group I ribozyme. Proc Natl Acad Sci USA, 96: 12299～12304

Stefani G, Slack F J. 2008. Small non-coding RNAs in animal development. Nat Rev Mol Cell Biol, 9 (3): 219～230

Strobel S, Donnelly L O. 1999. A hydrogen-bonding triad stabilizes the chemical transition state of a group I ribozyme. Chem Biol, 6: 153～165

Sudarsan N, Hammond M C, Block K F, et al. 2006. Tandem riboswitch architectures exhibit complex gene control functions. Science, 314 (5797): 300～304

Szewczak A A, Donnelly L O. 1998. A minor groove RNA triple helix within the catalytic core of a group I intron. Nat Struct Biol, 5 (12): 1037～1042

Toor N, Keating K S. 2008. Crystal structure of a self-spliced group II intron. Science, 320 (5872): 77～82

Toor N, Robart A R. 2006. Self-splicing of a group IIC intron: 5′ exon recognition and alternative 5′ splicing events implicate the stem-loop motif of a transcriptional terminator. Nucleic Acids Res, 34 (22): 6461～6471

Torres-Larios A, Swinger K K. 2005. Crystal structure of the RNA component of bacterial ribonuclease P. Nature, 437 (7058): 584～587

Tucker B J, Breaker R R. 2005. Riboswitches as versatile gene control elements. Curr Opin Struct Biol, 15 (3): 342～348

Ule J, Jensen K B, Ruggiu M, et al. 2003. CLIP identifies Nova-regulated RNA networks in the brain. Science, 302 (5648): 1212～1215

Walter N G, Burke J M. 1998. The hairpin ribozyme: structure, assembly and catalysis. Curr Opin Chem Biol, 2 (1): 24～30

Wang Z, Rolish M E, Yeo G, et al. 2004. Systematic identification and analysis of exonic splicing silencers. Cell, 119 (6): 831～845

Wank H, SanFilippo J. 1999. A reverse transcriptase/maturase promotes splicing by binding at its own coding segment in a group II intron RNA. Mol Cell, 4 (2): 239~250

Watanabe T, Sullenger B A. 2000. Induction of wild-type p53 activity in human cancer cells by ribozymes that repair mutant p53 transcripts. Proc. Natl Acad Sci USA, 97: 8490~8494

Watanabe T, Totoki Y, Toyoda A, et al. 2008. Endogenous siRNAs from naturally formed dsRNAs regulate transcripts in mouse oocytes. Nature, 453 (7194): 539~543

Watson J D, Baker T A, Bell S P, et al. 2007. Molecular Biology of the Gene. 6th edition. New York: Cold Spring Harbor Laboratory Press

Welch P J, Yei S. 1998. Ribozyme gene therapy for hepatitis C virus infection. Clin Diagn Virol, 10 (2~3): 163~171

Wu S, Romfo C M, Nilsen T W, et al. 1999. Functional recognition of the 3′ splice site AG by the splicing factor U2AF35. Nature, 402 (6763): 832~835

Xiao M, Li T. 2005. A peripheral element assembles the compact core structure essential for group I intron self-splicing. Nucleic Acids Res, 33 (14): 4602~4611

Xiao S, Scott F. 2002. Eukaryotic ribonuclease P: a plurality of ribonucleoprotein enzymes. Annu Rev Biochem, 71: 165~189

Yang J H, Zhang X C, Huang Z P, et al. 2006. snoSeeker: an advanced computational package for screening of guide and orphan snoRNA genes in the human genome. Nucleic Acids Res, 34 (18): 5112~5123

Yanofsky C. 2007. RNA-based regulation of genes of tryptophan synthesis and degradation, in bacteria. RNA, 13 (8): 1141~1154

Zamore P D, Haley B. 2005. Ribo-gnome: the big world of small RNAs. Science, 309 (5740): 1519~1524

Zhao T, Li G, Mi S, et al. 2007. A complex system of small RNAs in the unicellular green alga Chlamydomonas reinhardtii. Genes Dev, 21 (10): 1190~1203

Zhaxybayeva O, Gogarten J P. 2003. Spliceosomal introns: new insights into their evolution. Curr Biol, 13 (19): R764~R766

Chapter 3　RNA and Non-Coding RNA

The genetic information in the DNA genome of most organisms is first transcribed to a large number of RNA molecules. All the RNA transcripts from a genome are called transcriptome. RNAs are the final products of some genes, such as tRNA, rRNA, snRNAs, snoRNAs, and miRNAs as well. The messenger RNAs (mRNAs) are further translated into polypeptides, which are then modified and folded into functional proteins. RNA molecules are classified into mRNAs and non-coding RNAs based on the coding potential.

1.1　RNA Structure

1.1.1　The Primary Structure and Building Blocks of RNA

Bases, ribonucleosides and nucleotides

The primary structure of ribonucleic acids (RNA) is the sequence of bases attached to the sugar-phosphate backbone. RNAs are constituted by four ribonucleosides: adenine (A), guanosine (G), cytosine (C) and uracil (U). Nucleotides A and G are purines, C and U are pyrimidines. A ribonucleoside comprises a ribose sugar and a purine or a pyrimidines base attached to its $1'$ position. The ribonucleosides are linked together by phosphodiester bonds. The nucleoside and its corresponding phosphate group form a nucleotide, as shown in Figure 3-1A.

Base pairing

As are in DNA, Watson-Crick base pairs between G and C, A and U are also the most common and important base pairs in RNA. Another critical canonical base pair in RNA is G-U base pair, which is a wobble base pair and has comparable thermostability with Watson-Crick base pair (Figure 3-1B).

Figure 3-1　Composition and base pairs of RNA (see page 52)

Difference between RNA and DNA

Besides the difference in base pairs (A-U in RNA insteada of A-T in on A), RNA differs from DNA in that the sugar in RNA is a ribose, whereas it is a deoxy ribose in DNA. The $2'$ position of the ribose contains a hydroxyl group in RNA which is absent in DNA. The presence of the hydroxyl group forces RNA to form the A form geometry rather than the B form, which is common in DNA. Thus, RNA helix contains deep and narrow major groove but shallow and wide minor groove. Furthermore, the hydroxyl group can attack adjacent phosphodiester bond, leading to the cleavage of the RNA backbone. Therefore, RNA is usually less unstable compared to DNA.

1.1.2　Secondary Structures of RNA

RNA secondary structure is very different from that of DNA. The secondary structure of DNA is right-handed double strand helix, which consists of two antiparallel DNA strands. The helical backbone is made up of deoxyriboses and phosphates, which impose strong limitation to the base pairs in the helix, and only canonical Watson-crick base pairs G-C and A-T are allowed. RNA is usually single-stranded, whose backbone is highly flexible. This nature of RNA makes it easier to form more complicate base pairs.

The secondary structure of RNA is a planar representation of its base pairs and intervening unpaired regions. Watson-Crick base pairs and wobble base pairs are the fundamental constitution of RNA secondary structure. More than 60% nucleotides in large ribosomal RNAs form Watson-Crick base pairs and wobble base pairs. Diverse structural elements are defined in RNA secondary structure, for instance duplex, hairpin (stem loop), mismatches, internal loop, bulged loop and multi-branched loop, etc. The secondary structure of tRNA is a multi-branched loop consisting of three hairpins (D stem, T stem, and anticodon arms) and an acceptor stem, as shown in Figure 3-2a.

Figure 3-2　Structure of tRNA (see page 53)

The double-stranded helix of RNA is in A form geometry, which harbors a wide and shallow minor groove, and a deep and narrow major groove. The major groove in RNA helix is too narrow and deep to be accessed; the minor groove thus becomes more important in proteins-RNA interactions.

1.1.3　RNA Motifs and Tertiary Structure of RNA

Nearly all structural RNAs function with a well-defined tertiary structure. RNA tertiary structure is the compact assembly of the secondary structural elements. Figure 3-2 shows the secondary structure of a tRNA and its tertiary structure. The tertiary structure of a tRNA resembles an inverted letter "L". The acceptor stem and T-stem are coaxially stacked, generating the contiguous vertical stem. The horizontal stem is likewise formed by stacking of the D-stem and the anticodon stem. Loop-Loop interactions between the D-loop and the TψC loop stabilize the junction of letter "L". Complex tertiary interactions are involved in the formation of the tertiary structure, including: ① two helices; ② two unpaired regions; ③ one unpaired region and a helix.

1.1.3.1　Interactions between helices.

In an RNA helix, most of the functional groups involved in the formation of hydrogen bonds are embedded in the backbone, thus two helices lack the chemical diversity for their direct contact. However, inter-helical contacts can still be made by the interactions between functional groups on their backbones, primarily through water-mediated backbone-backbone contacts. Moreover, two helices can be packed together by the contacts between the backbone of one helix and the shallow minor groove of the other. This interaction is mainly mediated by the 2′hydroxyl groups of the RNA backbone.

Base stacking is an arrangement of RNA bases in which one base is on top of another. Base stacking is the most fundamental and important interaction stabilizing the structure of RNAs. For instance, 72 out of 76 nucleotides in tRNA^{Phe} are involved in base stacking interactions. Likewise, two helices can also be stacked on top of each other when the terminal base pairs of each helix are stacked together. The stacking of two helices like this is called coaxial stacking, which is a ubiquitous method utilized by large RNAs to achieve highly ordered structure. Coaxial stacking was firstly reported in the crystal structure of tRNA^{Phe}. In this crystal, D stem and anticodon stem stack upon one another, so as to the T stem and the acceptor stem, as can be seen in Figure 3-2b.

1.1.3.2　Interactions between helix and unpaired region

The A-minor motif

A-minor motif involves the insertion of adenines into the minor groove of its neighboring helices, preferentially at C-G base pairs, by hydrogen bonding with the 2′-OHs of those pairs. A-minor motifs can stabilize the contacts between RNA helices and between single-stranded regions and helices. The A-minor motif is the most ubiquitous and important long-range interaction in large RNA structures. It was found that 68 of the 106 Adenosines that are up to 95% conserved in 23S rRNA are involved in A-minor interactions. (Figure 3-3a)

Base triples

Base triples involve the formation of additional non-canonical base pairs between single-stranded bases and base of double-stranded helix (Figure3-3b). The single-stranded bases can then be packed onto the major or minor grooves of the helix. According to this, base triples can be classified as major-groove base triples and minor-groove base triples. By base triple interaction, the single stranded RNA can bring two helices together.

Tetraloop motif

Tetraloop is a terminal loop comprised of four nucleotides. Tetraloop is one of the most prevalent motifs observed in nearly all classes of structural RNAs. "UNCG" tetraloop and "GNRA" tetraloop (N for any nucleotides and R for purines) are the most encountered tetraloops. All GNRA tetraloops have a guanosine at position 1 and adenine at position 4. These two nucleotides can form sheared base pairs between N3(G)/N6(A) and between N2(G)/N7(A). The variability of the second and third nucleotides does not significantly alter their thermodynamic stability, but determines the diverse tertiary interactions they participated in. (Figure 3-3c).

1.1.3.3 Tertiary interactions between unpaired regions

Loop-Loop interactions

When two hairpins come together spatially, the complimentary terminal loops of the hairpins can form Watson-crick base pairs to create a new helix. (Figure3-3d).

The pseudoknot

A pseudoknot Contalns two stem-loop structures in which the loop of the first stem forms part of the second stem. The pseudoknot was first recognized in the turnip yellow mosaic virus in 1982. Pseudoknots fold into knot-shaped three-dimensional conformations but are not topologically real knots. (Figure3-3E). Pseudoknots are observed in a large class of RNAs, including viral RNA, rRNA, mRNA and ribozymes. Pseudoknot is an important structural motif for functional RNAs, for instance it's the catalytic core in group I ribozymes.

Figure 3-3 Illustration of RNA tertiary interactions (see page 55)

1.2 The Mechanisms of RNA Function

1.2.1 Base Pairing

A large variety of functional non-coding RNAs are trans-acting regulators. These RNAs function on their target main by base pairing. This kind of action can enrich the complexity of gene expression, i. e. , a single RNA can regulate the expression of multiple genes. Furthermore, the pairing between a functional RNA and its target could be either perfect or non-perfect base pairs. For instance, microRNA can regulate the target gene expression by translational inhibition. It can either form non-perfect Watson-Crick base pairs with the 3′-UTR region of the target gene, resulting in translation inhibition; or form perfect base pairs, which lead to the digestion of the target.

1.2.2 Forming Secondary Structure

Some RNAs are *cis*-acting regulators. These RNAs can usually regulate the only genes they reside in by forming secondary structures, to control the gene expression. An example is the riboswitch. A riboswitch can be divided into two parts, an aptamer and an expression platform. There are several kinds of expression platforms. First, it can form a hairpin structure to control the RNA level of the gene by rho-independent transcription termination. Second, it can fold into a certain structure which sequester the ribosome-binding site, and control the gene expression on protein level. Finally, it can also fold into certain structure which abolishes the splicing of pre-mRNA.

1.2.3 Forming Tertiary Structure

All structural RNAs have to fold into a compact structure which can fulfill their function. The best examples of this kind are catalytic RNAs, i. e. , ribozymes. The substrate and the active site of the ribozyme are usually remote in sequence. Thus the ribozyme has to fold into a compact three-dimensional structure that brings the substrate to the active site. In group I ribozymes, the substrate and the nucleophile that attack the substrate are on different molecules. The ribozyme has to be folded to a compact structure which can hold the nucleophile close enough to the substrate, to initiate the reaction.

1.2.4 Function with Proteins

Most of functional non-coding RNAs can't regulate target gene expression alone. They main act as bridges, bringing the protein factors they recruited to their target genes by base pairing. For instance, miRNA and siRNA can inhibit the translation or induce the cleavage of the target RNAs by bringing the RISC complex to these RNAs. The methylation guided by snoRNA is a similar case. SnoRNA first form base pairs with its target gene, and the methylation was accomplished by methylase it recruited.

1.3　Ribozymes

Ribozyme is the shortening of ribonucleic acid enzyme, referring to the RNA molecules that have catalytic ability. Tens of ribozyme species have been identified in the past twenty-six years. The ribozymes are wide spread in all three kingdoms of life, and are involved in diverse biological processes. The ribozymes can be grouped into small and large ribozymes according to the size of their primary sequences. The hammerhead, hairpin, varkud satellite (VS) and hepatitis delta virus (HDV) ribozymes are small ribozymes with the size no larger than 200 nt. Small ribozymes are mainly found in the viral, virusoid and satellite RNA genomes, in which they are responsible for the processing of rolling cycle replication intermediates to genome length. Small ribozymes catlyze site-specific cleavage reactions using their nucleotides as the attacking nucleophiles, and thus were also named nucleolytic ribozymes. Group I, group II, RNase P and ribosomes are large ribozymes, with the primary sequences from hundreds to thousands of nucleotides. Group I and group II introns which catalyze two consecutive estertransfer reactions, are abundant in orgnellar genomes, and the precise splicing of these introns are crucial for the maturation of their flanking exons. Ribonuclease P (RNase P) is a ubiquitous ribozyme found in all three kingdoms of life. It functions in the maturation of all tRNAs from their precursors using a hydrolysis reaction catalyzed by its RNA component. Moreover, it's found recently that the most abundant and important biological process, polypeptide synthesis is accomplished by the peptidyl transferase activity of the rRNA component of ribosome, which means that ribosome is also a ribozyme.

1.3.1　Nucleolytic Ribozyme

1.3.1.1　Reaction chemistry of nucleolytic ribozyme

The nucleolytic ribozymes all carry out a site-specific cleavage reaction to process the rolling cycle replication intermediates to genome length. The cleavage reaction is a transesterification reaction, in which the 2'-oxygen of the cleavage site attacks the adjacent 3'-phosphorus, producing a cyclic 2', 3'-phosphate on the 3'-end and a free hydroxyl group on the 5'end. The chemical reaction of the cleavage is a S_N2 reaction, as can be seen from Figure 3-4a. The reaction first involves the deprotonation of the 2'-hydroxyl group at the cleavage site. Then the deprotonated 2'-oxygen forms a weak bond with the adjacent 3'-phosphorus, and the cleavage site enters a transition state of a trigonal bipyramidal structure. In the transition state the 2'-oxygen nucleophile and the 5'-oxygen leaving group localize at the opposite apex, Lining up with the 3'-phosphorus (in-line). The reaction ends as the formation of the covalent bond between the 2'-oxygen and the 3'-phosphorus, and the rupture of 3'-5' phosphate bond. In general, these ribozymes can also perform the reversible ligation reaction, in which the 5'-oxygen functions as the nucleophile that attacks the cyclic phosphate.

Figure 3-4　Reaction mechanism of ribozymes (see page 57)

The reaction is an acid-base reaction, in which a general base is involved in the deprotonation of the 2'-hydroxyl to activate the 2'-oxygen nucleophile, and a general acid is involved in the protonation of the 5'-oxygen leaving group. Nucleolytic ribozymes are named as the fact that they all use nucleotide functional groups as the general acid and base, except HDV ribozyme, which may utilizes a metal ion in the active site to function as the general base.

Acid-base reaction is the universal mechanism of all ribozymes, there are two distinct features distinguish the small nucleolytic ribozymes from the rest. The major difference is that nucleolytic ribozymes use an inner-nucleotide group as the attacking nucleophile, instead of a remote (group II) or exogenous molecule (group I, RNase P and ribosome). Another distinctive difference lies in that the general acid and base are functional groups of the ribozyme, whereas other ribozymes utilize metal ions.

1.3.1.2　Structural features of small nucleolytic ribozymes and the formation of active site

Although the nucleolytic ribozymes have unrelated primary sequences and tertiary folds, they all need to form compact structures to bring the functional nucleotides into close proximity to the cleavage site, to serve as general base or acid to activate the nucleophile or stabilizing the leaving group, respectively.

HAMMERHEAD ribozyme is composed of a conserved core and three flanking helices assembling like a hammerhead. Hammerhead ribozyme is the first catalytic RNA structure solved at atomic resolution, which revealed that stem Ⅱ and Ⅲ are coaxially stacked, while stem Ⅰ lie at a sharp angle to stem Ⅱ (Figure 3-5a left). Unfortunately, the structural information inferred from the crystal structure can't fit the biochemical data in many aspects, of which a major problem is that the position of the scissile phosphate in the crystal structure is completely incompatible with "in-line" attack mechanism. It was suggested that the crystal structure was not the catalytically active structure, but represent a "ground state". This hypothesis was fully evidenced by the crystal structure of a full-length and active ribozyme in 2006 (Figure 3-5a middle). In this crystal, backbone distortions at the junction of helices Ⅱ and Ⅲ force the nucleotide C17 to stack on stem Ⅲ, placing it in the active site pocket at the three-helix junction. Nucleotide G5 hydrogen bonds to the furanose oxygen of C17, helping to position it for "in-line" attack. G12 is within hydrogen bonding distance to the 2′-O of C17, and the ribose of G8 hydrogen bonds to the leaving group 5′-O, suggesting that these two nucleotides function as the general base and acid, respectively. Nucleotide C3 forms a Watson-Crick base pair with G8, to help its position, as can be seen from Figure 3-5a right. This crystal structure thus addresses the major concerns that appeared irreconcilable with previous crystal structures.

Figure 3-5　Secondary, crystal structures of ribozymes and the formation of active sites (see page 58)

The **HAIRPIN ribozyme** consists of four helical stems, termed A, B, C and D, which form a four-way junction (Figure 3-5b left up). Stems A and B each contains a internal loop, and the cleavage site of the ribozyme is located in loop A. Nucleotides in these two loops are highly conserved and are critical for the catalysis, while the helical regions are highly variable, which are constrained only by maintaining the base pairs. In the active structure, stem A coaxially stacks on D, and stem B on C (Figure 3-5b right up). The two coaxial stacked domains are anti-parallel positioned to allow the contacts between the internal loops on stem A and B to form the catalytic center. The active site involves a complex ribose zipper formed between A10, G11 in loop A, and A24, C25 in loop B. Nucleotide G_{+1} is extruded from loop A, and inserted into a pocket within loop B by forming a Watson-Crick base pair with C25. This extrusion changes the backbone conformation and thus the 2′-O nucleophile of A_{-1} is well aligned for nucleophilic attack. Nucleotides G8 and A38 are closely juxtaposed at the active center and hydrogen-bond with the scissile phosphate, suggesting that they may function as the general base and general acid, respectively, to participate in general acid-base catalysis (Figure 3-5b down).

The **VS ribozyme** is about 150 nt in length and consists of 6 helices, which form two three-way junctions, the 2-3-6 junction and the 3-4-5 junction, as can be seen from Figure3-5d. Helix Ⅰ is the substrate helix, which contains the cleavage site in its internal loop. Helices Ⅳ, Ⅲ and Ⅵ are coaxially stacked to form a long column and helices Ⅱ and Ⅴ radiate from the column. The local sequences of the two three-way junctions are of critical importance to the stacking of helices. The catalytic center of VS ribozyme is suggested lying in the bulged loop, termed A730 loop of helix Ⅵ. The cleavage site of the ribozyme is after G620 in the internal loop of helix Ⅰ. Nucleotides GUC in the terminal loop of helix Ⅰ can form long-range base pairs with GAC in the terminal loop of helix Ⅴ. This interaction brings the cleavage site to the active site. Comparing the similarity between VS and hairpin ribozyme, together with the biochemical data, it is suggested that nucleotides A765 and G638 may function in the acid-based catalysis, but it's not clear yet which nucleotide acts as the general acid and which as the general base. A sheared base pair between G620 and A638, and the G-A sheared base pairs shared between G638 and A621, G638 and A622 help to hold the to be deprotonated 2′-oxygen close to the general base.

The **HDV ribozyme** consists of 5 helices, termed P1-P4 and a 2-base-pair stem P1.1 (Figure 3-5c). These helices form a double-nested pseudoknot: helix P1 and helix P2 form a single pseudoknot, P3 and P1.1 form another pseudoknot, and these two pseudoknots are further nested to form a third pseudoknot between stems P1 and P1. 1. The five helices are arranged into two parallel stacks, in which helices P1, P1. 1 and P4 are coaxially stacked, and helix P2 is coaxially stacked on P3 (Figure 3-5c). The two stacks are positioned side by side, linked by five strand crossovers and further constrained by the P1. 1 pairing. Helix P1 is the substrate helix that contains the cleavage site. The catalytic core of the ribozyme involves stems P1. 1, P3 and the junction regions, while stems P1, P2 and P4 are only structural elements stabilizing the catalytic core. Divalent cations are required for HDV ribozyme function, and it's believed that a hydrated metal ion is directly involved in the reaction chemistry. Another crucial factor in the catalysis is nucleotide C75. These two factors should act as the general acid and base in the catalysis, but it's yet unclear which factor act as the general base and which as the general acid (Figure 3-5c right). The first crystal structure of HDV ribozyme revealed that C75 is in close proximity with the leaving group, suggesting that it may function as the general acid. Multiple lines of biochemical evidences were subsequently found to support this notion. But a recent crystal structure revealed that C75 is closer to the nucleophile, indicating that it may function as the general base and the hydrated metal ion as the general acid. Thus further research is needed to tell the specific role of these two factors.

1. 3. 2　RNase P Ribozyme

1. 3. 2. 1　RNase P ribozyme catalysis

RNase P ribozyme is a universal ribozyme that is conserved in all three kingdoms of life. It functions in the maturation of tRNA genes utilizing a hydrolysis reaction. The reaction chemistry of the hydrolysis is also an acid-base reaction. Although several crystals of the ribozyme have been resolved, atomic view of the arrangement of the chemical groups is yet absent. The reaction mechanism is mainly inferred from biochemical data. The nucleophile in the reaction is proposed to be a water molecule activated by the general base. Three divalent metal ions are involved in the reaction, as can be seen from Figure3-4b. Metal A functions as the general base to deprotonated the nucleophile, resulting in an Mg^{2+}-hydrate to function as the nucleophile to start the reaction. Metal B is involved in the stabilization of the trigonal bipyramidal transition state. The $2'$-hydroxyl group of the precursor tRNA cleavage site (N_1) coordinates a third metal ion by inner-sphere water to protonate the $3'$-leaving group.

1. 3. 2. 2　RNase P ribozyme structure and catalytic core

The primary sequences and secondary structures of RNase P ribozymes are highly diverse, but all RNase P ribozymes share a set of highly conserved structural elements which make up the structural and catalytic core. The core consists of five Conserved Regions (CR for short), termed CRI to CRV which contain many highly conserved nucleotides (Figure 3-5e left). Besides the common core structures, each RNase P ribozyme contains its unique peripheral elements. The peripheral elements are phylogenetically conserved in each domain of life. The bacterial RNase P RNAs are typically 330~400 nt in length, and can be sub-typed into two classes, A type (Ancestral) and B type (Bacillus), according to the differences in the peripheral elements they harbor. Archaeal ribozymes are similar in size to the bacterial ribozymes but have different peripheral elements. Eucaryal ribozymes are two-thirds the length of the bacterial ribozymes and lack many peripheral elements present in bacterial ribozymes.

The RNase P ribozymes can be divided into two domains that are folded independently (Figure 3-5e left). The specificity domain (S-domain) is involved in the recognition of pre-tRNA substrate by interacting with its TΨC loop, whereas the catalytic domain (C-domain) is responsible for the recognition of the pre-tRNA acceptor stem and the $3'$-CCA sequence, and catalyzes the hydrolysis reaction. The overall structure of the ribozyme is a compact fold composed of coaxially stacked helices that are joined together by long-range docking interactions. These relatively spatial arranged helices form a remarkably flat surface responsible for the binding of pre-tRNA. The five conserved regions, together with helix P15 and three coaxially stacked helices P1-P4-P5, P2-P3 and P8-P9, form the catalytic core of the ribozyme (Figure 3-5e right). P4 is a long-range base pair formed between CR Ⅰ and CR Ⅴ, and is crucial for the catalysis of the ribozyme. Three divalent metal ions were found to bind on the major groove of helix P4 in the recent crystal, however, these metal ions are too far away from the active site, thus may not function in catalysis. Instead, two other metal ions in the junction regions are in close proximity to the active site, which may function in the catalysis. The less conserved peripheral elements are located on the surface and are away from the substrate binding site, which may stabilize the global structure by long-range docking interactions.

Although different RNase P ribozyme contain distinct peripheral elements, the tertiary structures are maintained considerably by emerging corresponding different interactions. For instance, docking between P12 and P13 in the A type RNA is replaced by a docking interaction of P10. 1 to P12 in the B type. Another example is the P16-P17-P6 structure in the A type, which is absent in B type. The tertiary interaction formed by this structure is rescued by the docking interactions between P5. 1 and P15. 1.

1. 3. 2. 3　Substrate recognition of RNase P ribozyme

Although several crystal structures of RNase P ribozyme have been obtained, there is no crystal of the ribozyme-substrate complex, thus the knowledge of substrate binding are mainly from biochemical data and modeling the tRNA on the ribozyme crystals. The determinates of pre-tRNA for its binding include the $5'$-leader sequence, $3'$-CCA and the TψC stem loop. The ribozyme folds contain a flat surface for the binding of the substrate, as mentioned previously. A universally conserved Adenosine in J5/15 junction of the ribozyme interacts with the N_{-1} nucleotide of the pre-tRNA $5'$-leading sequence. The recognition of the TψC loop of pre-tRNA is accomplished by the S-domain of the ribozyme, of which two highly conserved adenosines in P10/11 domain may participate in the docking of the TψC loop. The interaction between the $3'$-CCA of tRNA and L15 of ribozyme has long been found to be crucial for the selection of hydrolysis site. Recent biochemical data revealed that this interaction is far more complex than simple base pairing. For instance, the interaction calls for divalent cations and compensatory mutations which maintain the candidate "base pairing" still causes catalytic defects, indicating that this interaction is not simply base pairing.

1. 3. 2. 4　Protein function in RNase P ribozyme

RNase P ribozyme is a ribonucleoprotein complex, of which the protein component may function structurally or catalytically. The protein contents of RNase P are highly variable among the three kingdom of life. It appears that the protein content is correlated with the structural complexity of the RNA component of the ribozyme. Bacterial RNase P ribozymes have more complex RNA structures, and the RNA component alone can fulfill the features needed for catalysis. The protein component in these ribozymes makes up only about 10% of the total mass. The eukaryotic RNase P ribozymes have relatively simple RNA structures, and consist of over 70% proteins. Thus it is possible that the protein components in eukaryal RNase P ribozymes perform structural roles, functionally substituting for miss RNA substructures.

1. 3. 3　Group I Introns

1. 3. 3. 1　Group I intron catalysis

The self-splicing of group I intron is a two consecutive phosphotransesterification reaction, including the sequential excision of the intron at $5'$-and $3'$-splicing sites and the ligation of the flanking exons (Figure 3-4c left). In the first step cleavage reaction, RNA duplex containing $5'$-substrate docked into the catalytic core of the intron, positioning the $5'$-splicing site for being attacked. An exogenous guanosine bound to the active site to start the reaction, using its $3'$-hydroxyl group as the nucleophile to attack the phosphate of $5'$-splicing site. This leads to the broken of the phosphodiester bond at the $5'$-splicing site and result in a free $3'$-hydroxyl on the $5'$-exon and a $5'$-G linked intron. After the first step reaction, conformational changes occur to release the attacking guanosine and to bind the highly conserved guanosine (ωG) residue at the $3'$-end of the intron. The second step then can occur, using the free $3'$-hydroxyl of the $5'$-exon as the nucleophile to attack the phosphate of the $3'$-splicing site. This leads to a ligated exon and a released intron.

The chemical reaction of group I introns is an acid-base reaction. At least two Mg^{2+} are directly involved in the reaction. As can be seen from Figure 3-4c right, Mg^{2+} (MB) binds to the $3'$-hydroxyl group of the exogenous guanosine in the first step, functioning as a general base to activate the nucleophile. Another Mg^{2+} (M_A) stabilizes the $3'$-leaving group, acting as a general acid. These two metal ions exchange their roles in the second step reaction, of which M_A functions as the general base and M_B as the general acid, respectively. The transition states were suggested to be stabilized by the Mg^{2+}-coordinated 2′-hydroxyl group of the exogenous guanosine and the hydrogen-bond network, respectively, in the first and second step.

1.3.3.2 Group Ⅰ intron structure

Group Ⅰ introns vary significantly in their size, from hundreds to thousands in length. Nevertheless, all introns retain a central conserved secondary structure composed of 9 base-paired helices numbered P1 to P9 (Figure 3-5f left). The nine helices are organized into three domains, P1-P2, P4-P6, and P3-P9. Helices in each domain are all coaxially stacked. Helix P1 is the substrate helix containing 5′-splice site. Helices P3 and P7 form a pseudoknot critical for the function of the ribozyme. The tertiary structures of group Ⅰ introns reveal a compact fold in which domain P4-P6 function as the scaffold and domain P3-P9 tightly wrapped around P4-P6 domain (Figure 3-5f middle). The three domains are connected by conserved junction regions, which play critical roles in the formation of either global structure or the active site. It was shown that double-stranded junction segments are involved in domain-domain interactions by mediating minor groove-minor groove contacts. The single-stranded junctions line one side of a domain, providing a complementary surface for interdomain packing. For instance, J3/4 and J6/7 may form base triples at the P4-P6 interface, which facilitate the contact between P4-P6 and P3-P9 domains (Figure 3-5f left).

Besides the nine conserved helices, most group Ⅰ introns contain at least one additional structural elements, termed peripheral elements, such as the P5abc, P2.1, and P9.1-P9.2 segments in the *Tetrahymena* intron. These peripheral elements are mainly involved in forming tertiary interactions that stabilize the core structure, and deletion of these segments generally reduces but does not eliminate self-splicing. But still, some peripheral elements are critical for the catalysis the intron; deletions of such peripheral elements eliminate intron splicing. For instance, the root stem of the P2.1 peripheral element in *Candida* intron form a triple-helix structure with P3 and P6, and is essential for the formation of the compact core.

1.3.3.3 Substrate binding and recognition of the splicing site

Both 5′-substrate and 3′-substrate must be docked into the active site to position the scissile phosphate close enough to the attacking nucleophile. The 5′-splice site is fundamentally determined by the Internal Guiding Sequence (IGS), which form helix P1 with the substrate (Figure 3-5f left). The 5′-splice site is a GU wobble pair on the substrate helix P1, which is further determined by three elements. First, P2 is coaxially stacked to the substrate helix P1, thus P1 can be anchored by the docking of P2 into the minor groove of P8, or other unknown interactions. Second, J2/3 and J8/7 decorate the minor groove of the P1-P2 domain. Finally, the GU wobbles pair of the 5′-splice site is docked into the P4-P6 domain by an A-minor interaction with J4/5. These interactions positioned the 5′-splice site near the active site.

The 3′-end of group Ⅰ intron is always a guanosine residue. This residue has to be docked into the active site and being attacked during the second step reaction. Thus this guanosine is the determinate of the 3′-splice site. Two elements are involved in the choice of the 3′-splice site. The first is a short long rang base pair, P9.0, between the downstream nucleotides of P7 and nucleotides upstream and usually adjacent to the terminal guanosine. This interaction brings the terminal guanosine to close to P7, which participates in the active site. The second is P10, which is formed by the exon sequence adjacent to the terminal guanosine and part of IGS. This interaction further positioned the 3′-splice site to the active site.

1.3.3.4 Active site of group Ⅰ intron

The active site of group Ⅰ intron is located in the P7 helix, which can bind either an exogenous guanosine in the first step reaction to serve as the attacking nucleophile, or the terminal guanosine of the intron in the second step reaction to get attacked. The active site is composed of four layers of base triples. In Tetrahymena intron, the guanosine (either exogenous in first step or terminal in the second step) forms a coplanar base triple (G-triple) with the G264-C311 base pair (Figure 3-5f right). This triple is sandwiched by three other layers of base triples. Above the G-triple, C262 pairs with G312. A263, which is depicted as a bulged residue in secondary structure diagrams, forms two hydrogen bonds to the minor groove side of G312. The base of C262 is located right above the purine ring of ω G, presumably stabilizing its binding by base stacking. Below the G-triple, the A261 base contacts the major groove of the A265-U310 pair. In a further layer, the Hoogsteen surface of the A306 base contacts C266-G309 major groove and form the forth base triple. The active site of the ribozyme can bind two divalent cations to act as the general base and general acid, deprotonating the nucleophile and protonating the 3′-leaving group, respectively.

1.3.4　Group Ⅱ Introns

1.3.4.1　Group Ⅱ intron catalysis

The splicing mechanism of group Ⅱ introns is identical to the splicing of nuclear pre-mRNA introns, and it was suggested that group Ⅱ introns were the ancestors of spliceosomal introns. The splicing of group Ⅱ introns is also a two-step ester-transfer reaction (Figure 3-4d left). In the first step, the 2′-hydroxyl group of the branch-site adenosine attacks the 5′-splice site. This leads to the cleavage of the 5′-exon and the formation of a lariat intermediate, and results in a free 3′-hydroxyl on the 5′-exon. The 5′-exon is held into place after cleavage. In the second step, the 3′-hydroxyl of 5′-exon attacks the 3′-splice site and result in a ligated exon and a lariat intron.

The chemical reaction of group Ⅱ introns is nearly identical to that of group Ⅰ introns, except for two distinct features (Figure 3-4d right). First, group Ⅱ introns utilize an in-sequence adenosine to attack the 5′-splicing site, using its 2′-hydroxyl group as the nucleophile. Second, in contrast to group I introns, in which the 2′-hydroxyl group near the scissile phosphate is involved in stabilizing the transition site, it is free in group Ⅱ introns.

1.3.4.2　Group Ⅱ intron structure

The primary sequences of group Ⅱ introns are highly variable except the conserved sequence in domain Ⅴ, but they can all adopt into a highly conserved secondary structure which consists of six helical domains, termed domain Ⅰ～Ⅵ (D1～D6 for short) (Figure3-5g). D1 contains nucleotides that are essential for the recognition of both exons and the branch-site nucleophile. D2 and D3 are essential component for the full catalytic activity, and the junction regions between them (J2/3) is part of the active center. D4 usually encodes a large open reading frame which functions in intron mobility, and functions little in ribozyme folding or catalysis. D5 is the most conserved region in group Ⅱ introns. It directly involves the assembly of the active core and offers functional groups that mediate the chemical catalysis. D6 contains the branch site adenosine, of which the 2′-hydroxyl serve as the attacking nucleophile in the first step cleavage.

Group Ⅱ introns can be classified into three major subgroups, ⅡA, ⅡB and ⅡC according to their distinct structural features. The crystal structure of an ⅡC intron is resolved recently. As is shown in Figure3-5g middle, the active structure of group Ⅱ intron is a compact assembly of all six domains, of which domain Ⅰ functions as the scaffold, providing a preassembled external groove for the docking of the exons and the formation of catalytic core. On the top, ⅠA and ⅠB of D1 are coaxially stacked, as well as Ⅰ(i), Ⅰ(ii) of D1 and D2 on the left (front view, same below). IC of domain Ⅰ stays close together and is parallel with the Ⅰ(i)-Ⅰ(ii)-DⅡ stack. On the right, the substrate biding domain, ⅠD1 and ⅠD2 stay close together. D3 and D4 are coaxially stacked on the bottom. These domains hold the highly conserved domain Ⅴ in the center. Domain Ⅵ which contains the attacking adenosine, although included, can't be visualized in this crystal. But there is ample space for this domain within the structure, and it suggested to lie within the open cleft next to D5.

1.3.4.3　Substrate recognition

The substrate recognition of group Ⅱ introns is generally determined by base pairs between intron and exon sequences. ⅡA and ⅡB introns share more similarities in the recognition of substrate. The recognition of 5′-substrate in these two subgroups is fundamentally determined by two Watson-Crick base pairing interactions, IBS1-EBS1 (intron and exon binding site 1) and IBS2-EBS2 (intron and exon binding site 2) which are both about 6 bp base pairing. The 5′-splice site is precisely chosen by the intron core, which specifically targets the phosphodiester junction between single and double-stranded RNA at the EBS1 terminus (Figure 3-5g left). The recognition of the 3′-substrate are different between ⅡA and ⅡB introns. ⅡA introns use a Watson-Crick base pair called δ-δ′ to determine the 3′-splicing site, while ⅡB introns use an IB3-EB3 base pair. The 5′-substrate recognition of ⅡC introns are quite different from that of ⅡA and ⅡB introns. As mentioned above, the IBS1-EBS1 base pairs in ⅡC introns are quite short, and they have no EBS2-IBS2 interactions. The 3′-substrate recognition is similarly determined by EBS3-IBS3 base pairs (Figure 3-5g left), just as ⅡB introns.

1. 3. 4. 4　The active site

The active center of the ribozyme involves three main domains, D1, D5 and D6, and an important J2/3 linker, of which D1 binds the substrate, D6 provides the attacking nucleophile for the first-step cleavage while D5 and J2/3 are the catalytic region. The formation of the active center is the rearrangement of these elements to bring them together. Besides functioning as the scaffold of the ribozyme, D1 helps to anchor both 5′-splicing and the 3′-splicing site to D5 through κ-κ′, ζ-ζ′, λ-λ′ and ε-ε′ interactions. The positioning of the branch point adenosine is determined by the non-canonical base pairs between its flanking polypyrimidine sequence and the loop sequence in D1.

Domain V is the most conserved and catalytically important region of the ribozyme. Much of the D5 sequences are involved in the tertiary interactions guaranteeing the proper position of D5, for instance κ-κ′, ζ-ζ′ and λ-λ′. The functional groups for the chemical catalysis are the so called "catalytic triad" AGC sequence and the D5 bulge. These two groups are in close proximity in the folded ribozyme, and they are likely to function together. Domain V also contains the binding pocket for metal ions which participate in the catalysis (Figure 3-5g right).

The J2/3 junction between D2 and D3 is an indispensible component of the active site, but its concrete role is not clear until recently, after resolving the crystal structure of the *Oceanobacillus iheyensis* intron. J2/3 and D5 bulge forms a triple helix with the catalytic triad (Figure3-5g middle). It merged to the D5 stem completely, forming a triple helix that brings the catalytically essential residues together.

1. 3. 5　Ribosome is a Ribozyme

Ribosomes are the most important macromolecules that exist in all three kingdoms of life, which catalyze the peptidyl transfer reaction to translate messenger RNA to proteins. Ribosomes are composed of two-thirds ribosomal RNA and one-third ribosomal proteins. The crystal structures of the large ribosomal subunit complexed with substrate and product analogues show that only RNA is in a position to chemically facilitate peptide-bond formation. There are three RNA groups close enough to the reaction site to form a hydrogen bond with the attacking α-amino group: the 2′-hydroxyl group of A76 of the tRNA in the Psite, the N3 group of A2486 (A2451 in *E. coli*) of 23S rRNA, and the 2′-hydroxyl group of A2486. These hydrogen-bonds may be involved in positioning the α-amino group for nucleophilic attack, or more importantly, direct in catalysis. Great advances have been made in the past two years, and high resolution crystal structures of the ribosome complexed with mRNA and tRNA have been obtained, providing the structural basis of ribosome functioning. The crystal structure of the large subunit ribosome revealed that, no protein moiety lies closer than 18Å to the site of peptide bond formation, and excluded the possible function of proteins in this reaction. Thus, proteins contribute nothing to catalysis, and RNAs are the only components to accomplish the key peptidyl transferase reaction. So ribosome is an RNA based enzyme, i. e. ribozyme.

1. 4　RNA in Post-Transcriptional RNA Processing

1. 4. 1　snoRNAs and rRNA Modification

Small nucleolar RNA (snoRNA) is one large class of noncoding RNAs in eukaryotes. It mainly comprises two families , the C/D and the H/ACA snoRNAs, which are classified according to their characteristics of sequences and secondary structures. snoRNAs mainly function in ribosomal RNA (rRNA) modification. The two classes of RNAs guide different nucleotide modifications: the C/D class carries out ribose-2′-O-methylation, while the H/ACA class is responsible for pseudouridylation (conversion of uridine to pseudouridine). They modify functional key regions of rRNA, and both types of modification are essential for ribosome function. The site of modification is accurately specified by the formation of RNA duplex between the antisense element in snoRNA and the corresponding reverse complementary site in rRNA. The reaction is catalyzed by a methylase or pseudouridine synthase, which forms a RNP with the guide snoRNA and other specific proteins.

C/D and H/ACA RNAs have distinct conserved signature-sequence elements and characteristic secondary structures, as shown in Figure 3-6. The C/D snoRNAs are defined by two consensus motifs: box C and box D (Figure 3-6a). The RUGAUGA sequentce of box C is near the mature 5′ end of the RNA, while the CUGA sequence of box D is located near the 3′ end. These two motifs are brought together by base pairing between the 5′ and 3′ termini of the RNA. Most C/D snoRNAs contain a second set of conserved sequences, the C′ box and D′ boxe, which are located in the central region of the RNA and do not satisfy the consensus of C and D boxes. One or two antisense elements which locate in the 5′ terminus of box D or box D′ and are 10～20 nucleotides in length, form the gaide duplex with the reverse complemental sequenle in the rRWAs. The canonical methylation site is further defined as the length the of the rRNA which base pairs with the fifth nucleotide upstream the CUGA motif of the snoRNAs.

The H/ACA snoRNAs typically form a hairpin-hinge-hairpin-tail secondary structure (Figure 3-6b). The counecting hinge contains the "box H" sequence ANANNA. The single-stranded tail contains the "box ACA" which is 3 nt upstream of the 3′ end of the snoRNA. Each H/ACA snoRNA contains one or two bi-directional antisense elements in the internal loops of its hairpin host. The antisense elements can form 9-nucleotid to 13-nucleotide duplex with the substrate RNA sequences flanking the pseudouridylation site. The pseudouridylation site is usually located 14～16 nucleotides upstream of either the H or ACA box.

Figure 3-6 The guide snoRNAs and their target RNAs (see page 66)

The snoRNAs in each family interact with a set of highly conserved proteins to form the C/D and H/ACA RNPs, and the guide RNAs function as adaptors that link the catalytic protein component to their targets. Using archaeal naming convention, the core C/D ribonucleoprotein (RNP) proteins are: fibrillarin (the 2′-O-methyltransferase), L7Ae and Nop56/58. Eukaryotes have homologous proteins. The bridge formed by L7Ae and Nop56/58 is required for the association between fibrillarin and a C/D guide RNA. L7Ae nucleates the assembly of the C/D RNP. L7Ae binds a k-turn motif formed by the interaction between box C and D elements. The resulted complex undergoes a conformation change, which seems to create a new binding site for Nop56/58 protein. Nop56/58 binds to the complex and then recruits the catalytic protein fibrillarin to complete the assembly.

The core proteins of archaeal H/ACA RNP are: centromere binding factor 5 (Cbf5; the pseudouridine synthase), L7Ae, Gar1 and Nop10. These proteins interact with the guide RNA through their conserved features. Cbf5 contains a catalytic domain and a PUA domain. Box ACA and the lower stem of the guide RNA are bound by the PUA domain of Cbf5, anchoring the antisense elements close to the catalytic site. Gar1 and Nop10 bind to distinct sites in the catalytic domain of Cbf5. It has been shown that Gar1 is involved in the binding and/or release of the target RNA. As in C/D RNP, L7Ae binds directly with the kink-turn (k-turn) motif that is located in the terminal loop of the H/ACA RNA.

In addition to rRNA modification, the C/D and H/ACA RNAs have been recently found to have other functions and targets, and to be located in other cellular localizations outside of the nucleolus. They guide post-transcriptional modifications of snRNAs in eukaryotes, of transfer RNAs in archaea, and of spliced leader (SL) RNAs in trypanosomes. In addition, one H/ACA RNA is functional in telomere synthesis. The snoRNA HBⅡ-52 is shown to regulate alternative splicing of a pre-mRNA. Some snoRNAs are tissue-specific and parentally imprinted. Notably, an increasing number of orphan snoRNAs, which do not apparently target established substrates like rRNAs and snRNAs, have been experimentally and computationally identified from different eukaryotes. Human orphan snoRNA targets tend to locate close to alternative splice sites. All of these indicate the presence of other targets and novel functions of orphan snoRNAs.

1. 4. 2 snRNAs and Pre-mRNA Splicing

Precursor mRNA (pre-mRNA) splicing is a critical regulatory stage in gene expression which includes accurate recognition and removal of introns by splicing machinery, spliceosome. Spliceosome is a dynamic complex of five small nuclear ribonucleoproteins (snRNPs) —U1, U2, U4, U5 and U6—and many auxiliary proteins. The snRNPs are named after their corresponding uridine-rich snRNA components. U4 and U6 snRNAs can form extensive basepairs (Figure 3-7a).

**Figure 3-7　Pathways of assembly and catalysis of the major-class and minor-class
spliceosomes**（see page 68）

（Adapted from Patel A A, Steitz JA. 2003. Nat Rev Mol Cell Biol., 4（12）: 960-970）

U1 snRNA contains 163 nucleotides with a m3G cap at its 5′ end. In a cross species co-variation analysis, U1 snRNA was proved to have a common secondary structure of four stem-loops. The ACUUACCU sequence at U1 snRNA's 5′ end is involved in the binding of U1 snRNP to the conserved sequence at the 5′ splice site of pre-mRNA（AGGURAGU, R represents purine）. The Sm site（AUUUGUGG）located in a single-stranded region is bound by Sm proteins. The human U1 snRNP contains three specific proteins, named U1 70K, U1A and U1C, and Sm proteins are common to all U snRNPs. U1 70K and U1A bind to stem-loop 1 and stem-loop 2 of the U1 snRNA, respectively. According to the current model of spliceosome assembly, the snRNPs join the spliceosome in an ordered pathway. The U1 snRNP initially recognizes 5′ splice site by base pairing between U1 snRNA and the 5′ exon-intron junction. However, in the HeLa in vitro system, the canonical RNA-RNA base-pairing is not essential for either 5′ ss selection or splicing, which indicates that other factors, probably U1 snRNP proteins, make significant contributions to 5′ ss recognition and the splicing complex formation. It has been found that yeast U1 snRNP protein U1C recognizes the sequence of the 5′ splice site and has been argued that this RNA-protein interaction precedes base pairing with U1 snRNA. It is also possible that there is another snRNA basepairing with the 5′ splice site, which facilitates its recognition by the spliceosome. U6 snRNA has been shown to bind to a site very close to the 5′ end of the intron (Figure 3-7).

U2 snRNA consists of 187 nucleotides that form five stem-loop structures: stem-loop I, II a, II b, III and IV. U2 snRNP contains several proteins including two protein complexes known as splicing factors SF3a and SF3b. SF3b binds to U2 snRNP through its interaction with the 5′ end of U2 snRNA, and then SF3a joins in the complex. The U2 snRNA is extensively modified (2′-O-methyl groups and pseudouridines) within the 27 nucleotides at the 5′ end. These modifications are essential for the binding of SF3b and SF3a to the U2 snRNP. In spliceosome A complex, the conserved sequence at the branch point pairs with a complementary sequence within U2 snRNA. Genetic analysis has shown that base paring is essential for splicing, more precisely, this duplex is believed to specify the branch site adenosine as the nucleophile for the first step transesterification reaction. Besides base-pairing with the branchpoint, U2 also base-pairs with U6 snRNA (Figure 3-7a). This interaction is not essential in yeast, but is necessary in mammals, at least for high splicing efficiency.

U5 snRNA has no obvious complementarity with any other snRNAs or the conserved regions of a pre-mRNA, however, a highly conserved loop in U5 snRNA aligns the 5′ and 3′ splice site for the second step trans-esterification reaction by binding to the 3′-end of the first exon and the 5′-end of the second exon (Figure 3-7a).

As mentioned ealier, U4 and U6 snRNAs undergo extensive base-paring, forming two base-paired stems, called stem I and stem II. Upon spliceosomal assembly this base-pairing is unwound, and U6 snRNA subsequently base-pairs with U2 snRNA and the 5′ splice site, thereby displacing U1 snRNA. At this time, U4 dissociates from U6 and can then be removed from the spliceosome. Thus, U4 may function in binding and sequestering U6 until it is time for U6 to join in splicing by binding to the 5′ splice site. It is noteworthy that some U6 bases that participate in base pairing with U4 to form stem I are also involved in the essential base pairing with U2 as discussed earlier. This underscores the importance of removing U4 in allowing U6 to basepair with U2 and to form an active spliceosome subsequently.

Recently, a new class of rare introns (minor-class introns) have been found in metazoan genes, they have non-canonical consensus sequences and are excised by a low-abundance spliceosome. This low-abundance spliceosome includes four snRNPs (U11, U12, U4atac and U6atac) that are functionally analogous to the well-characterized U1, U2, U4 and U6 snRNPs, respectively. The U5 snRNP is shared by both spliceosomes. Because the excision of these minor-class introns is dependent on the U12 snRNP, they are called U12-type introns, whereas the canonical, major-class introns that require the analogous U2 snRNP are called U2-type introns (Figure 3-7).

Taken together, catalysis of the splicing reaction is proceeded by a coordinated series of RNA-RNA,

RNA-protein and protein-protein interactions, snRNAs function as "anchor" and "scafford" in splice site recognition and the two-step transesterification of precursor mRNA splicing.

1.4.3 Alternative Splicing

With the rapid development of sequencing technologies, it is found that the number of protein-coding genes in an organism does not correlate with its overall cellular complexity. For example, despite of the remarkble complexity, mammalian species have similar numbers of protein-coding genes as *Arabidopsis thaliana* (20 000~25 000) and are only four times of that in budding yeast *Saccharomyces cerevisiae* (~6000). These observations indicate that there must be some mechanisms acting to regulate and diversify gene functions with so few protein-coding genes. Alternative splicing (AS) is the process by which the exons of primary transcripts (pre-mRNAs) from genes can be spliced in different arrangements to produce distinct mRNAs and protein variants with different structures and functions. In human genome, it is estimated that 73% of human genes are alternatively spliced, which may represent one mechanism that accounts for the cellular complexity of higher eukaryotic organisms.

In higher eukaryotes, the coding sequence of most genes is splited to multiple exons by introns. For example, the average intron number per gene is approximately seven in human genome. In a typical multi-exon mRNA, the splicing pattern can be altered in many ways (Figure 3-8). Most exons are constitutive, as they are always included in the mature mRNA. A regulated exon that is sometimes included and sometimes excluded from the mRNA is called a cassette exon. Splicing of some cassette exons is mutually exclusive, producing mRNAs that contains one but not other cassette exons. In other cases, exons can also be lengthened or shortened by altering the position of one of their splice sites, which is called alternative 5′ splice site or 3′ splice site. Other splicing events are summarized in Figure 3-8.

Figure 3-8 Alternative splicing events (see page 70)
(Adapted from Matlin et al. 2005. Nat Rev Mol Cell Biol 6: 386-398)

Like other levels of gene expression control, the regulation of splicing involves both *cis* and *trans* components, which are composed of sequences in the pre-mRNA and cellular factors, respectively. As for the *cis* components, three sites, the 5′ splice site (5′ ss), the 3′ splice site (3′ ss), and the branch point sequence (BPS), participate in the splicing reaction and are present in every intron, and thus are known as the core splicing signals. However, these core signals lack sufficient information content for the splicing machinery to distinguish correct pairs of splice sites from cryptic splice sites, which are vastly more abundant than correct splice sites, implying the involvement of other *cis*-acting elements besides the core splice signals in splice site selection. These *cis*-elements include exonic splicing enhancers (ESEs) or silencers (ESSs), functioning to promote or inhibit the inclusion of the exon they reside in; and intronic splicing enhancers (ISEs) or silencers (ISSs), functioning to enhance or inhibit the usage of adjacent splice sites from an intronic location. In general, these splicing regulatory elements (SREs) function by recruiting trans-acting splicing factors that activate or suppress splice site recognition or spliceosome assembly.

ESEs include a diverse range of sequences, and many exons contain ESE sequences. Most ESEs function by recruiting members of the SR protein family. These factors usually regulate splicing by binding ESEs through their N terminal RRM domains and mediating protein-protein interactions that facilitate spliceosome assembly through C terminal RS domains. ESSs are often bound by splicing repressors of the hnRNP family, which is a diverse group of proteins containing one or more RNA-binding domains. HnRNPs function by a wide variety of mechanisms. For example, PTB (hnRNP I) can block essential interactions between U1 and U2 snRNPs, whereas hnRNP A1 can inhibit splicing by binding to the both sides of an exon and "looping it out" or by directly displacing snRNP binding. Both computational and experimental methods have been developed to identify ESEs and ESSs on a large scale. ESEs have also been identified experimentally by *in vitro* and *in vivo* SELEX approaches. 133 ESS decanucleotides active in human cells were identified by a cell-based fluorescence-activated screen (FAS) from a library of random sequences.

Large-scale screening has not yet been reported to identify ISE/ISS. One well characterized ISE is the G triplet (GGG) or G run, which often occurs in clusters and can enhance the recognition of adjacent 5′ ss or 3′ ss. Characterized ISSs include binding sites for the splicing repressors PTB and hnRNP A1, CA-rich sequences bound by hnRNP L, specific octamers flanking exon Ⅲb of the FGFR2 gene, and two elements in intron 7 of human SMN2 gene.

The protein factors involved in AS can be generally classified into two classes. One class of AS regulators consists of relatively widely expressed proteins, including SR proteins and hnRNP proteins (discussed above). Their cell-type-specific expression level changes or post-translational modifications can alter their activities or localization, providing a means of AS regulation. The other class of AS regulators consists of proteins with a much narrower job description; they have restricted expression patterns that are responsible for regulating tissue-specific AS events. These proteins include Nova-1/2, TIA1/TIAR, Fox-1/2 and nPTB, and all contain an RNA-binding domain of the KH or RRM type. The number of AS regulators identified so far is less than two dozen in mammals. Several genome-wide approaches have been recently developed to expand the list of known AS regulators.

Several microarray platforms have been designed to distinguish different splicing isoforms and to detect AS at a genomic scale, facilitating the recognition of genome-wide alternative splicing events in human, mouse, and chimp. Some new approaches, such as cross-linking/immunoprecipitation (CLIP), RNP immunoprecipitation (RIP) and genomic SELEX, have been developed to systematically identify RNA targets for different trans-factors at the genomic level. Analysis of the target sequences will help to define the sequence determinants of the splicing factor binding, and may also help to identify cooperative or antagonistic relationships between different splicing factors.

Comprehensive explanation of the complex network controlling constitutive and alternative splicing is a major challenge for post-genomic era. Based on the techniques mentioned above, the cis-acting elements, trans-acting factors and their regulator sequences, and the splicing events in different cell types can be systematically recognized. Intergating the information is expected to reveal the rules that govern the splicing mode of a pre-mRNA, and eventually decode the splicing code embeded in genome.

1.5 RNA Regulation of Gene Expression in Bacteria

1.5.1 Small RNAs

RNA regulators in bacteria have been traditionally denoted as "small RNAs" (sRNAs). Their size ranges from about 50~500 nucleotides, averages around 100 nucleotides, and is larger than that of those regulatory RNAs from eukaryotes (which range from 21~30 nucleotides). They are not generally processed from larger dsRNA precursors (as those eukaryotic RNA regulators are); instead, they are encoded in their final form by small genes. Many sRNAs have been identified in E. coli as well as other bacterial species by computational and experimental methods.

Based on the mode of action, there are two types of sRNAs: trans-acting and cis-acting regulatory RNAs. In the latter case, the gene regulation is primarily mediated through alternative RNA pairings that operate in cis, which means that the RNA regulatory elements control the expression of the genes they reside in. Riboswitch and attenuation belong to this type, which are illustrated in the following two sections. Here, we focus on the former, the trans-acting sRNAs. They mainly function by base pairing with target RNAs, and a few act by modulating the activities of proteins.

The sRNAs acting via base pairing could be called as antisense RNAs. After their binding with target mRNAs, they mainly promote mRNA degradation by ribonucleases, and repress or activate translation by inhibiting or facilitating ribosome binding. Some antisense sRNAs locate in the same DNA regions as their targets, and have long perfect complementarity with their targets. Several plasmid-encoded antisense sRNAs have been studied in detail, some of which affect replication and others repress the expression of proteins being toxic to the cell. However, in most cases, the pairing between sRNAs and their target is imperfect and weak, where the bacterial RNA chaperone Hfq is needed to facilitate the base pairing.

1.5.2 Riboswitch

Riboswitches are regulatory RNA elements that act as direct sensors of small molecule metabolites to control gene transcription or translation. These regulatory elements typically reside within the 5'-untranslated regions of the genes they control.

Riboswitches are typically composed of two functional components: the aptamer and the expression platform. The aptamer serves as a molecular sensor embedded within the riboswitch, and selectively recognizes its corresponding target molecule, for example, a specific metabolite. The expression platform generally locates immediately downstream from the aptamer. However, the two domains overlap with each other in many instances. The role of the expression platform is to transform metabolite-binding events into gene-control consequences by allosteric modulation of the structure of the 5'-UTR. The expression platform varies substantially between even closely related organisms except the ubiquitous mechanism: they undergo a conformational change in response to ligand binding to the aptamer, and hence influence some part of the gene expression process. In many cases, these changes involve the unpairing/pairing of Watson-Crick base pairs that render mutually exclusive alternative structures.

There are two common mechanisms of riboswitch mediated gene control. In one case, they operate at the level of transcription termination by using an anti-termination mechanism. In the other case, they operate at the level of translation by controlling the formation of an RNA structure that masks the ribosome binding site (RBS) on mRNA. They function through changes in RNA secondary structure as shown in Figure 3-9.

Figure 3-9　Common mechanisms of riboswitch gene control (see page 72)
(Adapted from Tucker, B J & Breaker, R R 2005. Curr Opin Struct Biol. 15: 342～348)

The proteins, whose expressions are controlled by riboswitches, often involve in the biosynthesis or transport of the metabolite being sensed. Thus, riboswitches are used as a form of feedback inhibition: when there are enough metabolites to bind with the riboswitches, the expression of the genes producing metabolites are repressed.

Currently, many riboswitches have been identified and they respond to a wide range of different metabolites, such as coenzyme thiamine pyrophosphate (TPP), flavin mononucleotide (FMN), S-adenosylmethionine (SAM), lysine and other amino acids, guanine, adenine, and Mg^{2+}. Although most prevalent in bacteria, riboswitches are also found in plants and fungi, where they are located near splicing junctions and appear to regulate mRNA splicing. Recently, it has been discovered that one gene could be regulated by two riboswitches in tandem. The effective control of gene expression by riboswitch is probably determined by a delicate balance among the rate of RNA folding, the rate of metabolite binding to the riboswitch, and the speed of transcription by RNA polymerase.

1.5.3　Attenuation

Transcription attenuation (Attenuation) can be defined as any mechanism that utilizes transcription pause or transcription termination to modulate expression of downstream genes. Several genes, especially genes for aminoacyl-tRNA synthetases are controlled by attenuation mediated by a 200～300 nucleotide long leader RNA, which directly and specifically interacts with cognate but uncharged tRNA. It is a kind of riboswitch which responds to uncharged tRNAs, rather than to small-molecule metabolites. The regulation relies on the coupling of transcription and translation in bacteria.

Here, the attenuation mechanism is explained by using the *E. coli* tryptophan biosynthetic operon (trpEDCBA). The 161-nucleotide *trp* leader transcript can form three overlapping RNA secondary structures referred to as the pause structure, the anti-terminator and a transcription terminator as shown in Figure 3-10.

Figure 3-10　Model of transcription attenuation of the trp operon

Attenuation is a typical negative feed-back regulation. Several other amino acid biosynthetic operons such as the *his*, *phe*, *leu*, *ilv*, and *pheA* operons are also regulated by transcription attenuation. Attenuation alone can provide robust regulation: other amino acids operons like *his* and *leu* have no repressors and rely entirely on attenuation for their regulation.

Transcription attenuation is a highly conserved regulatory strategy used by bacteria. The attenuation is mainly controlled by the translation of leader peptide, like in tryptophan operon. Moreover, RNA-binding proteins, transfer RNA, antisense RNA are also found to mediate transcription attenuation.

1. 6　Small RNA Regulation of Gene Expression in Eukaryotes

1. 6. 1　siRNA

In the 1990s, geneticists found an interesting phenomenon that both the endogenetic genes and exogenetic genes are co-suppressed, during contructing transgenetic plants and animals in order to overexpress a specific gene. This phenomenon is called co-suppression. There are two types of co-suppression: transcriptional gene silencing (TGS) and post-transcriptional gene silencing (PTGS). Previous studies have shown that TGS usually accompanies the increase in methylation and decrease in transcription efficiency in the regions homologous to the specific gene. On the other hand, PTGS is correlated to the decrease of the mRNA stability. It has been found that gene expression can be suppressed by negative strand RNA only under high concentration, but not suppressed by positive strand RNA. It is surprising to find out that double strand RNA could suppress the expression of specific targets even at a very low concentration, and the suppression could pass on to the next generation. This suppression effect caused by double strand RNA is called RNA interference (RNAi). The discovery of RNAi explains the questions which have puzzled the geneticists for many years, and promotes the study of its mechanism and application as a therapeutic drug. Andrew Fire and Craig C. Mello won the Nobel Prize for physiology and medicine in 2006 for their innovative discovery of RNAi.

RNA interference is conserved during evolution. In *Drosophila*, *C. elegans* and plants, dsRNA about 500 bp could silence its homologous genes. The exogenetic RNAs are processed into small interfering RNAs (siRNAs) by Dicer cleavage. The generated siRNAs have a phosphate group at the 5' end, and an overhang of two bases with a hydroxyl group at the 3' end, which are the characteristics of the RNAs processed by RNase Ⅲ. But in mammals, only the dsRNAs with a length of 20~30 bp have the RNAi effect. The siRNA is integrated into the RISC complex, which is composed of Dicer, TRBP and AGO2. The selection of the guide stand and excision of the passenger strand take place as RISC assembles. Then the mature RISC could specifically cleave its full complementary mRNAs and leads to their degradation.

In nature, RNAi is a conserved defense mechanism, which functions in preventing virus infection, in maintaining the stability of the genome. It is shown that endogenous siRNA could specifically suppress the activities of transposons. And studies in plants have demonstrated that endogeneous anti-virus siRNAs are synthesized after virus infection, to suppress the replication and further infection of virus. One recent study has shown that there are various siRNAs in mammalian cells, and these siRNAs primarily originate from the repeat sequences in genome, retrotransposons and some pseudogenes. This study also proves that there are endogeneous siRNAs even in the species which lack RdRP (RNA-dependent RNA polymerase), and they play important roles in maintaining the stability of the genome and in development.

There are 3 ways for carrying out RNAi in mammals. The first one is to synthesize siRNA using a chemical method. The second one is to use a vector containing H1 or U6 promoter to transcribe shRNA of about 60~70 nt, followed by Dicer processing. The third one is esiRNA, which is achieved by digesting long dsRNA with Rnase Ⅲ and generating a siRNA library (20~30 bp) to silence target genes.

RNAi, a popular experimental method, is broadly applied in the study of gene functions. SiRNA has been used as drugs in curing hepatitis breakout, virus and cancer. The high efficiency and specificity of siRNA promise it a bright future in application. But, now there are problems about targeted-delivery and off-target, which need to be solved.

1. 6. 2　miRNA

microRNAs (miRNAs) are a family of 20~25 nucleotide RNA molecules that negatively regulate gene expression by binding to complementary sequences in 3' untranslated region of messenger RNAs, leading to either degradation of the mRNAs or inhibition of their translation. In 1993, the first miRNA *lin-4* was discovered in *C. elegans*. Seven years later, the second miRNA was discovered in *Drosophila*. The conservation analysis revealed an amazing and astonishing fact that *let-7* was highly conserved among fruit fly, nematode and human beings. This suggests that miRNA might be a kind of highly-conserved noncoding RNA and play significant roles in gene regulation. This remarkable finding has intensively accelerated the discovery and identification of novel miRNAs. Upto June of 2008, the number of miRNAs deposited in miRBase has come to 6396, and 678 of which are in human. Now, it has been proved that miRNAs extensively exist in eukaryotic organisms, and are one of the biggest eukaryotic gene families.

Besides miRNA's significant role in eukaryotic gene regulation, they are also extensively encoded in unicellular organisms. Recent studies also reveal that even some viruses encode certain kinds of miRNAs, which play crucial roles in viral replication and infection.

miRNA is usually transcribed by RNA polymerase Ⅱ into a large primary RNA (pri-miRNA) molecule. This primary RNA molecule is processed by the complex composed of Drosha and a dsRNA binding protein DGCR8, into precursor miRNA (pre-miRNA) of about 70 nucleotides, which forms a stem-loop structure. Then pre-miRNA is transported into cytoplasm by a complex formed by RAN and the GTP-dependent protein Exportin-5, where the stem-loop structure is processed by Dicer into 22-nucleotide miRNA: miRNA* double strand RNA. This dsRNA is quickly loaded into the RISC complex, in which miRNA* is cleaved and miRNA stays in RISC complex, resulting in a functional RISC complex, named miRISC (Figure 3-11).

Figure 3-11　Biogenesis of miRNAs and their assembly into microribonucleoproteins
(Filipowicz, Bhattacharyya et al. 2008)

A class of newly discovered miRNA, which is called MiRtron, has a different processing pathway. The processing of primary MiRtron does not rely on Drosha but on the cellular splicing machinery. After being spliced out as a whole intron, the precusor MiRtron is firstly debranched and then refolds into pre-miRNA, then transported into cytoplasm and finally processed by Dicer.

Plant and animal miRNAs repress gene expression using different mechanisms. Plant miRNA forms perfect base pairs with its target mRNA and mediate target mRNA cleavage and degradation. By contrast, the miRNA in animal is partially complementary to its target mRNA, and it mainly functions by translational inhibition. Moreover, miRNA can inhibit translation by eliciting the drop off of the ribosome from mRNA. Besides translational inhibition, miRNA can also trigger the degradation of target mRNA or the newly synthesized polypeptide, yet the detailed mechanism remains unclear.

miRNA expression is remarkably spatial, temporal and tissue specific. This expression pattern results in gene inhibition at specific developmental stage or in specific tissues, thus determines the morphogenesis of tissue and organ. For instance, *lin-4* is expressed in L1-L4 stage, while *let-7* is expressed during L4 of adult stage, these miRNA expression determine the expression profile of target genes. miRNA-124, miRNA-122 and miRNA-1 are exclusively expressed in brain, liver and skeletal muscle, respectively. By this means they modulate the transcriptome of these tissues and determine their corresponding morphogenesis.

miRNA expression profiles are dramatically different between normal and carcinomatous tissues; In carcinomatous tissues, most miRNAs are downregulated while a few are upregulated, which indicates miRNAs play important roles in carcinogenesis. MiRNA location analysis reveals that 50% of the annotated miRNAs are located in the fragile sites of the chromosome, which tend to be deleted in cancer tissues. Recent studies reveal that some miRNAs function as oncogenes or tumor suppressor genes, among which miRNA-17-92 cluster is the first identified "oncomiR"; on the contrary, miRNA-16, *let-7*, miRNA-34a and miRNA-335 mainly function as tumor suppressors. The function of miRNA in carcinogenesis provides us promising targets for gene therapy. Besides, miRNAs also play crucial regulatory roles in cell proliferation and apoptosis, fat metabolism, axon guidance, differentiation of hematogenic cells, as well as development.

1. 6. 3　RasiRNAs

Studies in *Drosophila* have identified five Argonaute proteins: Ago1, Ago2, Ago3, Piwi, and Aubergine (Aub). Ago1 and Ago2 belong to the AGO subfamily of Argonaute proteins, which are ubiquitously expressed, whereas Piwi, Aub, and Ago3 belong to the Piwi subfamily, the expression of which is germ line specific. Ago1 and Ago2 are key players of miRNA-mediated and siRNA-mediated RNA-silencing pathways, respectively. Aubergine, Piwi and Ago3 bind to thousands of small non-coding RNAs, a family of RNAs called repeat-associated small interfering RNAs (rasiRNAs) with their sequences corresponding to repetitive elements and heterochromatic genome regions. These rasiRNAs are single-stranded and directly interact with Piwi proteins. They are slightly longer than Argonaute bound siRNAs/miRNAs (24~29 nucleotides in *D. melanogaster*), with a phosphorylated 5′ end and a 2′-*O*-methyl modification at their 3′ ends. RasiRNAs are involved in the establishment and maintenance of heterochromatin and transposon control.

1.6.4　piRNAs

Small RNA partners of Piwi proteins were firstly identified in Mouse testis, and later in other mammalian and Zebrafish testis. They are termed Piwi-interacting RNAs (piRNAs) and have several interesting characteristics. piRNAs are mainly found in animals' germline, and are longer than miRNAs and siRNAs (26~31 nt vs. 21~24 nt). Although piRNAs share some common features with rasiRNAs, there are also some differences between them. The percentage of piRNAs mapped into repetivive elements is smaller than that of rasiRNAs. Nonetheless, on the basis of their common features, rasiRNAs are deemed as a specialized subclass of piRNAs.

Sequence analysis of piRNAs revealed a strong bias for uridine residues at their 5′ termini, which is similar to siRNAs and miRNAs. The 5′ uridine bias is a characteristic of the RNAs processed from double-stranded precursors by RNase Ⅲ enzymes. However, a computational search failed to identify any secondary structures similar to pre-miRNAs in the regions flanking piRNAs, suggesting that piRNA processing is distinct from miRNA biogenesis. Besides, 2′-*O*-methyl modification at the 5′ end of piRNAs also exists in plant miRNAs.

In *Drosophila*, loss of Dicer, which functions in the miRNA pathway and siRNA pathway, has no effect on piRNA levels, suggesting piRNAs have an alternative biogenesis pathway. Sequencing data suggest that piRNAs interact respectively with three Piwi family proteins (Piwi, Aubergine, Ago3) in *Drosophila*, thus uncovered the mystery of piRNA biogenesis.

A new small RNA silencing pathway begins to be uncovered. Despite these interesting discoveries, several important questions remain to be answered. Is piRNA mediated regulation pathway conserved in evolution? What are piRNA's targets? We still do not know how a new piRNA response is initiated, and it is unclear why some regions in our genome are so enriched of piRNAs. The next breakthrough may come from further studies.

Summary

Ribonucleotide acid (RNA) is a single-stranded polymer that composed of four different ribonucleotides linked by phosphodiester bonds. Compared to its double-stranded DNA counterpart, RNA is much more capable in forming different structures through base-pairing and tertiary interactions, which credit it a large variety of biological functions. However, recent studies have showed that pure RNA molecules are prone to form local non-native base pairing, creating the thermodynamic and kinetic challenges for the correct folding of RNA molecules. Strikingly, RNAs in living cells are always associated with proteins, which may prevent the non-native structures of these flexible large molecules.

In this chapter, we gave brief introductions to the well characterized non-coding RNAs. Ribozymes are a large class of structural RNAs functioning in diverse biological processes. It can fold to distinct tertiary structures and catalyze different biochemical reactions in test tubes without the assistance of proteins, snoRNAs are responsible for the modification of ribosomal RNAs; snRNAs are the vital components for the maturation of mRNAs; small RNAs are important regulatory elements in both prokaryotes and eukaryotes, for example, microRNA and RyhB RNA. However, these non-coding RNAs make up only a small proportion of the transcriptome. How did the large proportion of the unannotated transcripts function remains a puzzle.

After the completion of Human Genome Project, it was found that the diploid human genome contains about 25 000 genes, far fewer than the estimated 100 000. These protein-coding sequences comprise no more than 1.5% of human genome size. However, it was found that about 10% of the human genome is transcribed into polyadenylated RNAs. This number is much larger than the constituents of protein-coding genes, which indicated that large portion of polyadenylated RNAs are instead non-coding RNAs. In addition to the polyadenylated transcripts, non-polyadenylated transcripts originated from intergenic regions were found making up the major proportion of the transcriptome when analyzing the transcription of 10 human chromosomes. Moreover, the antisense strands of about $15\% \sim 20\%$ protein coding genes are transcribed, this number can even reach 50% of the whole transcripts in certain tissues. It's becoming clear that the majority of human genome is transcribed, as well as other higher eukaryotes and even prokaryotes. For instance, at least 74.5% of yeast genome and 62% of the mouse genome are transcribed, respectively; and a substantial number of antisense and intergenic transcripts were detected in *E. coli*. Thus the so called "transcriptome" seems far more complicated than expected. Profited from the emergence of newly developed technologies (for example, Serial Analysis of Gene Expression (SAGE), Full Length cDNA Cloning, Cap-Analysis of Gene Expression (CAGE) and Tilling Array), the complexity of transcriptome becomes convinced. It is generally believed that the non-protein-coding RNAs must code something else, functioning either in gene regulation or in enriching the complexity of genome which contains such few genes.

Furthermore, the non-coding RNAs introduced in this chapter are generally small RNAs (except for some large ribozymes). Many of the non-polyadenylated transcripts originated from the intergenic regions can be as long as tens of kilobases. Did these large transcripts encode large functional non-coding RNAs? How did they function? We are pushed to the starting line again after the completion of genome sequencing.

(Yi Zhang)

第四章 蛋白质、蛋白质组与蛋白质组学

基因组学研究的深入开展已经推动了对复杂生物学过程的系统分析。通过比较不同发育时期的细胞或组织的基因表达，可以从总体上定量地描述这些系统的生物学过程。但是转录后加工和蛋白质半衰期机制的发现，说明 mRNA 水平的变化并不完全与蛋白质所决定的生物学过程相关。实验也证明，组织中 mRNA 丰度与蛋白质丰度的相关性并不好，尤其对于低丰度蛋白质来说，相关性更差。更重要的是，蛋白质复杂的翻译后修饰、蛋白质的亚细胞定位或迁移、蛋白质-蛋白质相互作用等几乎无法从 mRNA 水平来判断。毋庸置疑，蛋白质是生理功能的执行者，是生命现象的直接体现者，对蛋白质结构和功能的研究将直接阐明生命在生理或病理条件下的变化机制。蛋白质本身的存在形式和活动规律，如翻译后修饰、蛋白质间相互作用以及蛋白质构象等问题，仍依赖于直接对蛋白质的研究来解决。可以说蛋白质是细胞功能和疾病过程的中心环节，没有蛋白质组学的努力，基因组学的成果无法实现。

本章将从蛋白质的最小功能单位开始对蛋白质系统研究的理论和方法学做一简要介绍。

第一节 蛋白质加工、转运与降解

一、蛋白质的结构域

天然的蛋白质均包括一级、二级和三级结构。三级结构是特征性的，决定了该蛋白质的生物学功能。许多蛋白质都可以被分为多个结构组成单元，结构域就是这样一个组成单元。结构域不仅是结构单位，而且是功能单位。

1. 结构域的定义

结构域是蛋白质中的一类结构单元，是构成蛋白质三级结构的基本单元。结构域一般可以自稳定，且常常独立进行折叠，而不需要蛋白质其他部分的参与；同一个蛋白质的各个结构域之间是以肽链相互链接的，而链接两个结构域的绝大多数都是单股肽链，只有在极个别的情况下会有少数的双股肽链联系不同的结构域。很多结构域都有自己独特的生物学功能。很多结构域并不是一个基因或基因家族对应蛋白质的独特结构单元，而往往是许多类蛋白质的共同结构单元，与蛋白质完成生理功能有着密切的关系，有时几个结构域共同完成一项生理功能，有时一个结构域就可以独立完成一项生理功能，但是一个结构不完整的结构域是不可能产生生理功能的。因此结构域是蛋白质生理功能的结构基础。现在使用的结构域概念具有三种不同的含义，即独立的结构单位、独立的功能单位和独立的折叠单位。

2. 结构域和蛋白质结构与功能

结构域作为蛋白质结构与功能单位广泛存在于球状蛋白中，其域间联结从免疫球蛋白（IgG）的松散联结到木瓜蛋白酶、溶菌酶等小分子蛋白质的紧密联结。结构域作为

功能单位表现在不同的结构域具有特定的功能，如结合底物、催化反应、亚基间相互作用及活性调节等。丙氨酸 tRNA 合成酶是这种结构的典型例子（图 4-1）。

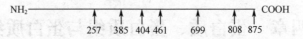

图 4-1　丙氨酸 tRNA 合成酶的结构域

257～385. 氨酰腺苷酸合成；406～461. 氨酰腺苷酸合成＋tRNA 氨酰化；699～808. 氨酰腺苷酸合成＋tRNA 氨酰化＋寡聚化

　　结构域作为结构功能单位，对于功能复杂的蛋白质的构成具有重要意义。在这些蛋白质中结构域往往以组件方式相互组合从而构成具有多种不同功能的分子。原核或真核生物的基因表达调控因子以及与信号转导有关的蛋白质因子往往以这种方式构成。表4-1 列举了一些结构域模块。

表 4-1　脊椎动物蛋白质中一些常见的结构域模块

结构域	特点	结构域	特点
Gla	含多个 γ 羧基谷氨酸残基	L	出现于 LDL 受体及补体蛋白中
G	类似表皮生长因子结构	T	发现于血小板结合蛋白和补体蛋白备解素中
K	"环饼"结构域	N	发现于一些生长因子受体中
C	经常出现在补体蛋白中	E	EF 手指形结构
F1、F2、F3	最早发现于纤维黏连蛋白	LB	一些细胞表面蛋白的凝集素模块
I	免疫球蛋白超家族结构域	SH2、SH3	2 类及 3 类 src 同源结构

3. 结构域与蛋白质折叠

　　蛋白质的二级结构堆砌具有一定的模式即折叠模式。对已知的蛋白质的结构研究表明它们可归属于数量有限的折叠模式，其数量预计为 700～1500。尽管众多蛋白质的结构在原子水平上错综复杂，在结构域层次上的折叠模式却非常简单和有规律。把蛋白质的折叠模式归纳为几大类：全 α、全 β、α/β 及 α＋β。对于小或中等规模的蛋白质，有限的折叠模式已能精确地描述其结构。对于大而复杂的蛋白质，其结构可进一步归结于亚结构——结构域层次的折叠模式，对于结构域层次的折叠模式的深入研究有助于我们对越来越多新发现的复杂蛋白的结构与功能进行研究。

　　蛋白质一级结构如何决定其高级结构，即所谓蛋白质折叠（protein folding）问题是现代分子生物学研究的热点。对于蛋白质如何形成其三维立体结构目前尚无定论。现在流行的融球态（molten globular state）模型等的主要实验依据是蛋白质体外变性/复性实验结果，且只适用于单域或简单蛋白。对于多域蛋白，现在倾向于所谓组件（modular）模型，即新生肽通过局部成核及组装逐渐形成空间结构，而成核中心或组装单位往往是结构域。结构域作为折叠单位已通过酶解或化学裂解获得的蛋白质片段的体外变性/复性实验证实。基因工程方法的发展已使定向分离多肽片段成为可能。应用这种技术研究分离的结构域的独立折叠已给出一些成功的例子。

　　对于结构域的组合规律和形成机制的研究有助于我们对以下问题的深入研究：①蛋白质结构域作为结构与功能单位，广泛存在于参与基因转录、翻译调控以及与信号转导有关的蛋白质因子等结构功能复杂的蛋白质中。对于复杂蛋白质，目前一种比较成功的研究战略是解剖完整蛋白质，将其按结构单元，主要是结构域，划分为相对独立的模块，研究这些模块的结构与功能及它们之间的相互作用。②如何从蛋白质的一级结构预测其三级结构一直是蛋白质研究的目标。拓扑学研究表明，结构域在蛋白质中的分布具有某种规律性。对其规律的深入研究必将为预测蛋白质的空间结构提供有益的帮助。③蛋白质工程可通过有控制的基因修饰和基因改造，产生符合人们意愿的蛋白质分子。定点突变技术为这一研究提供了强有力的工具，目前已发展得相当成熟。另一更为新颖的战略是利用蛋白质组合规律及结构与功能关系的研究结果，设计新的蛋白质分子。这种方式有些类似于建筑工艺中的预制件组装方式。通过不同的结构与功能元件的不同组合，可以形成数目众多、功能各异的蛋白质分子，从而为蛋白质分子设计提供更多的可能性。

二、新生肽链折叠与分子伴侣

长期以来关于蛋白质折叠，形成了自组装（self-assembly）的学说，因此，在研究新生肽链的折叠时，就很自然地把在体外蛋白折叠研究中得到的规律推广到体内，用变性蛋白的复性作为新生肽链折叠的模型，并认为细胞中新合成的多肽链，不需要别的分子的帮助，不需要额外能量的补充，就应该能够自发的折叠形成它的功能状态。其实，新生肽链的折叠在合成早期已开始，而不是合成完后才开始进行，随着肽段的延伸同时折叠，又不断进行构象的调整，先形成的结构会作用于后合成的肽段的折叠，而后合成的结构又会影响前面已形成的结构的调整。因此，在肽段延伸过程中形成的结构往往不一定是最终功能蛋白中的结构。这样，三维结构的形成是一个同时进行着的、协调的动态过程。20 世纪 90 年代一类具有新的生物功能的蛋白质，分子伴侣（molecular chaperone）的发现，以及在更广泛意义上说的帮助蛋白质折叠的辅助蛋白（accessory protein）的提出，说明细胞内新生肽链的折叠一般意义上说是需要帮助的，而不是自发进行的。

1. 蛋白质分子的折叠和分子伴侣的作用

蛋白质分子的三维结构，除了共价的肽键（peptide bond）和二硫键，还靠大量极其复杂的弱次级键共同作用。因此新生肽链在一边合成一边折叠过程中有可能暂时形成在最终成熟蛋白中不存在、不该有的结构，它们常常是一些疏水表面，它们之间很可能发生本不应该有的、错误的相互作用而形成非功能的分子，甚至造成分子的聚集和沉淀。按照自组装学说，每一步折叠都是正确、充分、必要的。实际上折叠过程是一个正确途径和错误途径相互竞争的过程，为了提高蛋白质生物合成的效率，应该有帮助正确途径的竞争机制，分子伴侣就是这样通过进化应运而生的。它们的功能是识别新生肽链折叠过程中暂时暴露的错误结构，与之结合，生成复合物，从而防止这些表面之间过早的相互作用，阻止不正确的非功能的折叠途径，抑制不可逆聚合物产生，这样必然促进折叠向正确方向进行。

分子伴侣的作用机制实际上就是它如何与靶蛋白识别、结合又解离的机制。有的分子伴侣具高度专一性，如一些分子内分子伴侣以及洋葱假单胞菌的酯酶，有它自己的"私有分子伴侣"。它是由 *limA* 基因编码的，与酯酶的 *lipA* 基因只隔三个碱基，可能是进化过程中造成的。而一般的分子伴侣识别特异性不高，它是怎样识别需要它帮助的对象的呢？现在只能说分子伴侣识别非天然构象，而不去理会天然构象。由于在天然分子中，疏水残基多半位于分子的内部而形成疏水核，去折叠后就可能暴露出来，或者在新生肽链的折叠过程中，会暂时形成在天然构象中本应该存在于分子内部的疏水表面，因此认为分子伴侣最有可能是与疏水表面相结合。只有β片层结构的蛋白质才可被分子伴侣识别。分子伴侣作用的第二步是与靶蛋白形成复合物。非常盛行的一种模型认为分子伴侣常常以多聚体形式而形成中心空洞的结构，用电子显微镜已经观察到由二层圆面包圈形组成的十四体分子和一个一层圆面包圈形组成的七体分子协同作用，形成中空的非对称笼状结构，推测靶蛋白可以在与周围环境隔离的中间空腔内不受干扰的进一步折叠。

2. 分子伴侣与蛋白质折叠酶

与分子伴侣不同，已经确定为帮助蛋白质折叠的酶目前只有两个，一个是蛋白质二硫键异构酶（protein disulfide isomerase，PDI）；另一个是肽基脯氨酸顺反异构酶（peptidyl prolyl cis-trans isomerase，PPI）。以PDI为例，众所周知，蛋白质分子中的二硫键与新生肽链的折叠密切相关，对维系蛋白质分子结构的稳定性和功能发挥也有重要作用。PDI定位在内质网管腔内，含量丰富，催化蛋白质分子内巯基与二硫键之间的交换反应。同时，它是目前发现的最为突出的多功能蛋白，除了二硫键异构酶的基本功能外，它是脯氨酸-4-羟化酶的α亚基，是微粒体内三酰甘油转移蛋白复合物的小亚基，还是一种糖基化位点结合蛋白等。其中，最引人注目的还是它有与多肽结合的能力，可以结合具有不同序列、长度和电荷分布的肽，特异性较低，主要是与肽的主链相互作用，但对巯基尚有一些偏爱。按照分子伴侣的定义，一般认为PDI和分子伴侣是两类不同的帮助蛋白，但是也有不同的看法，认为蛋白质二硫键异构酶也具有分子伴侣的功能。

蛋白质分子中天然二硫键的形成要求这些在肽链上往往处于不相邻位置的巯基，通过肽链一定程度的折叠，才能相互接近到可以正确形成二硫键的位置。肽链的自身折叠是一个慢过程，而蛋白质二硫键异构酶催化蛋白质天然二硫键的形成却是一个快过程。另外，蛋白质二硫键异构酶具有低特异性的与各种不同肽链相结合的能力，在内质网中以极高的浓度存在，是一个钙结合蛋白，是一个能被磷酸化的蛋白质，这些都已经符合了分子伴侣的条件。因此推测蛋白质二硫键异构酶很可能首先通过它与伸展的或部分折叠的肽链的结合，阻止错误的折叠途径，促进正确的中间物生成，帮助肽链折叠使相应的巯基配对，从而使正确的二硫键得以形成；然后催化巯基的氧化或二硫键的异构而形成天然二硫键。蛋白质二硫键异构酶的酶活性与它的分子伴侣功能不是相互排斥，而是密切相关，协调统一的。分子伴侣与帮助新生肽链折叠的酶之间，大概不应该，也不能够划一条绝对的分界线。

3. 分子伴侣研究的实际应用

分子伴侣的研究成果必然会大大加深我们对生命现象的认识，同时也一定会增加我们与自然斗争的能力和自身生存的能力。由于分子伴侣在生命活动的各个层次都具有重要作用，它的突变和损伤也必定会引起疾病，因此可以期望运用分子伴侣的知识来治疗所谓的"分子伴侣病"。另外，利用对分子伴侣的研究成果从根本上提高基因工程和蛋白质工程的成功率，也必将对大幅度提高人类生活水平起到重要作用。

三、分泌性蛋白

不同功能的蛋白质通过相应的分选信号与严格的运输途径，准确地定位到相应的细胞器中，或被分泌到细胞外发挥功能，其中分泌到细胞膜外的蛋白质称为分泌性蛋白（secreted protein），简称分泌蛋白。分泌性蛋白种类繁多，包括血液和胞外基质中的众多细胞因子、补体、降解酶类、肽类激素、免疫球蛋白等。分泌性蛋白通过内分泌、旁分泌、自分泌等途径发挥作用，全面系统的研究分泌性蛋白不仅有助于全面认识、分析和解释各种生理和病理现象，而且也为新药的开发提供更多的选择。分泌性蛋白是目前生物学界研究的热点和难点。

信号肽假说（signal hypothesis）认为分泌性蛋白 N 端的信号肽，指导着分泌性蛋白合成。虽然各种蛋白质中信号肽不具有同源性，但其结构框架和氨基酸残基类型有一定的规律。典型的信号肽结构通常分为三个区：n 区是一些带正电荷的残基，是信号肽变异的主要位置；h 区是信号肽功能核心区；c 区含信号肽酶切位点。信号肽基本功能是通过疏水作用和静电引力与质膜的相互作用，对含有信号肽序列的前体蛋白的胞内转运发挥作用。近年来随着生物信息学的发展，针对信号肽发展了多种遗传算法进行分泌性蛋白预测，但需与膜蛋白区别。因为膜蛋白也有信号肽序列，但其在信号肽之外还有一个以上的疏水跨膜区，由于该区的存在才使得膜蛋白停留在细胞膜中；相反，分泌性蛋白跨膜穿梭后，信号肽酶将信号肽切除，以完成其分泌的过程。

四、胞内蛋白的降解

蛋白质转换（protein turnover）的概念已经提出了 60 年。早先认为组成人体的蛋白质是稳定组成的，而来自饮食的蛋白质只是提供能量，独立于其结构与功能之外。使用 ^{15}N 标记的酪氨酸喂养大鼠，在其尿里只能回收到 50%。进一步研究发现，另一半掺入到体内蛋白质。这些实验表明体内的结构蛋白处于合成和降解两种动力学状态，并且个别氨基酸甚至处于动力学的转变状态。这个系列的发现颠覆了当时这个领域的观念。

我们现在认识到胞内蛋白存在广泛的转换，这个过程是特异性的。许多蛋白质的稳定性接受单独调节，在不同条件下变化很大。从不接受调节且非特异性的"清道夫"处理过程变成高复合物、动态控制及严格调控的细胞内蛋白质降解过程。蛋白质降解在广泛的基本代谢过程中扮演主要角色。这些接受蛋白质降解影响的过程包括细胞周期、发育、分化、转录的调节、抗原提呈、信号转导、受体介导的内吞等不同的代谢过程。它已经改变了人们的认识，即细胞代谢过程的调节大多数发生在转录和翻译水平，明白蛋白质降解的调控具有同等重要的地位。因此蛋白质降解的异常参与多种疾病的病理过程，如某些恶性肿瘤、神经退行性病变及免疫和炎症系统的紊乱。因此，蛋白质降解系统已经成为药物作用靶点的平台。溶酶体（lysosome）和泛素（ubiquitin）-蛋白酶体（proteasome）复合物是两大蛋白质降解系统。

1. 溶酶体与胞内蛋白的降解

在生物化学上，对于溶酶体的认识首先是作为一个囊泡结构，包含多种水解酶，以酸性 pH 为最适条件发挥作用。它被一层生物膜包围是为了保护细胞内蛋白质不被水解。溶酶体的定义已经得到很大扩展，这是因为认识到消化是动力学的过程并且在溶酶体成熟的多个时期通过受体介导的内吞作用（endocytosis）及胞饮作用（pinocytosis）消化外源蛋白质；经巨噬细胞的吞噬细胞作用消化外源颗粒；同样也可以经自吞噬作用消化内源的蛋白质和细胞器。溶酶体系统是一个不连续的异源系统，也包括不含水解酶的结构，例如，一个极端是早期溶酶体包含被内吞的受体和配体复合物和被吞饮/吞噬的外源物。另一个极端是它的残体：异体吞噬和自噬完全消化过程的终末产物。在这两个极端之间能观察到包含蛋白质降解任何阶段的初级和新生的溶酶体：早期自噬囊泡一定包含细胞器，中期和晚期吞噬囊泡（异噬囊泡）包含细胞外颗粒，在内吞囊泡与消化溶酶体之间转移囊泡的多囊体。

蛋白酶和他们的底物被膜分隔开控制降解，底物是如何被转进溶酶体囊腔，因溶酶体蛋白酶活性而被降解的？这方面一个重要的发现是溶酶体自噬作用（autophagy）机制的揭示。基础代谢条件下，细胞质内所有的蛋白质被一定的膜结构分离，然后融合到新生溶酶体被消化，这个过程被称为微量自噬。极端条件下，如饥饿，线粒体、内质网膜、糖原颗粒和其他的细胞内溶物也能被吞入，这个过程被称为巨量自噬。消化细胞内外蛋白质的溶酶体的作用是不同模式的。但是，基于溶酶体活性的胞内蛋白的降解有几个方面越来越无法得到合理解释：独立实验证据的积累表明至少某类胞内蛋白的降解是非溶酶体的。

蛋白质降解是靶底物与蛋白酶之间直接的相互作用，因此蛋白酶不能游离于细胞质，否则将导致细胞的毁坏。胞内蛋白降解存在物理的隔离机制，蛋白酶与其作用的底物仅在需要的时候发生联系。溶酶体膜提供了这样的保护机制。按照一种模式，不同蛋白质对溶酶体蛋白酶有不同的敏感程度，它们在体内的半衰期与它们对溶酶体蛋白酶的敏感程度相关。半衰期极长的蛋白质对溶酶体蛋白酶不敏感或不同生理条件下稳定性发生了变化。进入溶酶体的蛋白质只有短半衰期的蛋白质被降解，长半衰期的蛋白质则返回细胞质。按照另一种模式，选择性由不同蛋白质对溶酶体膜的亲和力决定，从而控制它们进入溶酶体的速率并决定它们降解的速率。最近大量的研究表明至少在应激诱导的巨量自噬的过程中，有一段氨基酸序列，赖氨酸—苯丙氨酸—苯丙氨酸—谷氨酸—精氨酸—谷氨酰胺，直接结合特异细胞质和溶酶体中的受体。调节许多细胞质蛋白进入溶酶体腔。进一步证实大量进入的蛋白质都包含这一同源序列。蛋白质降解直接或间接地需要能量，如蛋白质转运进入溶酶体膜。

如上所述，溶酶体作用的机制只能部分地解释蛋白质的降解，并且对于细胞内蛋白质降解的几个关键特征：个别蛋白质的异源稳定性，降解过程中营养与激素的作用，胞内蛋白降解的代谢能量依赖性，选择性抑制剂对不同类型蛋白质的效应都不能做出解释。因此，溶酶体假说受到挑战。

2. 泛素-蛋白酶体系统

泛素是一个小的、热稳定的、进化上高度保守的 76 个氨基酸的蛋白质。泛素研究领域的一个重要的进展是单一的泛素组分能共价地与组蛋白结合，特别是 H_2A 和 H_2B。泛素与 H_2A 通过泛素 C 端的甘氨酸（Gly76）与组蛋白分子中赖氨酸（Lys119）的氨基之间的异肽键连接。这个异肽键被认为是泛素与被降解的底物之间的连接，并且是泛素参与蛋白质降解下游分子 26S 蛋白酶体的识别信号。在这个多聚泛素链中，存在泛素的 Gly76 与底物 Lys48 的连接部分，仅仅基于 Lys48 的泛素链被 26S 蛋白酶体的识别参与蛋白质降解的信号转导。在最近几年，已经表明泛素的第一部分也能黏附待降解靶底物 N 端的线性部分。N 端泛素化是缺乏赖氨酸的蛋白质降解所必需的。但是，几个含有赖氨酸的蛋白质经同样的过程。在这些蛋白质链内赖氨酸可能不参与共价连接。

泛素与 H_2A 结合相似的高能异肽键的发现，解决了细胞内蛋白质降解需要能量的问题，并且为阐明异肽键形成的机制奠定了基础。利用泛素激活机制以及泛素的共价亲和结合，纯化了参与泛素降解蛋白质级联反应的三个酶：泛素激活酶（ubiquitin-activating enzyme，E1）、泛素交联酶（ubiquitin-carrier protein，E2）、泛素连接酶（ubiquitin-protein ligase，E3）。E3 是一个特异的底物结合组分，表明不同的蛋白质可能被特异地识别（图 4-2）。

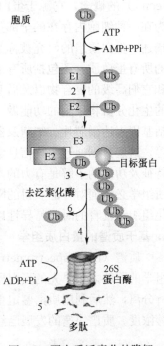

图 4-2　蛋白质泛素化的降解
(Ciechanover A, 2005)

被泛素标签的蛋白质可被 ATP 依赖大分子碱性蛋白酶降解。这个蛋白酶被命名为 26S 蛋白酶体；蛋白酶体是由两个 19S 和一个 20S 亚单位组成的桶状结构，19S 为调节亚单位，位于桶状结构的两端，识别多聚泛素化蛋白并使其去折叠。19S 亚单位上还具有一种去泛素化的同工肽酶，使底物去泛素化。20S 为催化亚单位，位于两个 19S 亚单位的中间，其活性部位处于桶状结构的内表面，可避免细胞环境的影响。20S 颗粒是一个由 4 个叠在一起的环组成的桶状结构。两个相同的外环称 a 环和两个相同的内环称 b 环。真核生物的 a 环和 b 环都由 7 个不同的亚基组成。催化位点位于 b 环。每个 20S 的桶都被由 17 个不同亚基的 19S 调节颗粒覆盖。9 个亚基是中心，8 个亚基是"唇"。19S 调节颗粒重要的功能就是识别泛素化的蛋白质和其他蛋白酶体潜在的底物。19S 调节颗粒的几个泛素结合亚基已经被鉴定，尽管生物学功能和作用模式尚不清楚。它的另外一个功能就是打开环的孔允许底物进入蛋白质降解的腔内。因此，如果一个折叠的蛋白质不适合通过窄窄的蛋白酶体通道，那么 19S 颗粒可使底物去折叠以便进入 20S 颗粒。通道打开和去折叠都需要代谢能量，而 19S 颗粒的确包含 6 个 ATPase 亚基为其提供能量。接着底物被降解，短肽从底物上释放，泛素重新被利用。

在很短的时期内，泛素标签假说得到一系列的支持。尽管细胞不提供毁坏过程中底物泛素化的直接证据，它们仍然提供了泛素连接与降解之间的最强的直接联系。泛素-蛋白酶体系统与蛋白质质量控制、细胞周期、DNA 修复、转录及免疫应答等密切相关，也与许多种疾病的发生相关。

第二节 蛋白质组与蛋白质组学

一、蛋白质组与蛋白质组学的概念

1994 年，澳大利亚麦考利大学的 Wilkins 和 Williams 首先提出了蛋白质组（pro-teome）的概念，它源于蛋白质（protein）与基因组（genome）两个词的杂合，其定义为在一种细胞内存在的全部蛋白质。蛋白质组首先指的是对基因组编码的所有蛋白质的描述。蛋白质组的研究被称为蛋白质组学（proteomics）。现在不仅指的是任何细胞内的所有蛋白质，还包括所有的同系物及修饰产物、它们之间的相互作用、蛋白质的结构和它们高级的复合物以及后基因组的每一个问题。现在希望高通量（high throughput）的生化分析将细胞的功能做一个直接的描述。蛋白质组学补充了功能基因组学方法，包括基因表达谱芯片、细胞及器官水平系统的表型谱、系统的遗传学分析以及小分子芯片等。这些数据通过生物信息学整合在一起将形成一个基因功能广泛的数据库，对蛋白质特征及功能的了解是有力的参考，并且对于研究中建立和检测假说是有用的工具。蛋白质组学必将面对样品获得的限制、样品的降解、巨大的动力学变化、翻译后修饰、多样化组织和发育的时间特异性以及疾病和药物的干预等问题。

1. 基于质谱的蛋白质组学

质谱（mass spectrometry）可从大量蛋白质混合物中鉴定微量的蛋白质，是蛋白质组学的原动力。早先的蛋白质组学依赖于二维电泳（2D‑electrophoresis）对蛋白质进行分离，然后利用质谱鉴定蛋白质点（图 4-3）。这个方法的限制是样品必须浓缩至足够浓度。质谱快速的发展已经可以直接分析样品。

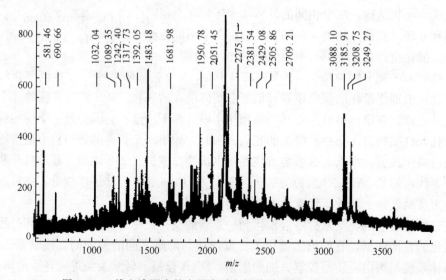

图 4-3   二维电泳图中某个蛋白质点的肽质谱图（张阳德，2004）

蛋白质组是动态的。对于这种蛋白质组动态调节判断的初始努力已经在酵母磷酸化组和泛素介导的降解组的单一实验中做了尝试。必须有更高通量和灵敏度来确保蛋白质组的动态以及细胞响应的实时扫描。

2. 基于芯片的蛋白质组学

不同的蛋白质芯片（protein chip）模式可以在蛋白质组的水平上进行快速的相互作用活性的检测。这些方法依赖重组蛋白质或者和蛋白质相互作用的特异性试剂，包括抗体、多肽及小分子。蛋白质检测方法来源于蛋白质的相互作用、蛋白质修饰以及酶活性。芯片的方法也能用于在体内检测，如通过 GFP 在细胞内进行蛋白质的定位。

3. 结构蛋白质组学

细胞行为的描述需要蛋白质结构信息。这就需要解析结构信息的技术，如蛋白质水平上的 X 射线晶体衍射技术、磁共振（nuclear magnetic resonance，NMR）、电子显微镜等。

4. 临床蛋白质组学

蛋白质组学的研究对临床诊断和药物的发现有深刻的影响。因为大多数药物的靶标是蛋白质，蛋白质组学的成果毫无疑问有助于临床试验。

与疾病状态相关的蛋白质谱的检测开始于蛋白质组学研究的早期，二维电泳首先使用的是临床样本。随着质谱技术的发展，体液中很多蛋白质和多肽也能够检测并作为基于蛋白质的诊断。质谱仪的高通量和高灵敏度适合于临床试验。标准化的样品准备、储存以及仪器使用是必需的。

蛋白质组学将毫无疑问地促进药物的发现，尽管在这个领域的进展比较缓慢。常常存在于细胞表面的新的疾病特异性靶标的鉴定在应用现代技术的基础上获得长足进步。生物学网络的理解将为靶点的重要性和适当性提供合理的基础。

5. 展望与挑战

蛋白质组学必须与其他技术结合才能准确地描述细胞的特征。所有的证据表明，现在的设备距离适合的分析还相距甚远，部分原因是制造商还没有制造机器和相关的硬件去完美地匹配蛋白质分析。质谱仪的改进可能有助于在任何复合物中检测蛋白质的修饰和相互作用。

在大量样品制备、蛋白质相互作用的选择性识别以及微量样品损失的减少方面，技术的发展将影响高效蛋白质组学的成功。同样重要的是，抗体的获得将改善和加快大、小通量蛋白质组学的进步。关于简单模式生物的蛋白质组学，相互作用的图谱还远没有成熟。当相互作用方面知识的增加时，可检测的假说就应该以相应的速度增加，特别是和基因组以及结构预测的数据结合在一起。一旦可以获得足够的动力学数据，就可以建立细胞行为的第一份草图，模型的进一步确认将需要大量的突变和药物处理来重复蛋白质组学的分析。所有这些信息都要以使用者可以获得的方式提供。然后更多的努力就要放在发展形象化的工具，包括和基因组数据的整合。

在临床方面，少量疾病组织的广泛的蛋白质组学的分析将有助于诊断与治疗的监控，特别是疾病预测的模式需要大量的临床数据。磷酸化蛋白质组学方法在临床样品上的应用使我们可以最大限度地了解和鉴别细胞的状态，有利于疾病诊断、药物的发现以及药物作用机制的阐明。宿主-病原相互作用的蛋白质组学是药物靶标研究的新领域。

人类蛋白质组组织（human proteome organization，HUPO）已经宣布了世界范围蛋白质组学研究的 5 个初级目标：血浆蛋白组的定义、对于特殊细胞类型的更加深入的蛋白质组学分析、对人类所有蛋白质的抗体的生产、新技术的发展以及信息学基础的建立。为了这些目标我们将增加所有蛋白质一级结构的分类，对所有被纯化的细胞器作图，建立模式生物的蛋白质相互作用的图谱，以及与功能基因组计划的整合。

二、蛋白质组学在人类疾病研究中的应用

蛋白质组学在人类疾病研究中的应用包括：对于蛋白质表达的描述不仅在细胞与组织水平，而且在亚细胞结构，在蛋白质复合物和生物体液；疾病早期诊断的新的标志物；治疗新靶标的鉴定，潜在的药物药效及毒理的评价等。

1. 超越二维电泳的疾病表达谱分析

常规的二维电泳可能是通量相当小的方法且需要大量的样品。但是临床样品的数量是有限的。此外，组织异质性（heterogeneity）使临床样品的分析复杂化。不同的组织微切片方法对减少异质性是有利的，但是这进一步减少了样品的数量。特别是激光微切片，可以从被分离的组织中得到特定的细胞类型，获得的蛋白质数量使用二维电泳分离就更困难。毫无疑问，多种依赖于液相的标记或非标记分离蛋白质或多肽的非凝胶技术有利于疾病蛋白质组学，特别是有自动化的潜力。

基于不分离样品的策略，包括直接的质谱分析或蛋白质芯片是重要的进展，质谱分析已经被应用于组织的原位蛋白质组学分析，可以同时检测正常和疾病组织的蛋白质表达。不同间隔获得的质谱进行比较，得到蛋白质在组织的空间分布。从正常和疾病组织获得的质谱图比较蛋白质表达的改变。使用这个方法进行肿瘤分析揭示了不同肿瘤类型患者的正常组织与肿瘤组织之间蛋白质表达差异的特异性。

2. 在疾病研究中 DNA 芯片与蛋白质芯片的比较

DNA 微阵列技术对于揭示基因表达模式以提供临床信息是非常有用的。对于疾病组织和细胞进行微阵列分析的一个重要的挑战是，疾病的临床特征与基因表达谱之间的关系无法建立在动态水平之上。另一个挑战是鉴定最微量但是与特异的临床特征最相关的标记，然后能使用临床试验技术进行检验。其他的挑战还有检测和预测基因的 RNA 与蛋白质水平之间缺乏相关性，可能暗示基因预测的特征是独立于基因功能的。例如，比较同一肿瘤的 mRNA 与蛋白质水平仅仅有少数的基因有统计学上的相关性。

DNA 微阵列技术继续在发展，有寡核苷酸阵列取代 cDNA 阵列的趋势。尽管完善的 DNA 微阵列及其分析工具完成疾病分析，但是它们不能减少、超越或替代 RNA 水平检测技术的需要，特别是相关疾病的调查。DNA 微阵列受到生物体液分析的限制，不能直接在体液中揭示可用的生物标记（体液中大多为蛋白质标记物）。大多数改变可能只发生在蛋白质水平而 RNA 水平没有改变，在蛋白质水平分析基因表达就必不可少。系统分析成千上万个蛋白质的芯片技术得到发展。不像 DNA 微阵列，蛋白质微阵列能检测疾病中发生改变的蛋白质的不同特征。这包括在生物样品中它们水平的测定，它们之间或与其他生物分子之间相互作用的检测，如其他蛋白质、抗体、药物或不同的小分子配体。针对蛋白质相互作用的检测，已经生产有相关的多肽和蛋白质芯片。多肽可以大量地直接合成在芯片上。纯化的重组蛋白可以大量地固定在芯片上。

应用于疾病组织研究的蛋白质芯片已经出现。作为一个模型可以较好地理解组织微环境中蛋白质的表达模式。固定一个组织的所有蛋白质的反向蛋白质芯片技术已经被发展，具有高灵敏度、准确和良好的线性关系，使定量人类组织细胞中信号蛋白的磷酸化

成为可能。蛋白质芯片的一个临床相关应用是在自体免疫紊乱中诱导抗体响应的蛋白质的鉴定。固定几百个蛋白质和多肽到玻片的表面制备成芯片。芯片与患者血清保温，荧光探针检测自体免疫疾病中结合特异蛋白质的自身抗体，包括系统性红斑狼疮以及风湿性关节炎。这种芯片代表了一个研究多种疾病免疫反应的强有力的工具，包括肿瘤。

制作芯片对蛋白质表达进行整体分析的主要挑战之一是现在缺乏广泛的捕捉工具，如抗体。对于蛋白质芯片另外一个重要的考虑是蛋白质发挥它们的功能是要经过大量的翻译后修饰，但是这些修饰一般使用重组蛋白或抗体因不具备识别的特异性而难以捕捉。

3. 蛋白质组学是发现疾病标志物的需求

应用蛋白质组学鉴定疾病标记物有着诱人的前景，包括对正常与疾病组织比较，鉴定异常表达的蛋白质可能代表新的标记物；细胞系以及原代培养细胞分泌蛋白质的分析以及直接血清蛋白的分析。质谱的潜在能力可以对生物体液蛋白不分离直接进行分析。原则上如此的方法因高通量与减少样品损失而非常适合标记物的鉴定。

这个方法现在非常受欢迎，特别对于血清分析。微升量的血清样品被结合到芯片上，被结合的蛋白质被处理分析。不同样品获得的质谱模式反映了这些样品的蛋白质和多肽的组成。癌症患者与正常个体之间的不同模式与区别已经在几种癌症中有报道。组织与生物体液的直接分析的主要缺点是偏向于低相对分子质量的蛋白质检测，鉴定修饰后的蛋白质仍有困难。

一个鉴定肿瘤标志物的富有成效的方法是分析血清中针对肿瘤蛋白的自身抗体。针对恶性肿瘤免疫反应的证据在增加，表明可以通过检测患者血清中针对细胞内以及细胞表面抗原产生的自身抗体来鉴定不同类型肿瘤。因此，引起抗体反应的肿瘤抗原的鉴定可以被用于肿瘤分析/诊断/预后评估以及疾病的免疫治疗。

诱导免疫反应的肿瘤抗原的检测有几种方法。通过扫描患者血清的表达文库已经检测了一组抗原，最近还使用了随机多肽文库的方法。使用二维电泳分离肿瘤蛋白，然后印迹并与患者血清保温，鉴定了许多不同类型肿瘤特异诱导产生自身抗体的蛋白质。对于大多数抗原蛋白都可以使用这种方法进行鉴定，包括翻译后修饰也归因于免疫反应。

4. 疾病相关的功能蛋白质组学

尽管通过多种表达蛋白质组学策略的数据能够揭示疾病蛋白质翻译修饰后的状态及改变水平与功能相关，但对于直接的功能分析还需要其他的技术。这需要分析蛋白质复合物在疾病状态下的异常，检测多种蛋白质的高通量模式，测定单个蛋白质的活性和水平，决定它们在不同生物学过程及疾病中的状态。

现在研究蛋白质复合物及蛋白质的相互作用有多种策略。如今，这些策略被用于有限范围内的疾病调查，即针对特定的复合物，如使用亲和共沉淀、二维电泳及质谱分析。现今大多数的蛋白质相互作用的研究只是比较了正常与疾病状态，即使在这个领域仍然处于早期，特别在疾病调查中，还没有得到全部的开发。

5. 蛋白质组学应用于病原微生物的研究

对于感染性疾病治疗的难点是对常用药抗性的增加。发展有效的新治疗方法迫在眉睫。对微生物的蛋白质组学分析要回溯至少 20 年，Fred Neidhardt 鉴定了不同生长状态下大肠杆菌的蛋白质表达模式。多数微生物基因组的完全测序已经提供了使用质谱鉴定这些基因组编码的蛋白质的基础，使了解其耐药产生机制和找到新的药物作用靶点成为可能。

6. 蛋白质组学应用于药物的研发

现在部分制药公司对蛋白质组学投入极大兴趣，许多蛋白质组学的研究由制药公司支持。鉴于很多药物的靶标为蛋白质，蛋白质组学对药物的发展应该有极大的应用价值。但是工业生产至今还是谨慎的态度，因为还缺乏针对蛋白质组学对药物发展的作用的严格评价程序。提供适当的技术平台是可行的，蛋白质组学可以在多方面支持药物的研发，如鉴定新靶标以及药物在临床前及临床中的药理及毒理评价。

第三节　蛋白质组学研究技术

一、蛋白质分离技术

蛋白质组学是对蛋白质组的分析与检测，即对生物蛋白质的大范围研究。从传统意义上讲，蛋白质组学是通过聚丙烯酰胺双向电泳对特定细胞或生物体内大量蛋白质的检测。随着质谱检测技术的出现，使蛋白质的鉴定变得准确而便捷。蛋白质组学并非只是蛋白质的生物化学分析，二者的区别在于蛋白质的生物化学分析侧重于单一蛋白质的序列、结构及功能的研究，它属于结构生物学范畴；而蛋白质组分析则侧重于蛋白质混合物的研究与检验，它通过对蛋白质部分序列的检测对它进行鉴定性分析，属于系统生物学的范畴。蛋白质组受基因组调控，而蛋白质组分析即是对基因组中特定基因所代表的相关功能的研究。

蛋白质组分析这一敏感而特异的蛋白质检测及鉴定方法，主要通过两道程序来完成。首先是蛋白质混合物的分离。其目的有两个，其一是将含有多种蛋白质的混合物分离为单个蛋白质或数个蛋白质的复合物，以便于后续的蛋白质质谱分析。其二是通过蛋白质的分离来比较不同样本蛋白质表达质与量的差别。作为蛋白质组分析核心技术之一的双向电泳已受到广泛的重视与应用。另外，蛋白质单向电泳、高效液相层析、毛细管电泳、等电聚焦和亲和层析等方法均可用于蛋白质组分析。而将不同方法结合起来的多维技术较单一方法更有优势。其次是质谱分析。质谱分析包括质谱检测、软件程序分析及数据库检索。质谱分析可以对蛋白质的相对分子质量进行准确测定。通过数据库检索，质谱分析可以比较被检蛋白的多肽质谱与数据库中蛋白质的多肽质谱，并以此为根据对被检蛋白进行鉴定。质谱分析所获得的蛋白质氨基酸序列是对蛋白质进行鉴定的最准确、有效的手段。蛋白质组分析主要在4个方面得到应用并发挥其优势。其一，样本全蛋白检测；其二，样本蛋白表达检测；其三，蛋白质相互作用检测；其四，蛋白质修饰检测（本章节涉及的技术方法参见第十章）。

1）等电聚焦电泳（isoelectric focusing，IEF）　蛋白质是兼性分子，它依环境 pH 的不同而携带正电荷、负电荷或不带电荷。pH 梯度对蛋白质等电聚焦技术而言至关重要。当存在 pH 梯度及电场的情况下，蛋白质分子将向其净电荷为零的 pH 处移动。当样本中的每一个蛋白质均达到其电荷为零的 pH 位置时，样本蛋白就根据其等电点而得到分离。

2）十二烷基硫酸钠（SDS)-聚丙烯酰胺凝胶电泳　SDS是阴离子洗涤剂，它同蛋白质混合后形成项链样结构，并以1.4∶1的比例结合，从而使蛋白质带有净负电荷。因此，电泳蛋白移动的距离同它的相对分子质量呈正比。在将胶条由第一相电泳移入第二相电泳之前，需先于平衡液中平衡30 min，使胶条浸透SDS缓冲液。

3）双向电泳　双向电泳是一种用于分析由细胞、组织或其他生物样本中提取的复杂蛋白质混合物的高效且广泛应用的检测方法。第一相为蛋白质等电聚焦电泳，它使蛋白质依各自等电点的不同进行分离；第二相为SDS-聚丙烯酰胺凝胶电泳，它使蛋白质依各自相对分子质量的不同进行分离。其所得到的蛋白质双维排列图中的每一点代表样本中的一种或几种蛋白质。采用两性电解质载体聚丙烯酰胺凝胶电泳法，限制了蛋白质双向电泳的稳定性及可靠性。固定pH梯度胶指的是，采用固定于塑料支撑条上的固定pH梯度胶代替过去的两性电解质载体毛细管胶进行蛋白质等电聚焦。大大地提高了蛋白质等电聚焦的分辨率与稳定性。

双向电泳也有缺陷，首先，双向电泳很难得到完全相同的分离结果，这使不同蛋白质双向电泳结果的比较存在误差。其次，蛋白质等电聚焦电泳法很难将一些大分子、疏水蛋白分离，降低蛋白质双向电泳结果的分辨率。双向电泳适合用来分析含量高的长效蛋白。在实际应用过程中有时采用其他的蛋白质分离技术，包括单向电泳、等电聚焦电泳、高效液相层析以及毛细管电泳等。

二、质谱分析

现今的质谱分析主要有两种方法：基质辅助激光解吸电离飞行时间质谱测量法（matrix-assisted laser desorption ionization）和电子喷雾电离质谱测量法（electrospray-ion-izationmassspectrometry）。两种方法操作方法完全不同，但其所获得的信息可相辅相成。

1. 基质辅助激光解吸电离飞行时间质谱测量法

基质辅助激光解吸电离飞行时间质谱测量仪是将多肽成分转换成离子信号，并依据质量ö电荷之比来对该多肽进行分析，以判断该多肽源自哪个蛋白质。仪器主要由两部分组成：基质辅助激光解吸电离离子源（MALDI）和飞行时间质量分析器（TOF）。待检样品与含有在特定波长下吸光的发光团的化学基质混合。样品混合物随即滴于一平板或载玻片上进行挥发。样品混合物残余水分和溶剂的挥发使样品整合于格状晶体中。样品然后置于激光离子发生器。激光作用于样品混合物，使化学基质吸收光子而被激活。此激活产生的能量作用于多肽，使之由固态样品混合物变成气态。由于多肽分子倾向于吸收单一光子，故多肽离子带单一电荷。这些形成的多肽离子直接进入飞行时间质量分析仪。飞行时间质量分析仪用于测量多肽离子由分析仪的一端飞抵另一端所需要的时间。而此飞行时间同多肽的质量ö电荷之比成反比，即质量ö电荷之比越高，飞行时间越短。最后，由计算机软件将探测器录得的多肽质量ö电荷比值同数据库中不同蛋白质经蛋白酶消化后所形成的特定多肽的质量ö电荷比值进行比较，以鉴定该多肽源自何种蛋白质。此法操作简便，敏感度高，同许多蛋白质分离方法相匹配。而且，现有数据库中有充足的关于多肽质量ö电荷比值的数据。

2. 电子喷雾电离质谱测量法

　　同基质辅助激光解吸电离飞行时间质谱测量法在固态下完成不同，电子喷雾电离质谱测量法（ESI-MS）是在液态下完成，而且多肽离子带有多个电荷。由高效液相层析等方法分离的液体多肽混合物，在高压下经过一细针孔。当样本由针孔射出时，喷射成雾状的细小液滴。这些细小液滴包含多肽离子及水分等其他杂质成分。去除这些杂质成分后，多肽离子进入连续质量分析仪。连续质量分析仪选取某一特定质量ö电荷比值的多肽离子，并以碰撞解离的方式将多肽离子分成不同电离或非电离片段。随后，依质量ö电荷比值对电离片段进行分析并汇集成离子谱。通过数据库检索，由这些离子谱得到该多肽的氨基酸序列。依据氨基酸序列进行的蛋白质鉴定较依据多肽质量指纹进行的蛋白质鉴定更准确、可靠。

3. 质谱在蛋白质组分析中的应用

　　1）样本全蛋白检测　样本全蛋白检测的目的在于发现并鉴定尽可能多的蛋白质组成分。但是，细胞或生物体在某一时段只有部分蛋白质组得到表达；而同时有许多蛋白质未得到表达。这使样本全蛋白质检测存在很大困难。

　　2）样本蛋白表达检测　蛋白质组的表达随发育、分化、疾病等生理和病理状态的不同而发生变化，故对样本蛋白表达的检测可了解与不同生理和病理状态相关的蛋白质表达状况，从而发现疾病的诊断和治疗方法。现多采用对比性蛋白双向电泳或对比性液相层析来首先比较不同状态下蛋白质表达的不同，然后再应用蛋白质谱分析来鉴定特异表达的蛋白质。

　　3）蛋白质相互作用检测　蛋白质通常经相互作用而形成多蛋白复合物来发挥其特殊功能。要了解蛋白质的特殊功能，就要首先了解这些多蛋白复合物的构成。而关键线索就是这些蛋白质是否同某些功能已知的蛋白质形成多蛋白复合物。现主要通过酵母双杂交系统与免疫共沉淀的方法进行蛋白质功能的研究。

　　4）蛋白质修饰检测　生物体中的大多数蛋白质以被修饰的形式存在，如翻译后裂解、磷酸化、糖基化与氧化等。以往采用单克隆抗体来检测蛋白质磷酸化等蛋白质修饰。由于修饰后的蛋白质，其质量ö电荷比值发生改变，因此蛋白质谱分析可作为一种便捷而高效的方法，用于检测作用于蛋白质序列上某一特定氨基酸的修饰方式。由于蛋白质谱检测可以检测蛋白质的全部氨基酸序列，所以用此法检测蛋白质修饰即准确又可靠。

三、蛋白质的相互作用研究技术

　　蛋白质组学是在蛋白质水平上动量、动态、整体性地研究生物体，不是一个封闭的、概念化的、稳定的知识体系，而是旨在阐明生物体全部蛋白质的表达模式和功能模式的领域。而蛋白质组功能模式的研究目前主要集中在蛋白质间相互作用的网络关系上，研究蛋白质之间相互作用的方法有如下几种。

1. 酵母双杂交系统

　　酵母双杂交系统（yeast two-hybrid system）又称蛋白质捕获系统，是研究蛋白质间相互作用的一种非常有效的分子生物学方法。真核生物细胞转录激活因子一般都含有两个不同的结构域：DNA 结合结构域（DNA binding domain，BD）和 DNA 转录激活结构域（transcription activation domain，AD）。这两个结构域相互独立但功能上又相互依赖，它们之间只有通过某种方式结合在一起才具有完整的转录激活因子的活性。将拟研究的靶蛋白基因与 AD 序列结合，编码诱饵蛋白（bait protein）的基因与 DNA BD 序列结合，形成两个融合基因。当这两个融合基因在同一菌株内表达，靶蛋白与诱饵蛋白在核内相互作用时，就重新形成完整的、有活性的转录因子，从而激活报告基因的转录。于是根据报告基因的表达与否，即可判断靶蛋白与诱饵蛋白是否相互作用。

　　酵母双杂交系统在研究蛋白质间相互作用时存在有一些缺陷，一个是"假阳性"问题。由于某些蛋白质本身具有激活转录功能或在酵母内表达时的转录激活作用，使靶蛋白 BD 融合基因与诱饵蛋白 AD 融合基因表达产物无需特异结合，就能启动转录系统，产生假阳性结果；另外一个问题是所分析的可能相互作用的蛋白质必须定位于核内，才能激活报告基因。但是很多蛋白质间的相互作用需在胞质内完成，这样的蛋白质就不适用此方法。针对"假阳性"问题，研究者在实践中通过对酵母双杂交系统进行改进出现了双筛选系统和假阳性显示分析法等。同时，为了克服酵母双杂交系统中存在的第二个问题，人们开始尝试使用其他蛋白质的结构特点来建立新的双杂交系统。于是，一些不依赖于转录因子活性的新型双杂交系统相继建立，如分离的泛素系统、蛋白质片段互补分析、阻抑物重构分析和 SOS 恢复系统等。由于这些新型双杂交系统的各自特点，使得它们成为酵母双杂交系统的有益补充和研究蛋白质间相互作用的有力工具。

2. 噬菌体展示

　　噬菌体展示（phage display）技术是一种噬菌体表面表达筛选技术，也是一种用于筛选和改造功能性多肽的生物技术。在编码噬菌体的外壳蛋白基因上连上外源蛋白或多肽的 DNA 序列（如单克隆抗体的基因序列），使外源基因随外壳蛋白的表达而表达，同时，外源蛋白随噬菌体的重新组装而展示到噬菌体表面。将噬菌体过柱，由于柱上含有目的蛋白，所以会特异性结合相应靶蛋白（抗体），将外源蛋白或多肽挑选出来。由于该技术的主要特点是将特定分子的基因型和表型统一在同一病毒颗粒内，即在噬菌体表面展示特定蛋白，而在噬菌体核心 DNA 中则含有该蛋白质的结构基因，因此，噬菌体展示技术使得表达蛋白（表达型）和编码基因（基因型）之间完美的结合，而且，它也能很好地解决酵母双杂交系统中存在的问题。噬菌体展示技术具有两点最显著的意义：一点是它引申出了分子库（molecular repertoires）的概念；另一点是此项技术克服了研究蛋白质相互作用时需基于对其结构的详尽知识的限制，这也是使得该项技术得以广泛运用的直接原因。目前，利用噬菌体展示技术构建随机肽库、蛋白质库和抗体库，研究受体或抗体的结合位置，改造和提高蛋白质、酶和抗体的生物学和免疫学属性，研究用于检测治疗用途的新型多肽药物、疫苗和抗体等，都具有广阔的应用前景。

3. 表面等离子体共振

　　表面等离子体共振（surface plasmon resonance，SPR）的应用是利用金属膜/液面界面光的全反射引起的物理光学现象来分析分子相互作用的技术。自 1982 年 Nylander 等首次将 SPR 技术用于免疫传感器领域以来，表面等离子体光学生物传感器得到了深入的研究和广泛的应用。SPR 生物传感器是利用表面等离子体共振现象和 SPR 谱峰对金属表面上电介质变化敏感的特点，通过将受体蛋白固定在金属膜上，检测受体蛋白与液相中配体蛋白的特异性结合。SPR 技术的特点是测定快速、安全，不需要标记物或染料且灵敏度高。除了应用于检测蛋白质外，还可检测蛋白质、核酸及其他生物大分子之间的相互作用，并且能对整个反应过程进行实时监测。目前，SPR 技术已成为当今一种全新的研究蛋白质之间相互作用的手段。

4. 蛋白质芯片

蛋白质芯片（protein chip）是高通量、快速、高效、微型化和自动化的蛋白质分析技术。其中一种蛋白质芯片类似于 DNA 芯片，即在固体支持物表面高密度排列的探针蛋白点阵，可特异地捕获样品中的靶蛋白，然后通过检测器对靶蛋白进行定性或定量分析。因此蛋白质芯片能够同时分析上千种蛋白质的变化情况，也使得在全基因组水平研究蛋白质的功能（如酶活性、抗体特异性、配体-受体交互作用及蛋白质与蛋白质或核酸或小分子的结合）成为可能。现阶段，蛋白质芯片除了用于研究蛋白质相互作用外，还广泛应用于疾病诊断、疗效判定、发现药物或毒物新靶点及其作用机制等方面。

5. 生物信息学

生物信息学重要研究领域之一就是蛋白质功能识别。通过同源性分析识别蛋白质功能已经被广泛使用，其依据是由同一祖先进化的同源蛋白家族，具有相似的序列和结构，相应的也具有相似的功能。最近，发展了几种计算机识别蛋白质功能的新方法，这些方法的依据不是蛋白质之间的同源性，而是蛋白质的相同特征，如系统发生模式、mRNA 表达模式及结构域融合模式等。相同特征的蛋白质之间具有功能上的关联或直接的作用，这种关联或作用指参与同一代谢途径或信号传递途径或参与形成同一结构复合体。应用这些方法可以对大量未知功能的蛋白质进行功能识别。今后的研究目标是进一步提高计算机识别的覆盖面、准确性和精确度（详见第十一章）。

四、蛋白质芯片应用于蛋白质组学分析

蛋白质组学的特点是采用高分辨率的蛋白质分离手段，结合高效率的蛋白质鉴定技术，全景式地研究在各种特定情况下的蛋白质谱。一个完整的蛋白质组分析将包括含量的测定、修饰、活性、定位以及样品中所有蛋白质的相互作用。在实践中，现在的技术限制我们的分析，只能获得部分蛋白质的其中一个或两个参数。这类型的实验等同于表达谱分析。称其为"定量蛋白质组学"（quantitative proteomics）。与之区别的"功能蛋白质组学"（functional proteomics），即定义研究样品中的每个蛋白质的功能。这可能包括酶与底物以及蛋白质之间的相互作用。芯片技术对"定量蛋白质组学"与"功能蛋白质组学"都有显著的影响，提供了一套快速蛋白质分析筛选方法。

1. 定量蛋白质组学

与基因组的普遍性和均一性不同，蛋白质组存在时间和空间上的多样性，即在同一物种的不同细胞中或同一细胞的不同生长时期，其蛋白质组在不断变化。因此精确描述细胞内存在的全部蛋白质的状态及相关的蛋白质组是不可能的。目前对疾病的研究主要采用比较蛋白质组学的方法，即通过比较正常与异常细胞或组织中蛋白质表达水平的差异，进而找到与人类疾病密切相关的差异蛋白，进而确定靶分子，定量蛋白质组学即对蛋白质的差异表达进行定量分析，是比较蛋白质组学的主要研究内容之一。利用蛋白质芯片技术来研究定量蛋白质组，会进一步促进蛋白质组的研究从静态的、定性的、组成分析过渡到动态的、定量的功能研究上来。

2. 功能蛋白质组学

基因组测序计划贡献了一列细胞或器官的蛋白质组成，而检测这些成分就是功能蛋白质组学的任务。酵母双杂交是已经被广泛使用的一项技术。这项体外技术尽管容易执行且相当有用，也存在一些限制。具有转录激活因子功能的蛋白质结合到 DNA 结合域时会产生假阳性，而蛋白质在酵母中展示不适当或没有正确的折叠就会产生假阴性。更显著的是，研究者既不能控制蛋白质翻译后修饰的状态，也不能控制研究相互作用的环境。

最近，免疫沉淀匹配质谱（immunoprecipitation and mass spectrometry）已经被用来大规模地鉴定蛋白质复合物。这项技术很容易适应高通量的调查，鉴定经标记的诱饵蛋白捕捉纯化的蛋白质。尽管这是很有效的鉴定蛋白质复合物的方法，但是在细胞中诱饵蛋白与其他蛋白质随机黏附存在，而非复合物，这是一个障碍。因此它可能共沉淀两个或更多实际上永远不可能共定位的蛋白质，得出蛋白质比实际更大、更复杂的错觉。作为这些方法的一个补充，蛋白质芯片技术提供了一个很好的控制，在体外研究系统的全基因组基础的功能。对于这种应用，蛋白质不是亲和而是固定在固相支持物上。为了将这种类型的芯片与抗原捕捉的蛋白质检测芯片区别，我们称其为蛋白质功能芯片。

尽管低密度的芯片是有用的，从芯片技术可以获得更大的利益。首先，原则上成千的蛋白质能被点在支持物上，微小的样品损耗能同时检测许多蛋白质的功能。其次，成百甚至上千的芯片拷贝使得相同的蛋白质可以重复地在不同条件下使用许多不同分子作为探针平行地进行分析。

蛋白质功能芯片将毫无疑问有利于改善制作、实验以及分析，但是现在扩展使用的最大障碍是纯的、重组蛋白的大量制备与收集制备的必要步骤是 cDNA 的克隆。使用高通量方法扫描 cDNA 文库的蛋白质表达克隆已经被描述。尽管这是一个对于大量收集蛋白质有潜在经济价值的方法，但是也有几个方面的限制。文库可能不包括低丰度表达蛋白质的基因，并且蛋白质常常开始或停止在编码序列的随机位置。随着整个基因组序列数据的获得，全长可读框的重组克隆载体构建。利用这些资源，配合高通量的重组蛋白表达及纯化方法，应该极大地促进芯片技术在蛋白质组学中的应用。每个方法都有其利弊。对于蛋白质功能芯片，蛋白质的纯度和完整需要被鉴定，它们的浓度要标准化。也许最大的挑战是发现一个可行的方法去检测被纯化蛋白质的功能状态。单一可溶性指标不能完全表明蛋白质是正确折叠的。通过对这些方面的紧密关注，系统定位分析与发现定位预期的数据将指导我们逐步接近完整地理解生物学过程。

小　结

本章从三个方面简要介绍了蛋白质组学的基本概念、研究内容以及研究方法。首先围绕蛋白质组学研究的主要内容介绍了蛋白质的基本知识，包括结构域、折叠、分泌以及降解；其次介绍了蛋白质组和蛋白质组学的基本概念以及在人类疾病研究中的应用；最后介绍了包括双向电泳、质谱、芯片等多种蛋白质组学研究中使用的重要方法的基本原理及应用。

（武军驻）

参 考 文 献

葛圣雷，陈主初，肖志强等. 2005. 分泌性蛋白质的研究策略. 生命的化学，25：476～477

何家田，王红霞，张学敏等. 2005. 膜蛋白组分析技术的研究进展. 军事医学科学院院刊，29：584～587

廉德君，许根俊. 1997. 蛋白质结构与功能中的结构域. 生物化学与生物物理进展，24：482～486

田云，卢向阳. 2003. 蛋白质间相互作用研究技术进展. 生物学通报，38：1～4

许克新，王云川. 2002. 双向电泳及质谱分析与蛋白组研究. 第四军医大学学报，23：2017～2022

Abbott A. 2001. And now for the proteom. Nature，409：747，748

Ciechanover A. 2005. Intracellular protein degradation：from a vague idea thru the lysosome and the ubiquitin-proteasome system and onto human diseases and drug targeting. Cell Death and Differentiation，12：1178～1190

Gygi S P, Corthals G L, Zhang Y, et al. 2000. Evaluation of two-dimensional gel electrophoresisbased proteome a-
  nalysis technology. Proc Natl Acad Sci USA，97：9390～9395

Hanash S. 2003. Disease proteomics. Nature，422：226～232

Jones S, Thornton J M. 1996. Principles of protein-protein interactions. Proc Natl Acad Sci USA，93：13～20

MacBeath G. 2002. Protein microarrays and proteomics. Nature Genetics，32：526～532

Tyers M, Mann M. 2003. From genomics to proteomics. Nature，422：193～197

Washburnl M P, Woltersl D, Yates Ⅲ J R. 2001. Large-scale analysis of the yeast proteome by multidimensional
  protein identification technology. Nature Biotechnology，19：242～247

White M A. 1996. The yeast two-hybrid system：Forward and reverse. Proc Natl Acad Sci USA，93：10001～10003

Zhu H, Snyder M. 2003. Protein chip technology. Current Opinion in Chemical Biology，7：55～63

Chapter 4   Protein, Proteome and Proteomics

The genomics has changed the paradigm for the comprehensive analysis of biological processes and systems. It was hypothesized that biological processes and systems can be described based on the comparison of global, quantitative gene expression patterns from cells or tissues representing different status. However, the discovery of posttranscriptional mechanisms that control rate of synthesis and half-life of proteins and the ensuing nonpredictive correlation between mRNA and protein levels expressed by a particular gene indicate that direct measurement of protein expression also is essential for the analysis of biological processes and systems. After that, proteomics is often considered the next step in the study of biological systems. It is more complicated than genomics because while an organism's genome is more or less constant, the proteome differs from cell to cell and from time to time. This is because distinct genes are expressed in distinct cell types. This means that even the basic set of proteins which are produced in a cell needs to be determined.

This chapter will focus on theories and experimental methods of protein, proteome and proteomics.

Section 1   The Processing, Transporting and Degradation of Protein

1.1   Domain of Protein

Most proteins fold into unique 3-dimensional structures, biochemists often refer to three distinct aspects of a protein's structure: primary, secondary and tertiary structure. Usually, the tertiary structure determines the protein's biological function. Now, a concept of domain is more and more often used to describe the structures and functions of a protein.

1.1.1   Definition

A domain is a part of protein sequence and structure that can evolve, function, and exist independently of the rest of the protein chain. Each domain forms a compact three-dimensional structure and often can be independently stable and folded. Thus, the current concept of domain should have three different meaning, that is, an independent structural unit, an independent functional unit and an independent folding unit. Many proteins consist of several structural domains. One domain may appear in a variety of evolutionarily related proteins.

1.1.2   Structure and Function of Domains

Domains generally present in the globular proteins as the structure and functional unit of proteins. The connection among domains include from a loose connection of IgG to the closely connection of some small molecules such as papain and lysozyme et al. The different domains have a specific function as a functional unit, such as the combination of substrate, catalysis reaction, interaction among subunits and regulation of activity and so on. Alanine tRNA synthetase is a classic example of this structure (Figure 4-1).

Figure 4-1   The structure of alanine tRNA synthetase domain (see page 104)

Domain is of great significance to those proteins which have complex functions as the structure and functional unit. Table 4-1 lists some of the domain modules.

Table 4-1　Some of the common domain module in vertebrate proteins

Domain	Characteristics
Gla	with a number of γ-carbonyl glutamic acid residue
G	similar with the structure of the epidermal growth factor
K	"Kringle" domain
C	often seen in the complement proteins
F1, F2, F3	first discovered in fibronectin
I	immunoglobulin superfamily domain
L	seen in LDL receptors and complement proteins
T	seen in platelets binding protein and complement protein properdin
N	seen in some of the growth factor receptors
E	EF-finger structure
LB	the lectin protein module on the surface of some cells
SH2, SH3	src homology structure

1. 1. 3　Domains and Protein Folding

The proper protein folding is important for the formation of protein secondary structure. The patterns of protein folding have been categorized as below: all-α, all-β, α/β and $\alpha + \beta$. For small or medium-sized protein, the limited folding model has been able to accurately describe its structure. In the case of large and complex proteins, its structure can be attributed to the folding model of domain level. The further studies for the folding model at domain level will help us understand the complex structures and functions of the proteins.

It is currently unclear about how proteins to form three-dimensional structures. The "molten globular state model" was based on the unfolding/refolding experiments of proteins in vitro, and it is only suitable to single domain or simple proteins. The multi-domain proteins are now tended to "modular model". Domain has been confirmed as a folding unit through the unfolding/refolding experiments of proteins in vitro. The genetic engineer helps us directly separate peptide fragments. Using this technology, the studies of separating and independent folding domain has been given a number of successful cases.

Domains have been widely found in complex proteins as structural and functional unit, which are involved in regulating transcription and translation of genes, as well as signal transduction. Topology studies benefit the studies on the distribution of domain in proteins and prediction of secondary and tertiary structures. The protein engineering and molecular design may help produce the expected protein molecule through the genetically modified methods.

1. 2　New Peptides' Folding and Molecular Chaperone

In fact, new peptides begin folding in the early stages of its synthesis, and not the end its synthesis. The conformation is continuously adjusted with extension and folding of the peptides. The formed structure will effect the folding of synthesizing peptide, and the forming structure can effect the adjustment of the formed structure. As a result, the formed structures are uncertainly the ultimate structure in the process of extension of peptides. In this way, the formation of three-dimensional structure is a dynamic process. The molecular chaperones and accessory proteins may help the peptide folding.

1. 2. 1　Protein Folding and the Role of Molecular Chaperone

The three-dimensional structure of protein molecules is maintained not only by the peptide bond and the disulfide bonds, but also a large number of extremely complex and weak secondary bonds. Therefore, the structure is likely formed in the folding process of the new synthetic peptides that should not have or does not exist in a mature protein. They are often some of the hydrophobic surface. Some error interactions are likely performed among them to form the non-functional molecules or even produce molecules' congregation and precipitation. In accordance with the theory of self-assembly, each step is correct, full and necessary in folding. In fact, the folding is a competitive and dynamic process between the right way and wrong way. In order to improve the efficiency of protein biosynthesis, the certain mechanism should help the right way to compete. Therefore, chaperones can functionally identify the wrong structure in the folded process of new peptide. Chaperones combine with the wrong structure to generate complex, thereby prevent them from the incorrect folding and the formation of irreversible polymer. So, the folding is inevitably carried out in the right direction.

The basic mechanism of how chaperones work is actually how to recognize, associate and dissociate with the their target proteins. Some of chaperones have a high specificity. For example, the onion pseudomonas esterase has its own "private chaperone". It is encoded by the gene *limA* which is three bases away from esterase gene *LipA*. This may be evolutionary process. The general chaperones have low specificity. How do they identify the targets? Generally, chaperones only recognize or identify non-natural conformation, and they do not care about the natural conformation. The most of hydrophobic residues are at the interior to form the hydrophobic core in natural molecules. The hydrophobic surface can be temporarily formed after the unfolding, or in folding of the new peptides. Chaperone is most likely to interact with the hydrophobic surface. Chaperone only identify the β-sheet structure of protein molecules. The role of the chaperone is to form complexes with the target protein at the second step.

1. 2. 2   The Difference Between Chaperone and Enzyme

At present, two enzymes have been identified as the helpers of protein folding. One is a protein disulfide isomerase, and another is peptidyl prolyl *cis*-trans isomerase. It is well known that the disulfide bonds are closely related with the folding of new peptide and is crucial to maintain the structural stability and develop function in protein molecules. For example, PDI is abubdant and locates in the lumen of endoplasmic reticulum and catalyze the exchange reaction of the disulfide and —SH in proteins. PDI is also one of the most prominent multi-functional proteins so far. In addition to the basic function as disulfide isomerase, PDI also serves as α-subunit of proline-4-hydroxylase and the small subunit of triglyceride transfer protein complex in microsomes and a glycosylation site-binding protein as well. The most striking thing is the binding ability with the peptides. PDI can interact with the peptides with the different sequences, length and charge distributions. These interactions are usually low-specific and mainly on the main peptide chain, they also occur to —SH groups. Thus, it is generally believed that PDI and chaperone are two different proteins to help the folding based on the definition of the chaperone, however sometimes one account PDI as a functional chaperone.

The natural disulfide formation requires the certain extent folding of peptides so that —SH group can be near to correctly form disulfide within or between protein molecules. The folding of peptides is a slow process by itself, but the natural disulfide formation is a fast process which is catalyzed by PDI. On the other hand, PDI has the capability to bind a variety of different peptides at a low specificity and it is abundant in endoplasmic reticulum, it also is believed a calcium-binding protein and can be phosphorylated. Thus, PDI has some common features of a chaperone.

1. 2. 3   The Practical Application of Chaperone Studies

Chaperone research achievements will be significantly deepen our understanding to life phenomenon and at the same time is bound to increase our survival capacity. As chaperone have played an important role at all levels of life activities, its mutation and damage will certainly cause the diseases, it could be expected to therapy the so-called "molecular chaperone disease" by using the knowledge of chaperone. On the other hand, chaperone research achievements fundamentally improved the success rate of genetic and protein engineering, it will play an important role for improving the living standards of human.

1. 3   The Secreted Proteins

Proteins with the different functions are accurately located at the corresponding organelles through protein sorting signals and strict transporting way, or secreted to the cellular outside. Proteins are named as the secreted proteins which are released into the outer membrane. Secreted proteins have a large variety, such as many cytokines in blood and extracellular matrix, complement, enzymes, peptide hormones, immunoglobulin and so on. Secreted proteins play a role through endocrine, paracrine and autocrine. The comprehensive and systematic studies on secreted proteins not only contribute to a comprehensive understanding, analysis and interpretation of a variety of physiological and pathological phenomena, but also provide more choices for the development of new drugs. Secreted protein is the hot and difficult point in biological field.

Blobel has received the 1999 Nobel Prize for Physiology or Medicine for his Signal Hypothesis. Although the signal peptides do not have homology in a variety of proteins, its structure and amino acid residues follow a certain rule. The typical structure of the signal sequence is usually divided into 3 areas: n is an area of positively charged residues and a main location of mutation of the signal sequence, h is the core functional area of the signal peptide, c area contains an identified site at which signal peptide is removed by enzymes. The basic function of the signal peptide is to interact with plasma membrane through hydrophobic and electrostatic attraction and to play a role for the transport of the protein precursor containing the signal peptide sequence in cells.

1.4　Intracellular Protein Degradation

The concept of protein turnover have become almost 60 years old. The most important systems in vivo for protein turn over are the lysosome and the ubiquitin-proteasome systems.

1.4.1　The Lysosome and Intracellular Protein Degradation

The lysosome was first recognized biochemically in rat liver as a vacuolar structure that contains various hydrolytic enzymes, which function optimally at an acidic pH. It is surrounded by a membrane that endows the contained enzymes latency that is required to protect the cellular contents from their action. The definition of the lysosome has been extended over the years. It has been known that the digestive process is dynamic and that all stages of lysosomal maturation involves digestion of both exogenous proteins (which are targeted to the lysosome through receptor-mediated endocytosis and pinocytosis) and exogenous particles (which are targeted via phagocytosis; the two processes are known as heterophagy), as well as digestion of endogenous proteins and cellular organelles (which are targeted by micro-and macroautophagy). The lysosomal/vacuolar system is a discontinuous and heterogeneous digestive system including some structures without hydrolases. For example, early endosomes which contain endocytosed receptor-ligand complexes and pinocytosed/phagocytosed extracellular contents. On the other hand, it also includes the residual bodies, they are the end products of the completed digestive processes of heterophagy and autophagy. Between these extreme cases, one can observe primary/nascent lysosomes that have not yet been engaged in any proteolytic process; early autophagic vacuoles that contain intracellular organelles; intermediate/late endosomes and phagocytic vacuoles (heterophagic vacuoles) that contain extracellular contents/particles; and multivesicular bodies (MVBs), which are the trafficking vacuoles between endosomes/phagocytic vacuoles and the digestive lysosomes.

In fact, the proteases are separated from their substrates by a membrane structure, however, how are the substrates translocated into the lysosomal lumen, and exposed to the lysosomal proteases for degradation? Autophagy may help a better undertsnading. Under basic metabolic conditions, cytoplasm proteins are segregated within a membrane-bound compartment, and are then fused to a primary nascent lysosome for digestion. This process is called microautophagy. Under some stress or extreme conditions (e. g., starvation), mitochondria, endoplasmic reticulum membranes, glycogen bodies and other cytoplasmic entities can also be engulfed by a process called macroautophagy. There are the different modes of action of the lysosome in digesting extra-cellular proteins and intracellular proteins. However, it has become gradually more and more difficult to explain several aspects of intracellular protein degradation based on the known mechanisms of lysosomal activity. The degradation of some cellular proteins must be nonlysosomal.

As mentioned above, the lysosome mechanism only partly explains the degradation of proteins and the explanation is not satisfactory to several key characteristics of protein degradation in cells. Many phenomena can not be explained, such as the heterologous stability of individual protein, the role of nutrition and hormones in the degradation process, the energy-dependent of intracellular protein degradation, and the effects of selective inhibitor on different types of proteins. Therefore, the lysosome hypothesis is being challenged.

1.4.2　The Ubiquitin-Proteasome System

Ubiquitin is a small, heat-stable and highly evolutionarily conserved protein of 76 residues. A single ubiquitin moiety can be covalently conjugated to histones, particularly to histones H2A and H2B. Two proteins are linked through a fork-like, branched isopeptide bond between the carboxy terminal glycine of ubiquitin (Gly76) and the NH2 group of an internal lysine (Lys119) of the histone molecules. The isopeptide bond in the histone-ubiquitin adduct has been known as the key connection between ubiquitin and the target proteolytic substrate. The target proteins (substrates) with the polyubiquitin chain can be recognized by 26S proteasome for degradation. The polyubiquitin chain functions as a proteolysis recognition signal in this processing. In recent years, it has been shown that the first ubiquitin moiety can also be attached in a linear mode to the N-terminal residue of the proteolytic target substrate. Yet, several lysine-containing proteins have also been described that traverse this pathway. In these proteins the internal lysine residues are probably not accessible to the cognate ligases.

Using the unraveled mechanism of ubiquitin activation and immobilized ubiquitin as a "covalent" affinity bait, the three enzymes that are involved in the cascade reaction of ubiquitin conjugation were purified. These enzymes are: (i) E1, the ubiquitin-activating enzyme; (ii) E2, the ubiquitin-carrier protein; and (iii) E3, the ubiquitin-protein ligase. The discovery of an E3, which is a specific substrate-binding component, indicated a possible solution to the problem of the varying stabilities of different proteins—they might be specifically recognized and targeted by different ligases.

The 26S proteasome is composed of two sub-complexes: a 20S core particle (CP) that carries the catalytic activity, and a regulatory 19S regulatory particle (RP). The 20S CP is a barrel-shaped structure composed of four stacked rings, two identical outer a rings and two identical inner b rings. The eukaryotic a and b rings are composed each of seven distinct subunits. The catalytic sites are localized to some of the b subunits. Each extremity of the 20S barrel can be capped by a 19S RP each composed of 17 distinct subunits, nine in a "base" sub-complex, and eight in a "lid" sub-complex. One important function of the 19S RP is to recognize ubiquitinated proteins and other potential substrates of the proteasome. Several ubiquitin binding subunits of the 19S RP have been identified, although their biological roles and mode of action have not been discerned. A second function of the 19S RP is to open an orifice in the ring that will allow entry of the substrate into the proteolytic chamber. Also, since a folded protein would not be able to fit through the narrow proteasomal channel, it is assumed that the 19S particle unfolds substrates and inserts them into the 20S CP. Both the channel opening function and the unfolding of the substrate require metabolic energy, and indeed, the 19S RP "base" contains six different ATPase subunits. Following degradation of the substrate, short peptides derived from the substrate are released, as well as reusable ubiquitin (Figure 4-2).

Figure 4-2　Degradation of ubiquitinated proteins (see page 109)

Ubiquitin-proteasome system is closely related with control of the protein quality, cell cycle, DNA repair, transcription and immune stress, as well as development of many diseases.

Section 2　Proteome and Proteomics

2.1　Concept of Proteome and Proteomics

The term proteome was first coined to describe the set of proteins encoded by the genome. The study of the proteome, called proteomics, now evokes not only all the proteins in any given cell, but also the set of all protein isoforms and modifications, the protein-protein interactions, the structural description of proteins and their higher-order complexes. It can be considered almost everything 'post-genomically'. We will use proteomics in an overall sense to mean protein biochemistry on an unprecedented, high-throughput scale. Proteomics complements other functional genomic approaches, including microarray-based expression profiles, systematic phenotypic profile at the cell and organism level, systematic genetics and small-molecule-based arrays. Integration of these data sets through bioinformatics will yield a comprehensive database of gene functions that serve as a powerful reference of protein properties and functions, and a useful tool for the individual researcher to both build and test hypotheses. Moreover, large-scale data sets will be crucial for the emerging field of systems biology. Proteomics would be impossible without the previous achievements of genomics, which provided the 'blueprint' of possible gene products that are the focal point of proteomic studies. Unlike the scalable exercise of DNA sequencing, with its attendant enabling technologies such as the polymerase chain reaction and automated sequencing, proteomics must deal with unavoidable problems of limited and variable sample material, sample degradation, vast dynamic range (more than 106 fold for protein abundance alone), a plethora of post-translational modifications, almost boundless tissue, developmental and temporal specificity, and disease and drug perturbations.

2.1.1 Mass Spectrometry-Based Proteomics

The ability of mass spectrometry to identify ever smaller amounts of protein from increasingly complex mixtures is a primary driving force in proteomics. Initial proteomic efforts relied on protein separation by two-dimensional gel electrophoresis, with subsequent mass spectrometric identification of protein spots. An inherent limitation of this approach is the depth of coverage, which is necessarily constrained to the most abundant proteins in the sample. The rapid developments in mass spectrometry have shifted the balance to direct mass spectrometric analysis, and further developments will increase sensitivity, robustness and data handling (Figure 4-3).

Figure 4-3　Mass-Spectrometry of a peptide (see page 110)

2.1.2 Array-Based Proteomics

More recently, various protein-array formats promise to allow rapid interrogation of protein activity on a proteomic scale. These arrays may be based on either recombinant proteins or, conversely, reagents that interact specifically with proteins, including antibodies, peptides and small molecules. Readouts for protein-based arrays can derive from protein interactions, protein modifications or enzymatic activities. A current challenge is to effectively couple highend mass spectrometry to array formats. Array-based approaches can also use in vivo readouts, for example in the systematic analysis of protein localization in the cell through green fluorescent protein.

2.1.3 Structural Proteomics

A full description of cell behaviour necessitates structural information. This all-encompassing structural endeavour spans several orders of magnitude in measurement scale and requires a battery of structural techniques, from X-ray crystallography and nuclear magnetic resonance (NMR) at the protein level, to electron microscopy of mega-complexes and electron tomography for high-resolution visualization of the entire cellular milieu.

2.1.4 Clinical Proteomics

Proteomics is set to have a profound impact on clinical diagnosis and drug discovery. Since most drug targets are proteins, proteomics will significantly benefit to drug discovery, clinical practice. An understanding of the biological networks that lie below the cell exterior will provide a rational basis for preliminary decisions on target suitability.

2.2　Disease Proteomics

Application of proteomics in human diseases may include: delineation of altered protein expression, not only at the whole-cell or tissue levels, but also in subcellular structures, in protein complexes and in biological fluids; the development of novel biomarkers for diagnosis and early detection of disease; and the identification of new targets for therapeutics and the potential for accelerating drug development through more effective strategies to evaluate therapeutic effect and toxicity.

2. 2. 1   Beyond Two-Dimensional Gels for Expression Profile

Even with the improvement, two-dimensional gels still have limitation with a rather low-throughput and requirement of a relatively large amount of samples. The latter is a particularly problem for clinical samples, which are generally procured in limited amounts. Furthermore, tissue heterogeneity complicates the analysis of clinical samples. Various tissue microdissection approaches are beneficial to reduce heterogeneity, but they further reduce the amount of sample available. In particular, the use of laser-capture microdissection, which allows defined cell type to be isolated from tissues, yields amounts of proteins that are difficult to reconcile with the need for greater amounts for two-dimensional gels. Undoubtedly, various non-gel-based schemas that rely on liquid-based separations of proteins or peptides, with or without tagging, will have utility for disease proteomics, particularly given their potential for automation.

Non-separation-based strategies, including direct profiling using mass spectrometry or the use of protein microarrays, are important developments. Mass spectrometry has been applied to the in situ proteomic analysis of tissues, an approach that allows imaging of protein expression in normal and disease tissues. By this method, frozen tissue is sliced and sections are applied on a matrix-assisted laser desorption/ ionization (MALDI) plate and analyzed at the regular spatial intervals. The mass spectra obtained at different intervals are compared, yielding a spatial distribution of individual masses across the tissue section. Mass profile of tissue sections obtained from normal and disease tissues may be compared to detect altered protein expression. Tumor analyses using this approach have uncovered differences in protein expression between normal and tumor tissues that may have specificity for different tumor types.

2. 2. 2   DNA Versus Protein Microarrays

The DNA microarray studies described above, as well as numerous others in the literature, indicate the great utility of DNA microarrays for uncovering patterns of gene expression that are clinically informative. An important challenge for microarray analysis of disease tissues and cells is to understand at a mechanistic level the significance of associations between subsets of genes and clinical features of diseases. Another challenge is to identify the smallest but most informative sets of genes associated with specific clinical features, which then can be interrogated using technologies available in clinical laboratories. Yet another challenge is to determine how well RNA levels of predictive genes correlate with protein levels. A lack of correlation may imply that the predictive property of the gene (s) is independent of gene function. For example, comparisons of mRNA and protein levels for the same tumors reported for lung cancer demonstrated that only a small percentage of genes had a statistically significant correlation between the levels of their corresponding proteins and mRNAs.

Technologies for DNA microarray analysis are still evolving. There is a tendency by manufacturers to favor oligonucleotide over cDNA-based microarrays. Nevertheless, however perfected DNA microarrays and their analytical tools become for disease profiling, they will not eliminate a pressing need for other types of profiling technologies that go beyond measuring RNA levels, particularly for disease-related investigations. DNA microarrays have limited utility for the analysis of biological fluids and for uncovering assayable biomarkers directly in the fluids. Numerous alterations may occur in proteins that are not reflected in changes at the RNA level, providing a compelling rationale for direct analysis of gene expression at the protein level. As a result, there is substantial interest in developing microarrays or biochips that allow the systematic analysis of proteins. Unlike DNA microarrays, which provide one measure of gene expression (namely RNA levels), there is a need to implement protein microarray strategies that address the many different features of proteins that can be altered in disease. These include not only determination of their levels in biological samples and, but also determination of their selective interactions with other biomolecules, such as other proteins, antibodies, drugs or various small ligands. For assays of protein interactions, biochips that contain either peptides or proteins are being produced. Peptides may be synthesized in very large numbers directly on the chip. Alternatively, recombinant proteins may be arrayed and effort is underway to assemble large sets of purified recombinant proteins for microarrays and other applications.

Profiling studies of disease tissues that have used protein microarrays are beginning to emerge. These may provide a better understanding to the expression patterns of proteins under the certain tissue microenvironment. A reverse-phase protein array approach that immobilizes the whole repertoire of a tissue's proteins has been developed. A high degree of sensitivity, precision and linearity was achieved, making it possible to quantify the phosphorylated status of signal proteins in subpopulations of human tissue cells. A clinically relevant application of protein microarrays is the identification of proteins that induce an antibody response in autoimmune disorders. Microarrays were produced by attaching several hundred proteins and peptides to the surface of derivatized glass slides. Arrays were incubated with the patient's sera, and fluorescent labels were used to detect autoantibody binding to specific proteins in autoimmune diseases, including systemic lupus erythematosus and rheumatoid arthritis. Such microarrays represent a powerful tool to study immune responses in a variety of diseases, including cancer.

One of the main challenges in making biochips for the global analysis of protein expression is the current lack of comprehensive sets of genome-scale capture agents such as antibodies. Another important consideration in protein microarrays is that proteins undergo numerous post-translational modifications that may be crucial to their functions. However, these modifications are generally not able to be captured using either recombinant proteins or antibodies that do not distinctly recognize specific forms of a protein.

2.2.3　The Quest for Disease Biomarkers Using Proteomics

There is substantial interest in applying proteomics to the identification of disease biomarkers. Approaches include comparative analysis of protein expression in normal and disease tissues to identify aberrantly expressed proteins that may represent new markers, analysis of secreted proteins in cell lines and primary cultures, and direct serum protein profiling. The potential of mass spectrometry to yield comprehensive profiles of peptides and proteins in biological fluids without the need to first carry out protein separations has attractive interest. In principle, such an approach is highly suited for marker identification because of reduced sample requirement and high throughput.

This approach is currently popular, particularly for serum analysis. Microlitre quantities of serum from many samples are applied to the surface of a protein-binding plate, with properties to bind a class of proteins. The bound proteins are treated and analyzed by MALDI. The mass spectra patterns obtained for different samples reflect the protein and peptide contents of these samples. Patterns that distinguish between cancer patients and normal subjects with remarkable accuracy have been reported for several types of cancer. The main drawbacks of direct analysis of tissues or biological fluids by MALDI are the preferential detection of proteins with a lower molecular mass and the difficulty in determining the identity of proteins owing to post-translational modifications obscuring the correspondence of measured and predicted masses.

A productive approach for the identification of cancer markers has been the analysis of serum for autoantibodies against tumor-specific proteins. There is increasing evidence for an immune response to cancer in humans, demonstrated in part by the identification of autoantibodies against a number of intracellular and surface antigens detectable in sera from patients with different cancer types. The identification of panels of tumor antigens that elicit an antibody response may have utility in cancer screening, diagnosis, prognosis, and in immunotherapy against the diseases as well.

There are several approaches for the detection of tumor antigens that induce an immune response. A number of antigens have been detected by screening expression libraries with patient sera by using a random peptide-library approach. Multiple proteins that induce cancer-specific autoantibodies have been identified using two-dimensional gels to separate tumor proteins, followed by western blotting and incubation with patient sera. For most antigenic proteins identified using this approach, post-translational modifications contributed to the immune response.

2.2.4　Disease Related Functional Proteomics

Although data obtained by various expression proteomics strategies have functional relevance by uncovering altered levels or posttranslational modification states of proteins in diseases, other technologies are needed for more direct functional analysis. This is exemplified by the need to analyze protein complexes and their disruption in diseases, to assay the activity of various classes of proteins in a high-throughput fashion, and to manipulate the levels and activities of individual proteins, in a cellular context, to determine their role in different biological processes and disease states.

Various strategies are currently in use for studies of protein complexes and protein-protein interactions. So far, such strategies have been applied to disease investigations on a limited basis, with a relatively narrow focus on particular complexes. For example, we can use affinity pull-down assays, two-dimensional gels and mass spectrometry for this strategy. Clearly, at the present time, most systematic studies of protein-protein interactions have dealt with normal or physiological states, as the field is still in its early stages and the merits of various technologies, particularly for disease investigations, have yet to be fully appreciated.

2.2.5 Contribution of Proteomics to Studies of Pathogens

A complicating factor in therapy for infectious diseases is the development of resistance to commonly used drugs, which heightens the need for developing effective new therapies. Interest in the application of proteomics to microbiology goes back at least two decades, with the pioneering work of Fred Neidhardt to characterize protein expression patterns in *E. coli* under different growth conditions. The complete sequencing of a number of microbial genomes has provided a framework for identifying proteins encoded in these genomes using mass spectrometry. A case in point is the sequencing of the genome of the malaria parasite *Plasmodium falciparum*, which has provided a basis for conducting comparative proteomic studies of this pathogen, leading to the identification of the new potential drug and vaccine targets.

2.2.6 Contribution of Proteomics to Drug Development

There is currently a burgeoning interest in proteomics on the part of the pharmaceutical industry, evidenced by implementation of proteomic programs by most major pharmaceutical companies. The notion has been advanced that, as the vast majority of drugs target proteins, proteomics should have substantial utility for drug development. But the industry has so far adopted a cautious attitude, and it is too early to make a critical assessment of the contribution of proteomics to drug development, relative to other approaches. Provided suitable technological platforms become available, the use of proteomics may permeate numerous aspects of drug development, by identifying new targets and facilitating assessment of drug action and toxicity in the preclinical and clinical phases.

Section 3 Techniques of Proteomics

3.1 Protein Separation Technology

Proteomics is the analysis and detection of proteome, that is, a large-scale biological research for proteins. The identification of proteins became accurate and convenient now with the emergence of mass spectrometry detection technology. Proteomics was mainly used in four aspects. The first is the detection of whole proteins of sample. The second is the detection of protein expression. The third is the detection of protein interactions. The forth is the detection of protein modifications. (The contents of this section see Chapter 10)

The first procedure is the protein separation. The samples for proteomics analysis are always a variety of multi-protein mixture. Mass spectrometry detectors can only be used to detect and identify peptides. Therefore, before mass spectrometry is carried out, protein mixture should be separated into a single protein or the complex containing several proteins. And then the separated proteins were digested to peptide in order to adapt to machine's working condition. Two-dimensional gel electrophoresis is one of the most effective methods for a separation of highly complex protein mixtures.

3.1.1 Isoelectric Focusing

Protein is the amphipathic molecule, which maybe carry a positive charge, negative charge or no charge in different pH environment. The gradient of pH is critical to isoelectric focusing technique. When the protein molecules exist in a pH gradient and electric field, they move to the place where its charge becomes to zero. When each protein of samples arrives at position where its charge becomes to zero, proteins has been isolated on the basis of their isoelectric point.

3.1.2   SDS Polyacrylamide Gel Electrophoresis

SDS is anionic detergent, which is mixed into proteins to form a necklace-like structure. SDS is bounded by proteins according to the ratio of 1.4 : 1 to make proteins load a net negative charge. Thus, the mobility of protein in electrophoresis is proportionally paralleled with its molecular weight.

3.1.3   Two-Dimensional Gel Electrophoresis

Two-dimensional gel electrophoresis is efficiently and widely used to analyze complex protein mixture in the cell, tissue or other biological samples. The first phase is isoelectric focusing electrophoresis, which allows proteins to be separated according their different isoelectric point. The second phase is the SDS polyacrylamide gel electrophoresis, which allows proteins to be separated according their different relative molecular weight. Every point in two-dimensional protein map represents one kind of protein in the sample or several. The polyacrylamide gel electrophoresis containing ampholyte limits the stability and reliability in two-dimensional gel electrophoresis. Instead of the ampholyte, the fixed pH gradient gel greatly enhances the resolution and stability of isoelectric focusing.

Two-dimensional gel electrophoresis has defects too. First of all, it is difficult to reproduce the same results in different two-dimensional gels. Secondly, it is hard to separate some macromolecules, hydrophobic protein. Two-dimensional gel electrophoresis is good at analysis of high content proteins. In the practical application, other protein separation technologies are sometimes used, which include one-way electrophoresis, isoelectric focusing electrophoresis, high performance liquid chromatography and capillary electrophoresis and so on.

3.2   Mass Spectrometry

Before mass spectrometry, proteins should be digested into small molecule polypeptide to improve the precise rate of mass spectrometry. The greater protein has the lower precise rate. The most common digestive enzyme is trypsin. At present, mass spectrometry has mainly two kinds of methods: Matrix-Assisted Laser Desorption, Ionization (MALDI-TOF-MS) and electrosprayion-izationmassspectrometry (ESI-MS).

3.2.1   MALDI-TOF-MS

MALDI-TOF-MS transfer polypeptide into ionic signal, and then the polypeptide are analyzed according to the ratio of quality and charge to judge this polypeptide. The instrument includes two parts: MALDI and TOF. This method is handy and sensitive, and can be used for and matched with many methods of protein isolation. Now, the database has enough data about the ratio of charge and quality of polypeptides.

3.2.2   ESI-MS

ESI-MS turns the analysis under liquid state, and polypeptide ion contains many charges. The liquid polypeptide mixture is isolated by using HPLC, and then passed through one fine pinhole below high pressure. When sample is shot by pinhole, it is become the petty droplet. These petty droplets include polypeptide ion, water and other foreign materials. After these foreign materials were removed, polypeptide ion enters Tan-Dem mass analyzer, which chooses some specific quality polypeptide ions to disintegrate them into different ionization or not ionization extract through colliding. Subsequently, Ion-spectrum of the above extract is drawn according to the ratio of charge and quality. The amino acid sequence is obtained according to the ion-spectrum by using database.

3. 2. 3　The Applications of Mass Spectrometry in Proteomics

3. 2. 3. 1　Detection of the whole proteins

This is to discover and identify proteins as much as possible. However, cells or organisms only express a part of proteome in a certain period of time, many proteins are not expressed. It makes the whole protein very difficult to detect.

3. 2. 3. 2　Detection of expression of proteins

The expression of proteome may change with the diversity of development, differentiation, disease and other physiological and pathological conditions. So, the detection of protein expression can help us discover the diagnostic and treatment approaches of diseases in different physiological and pathological conditions. Now people always compare the differences of protein expression at different states to identify the specially expressed protein by using the two-dimensional gel electrophoresis or liquid chromatography.

3. 2. 3. 3　Detection of protein interactions

Proteins usually form multi-protein complex by interactions for their special functions. To learn the special functions of proteins, it is necessary to firstly understand the composition of protein complexes. Now the yeast two-hybrid system and immunoprecipitation are primarily used.

3. 2. 3. 4　Detection of protein modifications

The proteins of organisms mainly exist in the modified formation, such as post-translational cleavage, phosphorylation, glycosylation and oxidation, etc. Because the modified proteins have a changed ratio of charge and quality, mass spectrum can be used as one convenient and efficient method for testing a specific amino acid modified manner in the protein sequences.

3. 3　The Technology of Protein Interactions

Proteomics is used to study the organism dynamically and wholly at the protein level. It is not a closed, conceptualizational and stable knowledge system, but rather a field which seeks to clarify the protein expression and functional patterns of the whole organism. The research of functional mode is currently focused on the network relations of protein interactions. Now the ways to study the protein interactions are as following.

3. 3. 1　Yeast Two-Hybrid System

Yeast two-hybrid system, also is called as protein capture system. It is a very effective method of molecular biology to study the protein interactions. Eukaryotic transcription activator generally contains two different domains: DNA binding domain (BD) and DNA transcription activation domain (AD). Two domains are independent of each other but also interdependent on function. They have the complete activity as a transcriptional activator when they combine with each other. The target protein gene is combined to AD sequences, while the gene encoding bait protein is combined to BD sequences. Then two fusion genes are formed. When the two fusion genes are expressed in one bacterium, the target protein would interact with bait protein and form a complete transcription activator with activation. As a result, according to the report gene expressing or not, we could determine if the target protein interact with bait protein.

3. 3. 2　Phage Display

Phage display technology is used to screen peptides which are expressed on the face of phage, and is also used to screen and transform the functional peptides. Using this technology, the exogenous protein or peptide gene sequence can be connected with the gene which encoded the envelope protein of the phage (for example: gene sequences of monoclonal antibody). So, the exogenous gene will be expressed with the expression of envelope protein, and the exogenous protein will be displayed on the face of phages. Because the target protein is fixed on the column, if the phages are flown through the column, the target proteins (antibodies) are combined specially, and the exogenous proteins or peptides will be picked out. As the main characteristics of the technology is that the genotype and phenotype of particular molecular are put into same virus particles, that is to display special protein on the face of phages and the target gene to be contained in the DNA of the phage. So the phenotype can be perfectly associated with the genotype by using this technology. And it can solve the problem of yeast two-hybrid system. Phage display technology have two significant meaning: One is that it derived out the concept of molecular repertoires. The other is that the technology overcomes the restrictions, which we must base on the detailed knowledge of their structure when the protein interactions are studied. It is why the technology is more widely used. At present, the phage display technology is used to construct a random peptide library, proteins library and antibody library. And it is also used to study the binding sites of receptor or antibody. It can be used to transform and improve the biological and immunological characters of proteins, enzymes and antibodies. And it is also used to study the new peptide drugs, vaccines and antibodies. Phage display has widely perspective.

3.3.3 Surface Plasmon Resonance

The application of Surface Plasmon Resonance (SPR) is that the interaction of molecules is analyzed by utilizing the physics optical phenomena that is produced by the total reflection of metal film/liquid interface. Since Nylander firstly used the technology of SPR in the field of immune sensor on 1982, surface plasmon optical biosensors has been deeply studied and widely applied. Currently, SPR technology has become a new way to study protein interactions. SPR biosensor is that the specific binding is detected between receptor protein and the ligand in the liquid by using the phenomenon of surface plasmon resonance and the characteristics that SPR peaks is sensitive to changes of electrolyte on the surface of the metal, after receptor is fixed in the metal membrane. SPR technology is characterized by rapid determination, security, no markers or dyes and high sensitivity. In addition to the detection of protein, SPR can detect the interaction between proteins and nucleic acids or other biological macromolecules, and monitor the whole process at real time.

3.3.4 Protein Chip

Protein chip is a high-throughput, rapid, efficient, miniaturizational and automated technology to analyze proteins. One of the protein chips is similar to the DNA chip, that is, probe proteins are high-density fixed on the face of solid support. It can capture the target protein from the samples, and then qualitatively or quantitatively analyze the target protein through the detector. Therefore, the protein chip can analyze the changes of thousands proteins at the same time, and also made it possible to study the function (for example, enzyme activity, specificity of antibodies, interactions of receptor and ligand , binding of between proteins and proteins or nucleic acids) of proteins in the level of total genome. At this stage, in addition to study protein interactions, protein chips are also widely used in disease diagnosis, gauging efficacy of therapy, finding a new target of drugs or poison and its mechanism and so on.

3.3.5 Bioinformatics

One of the important research areas of bioinformatics is to identify protein function. It has been widely used to identify the protein function through homologous analysis. It is based on that the family of homologous proteins from the same ancestor has similar sequences and structures, correspondingly also has similar functions. Recently, some new methods have been developed to identify the protein function using computer. The basis of these new methods is not the protein homology but the same features, such as the mode of system development, mRNA expression patterns and domain integration mode. The proteins with same characteristics have the association of the functions and direct effect. The association and effect are that the proteins are involved in the same metabolism pathways or signal transduction pathways. These methods can be applied to identify the function of a large number of unknown proteins. Future research goal is to further enhance the coverage, accuracy and precision of the computer-based analysis (You can see the detail on Chapter 11).

3. 4　Protein Microarrays and Proteomics

Proteomics is characterized by the use of high-resolution protein separation tools, combined with efficient protein identification technology, panoramicly researching the protein spectrum at various specific cases. A complete proteome analysis will include: determination, modification, activity, location, as well as the interactions of all the protein in samples. In practice, the current technology limits our assay, and we can only get one or two parameters of some proteins. This type of experiment is identical with expression spectra analysis. We call it as quantitative proteomics. Functional proteomics is different from it, that is, to define and study the function of each protein in samples. This could include the interactions of enzyme and substrate, as well as among proteins. Chip technology has a significant impact on quantitative proteomics and functional proteomics, and provides a method to analyze and screen proteins rapidly.

3. 4. 1　Quantitative Proteomics

Different from the universality and uniformity of the genome, there are diversity on the time and space for the proteome, that is, the proteome has variety in the different cells of same specie or in the different growth periods of same cell. Therefore, it is impossible to accurately describe the states of all proteins and the related proteome in the cell. At present, comparative proteomics is the key way to study diseases. Through comparing the differences of proteins expression between normal and abnormal tissue, the closely related proteins with human diseases are found, and then determine the targets. Quantitative proteomics is one of the main researches in comparative proteomics. The use of protein chip technology to study the quantitative proteomics, will further promote the study of proteome from the static, qualitative and constitutional analysis to a dynamic, quantitative and functional study.

3. 4. 2　Functional Proteomics

Genome sequencing projects contribute daily to lists of the protein components of cells and organisms; determining what these components do is the task of the functional proteomics. One technique that has been used extensively is the yeast two-hybrid system. This in vivo assay, although easy to implement and of considerable utility, has several limitations. Proteins that function as transcriptional activators yield false positives when fused to DNA-binding domains, whereas false negatives arise when proteins are displayed inappropriately or fail to fold correctly in yeast. More significantly, the investigator is unable to control either the post-translational modification state of the proteins under investigation or the environment under which the interactions are being studied.

Recently, immunoprecipitation coupled with mass spectrometry has been used to identify multi-protein complexes on a large scale in *Saccharomyces cerevisiae*. This technique can be adapted readily to high-throughput investigation, can identify interesting proteins associated with an epitope-tagged bait protein. Although this is a very efficient way to identify protein complexes, it has the caveat that a single bait protein may occur in more than one complex in a cell. It may therefore bring down two or more proteins that never actually colocalize, giving the illusion that protein complexes are bigger and more elaborate than they really are. As a complement to these approaches, protein microarray technology provides a well-controlled, in vitro way to study function on a system-wide or genome-wide basis. For this application, the proteins themselves, rather than affinity reagents, are arrayed on a solid support.

An advantage of studying proteins in an array format is that the investigator can control the conditions of the experiment. This includes not only factors such as pH, temperature, ionic-strength and the presence or absence of cofactors, but also the modification states of the proteins. Although low-density arrays such as these are useful, there is substantial benefit to be gained from using microarray technology. First of all, in principle thousands of proteins can be spotted on a single slide or similar support, enabling one to interrogate simultaneously the function of many different proteins with minimal sample consumption. Secondly, hundreds or even thousands of copies of an array can be fabricated in parallel, enabling the same proteins to be probed repeatedly with many different molecules under many different conditions.

Protein function microarrays will benefit undoubtedly from improved methods of fabrication, processing and analysis, but at present the greatest obstacle barring their widespread use is the production of large collections of pure, recombinant proteins. Such production requires, as a necessary first step, the cloning of cDNAs. One approach has been described, in which cDNA libraries are screened for protein-producing clones using high throughput methods. Although this is a potentially economical way to generate large collections of proteins and has been used for some discovery-oriented applications, it is limited in several respects. The resulting libraries are not normalized, they fail to contain genes that are expressed at low abundance, and the proteins often begin and end at random positions in the coding sequence. With the availability of whole-genome sequence data, systematic efforts are now underway to prepare perfectly normalized, indexed collections of full-length open reading frames in recombination-based cloning vectors. Resources such as these, coupled with high-throughput methods of expressing and purifying recombinant proteins, should accelerate greatly the application of microarray technology to functional proteomics.

Each approach comes with both advantages and limitations. For protein function microarrays, the purity and integrity of the proteins need to be determined, and their concentrations need to be normalized. Perhaps the biggest challenge is to find a feasible way to assess the functional states of purified proteins. Solubility alone, although informative, does not indicate that a protein is folded correctly or processed appropriately. By paying close attention to these issues, data generated by both systems-oriented analyses and discovery-oriented expeditions should lead us ultimately toward a more integrated understanding of biological processes.

Summary

This chapter gives some introductions about the basic concepts, the studies, the research methods of proteomics from the three aspects: first of all, focusing on proteomics research, introduced the main contents of the basic knowledge of the protein, including domain, folding, secretion and degradation; second, introduced protein and proteomics as well as the basic concepts in the study of disease; finally, introduced some basic principles and applications of some important methods used in the research of proteomics, including the two-dimensional gel electrophoresis and mass spectrometry and chips etc.

(Junzhu Wu)

第五章　基因表达的调控

近年来，生命科学研究揭示，一切生命现象从生物的遗传、变异到生物体的生长、发育、繁殖、分化以及包括癌变在内的许多疾病发生，都与基因表达调控有关，因此形成了众多的热点探索课题。在一定程度上可以说基因表达的调控是分子生物学的真谛所在。目前，总体来看对生物体复制—转录—翻译的过程是清楚的，然而这些过程是怎样调节控制的，并不十分清楚。真核生物同一机体的各种细胞都含有相同的遗传信息，即有相同的结构基因，但它们在各细胞中并非同时表达，而是按一定时间、空间、有序的表达，由此产生不同细胞的分化。分化本身就是基因表达调控的结果。

第一节　基因表达调控的基本概念

一、基因表达的概念

基因表达通常是指生物基因组中结构基因所携带的遗传信息经转录、翻译等一系列过程，合成特定的蛋白质，进而发挥其特定的生物学功能和生物学效应的全过程。但并非所有基因表达过程都产生蛋白质，rRNA、tRNA 的编码基因转录生成 RNA 的过程也属于基因表达。基因表达可以在转录、加工和翻译多个水平受到调控。转录水平的调控是基因表达的基本控制点。基因的转录调控是通过反式作用因子（*trans*-acting factor）和顺式作用元件（*cis*-acting element）之间的相互作用来进行。反式作用因子通常为蛋白质（也可能是 RNA），它可以在细胞内扩散，因此可以作用于任何合适的靶基因。顺式作用元件通常是 DNA，不需要转变为其他形式。一般它只影响与其邻近的 DNA 序列。

二、基因表达的特异性

基因表达表现为严格的规律性，即时间、空间特异性。生物物种越高级，基因表达规律越复杂、越精细，这是生物进化的需要及适应。基因表达的时间、空间特异性由特异基因的启动子（序列）和（或）增强子与调节蛋白相互作用决定。

1. 时间特异性

按功能需要，某一特定基因的表达严格按特定的时间顺序发生，这就是基因表达的时间特异性（temporal specificity）。在多细胞生物从受精卵到组织、器官形成的各个不同发育阶段，都会有相应基因严格按一定的时间顺序开启或关闭，表现为与分化、发育阶段一致的时间性。因此，多细胞生物基因表达的时间特异性又称阶段特异性（stage specificity）。

2. 空间特异性

在个体生长全过程中，某种基因产物在个体中按不同组织空间顺序出现，这就是基因表达的空间特异性（spatial specificity）。基因表达伴随时间或阶段顺序所表现出的这种空间分布差异，实际上是由细胞在器官中的分布决定的，因此，基因表达的空间特异性又称细胞特异性（cell specificity）或组织特异性（tissue specificity）。

在多细胞生物个体某一发育、生长阶段，同一基因产物在不同的组织器官表达多少是不一样的；在同一生长阶段，不同的基因在不同的组织、器官表达分布也不完全相同。

三、基因表达的方式

不同的基因对内、外环境信号刺激的反应性不同。按对刺激的反应性，基因表达的方式或调节类型存在很大差异。

1. 组成型表达

某些基因产物对生命全过程都是必需的或必不可少的，这类基因在生物个体的几乎所有细胞中持续表达，通常称之为持家基因（housekeeping gene）。这类基因在组织细胞中呈现持续表达，是维持细胞基本生存的需要，这类基因表达被称为细胞基本的或组成型基因表达（constitutive gene expression），其表达只受启动序列或启动子与 RNA 聚合酶相互作用的影响，而不受其他机制调节。例如，催化三羧酸循环各阶段反应的酶的编码基因就属这类基因。

2. 诱导和阻遏表达

与持家基因不同，另有一些基因表达极易受环境变化影响。在特定环境信号刺激下，相应的基因被激活，基因表达产物增加，即这种基因是可诱导的，该基因称为可诱导基因。可诱导基因在特定环境中表达增强的过程称为诱导（induction）。相反，在特定环境信号刺激下，如果相应的基因对环境信号应答时被抑制，即这种基因是可阻遏的，该基因称为可阻遏基因。可阻遏基因在特定环境中表达产物水平降低的过程称为阻遏（repression）。可诱导或可阻遏基因除受到启动序列或启动子与 RNA 聚合酶相互作用用的影响外，还受其他机制调节。这类基因的调控序列含有特异刺激的反应元件，如乳糖操纵子、色氨酸操纵子。诱导和阻遏是同一事物的两种表现形式，在生物界普遍存在，也是生物体适应环境的基本途径。

在一定机制控制下，功能上相关的一组基因，无论其为何种表达方式，均需协调一致、共同表达，即协调表达（coordinate expression），这种调节称为协调调节（coordinate regulation）。

四、基因表达调控的生物学意义

1. 适应环境、维持生长和增殖

生物体赖以生存的内、外环境是不断变化的。所有活细胞都必须对内、外环境变化作出适当反应，以使生物体能更好地适应变化的环境。通过一定的程序调控基因的表达，可使生物体表达出合适的蛋白质分子，以便更好地适应环境，维持生长和增殖。

2. 维持个体发育与分化

在多细胞生物生长、发育的不同阶段，细胞中的蛋白质分子种类和含量差异很大；即使在同一生长发育阶段，不同组织器官内蛋白质分子分布也存在很大差异，这些差异是调节细胞表型的关键。多细胞生物尤其是高等哺乳动物的各种组织、器官的发育与分化都是由一些特定基因控制的。当某种基因缺陷或表达异常时，则会出现相应组织、器官的发育与分化异常。

第二节　原核生物基因表达的调控

一、原核生物基因表达调控的特点

原核生物是单细胞生物，基因组由一条环状双链 DNA 组成，由于无核小体结构，无核膜，故 DNA 转录和 mRNA 翻译在同一时间和空间上进行（转录和翻译偶联）。原核生物与周围环境的关系非常密切，因本身无足够的能源储备，在长期的进化过程中演变出了高度适应性和高度的应变能力。原核生物必须不断地调节各种不同基因的表达，以适应周围环境、营养条件的变化（碳源、氮源等）和对付不利的理化因素（高温、射线、重金属、烷化剂等）。在反应中，细菌可迅速合成自身需要的酶、核苷酸和其他生物大分子，同时又能迅速地停止合成并降解那些不再需要的成分，使细菌的主要功能——生长、繁殖达到最优化。

原核生物细胞结构的特征及操纵子表达调控方式（下面将详细介绍）都与上述表达调控的特点相适应。转录的起始、终止和 mRNA 快速转换是细菌基因调控的三要素，细菌的大多数基因表达调控是在转录水平上进行的。

二、转录水平的调控

（一）RNA 聚合酶对转录起始的调控

转录的第一步是 RNA 聚合酶（RNA polymerase）与启动子结合。启动子（promoter）是 DNA 分子上 RNA 聚合酶识别、结合并起始转录的部位。原核生物只有一种 RNA 聚合酶，催化三种 RNA 合成。大肠杆菌 RNA 聚合酶由 5 个亚基组成，即 $\alpha_2\beta\beta'\sigma$，分子质量约为 500 kDa，$\alpha_2\beta\beta'\sigma$ 又称全酶（holoenzyme），5 个亚基中 σ 亚基（σ 因子）与其他亚基结合较松散，很容易从全酶上脱下来，$\alpha_2\beta\beta'$ 称为核心酶（core enzyme），核心酶具有催化活性，使合成的 RNA 链延长；σ 亚基本身没有催化活性，其作用是识别 DNA 分子上 RNA 合成的起始信号。细胞内哪条 DNA 链被转录，转录方向与转录起点的选择都与 σ 因子有关，因此，称 σ 亚基为起始因子。不同的 σ 因子可以竞争结合核心酶。环境变化可诱导产生特定的 σ 因子，从而开启特定的基因。例如，大肠杆菌在一般环境中发生作用的是 σ70，环境中温度改变可诱导产生 σ32，σ32 能识别热应激蛋白启动子，导致热应激蛋白的合成，产生热应激反应。

（二）操纵子水平的调控

原核生物基因表达调控主要发生在转录水平，而转录调控的基本单元是操纵子。所谓操纵子（operon）是指数个功能相关的结构基因串联在一起，受上游调控元件控制，形成的转录单位。操纵子转录的产物为 mRNA 分子，而这种 mRNA 分子上带有编码几种蛋白质的信息，可作为合成几种蛋白质的模板，所以这种 mRNA 也称为多顺反子 mRNA（polycistronic mRNA）。

在细菌生命周期的某一时刻，并非全部潜在的启动子都可以利用。RNA 聚合酶使用哪个启动子或哪个操纵子主要由细菌赖以生存的培养基里的营养成分决定。例如，将乳糖、半乳糖和阿拉伯糖转化为葡萄糖需要三组不同的酶类，这些酶分别由三个操纵子的基因编码。细菌根据培养基所含糖种类的不同，使用相应的操纵子。

由底物导致合成利用该底物的酶，这种现象称为酶诱导（enzyme induction），这个底物叫做诱导物（inducer），一旦除去诱导物，酶的合成就会很快终止。酶诱导在细菌中普遍存在，是生物进化过程中出现的一种经济、合理的利用有限资源的本能。细菌能合成超过千种酶，如果没有底物可以利用，合成这么多酶是浪费，而有底物没有酶，底物也得不到利用。1965 年 Monod 和 Jacob 深入研究酶诱导现象后首先提出了操纵子学说，用表达调控的原理，揭示了酶诱导的本质。

1. 乳糖操纵子

乳糖操纵子（lac operon）由结构基因和调控元件两部分组成。结构基因 Z、Y、A 分别产生 β-半乳糖苷酶（分解乳糖成为半乳糖和葡萄糖）；透过酶（使外界乳糖等透过大肠杆菌细胞壁进入细胞内）；乙酰转移酶（能将乙酰辅酶 A 上的乙酰基转到半乳糖上，形成乙酰半乳糖）。调控元件：启动子（P）和操纵基因（operator，O）。P 区段内有 RNA 聚合酶结合位点和 cAMP-CAP 结合位点；O 区段为阻遏蛋白（repressor）结合位点；P 区上游有阻遏基因（inhibitor gene，I），其能编码阻遏蛋白，阻遏蛋白对基因表达起抑制作用。

从基因表达的角度来看，乳糖操纵子的表达顺序首先是以 RNA 聚合酶与 P 结合，经过 O，到达首尾相连（串联）的三个结构基因（lacZ、lacY、lacA），转录出一条多顺反子 mRNA，最终产生三种不同的蛋白质。但从基因调控的角度出发，结构基因是否转录为 mRNA 要受调控基因的控制，而阻抑物是否与 O 结合，又决定该基因的关闭或开启。

在没有乳糖的条件下，阻遏蛋白能与操纵基因结合。只要具有活性的阻遏蛋白结合到 O 位点，就可以阻止 RNA 聚合酶的转录活动。这是由于 P 位点和 O 位点有一定的重叠序列，O 被阻遏蛋白占据后，抑制 RNA 聚合酶与启动子结合，从而抑制结构基因 lac Z、lac Y、lac A 的转录。在有乳糖存在时，阻遏蛋白与乳糖结合，使阻遏蛋白的构象发生改变，以致不能与操纵基因结合而失去了阻遏作用，于是 RNA 聚合酶便能结合于 P 位点，从而引起结构基因转录。乳糖能诱导基因表达，因此，称乳糖为诱导剂；在体外实验中常用的诱导剂是异丙基硫代半乳糖苷（isopropyl β-D-1-thiogalactopyrano-side，IPTG）。在这个调节系统中，阻遏蛋白是主要的作用因子，而诱导物可以影响阻遏蛋白的活性；只有阻遏蛋白被诱导失活，结构基因才得以表达，这是一种负调控方式。

原核基因表达的正调控或负调控是按照没有调节蛋白的存在下，操纵子对于加入调节蛋白的反应情况来定义的。正调控（positive control）是指没有调节蛋白操纵时，基因是关闭的，当加入调节蛋白分子后，基因活性开启，能进行转录；相反，在无调节蛋白时，基因表达具转录活性，一旦加入调节蛋白则基因被关闭，转录受到抑制，这便是负调控（negative control）。负调控中的调节蛋白称为阻遏蛋白或阻抑物。原核生物也存在比较复杂的 cAMP-CAP 正调控方式。详见 cAMP-CAP 正调控系统。

2. 阿拉伯糖操纵子

当细菌细胞以阿拉伯糖作为生长所需的能源时，能产生三种酶，催化阿拉伯糖转变为 5-磷酸木酮糖，后者进入糖酵解途径。这三种酶分别是核酮糖激酶、阿拉伯糖异构酶和磷酸核酮糖差向异构酶。编码基因分别为 $araB$、$araA$、$araD$；此外，调控元件有 I1、I2、O1、O2 和 P。就像乳糖代谢一样，细菌为了代谢阿拉伯糖而合成新的酶。但是，我们现已知道，这两个操纵子的调节途径是很不一样的。阿拉伯糖操纵子（araoperon）上游有一个 $araC$ 基因，$araC$ 基因的产物 AraC 蛋白不同于乳糖操纵子的阻遏蛋白，AraC 蛋白对阿拉伯糖操纵子具有正调控和负调控双重作用。AraC 有两个结合位点，一个在 I1、I2 区，另一个在 O1、O2 区（约 −280 位置）。当阿拉伯糖存在时，该糖与 AraC 蛋白一起结合在 I1、I2 区，有利于 RNA 聚合酶与启动子结合从而促进阿拉伯糖操纵子的 $araB$、$araA$、$araD$ 结构基因的转录，显示正调控作用；当阿拉伯糖缺乏时，AraC 蛋白既与 I1、I2 区结合，又与 −280 区 O1、O2 区结合，以致 DNA 发生扭曲，这样影响了 RNA 聚合酶与启动子接近、结合，阻止了结构基因的转录，是典型的负调控形式，见图 5-1。

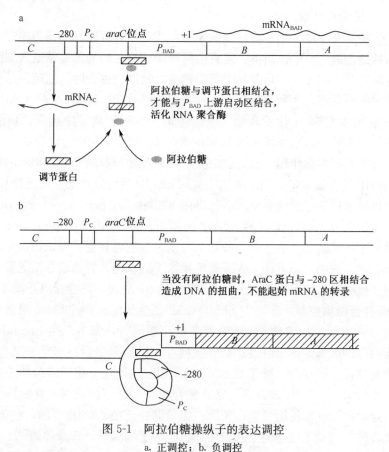

图 5-1　阿拉伯糖操纵子的表达调控
a. 正调控；b. 负调控

3. 色氨酸操纵子

前面所讨论的 lac 操纵子、ara 操纵子是编码分解代谢酶系的操纵子，编码的酶负责某一营养物的分解利用，是分解代谢过程，它们的表达只有在被分解的底物存在时才有意义。在细菌中还有负责某些物质合成代谢的操纵子，如色氨酸操纵子（trp operon），其编码的色氨酸合成酶负责细菌细胞内色氨酸的合成。色氨酸操纵子表达的调控有两种方式，一种是通过阻遏蛋白的调控；另一种则是通过衰减子作用（attenuation），在此仅介绍前者。

色氨酸操纵子在没有外源色氨酸（培养基中没有色氨酸）时，该操纵子表达，使细胞内有足够的色氨酸以进行蛋白质合成；而外源色氨酸存在（加入色氨酸后）时，细菌就不必自己合成了，这类操纵子就受到阻遏，则合成迅速停止。色氨酸操纵子的结构基因（A、B、C、D、E）编码 5 种酶，在色氨酸合成代谢中发挥作用；调控元件为 P、O。色氨酸操纵子阻遏蛋白是该操纵子 R 基因（repressor gene）的产物，它只有与色氨酸结合，才能成为有活性的阻遏蛋白，结合于 O 位点阻止转录。当色氨酸缺乏时，阻遏蛋白不能活化，阻遏解除。β-吲哚丙烯酸是色氨酸的竞争性抑制剂，它与阻遏蛋白结合后，阻止色氨酸与阻遏蛋白结合，因此解除阻遏而促进转录进行。在基因工程操作中，用色氨酸启动子（来自色氨酸操纵子）组建的载体表达目的基因时，用 β-吲哚丙烯酸可提高转录水平。

4. cAMP-CAP 正调控系统

许多微生物都专一地利用一种糖，但大肠杆菌等细菌可以利用葡萄糖也可以利用乳糖等。葡萄糖是细菌生长中最简单、最直接可利用的糖。因为葡萄糖进入细胞后，不需要产生任何新的酶。因此，当培养基中葡萄糖与乳糖同时存在时，细菌总是优先利用葡萄糖，直到葡萄糖耗竭，才利用乳糖。这种"葡萄糖效应"涉及 cAMP-CAP 的调控。

cAMP 是 20 世纪 50 年代发现的，现已清楚它在激素调节中起第二信使作用。60 年代，人们发现大肠杆菌培养液中葡萄糖的含量总是与 cAMP 的含量成反比。这并不是葡萄糖本身直接起抑制作用，而是它的分解代谢产物抑制腺苷酸环化酶的活性，进而使细胞内 cAMP 的含量降低。当培养基中加入 cAMP 后可以增加 β-半乳糖苷酶的产量。cAMP 的作用是通过和分解代谢基因活化蛋白（catabolite gene activitor protein，CAP）结合成复合体后完成的；CAP 又称 cAMP 受体蛋白（cAMP receptor protein，CRP），这种蛋白质是原核生物基因表达的一种正调控蛋白。它可将葡萄糖饥饿信号传递给许多操纵子，具有激活乳糖、半乳糖、麦芽糖等操纵子的功能，使细菌在缺乏葡萄糖的环境中可以利用其他碳源。从乳糖操纵子体外转录实验中发现，乳糖操纵子受 lac I 阻遏蛋白和 CAP 两种蛋白质控制，即处于 cAMP-CAP 复合体的正调控和 lac 阻遏蛋白负调控之中。在没有乳糖存在的情况下，不管葡萄糖存在与否，都不产生 Lac mRNA，这是因为阻遏蛋白与操纵基因结合所致；而有乳糖存在时，阻遏蛋白与乳糖结合，去阻遏，但如果有葡萄糖存在，cAMP 处于低水平，cAMP-CAP 与启动子结合的亲和力低，因此，只有很少量 Lac mRNA 被合成。当乳糖存在而葡萄糖缺乏时，乳糖操纵子转录达到最大量。这是因为乳糖使阻遏蛋白失活；葡萄糖缺乏使 cAMP 增加，它与 CAP 结合增加，其复合体与启动子结合的亲和力增加，激活了操纵子，转录起始。

阿拉伯糖操纵子体外转录要达到最大活性，也需要 AraC 蛋白和 CAP 蛋白。虽然色氨酸操纵子和阿拉伯糖操纵子转录的复合调控是复杂的，但对细菌是有益的。只要葡萄糖丰富，几乎没有 cAMP 产生，这样 CAP 未被激活，诱导消化其他糖的酶是不需要的。当葡萄糖缺乏而其他糖存在时，转录起始，其他糖被代谢。在研究中，科学家发现 CAP 和 RNA 聚合酶的结合位置相邻，并且在 DNA 螺旋的同一侧，因此他们推测 CAP 作用基础可能是 CAP 蛋白吸附 RNA 聚合酶从而促进转录起始。cAMP-CAP 的调控是极其广泛的，除了某些糖类的代谢酶外，细菌中许多其他功能的酶也表现为对葡萄糖效应的敏感，如三羧酸循环和呼吸链酶系统中的大多数酶、分解各种碳源底物的酶、降解某些氨基酸的酶、抗生素合成酶以及负责鞭毛形成的酶等。

综上所述，原核生物中，操纵子系统是最经济、最有效的。把功能相关的基因组织在一起，不必逐个进行调控，而是"一开俱开，一关全关"，达到快速调节的目的。

（三）RNA 聚合酶活性调控

RNA 聚合酶活性调控，又称魔斑（magic spot）核苷酸调节作用。其机制为当细菌细胞中缺乏氨基酸，即处于氨基酸饥饿时，会出现两种异常核苷酸：鸟苷四磷酸（ppGpp）和鸟苷五磷酸（pppGpp），他们都能与 mRNA 聚合酶结合形成复合物，使 RNA 聚合酶构象发生改变，活性降低。随后 rRNA、tRNA、mRNA 合成降低或停止。当氨基酸充足时，则不出现上述情况。

由于这两种异常核苷酸在层析谱上呈现斑点，所以称为魔斑，有人称之为警报素（alarmone）。当出现魔斑时，表示细胞内缺乏氨基酸。魔斑核苷酸出现的意义在于"让细菌知道：因氨基酸缺乏，蛋白质合成受限，不需要再生产 RNA 了"。于是细胞对这种情况可作出种种反应：抑制核糖体或其他大分子合成，活化某些氨基酸操纵子（如色氨酸操纵子）的转录，活化蛋白水解酶去抑制与氨基酸合成无关的转运系统等，从而节省能量和原料，帮助细胞渡过难关。

三、转录前水平的调控

转录前水平的调控是指发生在基因组内基因结构的变化，即通过 DNA 重排进行的调控。最典型的例子是鼠伤寒沙门氏菌两种鞭毛抗原的选择表达。鼠伤寒沙门氏菌有两种鞭毛抗原（两种血清型），分别由不连锁的两个基因 *H1*、*H2* 所编码。细胞中的两个基因在任何时候都只有一个基因表达，即 *H2* 基因表达则 *H1* 基因关闭；*H2* 基因关闭则 *H1* 基因表达，在 *H2*、*H1* 两相之间变换"开—关"。是什么因素使一个基因表达，又使另一个基因不表达呢？现已知，*H2* 基因和 *H1* 基因在染色体上相距很远，而 *H2* 基因和 *H1* 基因串联在一起。*H2* 基因的上游有一段 970 bp 序列，称 *hin* 基因——倒位基因，该基因内含有 *H2* 的启动子，两端各有一段 14 bp 倒转重复序列，即 IRL、IRR。基因表达有两种情况，当 *H2* 基因表达时，H1 阻遏蛋白基因也表达，编码的 H1 阻遏蛋白可以阻遏鞭毛抗原 *H1* 基因的表达，表现为 *H2* 基因表达，*H1* 基因抑制（Ⅱ相）；另外一种情况，*hin* 基因的表达产物——倒位蛋白可使 *hin* 基因及 IRL、IRR 发生倒位基因重排。这样 *H2* 的启动子移到另一头，且方向改变（转向）、远离结构基因，由此，*H2* 基因表达受抑制，同时 H1 阻遏基因也受抑制，不能对 *H1* 进行阻遏，而 *H1* 基因表达（Ⅰ相）。

倒位调控也许是使鼠伤寒沙门氏菌在感染过程中逃避免疫破坏的一种方式。例如，开始感染的细菌处于Ⅰ相，制造 H1 型鞭毛蛋白，随着细菌菌体的扩增，宿主细胞可能产生针对 H1 型鞭毛蛋白的抗体，这些抗体将消灭整个细菌菌体。若这时细菌以较高频转为Ⅱ相，细菌便存活并大量扩增。

四、翻译水平的调控

（一）反义 RNA 的调节

蛋白质作为阻抑物或激活物（诱导物）对转录进行调控的例子已屡见不鲜。然而现已发现有些 RNA 小分子也能调节基因表达，如反义 RNA（antisense RNA）的调控。

反义 RNA 是一类小的转录产物，长 70～200 bp，能通过互补的碱基与特异 mRNA 结合，从而阻断 mRNA 翻译成蛋白质，因为它们与 mRNA 之间的特殊关系，人们称之为反义 RNA。过去人们称这类 RNA 为 mRNA 干扰性互补 RNA（mRNA interfering complementary RNA，micRNA）。Mizuno 在有关渗透压变化对大肠杆菌外膜蛋白基因表达调控的研究中发现，大肠杆菌渗透压调节基因 *ompR* 的产物 OmpR 蛋白在不同的渗透压下有不同的构象，分别作用于渗透压蛋白 OmpF 和 OmpC 的调控区（两个基因不连锁）。低渗时，OmpR 蛋白对 *ompF* 基因起正调节作用，OmpF 合成增高，而 OmpC 合成抑制；高渗时，OmpR 蛋白发生构象改变，对 *ompC* 起正调节作用，OmpC 合成增高，而 OmpF 合成受抑。现已知，当 *ompC* 基因转录时，在 *ompC* 基因启动子上游方向有一段 DNA 序列——调节基因 *micF*，以相反的方向同时转录，产生一个 174 个核苷酸的 RNA——反义 RNA，这种 RNA 能与 *ompF* RNA 顺序中 5′端序列，包括 SD 序列以及编码区（包括 AUG）形成杂合双链，从而抑制 *ompF* 的翻译。所以，*ompC* 转录越多，*ompF* 反义 RNA 也就越多，OmpF 蛋白就越少，见图 5-2。

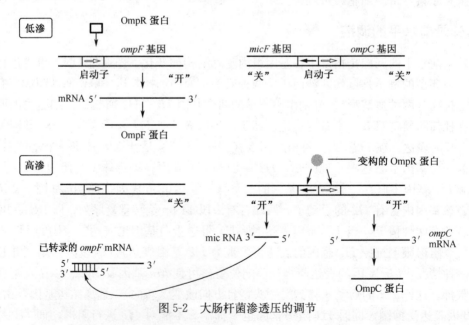

图 5-2 大肠杆菌渗透压的调节

反义 RNA 对基因表达的调控作用揭示了一种新的基因表达调控的机制。从目前对原核细胞研究表明，反义 RNA 作用的基本原理是通过碱基配对与特定的 mRNA 结合，形成二聚体，从而阻断后者的表达。

（二）mRNA 的稳定性

细菌的增殖周期是 20～30 min。代谢反应调控速度很快，这不仅要求有快速的转录起始和转录终止的调控，也需要有 mRNA 快速降解的调控，从而使 mRNA 保持较高的更新速度，mRNA 降解速度是翻译调控的另一重要机制。

原核生物中 mRNA 的半衰期相差较大，可以从几十秒到几十分钟（平均 2～3 min）。这与 mRNA 本身的结构、细胞生理状态和环境因素有关。mRNA 在其 5′端或 3′端的发夹结构可保护其不被外切核酸酶迅速水解，提高稳定性；而 RNaseⅢ能识别特殊的发夹结构，将其裂解，再使 RNA 被其他 RNA 酶降解。未被裂解的发夹结构，其他 RNA 酶不能将其破坏。如果这种发夹结构被保护，则 RNA 的寿命就延长了。有些特殊调控蛋白可以结合这种发夹结构，调节 mRNA 的稳定性。

在分子生物学发展进程中，原核生物基因表达调控的研究已取得了许多令人瞩目的成果，尤其是操纵子理论及其在代谢调节中的应用，不仅成为认识原核生物生命活动本质和改造原核生物为人类服务的重要环节，也对探讨真核生物基因调控机制有所启迪。

第三节　真核生物基因表达的调控

一、真核生物基因表达调控的复杂性及特点

真核生物尤其是高等生物的基因组不仅比原核生物大，而且结构、功能复杂。由此决定了其表达调控较原核生物范围更大、功能更复杂、更精细和微妙，同时给真核基因表达调控的研究带来困难（难以直接鉴定基因产物和基因控制的生化过程，难以直接选择出影响调节基因的突变体，难以直接通过改变外界环境条件来研究分析基因表达变化，难以直接操作等）。但是，随着分子生物学研究的深入，人类基因组计划及模式生物基因组计划的进展，人们在真核基因组结构、功能及调控研究方面已取得重要进展。当前，真核生物基因表达调控的研究，已成为探究生命奥秘的重要基础。

研究表明，绝大多数真核生物是多细胞的、复杂的有机体，基因表达调控的特点是能在特定时间和特定细胞中激活特定基因，从而实现"预定"的、有序的分化发育过程。一般来说，真核生物对外界环境条件变化的反应和原核生物十分不同，由于真核生物绝大多数细胞处于较恒定的环境中，一般能避免外界环境突然改变的影响，即对外界因素的变化通常不发生反应，由此保证生物体组织器官在千变万化的环境条件下维持正常功能。但真核生物的某些细胞例外，如肝细胞，这是由于肝脏本身的解剖特点所决定的。由肠道吸收的物质，包括营养物质和有毒物质，都经门脉系统首先进入肝脏，肝细胞中的一些基因表达可以因外界吸收进入体内的营养和物质毒性情况而受到调控。例如，低糖饮食时，哺乳动物肝细胞与糖异生有关的酶类基因被激活。镉及其他一些重金属能与肝细胞的金属硫蛋白结合，解除重金属毒性。有些药物，如苯巴比妥、可待因、吗啡等以及一些致癌物质能诱导肝细胞合成细胞色素 P450。P450 是一类加单氧酶，可使上述药物和致癌剂羟化，增加溶解度加速排出体外而解毒。这些反应的调控是真核生物基因表达调控的一种类型——瞬时调控，或称可逆调控，它相当于原核细胞对环境条件改变所做出的反应。瞬时调控包括某种底物或激素水平升降，或细胞周期不同阶段中

酶的活性和浓度的调节。真核生物基因表达调控的另一种类型——发育调控，或称不可逆调控，是真核生物基因调控的精髓。这是由于真核生物绝大多数细胞基因表达是与生物体的发育、分化有关。正常情况下，体细胞类型按一定计划严格调控，使个体发育顺利进行。不同细胞的基因表达依类型不同，所处发育阶段不同而异。因此，发育调控决定了真核生物细胞生长、分化、发育的全部过程。

真核基因表达调控是通过多阶段水平来实现的，即转录前、转录中、转录后、翻译和翻译后共 5 个水平。总的来说，与原核生物一样。真核生物转录水平的调控是最为重要的一环，但由于真核生物转录和翻译在时间和空间上完全分割，所以翻译水平上的调控对真核基因表达来说也是十分重要的。

二、转录前（基因组）水平的调控

转录前的调控指发生在基因组水平上基因结构的改变。这种调控方式稳定持久。

（一）基因扩增

基因扩增（gene amplification）是指细胞内某一基因的拷贝数高于正常的现象，是细胞在短时期内为满足某种需要产生足够产物的一种调控方式。细胞在发育分化时，对某种基因产物的需要量剧增，而单靠调控其表达不足以满足，只有增加这种基因的拷贝数来满足要求。例如，非洲爪蟾体细胞中 rRNA 基因拷贝数约为 500 个，而在卵细胞中拷贝数增加了 4000 倍。这是因为卵细胞的分裂需要合成大量蛋白质，而对 rRNA 的需要剧增。rRNA 基因扩增的结果，使细胞内迅速累积 10^{12} 个核糖体。如果没有这种扩增结果，则需 500 年才能积蓄到如此多的核糖体！

在肿瘤细胞中，某些原癌基因拷贝数异常增加，导致表达产物增加，使细胞持续分裂而致癌变。

（二）基因重排

基因重排（gene rearrangement）是指某些基因片段改变原来存在的顺序而重新排列组合。基因重排不仅可以形成新的基因，还可以调节基因的表达。以基因重排来调节不同基因表达的例子是哺乳动物免疫球蛋白各编码区基因的重排连接。已知当哺乳动物受到外界抗原刺激后会产生相应的抗体（免疫球蛋白）。粗略估计，免疫球蛋白可达几百万种。蛋白质都是由基因编码，哺乳动物总的基因数不过 3 万～5 万，怎么可能由这些基因来编码这么多种免疫球蛋白分子呢？也就是说决定多种多样抗体的基因库从何而来？研究表明，从胚胎细胞到 B 细胞（抗体形成细胞）分化过程中，抗体基因发生了两次重排。第一次发生在前 B 细胞中，由编码免疫球蛋白可变区的基因片段参与，使在种系中相互分离的片段经重排后相互连接在一起，称 V-D-J 复合体，即形成重链可变区（V 区）完整基因；重链重排后，接着是轻链 V 区基因重排，形成 V-J 复合体，即轻链 V 区完整基因。第二次重排发生在成熟 B 细胞经抗原刺激后出现重链改变的类别转换，其抗原特异性不变，B 细胞分化、发育成浆细胞。现已知 V、κ、λ、μ 基因有上百个，J、D 片段也有若干个，所以通过这种片段的组合重排，使基因组中有限的基因片段形成了抗体 V 区的多样性（抗体的多样性主要取决于 V 区）。

（三）DNA 甲基化

在脊椎动物中，DNA 上特定的 CG 序列处的 C 可发生甲基化修饰（DNA 中胞嘧啶环 C5 位甲基化，^{5}mC）。这种甲基化可以阻止某些基因的转录，并且能遗传到子细胞中去。研究表明，转录活跃的基因是低甲基化或未甲基化，而不表达的基因则高度甲基化，即基因的甲基化程度与基因表达呈反向平行关系。基因某一特定区域尤其是靠近 5′端调控序列的去甲基化可使基因转录活性增加。在正常情况下不表达基因，可因激素的变化，致癌物作用等使基因调控区去甲基化而重新激活。因此，DNA 甲基化（DNA methylation）异常可能为参与肿瘤和心血管疾病发生、发展的机制之一。

（四）染色体结构对转录激活的控制

真核基因的重要特征之一是基因组 DNA 与蛋白质结合，形成以核小体为基本单位的染色体结构而存在于细胞核内。这种结构特征产生了真核生物基因转录前在染色体水平上的独特的调控机制。换句话说，基因的转录是以染色体结构的一系列重要变化为前提的。基因的活跃转录是在常染色体上进行，转录发生时，带有编码基因的染色质首先发生构象可逆改变，由致密结构变为比较疏松的结构，以便于与转录有关的调控蛋白同 DNA 顺式作用元件结合，以及 RNA 聚合酶在 DNA 模板上的滑动。当染色质处于疏松结构时易被非特异性内切核酸酶，如 DNA 酶Ⅰ（DNaseⅠ）水解，形成了对 DNaseⅠ水解作用敏感区，称为 DNaseⅠ敏感位点（DNaseⅠ sensitive site）。当用极低浓度的 DNaseⅠ处理染色质时，水解将发生在少数特异位点上，这些特异位点即是活跃表达基因所在染色体上的对 DNaseⅠ的超敏感位点（hypersensitive site）。每个活跃表达基因都有一个或数个这类位点。这是活跃基因的共性，非活跃表达基因不表现 DNaseⅠ的超敏感性。因此，常将 DNaseⅠ超敏感性作为该基因转录活性的标志，即某些 DNA 序列上发现染色质 DNaseⅠ超敏位点，提示该段 DNA 序列可能在体内有重要生理作用。

三、转录水平的调控

真核生物基因表达在转录水平的调控，是各级调控中最重要的一步，主要涉及 RNA 聚合酶、顺式作用元件和反式作用因子三种因素的相互作用。

（一）RNA 聚合酶

基因转录是由 RNA 聚合酶（RNA polymerase）催化完成的。无论是原核生物还是真核生物，在转录过程中，RNA 聚合酶与启动子的结合是关键的一步。不同的是，细菌 RNA 聚合酶识别的是一段 DNA 序列，而真核生物 RNA 聚合酶识别的不单是 DNA 序列，而是 DNA-蛋白质复合物。即只有当一个或多个转录因子（transcription factor，TF）结合到 DNA 上，形成有功能的启动子时，才能被 RNA 聚合酶识别、结合。由于真核生物 RNA 聚合酶有三类，分别转录三类不同的 RNA，因此转录因子也有三类：TFⅠ、TFⅡ、TFⅢ。

RNA 聚合酶Ⅱ转录的编码蛋白质的基因称为Ⅱ类基因，此类基因品种多，与细胞生长、分化直接相关，其表达调控也最为复杂。

（二）顺式作用元件

顺式作用元件（cis-acting element）系与结构基因串联的特定的DNA顺序，包括启动子、增强子（沉默子）加尾及终止信号等（详见第二章）。

（三）反式作用因子

反式作用因子（trans-acting factor）是一类细胞核内的蛋白质因子，通过与顺式作用元件和RNA聚合酶的相互作用而调节转录活性。一个完整的反式作用因子含有两种结构域：DNA结合结构域和转录活化结构域。反式作用因子通过前者与DNA特定顺序结合，通过后者发挥转录活化功能。有些反式作用因子可能只含有两者之一，只有当互补的两个蛋白质存在于同一细胞时，才具有功能，这可能与基因表达组织特异性有关。

1. DNA 结合结构域

反式作用因子发挥其转录调节功能的首要条件是必须有一个与DNA特异结合的结构。对大量反式作用因子结构研究表明，DNA结合结构域多在100个氨基酸以下，大体有4种形式。

1）螺旋-转角-螺旋（helix-turn-helix，H-T-H）　这是研究得比较清楚的DNA结合区的结构模式。由约60个氨基酸残基组成，含有两个α螺旋，α螺旋之间有一个β转角。在两个α螺旋中，一个为识别螺旋（C端），其氨基酸残基直接与靶DNA双螺旋大沟特异性结合；另一个螺旋（N端）穿过大沟，与DNA中磷酸戊糖骨架形成非特异性结合。真核生物中，最早在控制果蝇早期发育的同源域蛋白中发现螺旋-转角-螺旋的结构模式，故也将此模式称为同源结构域（homodomain，HD）。具有这种结构的反式作用因子与机体发育、分化过程有关。

2）锌指结构（zinc finger）　锌指是由一小群氨基酸残基与一个锌原子结合，在蛋白质中形成相对独立的指结构，故而得名。锌指由约30个氨基酸组成。其特点是具有几个相同的指状结构，在Cys和His之间有12个氨基酸残基，其中数个为保守的碱性残基。4个Cys通过与锌离子络合而形成一个稳定的指状结构，也有Cys2/his2指状结构。锌指的共有序列：$Cys-X_{2-4}-Cys-X_3-Phe-X_5-Leu-X_2-His-X_3-His$。锌指对DNA的结合是必需的，但对结合的特异性并不重要。哺乳动物细胞的SP2以及其他蛋白质的DNA结合区也发现有类似的锌指结构。

3）亮氨酸拉链（leucine zipper）　亮氨酸拉链是反式作用因子DNA结合结构域的一种结构模式，大约由30个氨基酸残基组成。分成两部分，一部分以碱性氨基酸为主，为DNA结合部位；另一部分每间隔6个氨基酸出现一个亮氨酸，形成了两性α螺旋，即螺旋的一侧是以带电荷的氨基酸残基（如Arg、Gln、Asp）为主，具有亲水性；另一侧是排列成行的亮氨酸残基，具有疏水性，称亮氨酸拉链区，两个具有亮氨酸拉链区的反式作用因子靠疏水相互作用结合形成亮氨酸拉链。在许多反式作用因子之间存在这种特异序列。因此，该结构与反式作用因子之间的相互作用有关。

4）螺旋-环-螺旋（helix-loop-helix，H-L-H）　H-L-H是发现时间不长的一种DNA结合结构域的结构模式。H-L-H结构由三部分组成。①100～200个氨基酸残基

形成两个两性 α 螺旋的区域。两性 α 螺旋中以亮氨酸残基为主体形成疏水面，以亲水氨基酸残基为主体组成的另一侧亲水面。②两螺旋之间为长短不等的肽段（环）。③α 螺旋 N 端有一段碱性氨基酸区，带有大量正电荷，当与 DNA 相靠近时，这些正电荷被 DNA 的磷酸根离子所中和，形成稳定的 α 螺旋结构结合于 DNA 双螺旋的大沟。H-L-H 这种与 DNA 结合的性质与亮氨酸拉链结构相似（图 5-3）。免疫球蛋白基因增强子结合蛋白 E12、E47 等就发现有此结构。

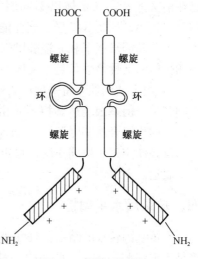

图 5-3　螺旋-环-螺旋

2. 转录活化结构域

在真核生物中，反式作用因子功能由于受蛋白质-蛋白质之间相互作用的调节，变得十分精密复杂。转录调控功能通常以复合体的方式来完成，这就意味着并非每一个反式作用因子都需要直接与 DNA 结合。因此，转录活化结构域就成为反式作用因子必须具备的物质基础。

反式作用因子功能具有多样性，其转录活化结构域也有多种，通常是由 20～100 个氨基酸残基组成。有时一个反式作用因子可含一个以上的转录活化结构域。转录活化结构域的结构特点，首先是含有很多带负电荷的酸性氨基酸残基，并能形成亲脂性的 α 螺旋，但在氨基酸序列上很少有同源性；其次是富含谷氨酰胺结构，与上述酸性结构域一样，此结构之间也无明显的序列同源性，也可能是可以相互替代的；最后是一些反式作用因子中富含脯氨酸残基。脯氨酸是一种亚氨基酸，可阻碍 α 螺旋的形成。

（四）反式作用因子的作用特点和规律

在基因转录的起始过程中，涉及反式作用因子与顺式作用元件、反式作用因子与反式作用因子之间的相互作用，而且不同基因的表达受到不同组合的反式作用因子协同调节，过程非常复杂，反式作用因子的作用有以下特点和规律。

1）一种反式作用因子能和一种以上的顺式作用元件结合　真核生物细胞的反式作用因子不像原核生物的调控蛋白与顺式作用元件结合有高度的专一性，有些反式作用因子可以和同源性很小的顺式作用元件结合。例如，酵母中的反式作用因子 HAP-1 能与 Cyc1 和 Cyc7 两种基因的顺式作用元件结合，而这两个结合位点的 DNA 序列没有同源性。C/EBP 可以识别 CAAT 盒，也能识别一些动物病毒的增强子序列。

2）一种顺式作用元件能和一种以上的反式作用因子结合　CAAT 盒能和多种反式作用因子结合，这些反式作用因子具有不同类型，如 CAAT 转录因子（CAAT transcription factor，CTF）族。CTF 族是能识别 CAAT 盒的一组转录因子；另一组 CAAT 结合蛋白被称为 CP，CP 族的 CP1 与 α 珠蛋白基因的 CAAT 盒有高度亲和力；CP2 与 γ 纤维蛋白基因的 CAAT 盒有高度亲和力；另外，CBP 因子、ACF 因子也能识别 CAAT 盒。

3）以二聚体或多聚体形式与顺式作用元件作用　反式作用因子二聚体形成的机制

已在前文述及，二聚体以不同因子形成的异二聚体为主，如 Fos/Jan 二聚体能识别 TGACTCA；E12/E47 异二聚体能识别免疫球蛋白 κ 链基因的增强子。一种能与免疫球蛋白重链基因增强子和 κ 轻链增强子结合的反式作用因子——μEBP-C（也称 NF-μE3）的四聚体的活性最强，是二聚体的 7 倍。

4）组合式调控　每一种反式作用因子作用于顺式作用元件后或者促进转录，或者抑制转录。但基因表达调控不是由单一因子完成的，而是几种因子组合发挥特定的作用，这称为组合式基因调控（combinatoral gene regulation）。结合到不同顺式作用元件上的反式作用因子协同作用，决定一个基因的转录活性，或者高表达，或者低表达，或者不表达。

四、转录后水平调控

真核基因转录的结果保证了遗传信息从 DNA 传递给 RNA，因此，转录水平的调控是决定细胞中 RNA 水平的一种十分重要的方式。然而，从诸多的实验观察发现，RNA 的数量、结构甚至种类在细胞核和细胞质中并非全然相同，而且在活跃生长的细胞与静止细胞之间，其 RNA 情况也存在差异。这表明遗传信息在转录后还有多种多样的选择性，即基因表达除 DNA 水平和转录水平调控外，转录后调控在决定细胞表型多样化方面也是十分关键的。

转录后水平的调控一般是指基因转录后对转录产物进行一系列修饰、加工的过程，主要包括 mRNA 前体修饰、剪接、mRNA 通过核孔在胞质内定位、RNA 编辑、mRNA 稳定性等多个环节。

（一）"加帽"和"加尾"的调控

真核生物 mRNA 在转录后要经过"加帽"（capping）反应，即在 5′端形成一个特殊结构，7-甲基鸟苷三磷酸（m^7G）。真核生物 mRNA5′端可有 3 种不同的"帽子"，其差异在于"帽子"中碱基甲基化程度不同。"帽子"结构与蛋白质合成效率之间的关系十分密切，以下几点可以说明：①"帽子"结构是前体 mRNA 在细胞核内和细胞质内的稳定因素，没有"帽子"的转录产物很快被核酸酶降解；②"帽子"为蛋白质合成提供识别标志，并促进蛋白质合成起始复合物的生成，因此提高了翻译强度；③没有甲基化的"帽子"（如 GpppN-）以及用化学或酶学方法脱去"帽子"的 mRNA，其翻译活性显著下降；④"帽子"结构的类似物，如 m^7G 等能抑制有"帽子"mRNA 的翻译，但对没有"帽子"mRNA 的翻译没有影响。

与核糖体结合的大多数真核生物 mRNA 在 3′端都含有 50～150 个腺苷酸，即 poly（A）尾。除组蛋白 mRNA 外，真核生物 mRNA 均有 poly（A），这也是在转录后加上去的——"加尾"（tailing），"加尾"信号是 AAUAAA。poly（A）的功能是保持 mRNA 稳定性，延长 mRNA 寿命。一些 mRNA 的 poly（A）长度是受控制的，细胞可以对不同 mRNA 的 poly（A）选择性加长或快速截短、去除。

（二）mRNA 选择性（可变）剪接对表达的调控

真核生物转录出的前体 mRNA 除了上述"加帽"、"加尾"修饰外，还要经过内含子的切除和外显子的拼接，这一过程即为剪接。关于 RNA 剪接的研究是 20 世纪 80 年代以来生物化学和分子生物学领域中最有生气的课题之一。

由于选择性剪接的多样化，一个基因可在转录后通过剪接加工而产生两个或两个以上的 mRNA，由此翻译成两个或更多的蛋白质。mRNA 选择性剪接的研究不仅解决了不连续基因转录产物的剪接问题，而且对于不连续基因的起源以及整个生命起源问题的探索也都是有力的推动（详见第三章）。

（三）RNA 编辑的调控

RNA 编辑（RNA editing）是一种较为独特的遗传信息加工的方式，即转录后的 mRNA 在编码区发生插入、删除或转换的现象，是在 RNA 分子上出现的一种修饰现象。mRNA 在转录后因核苷酸替换、插入、缺失等改变了 DNA 模板来源的遗传信息，从而编码出氨基酸序列不同的多种蛋白质。编辑的结果不仅扩大了遗传信息，也可能是生物适应中的一种保护措施。

1）核苷酸替换　最典型的例子是载脂蛋白 B 的 RNA 编辑。肠型载脂蛋白 B（Apo B）的组成占肝脏型 Apo B_{100} 分子的 48%，又称 Apo B_{48}，它是膳食中富含三酰甘油的脂蛋白与乳糜微粒分布到各组织所必需的。在蛋白质水平上，它只保留了 Apo B_{100} 分子 N 端脂蛋白装配结构域，缺少 Apo B_{100} 分子 C 端 LDL 受体结合区。在人、兔、鼠中都是由于第 26 个（最大的）外显子中第 6666～6668 个核苷酸由 Apo B_{100} 的 CAA 突变为 UAA，C→U 替换，由编码谷氨酰胺的密码子变为终止子，产生编码 Apo B_{48} 的 mR-NA。C→U 替换可能通过胞嘧啶脱氨酶（cytidine deaminase）脱氨作用来实现，见图 5-4。

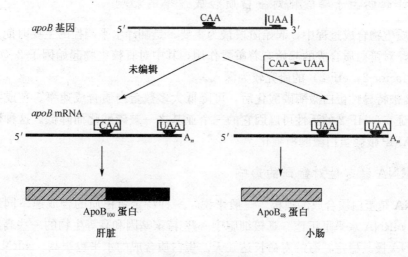

图 5-4　RNA 编辑

2）核苷酸的插入或缺失　椎虫中，RNA 编辑反应非常广泛，缺失、插入常发生，椎虫线粒体的细胞色素氧化酶亚基 Ⅱ 的基因与人类基因相比在相当于编码第 170 位氨基酸处有一个移码突变（frameshift mutation），然而在其基因转录产物中却由于 4 个 U 的插入恢复了相应可读框，因而产生出相应功能的蛋白质。提示 RNA 编辑的基因表达调控是一种重要的调控和补救机制。

由此可见，RNA 编辑不同于上述 mRNA 前体的剪接。剪接是在切除内含子后所得到的成熟 mRNA，其编码信息均可在原始基因中全部找到；但 RNA 编辑后所得的成熟 mRNA 编码区发生的碱基数量的改变，却并不出现在原始的基因中。

（四）mRNA 运输控制

成熟的 mRNA 并不是全部进入细胞质。有人通过放射性核素标记实验计算得到，大约只有 20% 的 mRNA 进入细胞质，留在核内的 50%mRNA 约在 1h 内降解，剩余的 mRNA 在某些类型细胞中是有功能的，而在另一些细胞中则功能不清。

虽然 RNA 运出核的控制机制目前尚不清，但至少有几点可以说明 RNA 出核是受控的：①RNA 通过核膜孔运出的过程是主动运输过程；②大多数的 RNA 需经过"加帽"、"加尾"，并在剪接完成后才能被运输。

mRNA 通过核膜孔从核中运出，它被运送的位置也有特异性。有的被直接运到内质网，在内质网膜上完成肽链的合成；有的则可能被运到细胞质，由细胞质中游离的核糖体进行翻译。

五、翻译水平的调控

蛋白质合成水平上的调控，也是表达控制的重要环节。同一细胞中同时出现不同的 mRNA，即使数目接近相等，产生蛋白质的多少可以相差很大，这主要取决于翻译的速率和 mRNA 寿命。

（一）翻译起始因子磷酸化对蛋白质合成速率的影响

蛋白质生物合成过程中，起始阶段最为重要，是翻译水平调控的主要时期。许多蛋白质因子，对蛋白质合成的起始有着重要作用，其中对真核生物起始因子-2（eukaryotic initiation factor-2，eIF-2）的研究较为深入。

eIF-2 被特异性蛋白激酶磷酸化后，可降低大多数蛋白质合成速率。在成熟前的红细胞中已证实，eIF-2 的活性可以因它的三个亚基之一被磷酸化而降低。这种磷酸化是由一种 cAMP 依赖蛋白激酶所催化。

（二）mRNA 稳定性对翻译的影响

mRNA 是蛋白质合成的模板。一般来说，一种特定蛋白质的合成速率同细胞质内编码它的 mRNA 水平呈正比。真核细胞中一些 持家基因和高等生物的一些高度分化细胞的 mRNA 极其稳定，有的寿命长达几天。蛋白质合成的水平也很高。mRNA 的稳定性既取决于自身的二级结构，又决定于转录后的修饰。

六、翻译后水平的调控

（一）信号肽在蛋白质翻译后的作用

信号肽（signal peptide）由 15~30 个疏水氨基酸残基组成，其特点为疏水性。它的作用是使蛋白质从内质网膜进入高尔基体。一旦蛋白质进入高尔基体，信号肽就被信号肽酶水解。切去信号肽后，前蛋白就变为有生物学活性的蛋白质了。有些蛋白质经过一次加工切去信号肽还不够，还要经过第二次。例如，胰岛素由 51 个氨基酸残基组成，但胰岛素 mRNA 的翻译产物在兔网织红细胞无细胞翻译系统中为 86 个氨基酸残基，称胰岛素原（proinsulin）。在麦芽无细胞翻译系统中为 110 个氨基酸残基组成的前胰岛素原。后来证明，在前胰岛素原的 N 端有一段富含疏水氨基酸残基的肽段作为信号肽，使前胰岛素原能穿过内质网膜进入内腔，在内腔壁上信号肽被水解。所以在哺乳动物细胞内，当合成完成时，前胰岛素原已成为胰岛素原，然后胰岛素原被运到高尔基复合体，切去 C 肽成为成熟的胰岛素，最终被排出胞外。后来发现，几乎各种分泌性蛋白均含有信号肽。

（二）新生肽链中氨基酸的修饰

从核糖体上最终释放出的多肽链，还不是具有生物活性的成熟蛋白质，需要进行氨基酸的修饰（包括磷酸化、羟基化、糖基化、乙酰化等）和肽链的正确折叠与装配。

为什么真核生物不直接产生有功能的蛋白质而要采取翻译后加工呢？原因是多样的。例如，密码子只有 20 种，只能编码 20 种氨基酸，而已知的修饰氨基酸不下 100种，它们只通过翻译后加工得到，而且它们并不是每一种蛋白质所必需，因此通过翻译后加工得到比较合理。某些修饰是为了暂时需要，例如，高等动物的消化酶先以酶原形式翻译出来，等到需要时，才加工成为有活性的酶；不少蛋白质的磷酸化-脱磷酸化，或乙酰化-脱乙酰化作用则起到调节作用。总之，生物在翻译后对蛋白质所起的某种修饰作用也是基因调控的一种方式，增强了生物对环境的适应性。

（三）新生肽链的正确折叠

新生肽链的一级结构是遗传信息所决定的，是蛋白质最基本的结构，它决定着蛋白质的空间结构。而蛋白质的空间结构则是其生物学功能的基础，即空间结构决定着蛋白质的生物学功能。空间结构就涉及肽链的正确折叠。有些基因工程蛋白质产物，其一级结构与天然蛋白质相同，但功能却与天然蛋白质有差异，这里也涉及新生肽链正确折叠的问题。

现已知和新生肽折叠有关的蛋白质大体上可分为两大类。一类是直接催化和蛋白质折叠有关特定反应的酶，如蛋白质二硫键异构酶、脯氨酸顺反异构酶；另一类则是帮助新生肽的折叠，使之成为成熟的蛋白质，但本身并不参与共价反应，称为分子伴侣（molecular chaperone）。分子伴侣的概念由 Lasky 于 1978 年首先提出，现已应用于许多蛋白质。分子伴侣在原核生物和真核生物中广泛存在。分子伴侣可调节和稳定未折叠或部分折叠的多肽，并防止不适当的多肽链内或链间相互作用；有些分子伴侣也可与天然的蛋白质相互作用以促使寡聚肽发生结构重排。它们还具有介导线粒体蛋白跨膜转运，调控信息传导通路和参与微管形成与修复等功能。目前已鉴别出来的分子伴侣主要属于伴侣素 60 家族、热休克蛋白 70、热休克蛋白 90 家族等几类高度保守的蛋白质家族。分子伴侣具有酶的特征但又和酶很不同，其作用机制现在还没有一致认识，但已越来越受到人们的广泛重视。

第四节　RNA 干扰与基因沉默

一、RNA 干扰概念与特点

RNA 干扰（RNA interference，RNAi）是正常生物体内抑制特定基因表达的一种现象（详见第三章）。RNA 干扰被美国科学杂志评为 2001 年十大科技突破之一，科学家对 RNA 干扰现象之所以表现出极大关注在于，RNA 干扰在基因功能和相关方面的研究中具有许多传统方法无法比拟的特点和优势。

1）特异性　RNA 干扰的最显著特征就是只引起与 dsRNA 同源的 mRNA 降解。实验表明，在果蝇细胞中 dsRNA 能特异性的短暂性抑制外源转基因和细胞内源基因表达，而对其他无关基因的表达不产生影响。

2）高效性　无论是在体内还是体外实验中，仅需少量的 dsRNA（每个细胞几个分子）就能有效的抑制靶基因表达，抑制的效率在低等动物中大于 90 ％。这表明 dsRNA 介导的 RNA 干扰是一个以催化放大的方式进行的过程。

3）dsRNA 长度限制性　引发有效 RNA 干扰的 dsRNA 最短不小于 21nt，如果是长链 dsRNA，也会在细胞内被 Dicer 酶（dsRNA specific endonuclease）切割为 21nt 左右的 siRNA，并由 siRNA 来介导 mRNA 切割。

4）ATP 依赖性　在去除 ATP 的样品中 RNA 干扰现象降低或消失，显示 RNA 干扰是一个 ATP 依赖的过程。可能是 Dicer 酶和 RNA 诱导沉默复合物（RNA-induced silencing complex，RISC）的酶切反应必须由 ATP 提供能量。

5）可传播性　RNA 干扰有一种令人惊奇的跨越细胞界限的能力，即 siRNA 的沉默信号可以被放大和传播，siRNA 可以作为引物以靶 RNA 为模板在 RNA 依赖的 RNA 聚合酶（RdRp）的作用下合成更多的 dsRNA，并被 Dicer 酶剪切成更多的 siRNA，使 RNA 干扰作用被放大。果蝇细胞中 RNA 干扰作用可在细胞群落之间传播。线虫中可在局部注射 dsRNA 而传播到整个机体。

二、基因沉默

基因沉默（gene silencing）是生物体生命进程中普遍存在的一种现象，是指生物体中特定基因由于某种原因，表达被抑制，是基因表达调控的一种重要方式。基因沉默现象广泛存在于生物界中。根据其作用机制和水平不同，可分为三种，转录水平的基因沉默（transcriptional gene silencing，TGS）、转录后水平的基因沉默（post-transcriptional gene silencing，PTGS）和位置效应（position effect）。

TGS 是发生在核内的事件，而 PTGS 发生在细胞质中。位置效应是指基因在基因组中的位置对其表达的影响。外源基因进入细胞核后先整合到染色质上，其整合位点与表达有密切关系。如果整合到甲基化程度高、转录活性低的异染色质上，一般不能表达。

基因沉默最初被认为是一种不幸的偶然事件，因为当人们将外源基因转入受体以后，总是希望外源基因能高效表达，产生新的性状。然而事与愿违的是，在高等植物中大约 30 ％的转基因实验都导致了基因沉默。现在普遍认为 PTGS 是生物在长期进化过程中所形成的对病毒、转座子和其他可转移核酸的防御系统。

三、RNA 干扰及基因沉默的研究意义及存在的问题

RNA 干扰作为一种非常保守的机制在真核细胞中广泛存在。对其研究将大大促进对基因功能的研究。RNA 干扰可高度专一性地抑制生物体内基因的表达，因此可以作为一种十分有效的手段，帮助我们进行基因功能的研究。可以通过编码区 RNA 干扰技术，研究某一基因家族和某一具体基因的功能。它比传统的反义 RNA 技术和基因敲除技术更迅速和有效。还可以通过与细胞特异性的启动子及可诱导系统的结合使用，在不同的发育时期或不同的器官中有选择地进行，可以克服由于同源重组导致胚胎死亡而无法观察其表型的缺点。另外 RNA 干扰引起的基因沉默现象与诱导系统的结合，可随意控制基因的表达或抑制，这更利于分析目的基因的功能。更有意义的是，RNA 干扰技术很有可能用于人类疾病的基因治疗。例如，肿瘤是一种多基因遗传病，是多个基因相互作用形成的基因网络调控的结果，单一基因的阻断不一定能完全抑制或逆转肿瘤的生长。而 PTGS/RNA 干扰可能在体内同时特异地抑制多个癌基因表达，而且不影响其他基因的功能。

需要说明的是，由于 dsRNA 抑制基因表达具有潜在高效性，任何导致正常机体 dsRNA 形成的情况都会引起相应基因沉默，所以正常机体内有一套严密防止 dsRNA 形成的机制。因为生物体是一个复杂的系统，存在多种机制相互作用以调控基因的表达。在体外 RNA 干扰实验中，我们看到 RNA 干扰能否成功地抑制基因表达以及抑制的程度还取决于细胞类型。对线虫来说，可以采用注射、浸泡后喂食的方法转入 dsR-NA，而对哺乳动物来说，寻找高效的方式来转入 siRNA 以及快速的方式来筛选 RNA 干扰尚需进一步探索。

RNA 干扰在抗病毒感染中的应用令人振奋，但能否成功还无法保证。在研究中也有诸多困难：①并不是所有的病毒 RNA 序列都能比较容易被 siRNA 接近、识别和切割。有些病毒靶序列可能隐藏在二级结构下，或者位于高度折叠的区域中，而有些病毒序列可能与蛋白质形成紧密的复合物，阻碍了与 siRNA 的识别。②病毒的复制过程并不十分严格，随意性很强，因此往往产生突变的子代。这一特点能够帮助病毒躲避机体的免疫监控和药物的抑制作用，同时也会干扰 siRNA 的识别。③病毒编码 RNA 干扰抑制因子。截至目前，已经从植物病毒中鉴定出超过 20 种 RNA 干扰的抑制因子，同时还在脊椎动物和昆虫中发现了多种抑制因子。这些抑制因子可以增强病毒的病原性并且促进病毒复制。④细胞内的 RNA 酶会降解 siRNA，故怎样保证 siRNA 的稳定性仍是一个亟须解决的问题，并且在 siRNA 的导入问题上，同样也面临着技术上的困难。

小　　结

遗传信息传递的"中心法则"是生命过程的基本规律，也是原核生物和真核生物中普遍存在的规律。遗传信息传递的整个过程中存在着调控机制，可以说生命现象的基础是基因表达的调控。

原核生物是单细胞生物，与外界环境关系密切，它们必须快速调节各种基因的表达，以适应营养条件的改变和应付不利的理化因素。原核生物细胞结构的特点和调控方式都与此相适应。发生在转录水平的操纵子系统是原核生物最重要的调控模式。将功能相关的基因组织在一起，受同一组调控单元控制，一开俱开，一关俱关，最为经济有效。操纵子学说的提出扩大了基因的概念，人们开始认识除了有能编码蛋白质一级结构的这么一类基因外，还有具备其他功能的基因，使人们建立了调控的思想，有了调控的概念。

真核生物绝大多数是多细胞的、复杂的有机体，基因组结构的特点决定了真核生物表达调控较之原核生物更复杂、范围更广泛。真核基因表达主要与生物体发育、分化有关，故细胞分化本身就是不同基因表达调控的结果。真核生物基因表达又有多层次调控，在每个水平上都存在对信息补充、修正的可能性，这就是真核基因调控的复杂性所在。可以说真核生物生命现象的基础是基因调控，而转录及转录后遗传信息扩展的各种方式又是表达多样性的基础。

在各种研究基因功能的方法中，RNA 水平使目的基因沉默无疑是目前研究的新点和热点，特别是双链 RNA 介导的转录后水平的基因沉默，更是快速关闭基因的途径。这一现象的发现和应用将有助于基因功能及其在信号转导、病毒抵御机制等方面的研究，并有可能开创基因治疗的新模式。

（伍欣星）

参 考 文 献

邓耀祖，屈伸. 2002. 医学分子细胞生物学. 北京：科学出版社

冯作化，皇甫永穆. 1998. 医学分子生物学. 武汉：武汉出版社

伍欣星，聂广，胡继鹰. 2000. 医学分子生物学原理与方法. 北京：科学出版社

朱玉贤，李毅. 1998. 现代分子生物学. 北京：高等教育出版社

Gilmore I R, Fox S P, Hollins A J, et al. 2006. Delivery strategies for siRNA-mediated gene silencing. Curr Drug Deliv, 3 (2)：147~155

Vaucheret H. 2006. Post-transcriptional small RNA pathways in plants：mechanisms and regulations. Genes Dev, 20 (7)：759~771

Wang Haifang, Luo Xiaoxing. 2003. A-to-I RNA editing：A new mechanism of genomic information modification. Chinese Science Bulletin, 48 (12)：1183~1187

Chapter 5   Regulation of Gene Expression

Recently, it was revealed by that, every biological phenomenon—from heredity, mutation to growth, development, breeding, differentiation and many diseases generation such as tumorigenesis—are intertwined with the regulation of gene expression. Many hot spot topic emerged ever since. As a matter of fact, the regulation of gene expression sits at the hub of genetics and molecular medicine. Generally speaking, the process of replication-transcription-translation is basically clear while the regulations of these processes are not very distinct. All the cells from a single multi-cellular organism bear the same genetic information with the same structure of genes. However, the genes were not expressed necessarily at the same time. The expression of genes was controlled temporarily and spatially, which laid the foundation of differentiation and tissue specific functions.

Section 1   The Basic Concept of Regulation of Gene Expression

1. 1   Gene Expression Overview

Gene expression usually means the process by which the genetic information carried by genome was presented into specific proteins for certain biological function via transcription and translation. Nevertheless, not all gene transcription generates proteins. Some encode rRNA and tRNA. Gene expression can be regulated at the levels of transcription, processing, and translation. The regulation of transcription is the key point, which is executed by the interaction between *trans*-acting factor and *cis*-acting element. *Trans*-acting factors are usually proteins (sometimes can be RNA). They could diffuse in cells and react with every appropriate target gene. *Cis*-acting elements are often DNAs, they usually just interact with adjacent DNA sequences.

1. 2   Specificity of Regulation of Gene Expression

Gene expression (no matter virus, bacteria or multi-cellular organism, or mammals and human beings) was tightly controlled temporarily and spatially. The higher in the evolutionary order, the more sophisticate control of gene expression is. This caters for the species evolution. The temporary and spatial control of gene expression is determined by the interaction between promoters (sequences) and (or) enhancers, and regulatory proteins.

1. 2. 1   Temporary Specificity

To fulfill certain function, the expression of a particular gene happens in an order of timing, which is so-called temporary specificity. In the biological processes of multi-cellular organism, to follow a sequential order of fertilizing eggs to the formation of tissues and organs, relevant genes will open or close in strict accordance with time order, and also manifest as temporary concordance with the stages of differentiation and development. Thus, the temporary specificity of multi-cellular organism gene expression is also known as stage specificity.

1. 2. 2   Spatial Specificity

In the process of organism development, certain gene expression happens at specific spatial organization, which was so-called spatial specificity of gene expression. The spatial and temporary distribution of gene expression in certain developmental stage is actually determined by the tissue specific distribution of cells. Therefore, the spatial specificity of gene expression is also known as cell specificity or tissue specificity.

With respect to the specific stages of growth and development of multi-cellular organism, a given gene is expressed diversely in different tissues and organs; at the same stage of growth, different genes express differently in various tissues and organs.

1.3 Types of Gene Expression

Gene expression responses to internal or external stimuli differently. Tremendous diversities exist in the types of gene expression as a response to the stimuli.

1.3.1 Constitutive Expression

Certain genes express is indispensable to the entire process of organism and is expressed in almost all the cells. Such genes are called house-keeping gene. These genes express continuously in the cells, which is required for the basic function of cells. Such gene expressions are regarded as elementary or constitutive gene expression. This type of expressions is only regulated by the promoter sequences and RNA polymerase. For example, the gene expression of enzymes involved in TAC (tricarboxylic acid cycle) is considered constitutive.

1.3.2 Inducible and Repressible Expression

Different from house-keeping genes, the expressions of some genes are extremely vulnerable to the changes of environment. Under the stimulation of certain environmental signals, genes which can be induced to express are called inducible genes. The elevated expression an inducible gene in special environment is called induction. On the contrary, if the expression of certain gene was suppressed by certain environmental signals, this process is called suppression and these genes are called repressible genes. Apart from the promoter sequence and RNA polymerase, inducible or repressible genes are regulated by additional mechanisms. The regulatory regions of genes, such as lactose operon and tryptophan operon, contain specific response elements to specific stimuli. Induction and repression are two sides of one coin, which are ubiquitously utilized by the organisms to adapt to the environment.

Under the control of certain mechanisms, no matter what the expression type is, a group of function-related genes are all needed to be coordinated and co-expressed, which is called coordinate expression. This regulation is known as coordinate regulation.

1.4 Biological Significance of Regulation of Gene Expression

1.4.1 Acclimation, Growth and Proliferation

The internal and external environments of the organisms are subjected to continuous changes. All living cells must response appropriately to those changes. In order to adapt to the changing environment, the levels of genes expression depend on quantitative alternation of gene expression. Through regulated gene expression, the organism will express appropriate genes for a better adaptation to the environment for the growth and proliferation.

1.4.2 Development and Cell Differentiation

At different developmental stages of multi-cellular organism, the types and amounts of proteins expressed vary tremendously. Even at the same stage of development, significant difference exists as to the proteins expressed in different tissues and organs, which constitute the key knot in the regulation of cellular phenotypes. The development of multi-cellular organism, particularly mammal, is controlled by a series of genes. If certain gene expressed aberrantly, abnormal development might be expected in corresponding tissues and organs.

Section 2   Regulation of Gene Expression in Prokaryotes

2. 1   Characteristics of Regulation of Prokaryotic Gene Expression

Prokaryote is a unicellular organism, the genome of which is composed of circular double-stranded DNA. Since there is no nucleosome nor nuclear membranes, gene transcription and translations happen at the same location simultaneously (the transcription and translation are coupled). The prokaryotes have a very close interaction with their environment. Due to the shortage of energy reserves, the gene expression pattern of prokaryotes is highly adaptive and versatile. They have to constantly adjust the expression of various genes to adapt to the changes of environment and nutrition demands (carbon, nitrogen, etc) and adverse physical and chemical factors (hyperthermia, ray, heavy metals, alkylating agents, etc). In the reactions, the bacteria can make fast synthesis of enzymes, nucleotides and other biomacromolecules, and at the same time fleetly cease synthesis and decompose the elements which are no longer required, so as to optimize the growth and reproduction of bacteria.

The above-mentioned characteristics of gene expression regulation in prokaryotes are consistent with its cellular structure. Initiation and termination of transcription and fast switch of mRNA are the three factors in bacterial gene regulation. Regulations of expression of most bacterial genes are achieved at the transcriptional level.

2. 2   Regulation at Transcriptional Level

2. 2. 1   Regulation of Transcription Initiation by RNA Polymerase

The first step of transcription is that RNA polymerase recognizes and binds promoter to initiate transcription of a gene. Prokaryotes have only one type of RNA polymerase to synthesize three types of RNAs. The RNA polymerase of *E. coli* composes of five subunits, namely $\alpha_2 \beta\beta' \sigma$, with a molecular weight of 500kDa. $\alpha_2 \beta\beta' \sigma$ is also so-called holoenzyme. σ- subunit binds loosely with other subunits and quite often falls from the holoenzyme. The $\alpha_2 \beta\beta'$ is so-called core enzyme, possess the capacity to catalyze the extension of RNA strand; σ-subunit cannot catalyze the RNA synthesis, but to recognize the location at the DNA for the initiation of RNA synthesis. σ-subunit was called initiation factor, as it determines which stand of DNA to be transcribed, direction and start site of transcription. Different σ factor could bind to the holoenzyme competitively. The expression of σ factor is inducible by the environmental changes. For example, $\sigma 70$ gene was open in the normal condition. $\sigma 32$ expression is induced upon the change of environmental temperature and further recognizes and binds to the start site of heat shock protein (HSP) promoter with resultant HSP expression.

2. 2. 2   Transcriptional Regulation Via Operon

Prokaryotic regulation of gene expression mainly occurred on the transcription level, and operon was the basic unit of transcriptional regulation. Genes devoted to a single metabolic goal, such as the synthesis of the amino acid tryptophan, are most often found in a contiguous array in the DNA. Such an arrangement of genes in a functional group is called an operon, because it operates as a unit from a single promoter. The transcriptional product of the operon is mRNA, which encodes a few types of proteins, and thus this mRNA was also known as polycistronic mRNA.

In any single moment of the life cycle of bacteria, not all of the potential promoters could be used. Which promoter or operon the RNA polymerase used depended on the nutrition in the medium, which was indispensable for bacteria. For example, to make the lactose, the galactose and the arabinose into glucose needed three different groups of enzyme, and they were encoded by three operon genes. Bacteria use correspondent operon according to the available sugar in the medium.

Increase in the rate of synthesis of an enzyme due to the presence of an inducer which acts to derepress the gene responsible for enzyme synthesis. This phenomenon is so-called enzyme induction. Once the inducer was withdrawn, the enzyme production will cease soon. Enzyme induction widely existed in bacteria as a manner of using limited resources economically and reasonably, which formed along with biological evolution path. The bacteria could produce thousands kinds of enzymes, but if there was no substrate to use, this action was just a tremendous waste; or if there was much substrate but no enzyme, the substrate could not be used too. Monod and Jacob investigated the enzyme induction thoroughly in 1965, and then prompt operon doctrine, which revealed the essence of the enzyme induction in the context of transcriptional regulation.

2.2.2.1 Lactose operon (Lac operon)

The lac operon is an operon required for the transport and metabolism of lactose in E. coli and some other bacteria. It consists of three adjacent structural genes, a promoter, a terminator, and an operator. The lac operon is regulated by several factors including the availability of glucose and of lactose. The three structural genes are: lacZ, lacY, and lacA. LacZ encodes β-galactosidase (LacZ), an intracellular enzyme that cleaves the disaccharide lactose into glucose and galactose. LacY encodes β-galactoside permease (LacY), a membrane-bound transport protein that pumps lactose into the cell. LacA encodes β-galactoside transacetylase (LacA), an enzyme that transfers an acetyl group from acetyl-CoA to β-galactosides. As for the control component, there were two kinds: promoter P and operator O. In the P fragment there were RNA polymerase binding site and cAMP-CAP binding site; O fragment was repressor binding site; in the upper reach of the P fragment, there was inhibitor gene, which encoded repressor, and the repressor could inhibit the gene expression.

From the point of view of gene expression, the action of Lac operon was: RNA polymerase binding P, via O, reaching the adjacent three structure gene Z, Y, and A, making a polycistronic RNA via transcription, at last producing three different kinds of proteins. And from the point of gene regulation and controlling, if the structure gene would transcribe into mRNA or not, it would all depend on the controlling of the regulating gene; as for whether the repressor binds O, it would determine the just gene in action or not.

If there was no lactose, the repressor would bind the operator gene. Once the active repressor binds O, the transcription of the RNA polymerase would be interfered. Between the P and O, there was some overlap. If O was occupied by the repressor, it would inhibit the binding of RNA polymerase to the promoter, and then inhibits the transcription of the structure gene Z, Y and A. In presence of lactose, the repressor would bind to lactose, and further alters the conformation of the repressor. As a result, the repressor could no longer bind to the operator region. When repressor loses its function, RNA polymerase could bind to P to induce transcription of the structure gene. Lactose was therefore called inducer since it can induce gene expression. The most common inducer used in vitro study was isopropyl thiogalactoside b (IPTG). The repressor was the main factor, and the inducer could affect the activity of the repressor; only when the repressor was induced to lose its activity, the structure gene could express. Obviously it is negative regulation.

To determine whether a certain prokaryotic gene regulation is positive or negative, we could add certain regulatory factor to the system to see how the operator works. When gene stays closed in absence of regulatory factor means positive control. Transcription starts when regulatory factor is added. In contrast, when gene kept open in absence of regulatory factor means negative control. The regulatory factor in negative control is also called inhibitor or repressor. More complicated cAMP-CAP positive control also existed in prokaryotic organism. See follows.

2.2.2.2 Arabinose operon (Ara operon)

When the bacteria used arabinose as its energy resource, it could produce three enzymes. Arabinose was catalyzed to generate xylulose 5-phosphate, which further enters the embden-meyerhof-parnas pathway. These three enzyme are ribulokinase, arabinose isomerase and phosphoribulose epimerase. The coding genes are AraB, AraA and AraD, respectively. The regulatory elements are I1, I2, O1, O2 and P. Just like the lactose metabolism, the bacteria produces new enzyme to make use of arabinose. Whereas, as we know, the regulation pathway of these two sugar operons are quite different. There was an AraC in the upper reach of the arabinose operon, and the product of the AraC. AraC was different with the repressor of the lactose operon, AraC had positive regulation and negative regulation, dual effects to the arabinose operon. AraC had two binding sites, one was in the I1 and I2 area, and the other was in the O1 and O2 area (about 280 in location). In presence of arabinose, it binds to the I1 and I2 area with the *AraC*, and this procedure conduces to the binding between the RNA polymerase and the promoter. In this scenario, the structure gene *B*, *A*, *D* of the arabinose operon transcribed normally. The positive regulation is apparent. But the *AraC* could bind to the I1, I2 area and at the same time bind to the 280 O1, O2 area in absence of arabinose. Consequently the RNA polymerase can not get access to promoter. Therefore, the transcription of structure gene was blocked. This process was a typical example of negative regulation (Figure 5-1).

Figure 5-1　Arabinose operon (see page 139)

2. 2. 2. 3　Tryptophan operon (Trp operon)

We have mentioned the Lac operon and the Ara operon, which encodes enzymes involved in catabolism. These enzymes catalyzed the metabolism of nutrition. Moreover there were a lot of operons for substance synthesis in bacteria. Just like the tryptophan operon. It encodes the tryptophan synthetase, which is responsible for the synthesis of tryptophan in bacteria. This process was an anabolic. There are two types of regulation of tryptophan operon: via repressor or attenuation. We would just introduce the former one herein.

The tryptophan operon expressed normally in absence of exogenous tryptophan (no tryptophan in medium), which supply sufficient amount of tryptophan for protein synthesis; When exogenous tryptophan is introduced, the bacteria no longer need tryptophan synthesis by themselves, thus the operon is blocked and the synthesis ended immediately. The structure gene of the tryptophan (*A*, *B*, *C*, *D* and *E*) encodes 5 enzymes needed for tryptophan synthesis, the regulatory factors for which were *P* and *O*. The repressor of tryptophan operon is transcribed from *R* gene (repressor gene). Only when it binds to the tryptophan, it acts an active repressor factor to prevent the transcription by binding to the *O* gene. Lacking of tryptophan, the repressor factor is not active. Thus the blocking would be ceased. β-indole acrylic acid is one type of competitive inhibitors of tryptophan. Once it binds to the repressor, the binding between the tryptophan and the repressor was abrogated, so the repression was relieved and the transcription was facilitated. In genetic engineering, when we used the vector with the Trp operon (from the tryptophan operator), we usually used β-indole acrylic acid to enhance the transcriptional level.

2. 2. 2. 4　cAMP-CAP positive regulation system

Many types of microbes used only one type of sugar, but some other bacteria, like *E. coli*, could use both glucose and lactose. Glucose is the sugar most ready for bacteria growth. When glucose enters into the cells, no new enzyme must be produced. Therefore, when both glucose and lactose are in the medium, the bacteria always prefer glucose. When no glucose remains, the bacteria began to use lactose instead. This phenomenon was called "glucose effect", which associates the regulation of cAMP-CAP.

cAMP was discovered in the 1950s, and we have known it as the second messenger in hormonal regulation. In 1960s, scientists noted that the content of glucose and cAMP in the medium of *E. coli* were inversely correlated. This was not the direct inhibitory effect of the glucose, but the metabolites of glucose inhibits the activity of adenylate cyclase, which results in the reduction of cAMP level. When cAMP is added into the medium, the expressed β-galactosidase will be released. cAMP works by its association with CAP (catabolite gene activator protein). CAP was also called CRP (cAMP receptor protein), which positively regulates prokaryote gene expression. CAP could convey the glucose starvation signal to many operons, so it could activate the lactose operon, galactose operon and maltose operon, etc. Without glucose, bacteria will use other carbon resources. *In vitro* transcription of lactose operon showed that lactose operon is controlled by LacI (repressor) and CAP. In another word, the lactose operon is positively regulated by the cAMP-CAP complex and negatively regulated by the Lac repressor. If there is no lactose, whether there is glucose or not, it could not produce Lac mRNA, because the operon is bound by the repressor. If there is lactose, the repressor will bind to stop the repression. But if there is some glucose, cAMP will be maintained at low level, thus the affinity between cAMP-CAP and promoter is low. Consequently, only very low level of Lac mRNA was produced. If there is some lactose but no glucose, the transcription of Lac operon will reach its maximum. In this case, lactose renders the repressor inactive; Glucose deficiency increases cAMP level via binding CAP. The affinity between the cAMP-CAP and the promoter increased either. The operon is then triggered to start transcription.

AraC and CAP is needed for the maximal activation of the transcription of arabinose operon. Though the regulation of the Lac and Ara operon is very complicated, it is beneficial for the bacteria. As long as glucose is sufficient, hardly any cAMP is produced. So the CAP is inactive, and the enzymes to digest other types of sugar are futile.

When other sugar but glucose is present, transcription will be activated to make use of those sugars. Scientists found that CAP is close to the binding site of the RNA polymerase, so they inferred that CAP could attract the RNA polymerase for the start of transcription. The regulation by cAMP-CAP is a very common process. Except for some special enzymes to digest some sugar, many processes in bacteria are sensitive to the glucose effect. For example, many enzymes involved in tricarboxylic acid cycle and respiratory chain, digestion of various carbon sources, degradation of some amine acid, antibiotic synthetase, etc.

In summary, the operon system was most economical and efficient way for the prokaryotic organism. Binding to a group of related genes, but not one by one, makes the regulation agile and consistent.

2.2.3　RNA Polymerase Activity Modulation

RNA polymerase regulation is also known as magic spot nucleotide modulation. Guanosine tetraphosphate (ppGpp) and Guanosine Pentaphosphate (pppGpp) appears to mediate the cessation of production of rRNA in the absence of an essential amino acid, a phenomenon present in stringent bacteria. ppGpp form a complex with the RNA polymerase and change the conformation of the polymerase, which, in turn, lead to the decreased polymerase activity. As a result, the synthesis of rRNA, tRNA and mRNA is reduced or stopped. But it does not happen when the amino acids were ample.

Abnormal nucleotides result in the appearance of spots on the chromatography. This is so-called alarmones, it signals a cell's response to deficiency of amino acids. The magic spots signaling bacteria for the synthesis of rRNA and tRNA was no longer needed because protein synthesis was abrogated due to lack of amino acids. Consequently, cells then responses as follow, inhibit the synthesis of ribosomal and other macromolecular, activate the transcription of some amino acid operator (such as tryptophan operon), activate the proteinase to inhibit some unnecessary transportation systems, thereby cells could live by the difficult situation.

2.3　Pre-Transcriptional Regulation

It occurs within the genome and is regulated through DNA rearrangement. The most typical example is the alternative expression of two flagella antigen in *Salmonella*. There were two flagellum antigens in *Salmonella typhimurium*, which are expressed by two unlinked gene *H1* and *H2*. Only one gene is expressed at one time, it means that when gene *H2* is expressed, *H1* switches off vice versa. There is a switch between *H1* and *H2*. What choice of gene expression to make? Now we know that gene *H2* is far away from *H1* in chromosome but is coupled with the *H1* repressor gene. There is a 970bp sequences upstream the gene *H2* which is called hin, an inversion gene, where *H2* promoter was. Two 14bp reverse repeat sequences which are called IRL and IRR located at the two end of hin. When *H2* gene is expressed, *H1* repressor gene is also be expressed, this could repress the expression of *H1*. It means *H2* is expressed and *H1* is inhibited (Phase Ⅱ). Another situation is that reversion protein could reverse the sequence of hin, IRL and IRR. So the *H2* promoter is shifted to another end with altered direction far away from structure gene. Thus, *H2* gene is suppressed and *H1* repressor gene is also inhibited. This could not repress the *H1* any longer. Accordingly *H1* is expressed (Phase Ⅰ).

Inverted modulation is perhaps a way utilized by almonella to evade from immune destruction during the infection. For example, the bacterial which were in the phase Ⅰ produce H1-flagellum initially. But host cells produced antibodies specific to H1-flagellum which will eradicate them as amplification of bacteria. At this time if the bacteria shift to the phase Ⅱ at a high frequency, a large number of bacterial may survive and expand.

2.4　Translational Regulation

2.4.1　Regulation by Anti-Sense RNA

It is well known that proteins regulate transcription as repressors or activators. It was found recently that some small regulatory RNAs could modulate gene expression, such as antisense RNA.

Antisense RNA is small single-stranded RNA transcripts, usually rather short (70~200 bp), which binds to specific mRNA via its complementary region. As a result, such binding will prevent the translation of mRNA. Due to the specific relationship with mRNA, it is named antisense RNA. It was used to be called mRNA interfering complementary RNA. Mizuno found that osmotic pressure controlling OmpR protein had different conformations under different osmotic pressures in the study of the osmotic pressure regulated outer membrane protein gene expression in *E. coli*, which had different effects on the modulatory region of osmotic pressure protein OompF and OmpC expression. Osmotic pressure gene *ompF* and *ompC* aren't linked. Upon low osmotic pressure, OmpR protein positively regulates *ompF* gene expression. OmpF synthesis is therefore increased and OmpC synthesis is inhibited. Upon high osmotic pressure, OmpR protein conformation changes and positively regulates *ompC* expression. OmpC synthesis increased whereas OmpF synthesis is inhibited. As it is aware that there is a DNA sequence transcribed at the reverse direction when *ompC* gene transcribes at the upstream of OmpC gene promoter which regulates *micF* gene expresion. This process produces a 174 bp antisense RNA which can form the hybrid double strands with 5′ terminal in ompF RNA including SD sequence and coding region (including AUG) to inhibit the translation of ompF. As a result, the more OmpC transcribes, the more ompF antisense RNA produced, the less OmpF protein generated (Figure 5-2).

Figure 5-2 **E. coli osmotic pressure regulation** (see page 142)

The gene expression regulation by antisense RNA reveals a new mechanism of gene regulation. The research in prokaryotic cell indicates the basic effect of antisense RNA is to form a dimmer with specific mRNA through base pairing thereby blocks the expression of mRNA.

2.4.2 Stability of mRNA

The proliferation cycle of bacteria is 20~30min. The regulation of metabolic reaction is also very fast. Fast regulation of transcription initiation and transcription termination, as well as fast degradation of mRNA is needed to keep high speed renewal of mRNA. The degradation speed of mRNA is another important mechanism of translation control.

The half-life of prokaryote mRNA is in the big difference, with variation from seconds to minutes (2~3 minutes by average). It highly correlates with the structure of mRNA, physiological state and environmental factors. The hairpin structure of 5′-terminal or 3′ terminal of mRNA can protect mRNA from rapid enzyme hydrolysis by exonucleases to sustain its stability. RNase Ⅲ can recognize the specific hairpin structure and hydrolysize it and the rest of part will be degraded by other RNases. The life RNA will be extended if hairpin structure is protected. Some special regulatory proteins can bind to hairpin structure to regulate mRNA stability.

In the development of molecular biology, prokaryotic regulation of gene expression has made many remarkable achievements, especially in the operon theory and its applications such as metabolism regulation. These have become important links in the understanding of the nature in the prokaryote for the benefit of human. It also inspires further investigation of eukaryotic gene regulation.

Section 3   Regulation of Gene Expression in Eukaryotes

3.1 Overview of Regulation of Gene Expression in Eukaryotes

Eukaryotes genome is significantly larger than prokaryotic genome and more complicated in structure and function. Therefore its regulation of expression is more sophisticated and refined, which bring the difficulty for research in the gene regulation of eukaryotes (difficult to directly identify the biochemical process of gene regulation; difficult to directly select mutants; difficult to study the diversity of gene expression by directly changing outer environment conditions; difficult in handling). However, with the recent progress of molecular biology, and completion of human genome project and model organism genome project, important progresses have been made in the structure, function and regulation of eukaryotic genome.

The research on gene expression of eukaryotes is critical to decipher the secrets of living organism.

Most eukaryotes are multi-cellular organism and activate genes transcription in the specific cells on the timely basis to direct predetermined differentiation and development via regulation of gene expression. Generally speaking, the reactions to environmental between prokaryotes and eukaryotes are very different. Most of eukaryotic cells stay in relatively constant environment and sudden changes of outer environment is not generally expected. This is the way to maintain the functional stability at different levels upon myriads of changes. But there are some exceptions. For example, gene expression in hepatic cells could be regulated by the condition of intern nutrients and toxic substances absorbed from outside. For instance, the genes of enzymes related to glyconeogenesis in mammal hepatic cells can be activated upon low carbohydrates diet; Chromium and some other heavy metals could combine with metallothionein in hepatic cells to remove the toxicity from the heavy metal; some medicine such as phenol barbital, codeine, morphine, and some carcinogens could induce the synthesis of cytochrome P450 in the hepatic cells—a kind of monooxygenase which could increase the dissolubility of these medicines and carcinogen by hydroxylation and accelerate the excretion of poisons. The above mentioned regulation of gene expression in eukaryotes is called transient regulation, or reversible regulation, which is similar to the reaction to those in prokaryotes in response to the environmental changes. Transient regulations include change of levels of certain substrates or hormones, and regulation of activity and concentration of enzymes in different stages of cell cycles. Developmental regulation (irreversible regulation) is another type, also the essence of gene expression regulation of eukaryotes because the gene expressions in most cells of eukaryotes are related to the development and differentiation of organism. Normally, somatic cells strictly follow certain track for individual development. Therefore, the regulation of development is critical in the progress of cell growth, differentiation and development of eukaryotes.

Gene expression regulations of eukaryotes are implemented on five stages, pre-transcription, transcription, post-transcription, translation, post-translation. Generally speaking, similar to prokaryotes, transcriptional regulation in eukaryotes is the most important stage. But translational regulation is also very important for eukaryotes since the transcription and translation are completely separated in spatially and temporally.

3.2 Pre-Transcriptional Regulation

Pre-transcriptional regulation means alteration of genetic structure at the level of genome. This regulation is stable and lasting.

3.2.1 Gene Amplification

Gene amplification is a phenomenon that copy number of certain gene is higher than normal in cells. It is a regulatory approach that cells meet a need to generate sufficient product in a short period of time. There is intense demand of a gene product in certain period of differentiation and development. Increased gene copies can meet this demand in case elevated gene transcription alone is insufficient. For example, the copy number of rRNA gene increases by 4000 times in ovocytes which is about 500 in the Xenopus larva's somatic cell. It is because the split of ovocyte needs a large amount of proteins. As a result, the need of rRNA is increases rapidly to accumulate 10^{12} ribosomes. 500 years will be needed to get so many ribosomes without gene amplification.

In tumor cells, copy number in some oncogene increases aberrantly, leading to the aberrant gene expression and proliferation.

3.2.2 Gene Rearrangement

Gene rearrangement refers to alteration of some gene fragments in sequential order. Gene rearrangement can not only create new genes, but also regulate gene expression. One of the best examples of gene rearrangement is the rearrangement of the encoding region of mammalian immune globulin gene. It's known that mammals will produce antibody (immune globulin) after challenge with antigens. Roughly estimated, the types of immune globulin could be up to several million. The maximal number of protein coding genes in the genome is only 30 000 to 50 000. How does it happen to code so many immune globulins? Where was the antibody library from? It was found that antibody genes rearrangement happened during the differentiation from embryonic cell to B cell (antibody-producing cell) twice. The first one occurred in pre-B cell stage. The genes encoding immunoglobulin variable region are involved to make mutual separate regions connect with each other after the rearrangement, said VDJ complex, a complete gene of Variable region of heavy chain. The heavy chain rearrangement is followed by light chain V region gene rearrangement to form a VJ complex, which is a complete gene of V region of light chain. The second one occurred in the mature B cells upon antigen challenge, which modify heavy chain to form class switch. The antigenicity remains the same and B cells become plasma cell with differentiation and development. It's known that there are several hundred genes for V, κ, λ, μ and several fragments for J and D. Thus a limited fragment of genes in the genome form numerous types of antibody V regions through gene rearrangement.

3.2.3 DNA Methylation

In vertebrates, c on a specific CG sequence of DNA can be modified via methylation (methylation is located at C5 cytosine on DNA, $5'^mC$). The methylation can prevent the transcription of some genes and could also be inherited to daughter cells. The research shows that low or no methylation correlate with active gene and a high degree of methylation correlates with gene expression silence. Apparently gene expression and the degree of DNA methylation negatively correlate with each other. The demethylation on a particular region of DNA, especially the regulation sequence around $5'$ terminal can increase transcriptional activity. Genes expressed at specific tissue, such as the promoter region of house keeping genes, show the high level of methylation. The silent gene can be re-activated by DNA demethylation due to stimulation with hormone or carcinogens. Therefore, aberrant DNA methylation may be another mechanism involved in cardiovascular disease and cancer development.

3.2.4 Chromosome Modification

One of the important characteristics of eukaryotic gene is combination of genomic DNA and proteins to form the nucleosome as the basic unit of chromosome structure in the nucleus. Such a structure forms the distinct mechanism at the chromosome level before the eukaryotic gene transcription. In other words, the precondition of gene transcription is a series of important changes of chromosome structure. The active transcription of gene is in the somatic chromosomes. At the time of transcription, chromatin forms a loose structure from a dense structure in order to make DNA *cis*-acting element accessible to RNA polymerase and transcription factors. The chromatin is easily hydrolyzed by the non-specific DNA endonuclease such as DNase I when it's in a loose structure. The formation of the DNase I hydrolysis sensitive areas are known as DNase I sensitive sites. When apply a very low concentration of DNase I to chromatin, the hydrolysis would occur in a few site-specific locations. These are where hypersensitive sites on the chromosome with active gene expression. Each active gene at chromatin has one or more such sites, but not for non-active genes. For this reason, the ultra-sensitivity to DNase I is the sign of active transcription of genes.

3.3 Regulation at Transcriptional Level

Regulation of eukaryote gene expression at transcriptional level is probably the most important step in gene regulation. The major components involved in the process include RNA polymerase, *cis*-acting element and *trans*-acting element.

3.3.1 RNA Polymerase

The gene transcription is catalyzed by RNA polymerase. For both prokaryotic and eukaryote, binding of RNA polymerase to promoter is the key point. The difference between them is that bacteria RNA polymerase identifies DNA sequence, but eukaryote RNA polymerase identifies DNA-protein complex. Only when one or more transcription factors bind to DNA to form functional promoter, the promoter can be recognized by RNA polymerase. There are three types of Eukaryotic RNA polymerases for three different types of RNAs. Transcript factors are therefore classified by TF Ⅰ, TF Ⅱ, and TF Ⅲ.

The gene of transcripts by RNA polymerase Ⅱ is called class Ⅱ gene, which is in largest number and exert many different functionality.

3.3.2 *Cis*-Acting Element

Cis-acting element is the DNA sequence with structural genes, including promoter, enhancer, silencer, termination signal, and so on.

3.3.3 *Trans*-Acting Factor

Trans-acting factor is one group of nuclear protein factors, which interact with *cis*-acting elements and RNA polymerase to regulate gene transcription. A complete *trans*-acting factor includes two types of domains: DNA binding domain and transcriptional activation domain. *Trans*-acting factor binds to special DNA sequence via the former, and then activate transcript via the later. Some may have only one domain, and only when two complementary proteins exist in the same cell it might be functional. This may be connected with tissue specific gene expression.

3.3.3.1 DNA binding domain

To play a role of transcription regulation, *trans*-acting factor must possess DNA binding domain. Most of DNA binding domains have less than 100 amino acids. Four types of those are listed below:

a) helix-turn-helix (HTH): This is a major structural motif capable of binding DNA. It is composed of two α helices joined by a short strand of amino acids and is found in many proteins that regulate gene expression. Of the 2 α-helix, one is recognition helix (carboxyl terminal), whose amino acid residue binds to major groove of target DNA; the other (amino terminal) cuts through the major groove and non-specific binds to the phosphoric acid pentose frame of DNA. The first helix-turn-helix found in eukaryotes was from Homeo-box protein which controls the early development of *Drosophila*. So this model is also named homodomain (HD). The *trans*-acting factors with this kind of structure connect to development and differentiation.

b) Zinc finger: Zinc finger is a large superfamily of protein domains that can bind to DNA. A zinc finger consists of two antiparallel β-strands, and a α- helix. The zinc ion is crucial for the stability of this domain type, in the absence of the metal ion, the domain unfolds as it is too small to have a hydrophobic core. The structure of each individual finger is highly conserved and consists of about 30 amino acid residues, constructed as a ββα fold and held together by the zinc ion. The α-helix occurs at the Cterminal part of the finger, while the β-sheet occurs at the Nterminal part. The consensus sequence of a single finger is: $Cys-X_{2-4}-Cys-X_3-Phe-X_5-Leu-X_2-His-X_3-His$. It is found that there are similar Zinc finger in DNA binding domain of SP2 and other proteins in mammalian cell.

c) Leucine zipper: Leucine zipper is a DNA binding motif of the trans-acting factors. It's made up of about 30 amino acid residues. It has two parts: the part with mainly basic amino acids binds to DNA, the other part has a leucine every 6 amine acids, forming amphoteric α-helix. One side of the α-helix is mainly made up of charged amino acid residue such as Arg, Gln, Asp, so it is hydrophilic, the other side consists of leucine amino acid residues, so it is hydrophobic. The latter is called leucine zipper. Two tans-acting factors with leucine zipper domain form leucine zipper. Via hydrophobic interaction, there are many such special sequences between *trans*-acting factors. So this structure is related to the interaction among *trans*-acting factors.

d) Helix-loop-helix (HLH): H-L-H structure consists of three parts: ①100~200 amine acids form two amphiprotic α-helix domains. The amphiprotic α-helix includes hydrophobic surface which mainly consists of leucine and hydrophilic surface that consists of hydrophilic amino acids. ②Between the two helixes there are peptides of different length. ③The N terminal of α-helix consists of basic amino acids with positive charges. When closing to DNA, these positive charges are neutralized by the phosphate radicals of the DNA, and form a stable α-helix similar to the major groove of DNA. This type of structure is discovered in the enhancer binding protein E12 and E47 of immunoglobulin genes (Figure 5-3).

Figure 5-3 Helix-loop-helix (see page 147)

3.3.3.2 Transcriptional activation domain

In eukaryotes, the function of *trans*-acting factor becomes very intricate because of protein-protein interactions. The transcriptional regulation is quite often accomplished via forming a complex. Thus not every *trans*-acting factor needs to bind to DNA directly. Transcriptional activation domain is therefore indispensable for *trans*-acting factor.

Trans-acting factor is highly diversified and so is its transcriptional activation domain, which is made up of 20~100 amino acids. Sometimes a *trans*-acting factor has more than one transcriptional activation domain. The structural characteristics of transcriptional activation domain are: First, there are many negatively charged acidic amino acids to form lipophilic α-helix that exhibits hardly any homology. Second, it is rich in glutaminate also with no significant homology to each other. Third, some *trans*-acting factors are rich in proline residue. Proline is an amino acid that could hinder the formation of α-helix.

3.3.4 The functional manner of trans-acting factor

In the start of gene transcription, interactions between *trans*-acting factor and *cis*-acting factor, *trans*-acting factor and *trans*-acting factor exist. The expression of different genes is regulated by different combination of *trans*-acting factors, which make the process extremely complicated. The *trans*-acting factors exhibit the following functional manners.

3.3.4.1 One *trans*-acting factor to more than one *cis*-acting element

Unlike the interaction between *cis*-acting elements and modulator proteins in prokaryote, the function of *trans*-actings in eukaryote is not highly specific. Some *trans*-acting factors can bind to *cis*-acting elements that are rarely homologous. For example, the *trans*-acting factor HAP-1 in Yeast could bind to *cyc1* and *cyc7*, whose DNA sequences are not homologous. C/EBP could recognize both CAAT box, and some enhancer sequences of animal virus.

3.3.4.2 One *cis*-acting element to more than one *trans*-acting factor

CAAT box could bind to many types of *trans*-acting factors, such as CTF family. CTF could bind to CAAT, and it can also work in the replicating adenovirus. The CP1 of CP family has high affinity to the CAAT box of α-globin gene; CP2 has high affinity to CAAT box of γ-fibrin gene. In addition, CBP factor and ACF factor could recognize CAAT box, too.

3.3.4.3 Dimer or polymer interacts with *cis*-acting element

Most of dimerization forms heterodimers. For example, Fos/Jan heterodimers could recognize TGACTCA; E12/E47 heterodimers could recognize the enhancer of κ chain gene of immunoglobulin. There is one type of trans-acting factor, μEBP-C (also named NF-μE3), which binds to the enhancer of immunoglobulin heavy chain genes and the enhancer of kappa light chain. The tetramer exhibits the highest activity of all the forms, as high as 7 times as that of dimer.

3.3.4.4 Combinatorial gene regulation

Every *trans*-acting factor promotes or suppresses gene transcription via its binding to the *cis*-acting element. However, gene expression regulation is controlled by several factors combined instead of just one factor. This type of regulation is called combined gene regulation. *Trans*-acting factors binding to different *cis*-acting factors work together to decide the transcription activation levels of the genes.

3.4 Post-Transcription Regulation

Transcription in eukaryotes ensures the transfer of inheritable message from DNA to RNA. As a result, regulation of transcription is critical to control the amount of RNA in the cells. However, the amounts, structures, or even types of RNA are not necessarily the same in nucleus or cytoplasm. Moreover, differences also exist between active and inactive cells. These suggest that post-transcriptional control can also affect gene expression for different cellular phenotypes besides regulation at DNA and transcriptional levels.

Post-transcriptional regulation generally refers to processing of transcript, including modification of pre-mRNAs, splicing, mRNA transport, RNA editing, mRNA stability.

3. 4. 1　Regulation of Capping and Tailing

mRNA in eukaryotes would be capped after transcription, forming a special structure of 7-methyl guanosine triphosphate at $5'$ end. Those three types of caps have different levels of methylation. Its structure can also closely affect the efficiency of protein expression: ①the structure of cap is critical to stabilize pre-mRNA in nucleus as well as cytoplasm; transcripts without caps will be degraded by nuclease rapidly; ②the cap can be identified by the initiation complex of protein synthesis to facilitate translation; ③Caps (e. g. GpppN-) without methylation (m^7G) or removal of caps (by chemicals or enzymes) significantly abrogate the translation; ④analogs of cap such as m^7GMP, etc. can inhibit the translation of capped mRNA but have no impact on mRNA with no caps.

Most eukaryotic mRNAs attached to ribosome would have a polyA tail at $3'$ end, with about $50\sim150$ A nucleotides. Except mRNA for histones, eukaryotic mRNAs contain polyA, which is the result of tailing after transcription. Generally, the signal sequence for tailing is AAUAAA. The poly (A) tail is formed to stabilize mRNA. The length of poly (A) tail of some mRNAs is manipulated so that it can be alternatively shortened or elongated.

3. 4. 2　Alternative Splicing on Expression Regulation

The pre-mRNA of eukaryotes needs further cleavage of introns and splicing of exons in addition to capping and tailing. This process called splicing. Research of RNA splicing is one of the most spiritoso subject in the area of biochemistry and the molecular biology in the 1980s.

Since one gene can produce at least two transcripts for at least two different proteins, the traditional concept of one gene to one peptide will have to be extended. Alternative splicing solves the problem of the interrupted gene splicing and also be invaluable to the research of the origin of life. (For details sees chapter 3)

3. 4. 3　RNA Editing

The RNA edition (RNA editing) is a unique way of genetic information processing, that is to say the insertion, the deletion or the transformation in the code area happens after transcription—beautification appears on RNA. Because of nucleotide replacement insertion flaw and so on, mRNA changes genetic information which DNA template originated after transcription, thus it codes many kinds of protein with different amino acid sequences. The edition result not only expanded the genetic information, but also one kind of possible protective measures in biotic adaptation.

a) Nucleotide replacement: An important example of RNA editing in mammals involves the *apoB* gene, which encodes two alternative forms of the serum protein apolipoprotein B (ApoB): ApoB-100 expressed in hepatocytes and ApoB-48 expressed in intestinal epithelial cells. Both ApoB proteins are components of large lipoprotein complexes that transport lipids in the serum. However, only low-density lipoprotein (LDL) complexes, which contain ApoB-100 on their surface, deliver cholesterol to body tissues by binding to the LDL receptor present on all cells. The cell-type-specific expression of the two forms of ApoB results from editing of *ApoB* pre-mRNA so as to change the nucleotide at position 6666 in the sequence from a C to a U. This alteration, which occurs only in intestinal cells, converts a CAA codon for glutamine to a UAA stop codon, leading to synthesis of the shorter ApoB-48. Studies with the partially purified enzyme that performs the posttranscriptional deamination of C6666 to U shows that it can recognize and edit RNA as short as 26 nucleotides with the sequence surrounding C6666 in the *ApoB* primary transcript.

b) Nucleotide insertion or deletion: The nucleotide sequence of mitochondrial pre-mRNAs in trypanosomes is posttranscriptionally edited by the insertion and deletion of uridylate (U) residues. In some RNAs editing is limited to small sections but in African trypanosomes, such as Trypanosoma brucei, 9 of the 18 known mitochondrial mRNAs are created by massive editing which can produce more than 50% of the coding sequence. In all cases, however, RNA editing is a key event in gene expression during which translatable RNAs are generated. The information for the editing process and possibly also the inserted

Us are provided by small guide RNAs, which are encoded in both the maxicircle and minicircle components of the trypanosome The information for the editing process and possibly also the inserted Us are provided by small guide RNAs, which are encoded in both the maxicircle and minicircle components of the trypanosome mitochondrial DNA (Figure 5-4).

Figure 5-4 RNA editing (see page 149)

3.4.4 mRNA Transport

Not all the mature mRNA will enter the cytoplasm. Only 20% of mRNA gets into the cytosol and those in the nucleus will be degraded in around one hour.

The mechanism of RNA export is not revealed yet. But RNA export to cytoplasm is controlled at least by ①Active transportation via nuclear pore; ②Most of RNAs have to be capped, tailed and spliced before the export.

mRNA will be exported from nuclear pore to the specific locations; some will be transported to endoplasmic reticulum for the synthesis of peptides; whereas some will be transported to the ribosome in the cytosol for translation.

3.5 Translational Control

Translational regulation is also a critical step in the expression control. Different mRNA species exist in the same cells, but their comparable protein products are not necessarily the same. This is mainly due to the differential translational control over the speed of synthesis and mRNA half-life.

3.5.1 Phosphorylation of Initiation Factor

The initiation stage is the most critical stage in the protein synthesis. Many protein factors exert import effects herein. Eukaryotic initiation factor, eIF-2 is probably the most thoroughly investigated one.

eIF-2 inhibits the protein synthesis of most protein after its own phosphorylation. The activity of eIF-2 in pre red blood cells can be inhibited by phosphorylation of one of its three subunits. This phosphorylation is catalyzed by cAMP dependent kinase activation.

3.5.2 mRNA Stability

mRNA is the template for protein synthesis. Generally, the synthesis of certain protein in the cytoplasm was directly proportional to its mRNA level. A number of house-keeping genes in eukaryotic cells and some mRNA species in the highly differentiated cells are extremely stable with a life up to a few days. The level of protein synthesis is also very high. mRNA stability depends both on its own second structures and post-transcriptional modification.

3.6 Post-Translational Control

3.6.1 Signal Peptide

Signal peptide (signal peptide) is usually made of 15~30 hydrophobic amino acids, which enables the proteins from the reticulum to enter the Golgi body. Once in the Golgi body, the signal peptide will be cleaved to form biologically active proteins. Further processing might be needed. For example: Proinsulin is the prohormone precursor to insulin made in the beta cell of the islets of Langerhans. It is synthesized in the endoplasmic reticulum, where it is folded and its disulfide bonds are oxidized. It is then transported to the Golgi apparatus where it is packaged into secretory vesicles, and where it is processed by a series of proteases to form mature insulin. Mature insulin has 39 fewer amino acids; 4 are removed altogether, and the remaining 35 form the C-peptide. The C-peptide is abstracted from the center of the proinsulin sequence; the two other ends (the B chain and A chain) remain connected by disulfide bonds.

3. 6. 2　Amino Acid Peptide Modification in the Neonatal Peptides

Released from the ribosome, the peptide chain is not the mature protein with biological activity. Further modification (e. g. phosphoric acid, hydroxyl, glycosylation, acetylation, etc.) are needed for the correct folding and assembly of proteins.

Why eukaryotes not synthesize functional proteins directly but take post-translational processing? There are only 20 protein codons for 20 amino acids, whereas more than 100 amino acid species are needed to produce via post-translational modification and processing. Not all the amino acid is essential for the proteins. Post-translational modification should be more reasonable. Some modification is needed only temporarily. Phosphorylation/dephosphorylation, or acetylation/deacetylation of lots of protein plays an important role. In short, the protein post-translational modification is an important way of gene regulation, this, in turn, enhances the adaptability of the organisms.

3. 6. 3　Correct Folding of Neo-Peptides

As the most basic structure of the proteins, the primary structure of the neo-peptide was determined by the genetic information. The spatial structure of protein is the basis of their biological functions. Correct folding of peptide chains is needed for the correct spatial structure. Some genetically engineered protein product is functionally quite different from the natural proteins due to a difference in neo-peptide folding.

Two types of proteins are involved in neo-peptide folding: The specific enzymes directly catalyze protein folding, such as protein disulfide isomerase, proline cis-isomerase; The ones assist the folding of the neo-peptides into mature proteins, known as molecular chaperone (molecular chaperone). Molecular chaperone proposed by Lasky in 1978. Widely existed in prokaryotes and eukaryotes, molecular chaperones regulate and stabilize folded or unfolded part of the peptide, and to prevent inappropriate interaction between chains and inter-chains. They have also mediated mitochondrial transmembrane protein functionality, regulate signal transduction and participate in microtubule formation and repair. Heat shock protein family 60, 70, 90 are most common highly conservative proteins. Their functional mechanism has not been unanimously identified and further investigation is warranted.

Section 4　RNA Interference and Gene Silence

4. 1　The Characteristics of RNAi

RNA interference (RNAi) is a process of post-transcriptional gene silencing (PTGS) by which double stranded RNA (dsRNA), when introduced into a cell, causes sequence-specific degradation of endogenous homologous mRNA sequences. It is wildly occurred in fungus, plant, invertebrate and mammal (not found in Prokaryote).

RNAi was evaluated as one of ten big technical breakthroughs by the Science magazine in 2001. The reason that the scientists display the enormous attention to the RNAi phenomenon lies in that RNAi has its characteristics and superiority in the gene function and in the related aspect research in which many conventional routes are unable to compare. ①Specificity: The most dominant character of RNAi is to cause mRNA degradation which is only homologous to dsRNA. dsRNA can specifically suppress the transient expression of external transgene or endogenous gene in the Drosophila cell without affecting other irrelevant gene expression. ②Highly effective: Regardless in vivo or in vitro, only a few of dsRNA molecules in each cell can effectively suppress the target gene expression. The efficiency of suppression is more than 90% in primitive animals. This indicated that dsRNA mediated RNAi is a catalytically amplified process. ③Length restriction: dsRNA which can induce effective RNAi should not be at least 21 bp by length. Moreover long dsRNA is also cut by Dicer into about 21nt siRNA in the cell to mediate mRNA degradation. ④ATP dependence RNAi will be affected or even abrogated upon ATP deprivation in the system, indicating that this process is ATP dependent. Dicer and RISC catalyzed reactions are most likely ATP dependent. ⑤Transmissibility RNAi possess astounding ability to trespass cell boundary, thereby the siRNA silencing signal may be enlarged and transferred. siRNA can serve as the primers for the target sequence to synthesize more dsRNA under polymerase (RdRp) which is RNA dependent, which is further cut by dicer into more siRNA to amplify the effect of RNAi. RNAi may disseminate between the *Drosophila* cell communities. Local injection of dsRNA in the *C. elegans* can be propogated to the whole.

4. 2　Gene Silencing

As a very important approach of gene regulation, gene silencing is a general phenomena in the evolution in which expression of a specific gene is suppressed. According to mechanisms and levels, it can be grouped into three categories: transcriptional gene silencing (TGS), post-transcriptional gene silencing (PTGS) and position effect.

Later, several groups showed that PTGS can also affect transgenes that are not homologous to endogenous genes, suggesting that this phenomenon is not a simple regulatory mechanism that controls the expression of endogenous genes. RNA interference (RNAi) strongly suggests that a mechanistic link between PTGS, quelling and RNAi exists. Currently studies suggest a model of PTGS whereby long dsRNA is processed by RNase Ⅲ-like enzyme (Dicer) into small dsRNA molecules called small interfering RNAs (siRNAs). siRNAs direct a nuclease to the target mRNA, but degradation of the mRNA by the small RNA guided nuclease most likely requires other enzymatic activities. The study of how PTGS is triggered has revealed the existence of at least three steps: initiation, propagation and maintenance.

Recovery can be induced after viral infection, indicating that viruses can trigger PTGS. The induction of PTGS by viruses is confirmed by the observation that endogenous genes or transgenes can be silenced after infection with recombinant viruses carrying partial gene sequence. This phenomenon is called virus-induced gene silencing. In the case of VIGS directed against transgenes, plants recover from virus infection, but transgene silencing and methylation persist in the absence of the virus. This suggests that VIGS induces the production and the propagation of a silencing signal in the uninfected parts of the plant, which triggers transgene silencing and immunity against the virus. In contrast, in the case of VIGS directed against endogenous genes, the virus persists in plants, and the endogenous gene remains unmethylated. This suggests that endogenous genes cannot produce the silencing signal.

4. 3　Some Considerations of RNAi and Gene Silencing

RNAi has been used to study the function of candidate genes. The development of stable and inducible expression vectors driving the expression of short hairpin RNAs (shRNAs) has further expanded the application of RNAi both in tissue culture and in animal models. Firstly, we can study a gene family and a concrete gene function through RNA interference technology in coding region. It is more quickly and effectively than the traditional anti sense RNA technology and gene knockout technology. And also through the combination of cell-specific promoter and inducible system, we have a choice to use RNAi in different growth time or in the different organ, the shortcomings that we cannot observe the phenotypes due to the death of embryos caused by homologous recombination can be overcome. Furthermore, a number of successful animal trials indicate that RNAi might ultimately become a potent therapeutic approach for the treatment of various human diseases. For instance, the tumor is one kind of polygenic hereditary disease, and the results of interaction of multiple genes forming gene regulation network. But the PTGS/RNA interference possibly specially suppresses many cancer genes expression in vivo, moreover does not affect the functions of other genes.

It is worth noting that the cause that dsRNA forms to lead corresponding gene silence via dsRNA suppressing gene expression has highly latent effective, therefore there has been a set of mechanisms that strictly prevent dsRNA formation in the normal organism. Because the organism is a complex system, interaction between many kinds of mechanisms exist to control gene expression. In vitro RNA interference experiment, we see whether RNA interference suppress gene expression successfully or not as well as the degree of inhibition depends on the types of cell. For *Caenorhabditis elegans*, we can transfer dsRNA using injection, soaking and feeding, but to the mammal, finding the highly effective way to transfer siRNA as well as the fast way to screen the RNA interference needs further exploration. The application of RNA interference in anti-virus infection is exciting, but it cannot guarantee success. There have been many difficulties in the research: firstly. it is not all viral RNA sequence that can quite easy close siRNA, recognized and cut. Some viral target sequence possibly hides under the secondary structure or is located in the region that is highly folded. Furthermore, some viral sequence and the protein may form the close compound which hinder siRNA recognize viral sequence. Secondly, the process of viral replication is not strictly strong capriciousness, so often generating offspring with mutation. This characteristics can help the virus evade the body's immune surveillance and drug inhibition. It will also interfere with recognition of siRNA. Thirdly, the virus codes the factors which inhibit RNAi. Up to now it has been identified from Plant viruses to more than 20 kinds of RNAi inhibitor. At the same time a variety of inhibitory factor has been found in vertebrate animals and insects. This inhibitory factor may enhance virus's pathogenicity and promote virus replication. Lastly, RNA enzyme can degrade siRNA in cells, so how to ensure the stability of siRNA is still an urgent need to solve the problem. Furthermore, we also face technical difficulties in the introduction of siRNA.

Summary

The Central Dogma of genetic information transfer is one of the Basic Rules of Life, and is also considered universal in prokaryotes and eukaryotes. Mechanism of gene expression regulation is involved in the whole process of genetic information passage. That is to say, life phenomena are based on gene expression regulation.

Prokaryotes are unicellular organisms. They depend on the surrounding environment. Therefore, they need to regulate gene expression swiftly in order to adapt to the environment and survive in nature. The characteristics of architecture and ways of gene expression regulation in prokaryotes are adjusted to the environment. As is well known, the operon is a basic unit of gene expression regulation in prokaryotes. Prokaryotes may turn genes on and off (or more finely regulate gene expression) in economical way. The emergence of the operon theory helps us enlarge the concept of gene. Besides the genes that encoding primary structure of protein, there are many other genes that have specific functions, which helps us built the idea and concept of gene expression regulation.

Eukaryotic gene regulation, especially in multicellular organisms, is complicated by the process of development unique to the organisms. The complexity of regulation of gene expression is determined by the characteristic of the eukaryotic genome. Gene expression regulation is involved in organism development and differentiation, so cell differentiation itself is the result of gene expression regulation. The gene expression regulation is multi-level, and the genetic information is repaired or modified at every level in eukaryotes. Like prokaryotes, life phenomena are also based on gene expression regulation. Moreover, multiform transcriptional regulation and post-transcriptional regulation are the basis of multiform expression to propagate genetic information in eukaryotes.

Among the study methods of gene function, RNA interference technique is undoubtedly becoming the method of choice for inactivating eukaryotic genes. RNA interference, which occurs post-transcriptionally, mediated by dsDNA, is a rapid approach to turn off target genes. It is likely that RNA interference will become the appropriate choice for studies in fields of gene function, signal transduction, defense strategy against viruses, with a potential to develop novel gene therapy regime.

(Tao Zhu)

第六章　基因工程及其应用

20世纪70年代以来，生命科学领域形成了一个融科研和生产为一体的新兴技术领域——生物技术（biotechnology）。生物技术中将基因进行克隆，利用克隆的基因表达、制备特定的蛋白质（多肽）或定向改造基因结构所用的方法和相关工作统称基因工程（genetic engineering）。基因工程中最主要的工作（技术）是基因克隆（gene cloning），也称分子克隆（molecular cloning）、重组DNA（recombinant DNA）。

第一节　工　具　酶

基因工程中许多工作涉及对DNA进行操作，包括切割、连接、合成、修饰等，而这些都是通过酶的作用来完成，因此，各种酶就成为人们对DNA进行操作的基本工具，故称工具酶。

一、限制性内切核酸酶

限制性内切核酸酶（restriction endonuclease）是一类能够识别双链DNA分子中的某种特定核苷酸序列，并由此水解其双链DNA内部特异位点——磷酸二酯键的核酸酶。这些限制性内切核酸酶，最初是从原核生物中分离纯化出来的。限制性内切核酸酶的发现与应用极大地促进了体外重组DNA技术的发展，使人们对真核染色体的结构与组织、基因的表达与进化等问题进行深入研究成为可能。

（一）寄主的限制与修饰现象及限制性内切核酸酶的发现

1952年Luria和Human在研究T偶数噬菌体过程中，1953年Weigle和Bertani在λ噬菌体对大肠杆菌的感染实验中发现了细菌的限制和修饰现象，导致了限制性内切核酸酶的发现。寄主控制的限制和修饰现象同免疫体系相类似，它能辨别自身的DNA和外来的DNA，保护自己的DNA不被改变，通过降解外源DNA以减少其对自身遗传信息传递的影响。

事实上，噬菌体在某一特定细菌宿主中生长的能力，取决于它最终在其中繁殖的细菌是什么菌株。例如，将λ噬菌体从一株大肠杆菌（K）转移到另外一株大肠杆菌（J），其生长效率往往会削弱，两个菌株的滴定度相差可达几个数量级。第二个菌株（J）释放的噬菌体能百分之百再感染同类菌株（J），但是若先将它们感染原来的宿主菌（K），再将释放的子代噬菌体重新感染第二个菌株（J）时，感染率就大大下降。此种现象即为宿主控制的限制作用（host-controlled restriction）。用放射性核素标记噬菌体进行的实验结果表明，在受感染的宿主细胞中，噬菌体生长的限制伴有噬菌体DNA的迅速降解，然而，用作繁殖噬菌体的感染宿主菌株并不导致类似的噬菌体DNA的降解。如果某一细菌细胞含有一种能选择性降解来自侵染病毒（或其他来源）DNA的核酸酶，那么，它必须能将这种外来 DNA 同它自己的 DNA 区分开来，之所以能够如此，是通过被称之为宿主控制的修饰作用（host-controlled modification）所致。

20 世纪 60 年代中期，科学家推测细菌中有限制-修饰系统（restriction-modification system，R-M system）。限制-修饰系统中的限制作用是指一定类型的细菌可以通过限制酶的作用，破坏入侵的噬菌体 DNA，导致噬菌体的寄主范围受到限制；而寄主本身的 DNA，由于在复制过程中通过甲基化酶的作用得以甲基化，使 DNA 得以修饰，从而免遭自身限制酶的破坏，这就是限制-修饰系统中修饰的含义。限制作用是指细菌的限制性核酸酶对 DNA 的分解作用，限制一般是指对外源 DNA 侵入的限制。修饰作用是指细菌的修饰酶对于 DNA 碱基结构改变的作用（如甲基化），经修饰酶作用后的 DNA 可免遭其自身所具有的限制酶的分解。

1968 年，Meselson 从大肠杆菌 K 株中分离出了第一个限制酶 *Eco*K，同年 Linn 和 Aeber 从大肠杆菌 B 株中分离到限制酶 *Eco*B。遗憾的是，由于 *Eco*K 和 *Eco*B 这两种酶的识别和切割位点不够专一，在基因工程中意义不大。1970 年，Smith 和 Wilcox 从流感嗜血杆菌中分离到一种限制酶，能够特异性地切割 DNA，这个酶后来命名为 *Hind* Ⅱ，这是第一个分离到的 Ⅱ 类限制性内切核酸酶。由于这类酶的识别序列和切割位点特异性很强，对于分离特定的 DNA 片段具有特别的意义。

（二）限制性内切核酸酶的命名和分类

1. 限制性内切核酸酶的命名

按照国际命名法，限制性内切核酸酶属于水解酶类。由于发现的限制酶的数量众多，而且越来越多，并且在同一种菌株中发现了几种酶。为了避免混淆，1973 年 Smith 和 Nathans 对内切核酸酶的命名提出建议，1980 年，Roberts 对限制酶的命名进行分类和系统化。限制性内切核酸酶采用三字母的命名原则，即属名 ＋ 种名 ＋ 株名的第一个首字母，再加上序号，表 6-1 为限制性内切核酸酶的命名原则。

表 6-1　限制性内切核酸酶的命名原则

表达方式	基本原则
组成结构	3 或 4 个字母组成，方式：属名＋种名＋株名＋序号
首字母	取属名的第一个字母，大写，斜体
第二字母	取种名的第一个字母，小写，斜体
第三字母	①取种名的第二个字母，小写；②若种名有词头，且已命名过内切核酸酶，则取词头后的第一字母代替，斜体
第四字母	若有株名，株名则作为第四字母，一般为大写，正体。是否大、小写，根据原来的情况而定
顺序号	若在同一菌株中分离了几个限制性内切核酸酶，则按先后顺序冠以大写的罗马数 Ⅰ、Ⅱ、Ⅲ 等，正体

例如，大肠杆菌 *Eco*RI 命名方式如下：

E	*Escherichia*	属名
co	*Coli*	种名
R	RY13	株系
I	First identified	鉴定顺序

2. 限制性内切核酸酶的分类

目前发现的限制性内切核酸酶有数百种以上。根据酶的组成、所需因子及裂解DNA方式的不同，可将限制性内切核酸酶分为三大类。重组DNA技术中常用的限制性内切核酸酶为Ⅱ类酶（如 $EcoR$Ⅰ、BamHⅠ等）。

（1）Ⅰ类限制性内切核酸酶

Ⅰ类限制性内切核酸酶的分子质量较大，一般在 3.0×10^5 Da 以上，其结构都是多聚体蛋白，通常由三个不同的亚基所组成，具有切割DNA的功能。切割反应时除了需要 Mg^{2+} 作辅助因子外，还要求ATP和S腺苷甲硫氨酸（SAM）的存在。Ⅰ类酶不仅是一种内切核酸酶，还具有甲基化酶和ATPase的活性，它们是一类具有多种酶活性的复合酶。Ⅰ类酶特异的识别序列大约为15个碱基对，此类酶虽然能够在一定序列上识别DNA分子，并作用于DNA分子，但因其识别DNA后，要朝一个方向或两个方向移动一段距离（通常为1000个碱基左右），并且要形成一个环才能切割DNA，所以识别位点和切割位点不一致，产生的片段较大（图6-1）。由于这类酶不能专门切割DNA的某种特殊位点，因此在基因工程中用处不大。

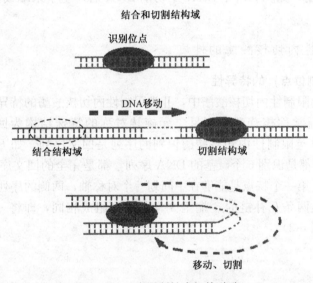

图 6-1　Ⅰ类限制性内切核酸酶

（2）Ⅲ类限制性内切核酸酶

Ⅲ类限制性内切核酸酶也是基因工程中不常用的酶类，相对分子质量和亚基组成类似于Ⅰ类酶，与Ⅰ类酶不同的是Ⅲ类酶有特异性的切割位点。作用方式基本与Ⅱ类酶相似。例如，$EcoP$Ⅰ是由两个亚基组成，一个亚基（M亚基）负责位点识别和修饰；另一个亚基（R亚基）具有核酸酶的活性（图6-2）。切割DNA时需要ATP、Mg^{2+} 参与，但并非必需条件。

（3）Ⅱ类限制性内切核酸酶

第一个被分离的Ⅱ类限制性内切核酸酶是 $Hind$ Ⅱ。这类酶的分子质量较小，一般为 $2 \times 10^4 \sim 4 \times 10^4$ Da，通常由 $2 \sim 4$ 个相同的亚基所组成。作用时需要 Mg^{2+} 作辅助因子，但不需要ATP和SAM。它们的作用底物为双链DNA，极少数Ⅱ类酶也可作用于

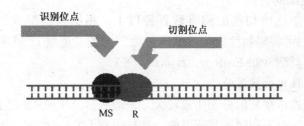

图 6-2　Ⅲ类限制性内切核酸酶

单链 DNA，或 DNA/RNA 杂种双链分子，尚未发现Ⅱ类酶与对应的甲基化酶在蛋白质亚基上有什么关联。与Ⅰ类酶的不同之处是这类酶的专一性强，它不仅对酶切位点邻近的两个碱基有严格要求，而且对更远的碱基也有要求，因此，Ⅱ类酶既具有切割位点的专一性，也具有识别位点的专一性，一般在特异性识别序列内切割。切割的方式有平切和交错切，产生平末端的 DNA 片段或具有突出黏性末端的 DNA 片段（5′或 3′黏性末端）。而另有一些酶，如 *Hpa*Ⅰ、*Eco*RⅤ切割 DNA 后产生平末端或钝性末端（blunt end）。

（三）Ⅱ类限制性内切核酸酶的性质

1. 识别序列（切割位点）的特异性

在三种类型的限制性内切核酸酶中，Ⅱ类限制性内切核酸酶的特异性最强。在分子生物学术语中，通常称核苷酸序列呈二元旋转对称的特殊结构为回文结构（palindrome）。大多数Ⅱ类限制性内切核酸酶识别的序列是回文序列，如 *Eco*RⅠ、*Bam*HⅠ、*Eco*RⅤ和 *Bgl*Ⅰ都是识别 6 个碱基的 DNA 序列，都是完全的回文序列。这段序列有两个基本的特征，第一个特点是能够在中间划一个对称轴，两侧的序列两两对称互补配对，第二个特点是两条互补链的 5′端至 3′端的序列组成相同，即将一条链旋转 180°，则两条链重叠（图 6-3）。

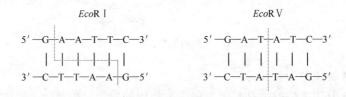

图 6-3　Ⅱ类限制性内切核酸酶的识别序列和切割位点

2. 限制性内切核酸酶的切割频率与产生的末端

切割频率是指限制性内切核酸酶在某 DNA 分子中预测的切点数。由于 DNA 是由 4 种类型的单核苷酸组成，假定 DNA 的碱基组成是均一的，而限制性内切核酸酶的识别位点是随机分布的，那么对于任何一种限制性内切核酸酶的切割频率，理论上应为 $1/4^n$，n 表示该限制性内切核酸酶识别的碱基数，如识别 4 个碱基的限制性内切核酸酶，其切割频率应为每 256 个碱基有一个识别序列和切点（$1/4^4=1/256$），识别 5 个碱基的限制性内切核酸酶，其切割频率应为每 1024 个碱基有一个识别序列和切点，以此类推。

事实上因 DNA 的分布是不均一的，且有大量的重复序列，加上限制性内切核酸酶的切点具有 GC 倾向，所以实际的频率偏低。例如，同是识别 6 个碱基的限制性内切核酸酶，切割的频率相差很大：$EcoR$ I 为 4000；$BamH$ I 为 6000；Sal I 为 8000；Hpa II 为 200。

根据限制性内切核酸酶切割 DNA 所产生的产物末端，发现限制性内切核酸酶对 DNA 的切割有两种方式，即平切和交错切。所谓平切，就是限制性内切核酸酶在 DNA 双链的相同位置切割 DNA 分子，这样产生的末端就是平末端。交错切就是限制性内切核酸酶在 DNA 双链的不同位置切割 DNA，产生的 DNA 片段的末端不是平齐的。

II 类限制性内切核酸酶的切割产物有平末端和黏性末端。黏性末端是指 DNA 分子的两端具有彼此互补的一段突出的单链部分，这一小段单链部分和同一分子的另一端或其他分子末端的单链部分如果互补的话，则能通过互补碱基之间的配对，形成双链。并在 DNA 连接酶的作用下，使同一 DNA 分子的两端连接成环状，或使两个小的短片段的分子连成一个大的长片段的线状分子。不同限制性内切核酸酶切割 DNA 产生三种不同类型的末端（表 6-2，图 6-4）。

表 6-2　一些常用限制性内切核酸酶及产生的末端

5′黏性末端		3′黏性末端		平末端	
酶	识别序列	酶	识别序列	酶	识别序列
Taq I	T/CGA	Pst I	CTGCA/G	Alu I	AG/CT
Cla I	AT/CGAT	Sac I	GAGCT/C	$FnuD$ II	CG/CG
Mbo I	/GATC	Sph I	GCATG/C	Dpn I	GA/TC
Bgl II	A/GATCT	Bde I	GGCGC/C	Hae III	GG/CC
$BamH$ I	G/GATCC	Apa I	GGGCC/C	Pvu II	CAG/CTG
Bcl I	T/GATCA	Kpn I	GGTAC/C	Sma I	CCC/GGC
$Hind$ III	A/AGCTT			Nae I	GCC/GGC
Nco I	C/CATGG			Hpa I	GTT/AAC
Xma I	C/CCGGG			Nru I	TCG/CGA
Xho I	C/TCGAG			Bal I	TGG/CCA
$EcoR$ I	G/AATTC			Mst I	TGC/GCA
Sal I	G/TCGAC			Mha III	TTT/AAA

有些限制性内切核酸酶虽然识别序列不完全相同，但切割 DNA 后产生相同类型的黏性末端，可进行相互连接；产生平末端的酶切割 DNA 后，也可彼此连接。这种酶称为同尾酶（isocaudamer）。它们虽然来源各异，识别的靶序列也各不相同，但都产生相同的黏性末端。常用的限制酶 $BamH$ I、Bcl I、Bgl II、Sau3A I 和 Xho II 就是一组同尾酶，它们切割 DNA 后都形成由 GATC 4 个核苷酸组成的黏性末端。显而易见，由同尾酶产生的 DNA 片段是能够通过其黏性末端之间的互补作用而彼此连接起来的，因

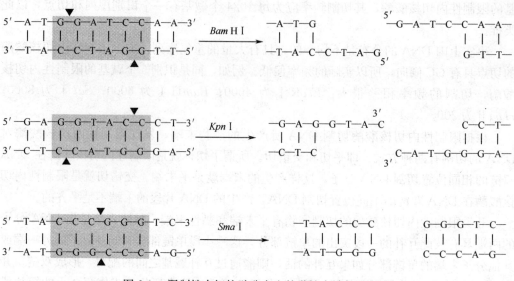

图 6-4　限制性内切核酸酶产生的黏性末端与平末端

此在基因工程操作中很有用处。

　　还有一类酶称为同裂酶（isoschizomer），同裂酶是与同尾酶对应的一类限制酶，这类酶来源不同，识别的是同样的核苷酸靶序列，同裂酶产生同样的切割，形成同样的末端。例如，$Hind$Ⅲ和HsuⅠ就是同裂酶，它们识别 DNA 的位置相同，切割后产生的末端是一样的。另外存在一种称为不完全的同裂酶，即几种酶可识别相同的 DNA 序列，但是在序列中切割位置却不同。

二、其他常用工具酶

　　基因的分子手术是相当复杂的过程，除利用限制性内切核酸酶对其目的基因片段进行有效的切割外，还需要一些其他重要的工具酶进行基因操作，以达到重组目的 DNA 的目的。这些工具酶包括连接酶、DNA 聚合酶、RNA 聚合酶、核酸酶、末端修饰酶等，这里简要列出一些主要工具酶的特性与功能，见表 6-3。

表 6-3　重组 DNA 技术中常用工具酶类

酶种类	功能
DNA 连接酶	催化 DNA 中相邻的 5′-磷酸基和 3′-羟基之间形成磷酸二酯键，使 DNA 切口封合或使两个 DNA 分子或片段连接
DNA 聚合酶Ⅰ	①合成双链 cDNA 的第二条链；②缺口平移制作高比活探针；③DNA 序列分析；④填补 3′端
反转录酶	①合成 cDNA；②替代 DNA 聚合酶Ⅰ进行填补、标记或 DNA 序列分析
多聚核苷酸激酶	催化多聚核苷酸 5′端磷酸化或标记探针
末端转移酶	在 3′-羟基末端进行同质多聚物加尾
碱性磷酸酶	切除末端磷酸基

第二节　基因克隆的基本过程

一、目的基因制备

利用重组 DNA 技术构建嵌合 DNA 时，欲插入载体 DNA 的外源 DNA 片段中即含有研究者感兴趣的 DNA 序列——目的基因。目前获取目的基因大致有如下几种途径或来源。

1. 化学合成法

这种方法是建立在 DNA 序列分析基础上的。如果已知某种基因的核苷酸序列或根据某种基因产物的氨基酸序列推导出该多肽编码基因的核苷酸序列后，再通过 DNA 合成仪利用化学合成原理合成目的基因。利用该法合成的有人生长激素释放抑制因子、胰岛素原、脑啡肽及干扰素基因等。但是这种方法目前仅限于合成核苷酸数量较少的一些简单基因，而且事先必须搞清楚它们的核苷酸序列。这种合成基因的方法有一个很大的优点，就是可以人工合成自然界不存在的新基因，使生物产生新的性状以满足人类需求。因此，这一方法今后将随着技术的不断改进而得到越来越广泛的应用。

2. 基因组 DNA 分离

分离组织或细胞染色体 DNA，利用限制性内切核酸酶将染色体 DNA 切割成基因水平的许多片段，其中即含有研究者感兴趣的基因片段。将它们与适当的克隆载体拼接成重组 DNA 分子，继而转入受体菌扩增，使每个细菌内都携带一种重组 DNA 分子的多个拷贝。不同细菌所包含的重组 DNA 分子内可能存在不同的染色体 DNA 片段，这样生长的全部细菌所携带的各种染色体片段就代表了整个基因组。存在于细菌内、由克隆载体所携带的所有基因组 DNA 的集合体称基因组 DNA 文库（genomic DNA library）。基因组 DNA 文库就像图书馆库存万卷书一样，涵盖了基因组全部基因信息，也包括研究者感兴趣的基因。建立基因文库后需要结合适当筛选方法从众多转化子菌落中筛选出含有某一基因的菌落，再进行扩增，将重组 DNA 分离、回收，获得目的基因的无性繁殖系——克隆。

3. cDNA 合成

这种方法是把含有目的基因的 mRNA 的多聚核糖体提取出来，分离出 mRNA，以 mRNA 为模板，利用反转录酶合成与 mRNA 互补的 DNA，即互补 DNA 单链（complementary DNA，cDNA），再复制成双链 cDNA 片段，与适当载体连接后转入受体菌，扩增出 cDNA 文库（cDNA library），然后再采用适当方法从 cDNA 文库中筛选出目的 cDNA。与基因组 DNA 文库类似，由总 mRNA 制作的 cDNA 文库包括了细胞全部 mRNA 信息，自然也含有人们感兴趣的编码 cDNA。当前发现的大多数蛋白质的编码基因几乎都是这样分离的。这种方法专一性强，但是操作过程比较麻烦，特别是 mRNA 很不稳定、生存时间短，所以要求的技术条件较高。

4. 聚合酶链反应扩增

目前，采用聚合酶链反应（PCR）获取目的 DNA 十分普通，应用这一技术可以将微量的目的 DNA 片段在体外扩增 100 万倍以上。它不仅可以用于基因的分离、克隆和核苷酸序列分析，还可以用于突变体和重组体的构建，基因表达调控的研究，基因多态性的分析，遗传病和传染病的诊断，肿瘤机制的探索，法医鉴定等诸多方面。

PCR 的基本工作原理是以拟扩增的 DNA 分子为模板，以一对分别与模板 $5'$ 端和 $3'$ 端互补的寡核苷酸片段为引物，在 DNA 聚合酶的作用下，按照半保留复制的机制沿着模板链延伸直至新的 DNA 合成，重复这一过程，可使目的 DNA 片段得到扩增。组成 PCR 反应体系的基本成分包括：模板 DNA、特异性引物、DNA 聚合酶（具耐热性）、dNTP 以及含有 Mg^{2+} 的缓冲液。

PCR 的基本反应步骤包括：①变性，将反应体系加热至 95℃，使模板 DNA 完全变性成为单链，同时消除引物自身和引物之间存在的局部双链；②退火（anneal），将温度下降至适宜温度（一般较 T_m 低 5℃）使引物与模板 DNA 退火结合；③延伸，将温度升至 72℃，DNA 聚合酶以 dNTP 为底物催化 DNA 的合成反应。上述三个步骤称为一个循环，新合成的 DNA 分子成为下一轮合成的模板，经多次循环（25～30 次）后即可达到扩增 DNA 片段的目的。

二、载体类型与选择

在基因工程操作中，常常把外源 DNA 片段利用运载工具导入宿主细胞，并使之在细胞内建立稳定的遗传状态，在细胞内繁殖、传代或进行表达。我们把能携带外源基因进入受体细胞的、有特殊结构的分子叫做载体（vector）。载体的本质是 DNA。经人工构建的载体按照其主要用途可分为克隆载体、表达载体。克隆载体的主要用途是克隆和扩增某一 DNA 序列，为研究提供材料及建立基因库；表达载体主要用途是得到基因表达的产物——蛋白质，从而研究其性质和功能。在基因工程中所用的克隆载体主要有 3 种类型：①质粒载体，②噬菌体与病毒载体，③柯斯质粒。虽然这些载体的来源不同，在大小、结构与复制方面的特性各异，但作为基因工程克隆载体，应当具备如下条件：①在宿主细胞株能自主复制；②载体具备可供选择的遗传标记，借助这些标记便于进行重组体筛选和鉴定；③载体的合适位置上必须提供外源基因的插入位点，即克隆位点，插入其中的外源基因可以像载体的正常组分一样进行复制和扩增。但作为基因工程表达载体，除满足上述条件外，还应具备表达所需的启动子、核糖体结合位点、增强子、终止子等调控元件。由于各类载体具有各自独特的性质，故实验者可根据自己的目的选择合适的载体，以满足不同克隆载体构建与应用的需要。以下分别介绍常用的一些克隆载体。

1. 质粒载体

质粒存在于多种细菌中，是独立于染色体外的遗传因子，由双链环状 DNA 分子组成，本身含有复制功能的遗传结构，能在宿主细胞独立自主地进行复制，并在细胞分裂时保持恒定地传给子代细胞。质粒带有某些遗传信息，在细菌内存在会赋予宿主细胞一些遗传性状，如对某些抗生素或重金属的抗性等。根据质粒赋予细菌的表型可识别质粒的存在，这是筛选转化子细菌的依据。因此，质粒 DNA 的自我复制功能及所携带的特殊遗传信息在重组 DNA 操作中（如扩增、筛选过程）是极其有用的。

根据质粒的拷贝数将质粒分为松弛型质粒和严谨型质粒。质粒拷贝数（plasmid copy number）是指细胞中单一质粒的份数同染色体数的比值，常用质粒数/每条染色体来表示。不同的质粒在宿主细胞中的拷贝数不同，松弛型质粒（relaxed plasmid）的复制只受本身遗传结构的控制，而不受染色体复制机制的制约，因而有较多的拷贝数，通常可达10～15个/每条染色体，并且可以在氯霉素作用下进行扩增，基因工程中使用的多是松弛型质粒。严谨型质粒（stringent plasmid）在宿主细胞内的复制除了受本身的复制机制控制外，还受染色体的严谨控制，因此拷贝数较少，一般只有1个或2个/每条染色体。这种质粒一般不能用氯霉素进行扩增。严谨型质粒多数是具有自我传递能力的大质粒。质粒的复制特性是受复制子控制的（图6-5）。

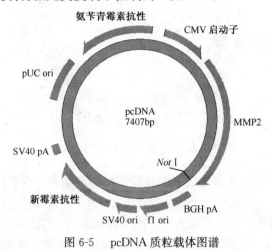

图 6-5　pcDNA 质粒载体图谱

目前，基因工程中应用最广泛的质粒载体多为 pcDNA 载体及其衍生物系列。在正常情况下，质粒载体的拷贝数可达上千个，所以不需要用氯霉素进行扩增。

2. 噬菌体载体

细菌质粒载体为基因克隆中应用最为普遍的一类载体。但由于这类载体易受到自身条件的限制，常常只能携带 10 kb 以下的外源基因。在基因工程中，由于不同的目的要求及实验操作，克隆的目的基因片段通常超出这个长度，如构建基因组文库等，噬菌体载体就显得更为有用。这里主要介绍用途较为广泛的 λ 噬菌体及丝状噬菌体载体。λ 噬菌体是最主要的载体。自 1974 年以来，已用野生型 λ 噬菌体改造和构建出一系列噬菌体载体。

（1）λ 噬菌体载体

λ 噬菌体是一种温和噬菌体，即在感染其寄主细胞后，呈现溶源及溶菌两种方式。λ 噬菌体基因组为双链线状 DNA 分子，总长度为 48 502 bp。DNA 分子的两端，各有一条由 12 个核苷酸组成的彼此完全互补的 5′ 单链突出序列，即黏性末端。λ 噬菌体 DNA 进入宿主细胞后，通过两个末端互补结合而形成黏性末端位点使 DNA 分子环化。根据对 λ 噬菌体基因结构与功能研究（图6-6），已阐明控制 λ 噬菌体复制转录的基因。

野生型 λ 噬菌体 DNA 上有 65 种限制酶的酶切点，除 Apa Ⅰ、Nae Ⅰ、Nar Ⅰ、Nhe Ⅰ、Sna BⅠ、Xba Ⅰ 和 Xho Ⅰ 限制酶各有一个切点外，其余都多于两个。有些酶切点在 λ 噬菌体增殖所必需的基因区域内。因此，λ 噬菌体必须经过改造才能用作载

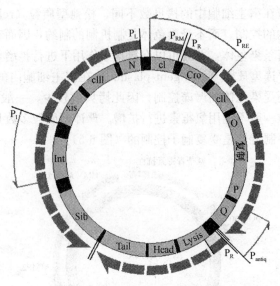

图 6-6 λ噬菌体的基因组结构

体。现在用的 λ 噬菌体载体大都除去了某种限制酶的酶切点。因此，作为载体的 λ 噬菌体都比野生型短。

　　λ 噬菌体载体有两种类型：①插入型。由于改建后的 λ 噬菌体 DNA 都比野生型短，所以可插入外源 DNA，只要插入的位置不影响噬菌体增殖。噬菌体 DNA 缺失越长，插入片段就越大。②置换型。λ 噬菌体的基因组有三个区域，左侧区包括使噬菌体 DNA 成为一个成熟的、有外壳的病毒颗粒所需的全部基因，全长约 20 kb。右侧区内包含所有的调控因子、与 DNA 复制及裂解宿主菌有关的基因，这个区域约长 12 kb。中间区域约长 18 kb，这一段 DNA 可以被外源 DNA 置换而不会影响 λ 噬菌体裂解生长的能力。置换型 λ 噬菌体是使用最广泛的载体。

　　λ 噬菌体裂解生长的能力同包装在头蛋白中的 λDNA 的大小有关。当 DNA 的长度小于野生型的 78％或超过 105％时，噬菌体的活性就急剧下降。因此要求 λ 噬菌体 DNA 和外源 DNA 长度之和为 39～53 kb。置换型载体 DNA 用选定的限制酶完全酶切后，要设法除去中间片段，只留下左臂和右臂以便用外源 DNA 片段连接，包装成重组噬菌体。这是因为左臂和右臂与中间片段间都有单链"黏性末端"，在连接酶作用下可以重新恢复原来的结构，从而影响了同外源 DNA 的连接。区分重组噬菌体形成的噬菌斑和重新恢复的载体噬菌体形成的噬菌斑的方法是用 X-gal（5-溴-4-氯-3-吲哚-β-D-半乳糖苷）蓝色噬菌斑实验。有些噬菌体载体的中间片段带有编码 β-半乳糖苷酶的基因。当这种噬菌体感染 lac 宿主细胞并在含有 X-gal 的培养基上生长时，半乳糖苷酶与 X-gal 反应的产物为不溶性的靛蓝染料。因此，含有中间片段的重新恢复的噬菌体形成蓝色噬菌斑；而同外源 DNA 连接的重组噬菌体形成的噬菌斑是无色透明的。

　　（2）丝状噬菌体载体

　　丝状噬菌体又称杆状噬菌体，M13、fd 和 f1 均为丝状噬菌体，其核心部分由单链环状 DNA 组成。M13、fd 和 f1 噬菌体基因组的结构和组成十分相似，它们之间有98％以上的核苷酸序列是同源的，个别序列差异散布于整个基因组中，所以 M13、fd

和 f1 基因组之间可以产生互补，并发生重组。

  M13 噬菌体通过吸附到细胞壁的性纤毛上，感染雄性大肠杆菌细胞。一经感染，基因组的单链 DNA 便转变成双链复制型 DNA（replicative form，RF），故 M13 能以双链 DNA 的形式从细胞中分离作为克隆载体。M13 基因组为含有 6407 个核苷酸的合环状正链单链 DNA，由 10 个基因组成（其基因结构见图 6-7），其基因组有 90% 以上的序列编码蛋白质，除了在基因 2 与基因 4 以及基因 3 与基因 4 之间是有两个较长的基因间隔区（intergenic region，IR）外，其他基因之间的间隔仅有几个核苷酸。基因 2 与基因 4 之间的 IR 不仅有 DNA 复制原点（ori），而且在 5480～5820 位核苷酸还存在 5 个发夹结构，其中以第三个发夹结构最为重要，这一发夹结构含有 59 个碱基，与转录终止有关。在 M13 基因组中还有许多转录启动子，基因组的转录可以从任意一个启动子开始。

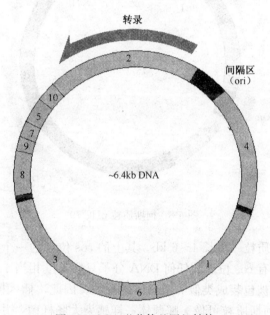

图 6-7 M13 噬菌体基因组结构

  野生型 M13 噬菌体不适于作克隆载体，在对 M13 噬菌体基因组间隔序列进行改造后，可组建成一系列克隆载体。M13 噬菌体作为载体具有以下几个重要的特点：①M13噬菌体的感染与释放不会杀死宿主菌，仅导致宿主菌生长缓慢；②M13 噬菌体 DNA 在宿主菌内既可以是单链也可以是双链，通过感染或转化的方法能将 M13 噬菌体 DNA 导入宿主菌中；③M13 噬菌体的包装不受 DNA 大小的限制，其噬菌体颗粒的大小可随 DNA 的大小而改变，即使 DNA 的大小超出本身 DNA 的大小 6 倍，仍能进行包装。由于 M13 噬菌体的这些特性，使其广泛地用于 DNA 重组。M13 噬菌体载体的主要用途是为 Sanger ddNTP 链终止法测定 DNA 序列制备单链 DNA 模板。噬菌体表面展示系统，就是利用大肠杆菌丝状噬菌体（M13 或 fd）作为载体，在重组噬菌体颗粒表面展示外源多肽。通过对这一系统的改进，证明其在 cDNA 文库的筛选上同样适用。

3. 柯斯质粒

1978 年 Collins 和 Hohn 构建了一种新型的大肠杆菌克隆载体，命名为柯斯质粒（cosmid）。它是用正常的质粒同 λ 噬菌体的 cos 位点构成。例如，柯斯质粒 pHC79 由质粒 pBR322 和 λ 噬菌体的 cos 位点的一段 DNA 构成（图 6-8），全长 43 kb。在包装时，cos 位点打开并产生 λ 噬菌体的黏性末端。由于 pHC79 有 pBR322DNA，所以也就有氨苄青霉素抗性和四环素抗性两个标记。图 6-8 展示了克隆基因时有用的限制酶切点。所以可以说柯斯质粒是一类特殊的重组质粒。

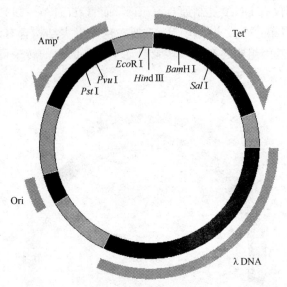

图 6-8　柯斯质粒 pHC 79

设计构建的柯斯质粒一般长 4～6 kb。其上的 cos 位点的一个重要作用是识别噬菌体的外壳蛋白。凡具有 cos 位点的任何 DNA 分子只要长度相当于噬菌体基因组，就可以同外壳蛋白结合而被包装成类似 λ 噬菌体的颗粒。因此，插入柯斯质粒的外源 DNA 可＞40 kb。重组的柯斯质粒可像 λ 噬菌体一样感染大肠杆菌，并在细菌中复制，如宿主菌在含氯霉素的培养基中生长，柯斯质粒可以扩增到宿主细胞 DNA 总量的 50% 左右。

采用柯斯质粒作载体的困难：①载体自身只相当于可以插入片段的 1/10 左右，因此往往会出现载体同载体自身连接，结果在一个重组分子内可有几个柯斯质粒载体连在一起。但用碱性磷酸酶处理，可阻止载体分子自身连接。②大小不等的外源片段相互连接后插入同一个载体分子，结果使在基因组内本来不是相邻的片段错乱地连接成一个片段，会影响实验结果的分析，后来专门选出 30～45 kb 的外源 DNA 插入载体 DNA，此时，每个载体只可能插入一个外源片段，因为如果两处均插入片段，则将超过包装成噬菌体颗粒的限度；③细菌的菌落体积远大于噬菌斑，因此如用柯斯质粒制备基因文库，则筛选所需的含某一 DNA 片段的菌落很费时间。现虽建立了高密度菌落筛选法，但由于柯斯质粒制成的基因文库常常不太稳定，插入的大片段外源 DNA 有可能通过同宿主基因组交换而致丢失等，所以最常使用的还是噬菌体载体。

三、目的基因与载体的体外重组

通过不同途径获取目的基因 DNA、选择或改建适当的克隆载体后，下一步工作是如何将目的基因 DNA 与载体 DNA 连接在一起，即 DNA 的体外重组。这种 DNA 重组是靠 DNA 连接酶将外源 DNA 与载体共价连接的。由于 DNA 连接酶具有修复单链或双链的能力，因此它在 DNA 重组、DNA 复制和 DNA 损伤后的修复中起着关键作用。特别是 DNA 连接酶具有连接平末端或黏性末端 DNA 的能力，这就促使它成为重组DNA 技术中极有价值的工具。改建载体、着手进行外源基因与载体连接前，必须结合研究目的及感兴趣基因的特性，认真设计最终构建的重组体分子。应该说，这是一件技术性极强的工作，除技巧问题，还涉及对重组 DNA 技术领域深刻的认识。下面仅就连接方式做扼要介绍。

1. 黏性末端连接

同一限制内切核酸酶切割位点连接：由同一限制性内切核酸酶切割的不同 DNA 片段具有完全相同的末端。只要酶切割 DNA 后产生单链突出（5′突出及 3′突出）的黏性末端，同时酶切位点附近的 DNA 序列不影响连接，那么，当这样的两个 DNA 片段一起退火时，黏性末端单链间进行碱基配对，然后在 DNA 连接酶催化作用下形成共价结合的重组 DNA 分子。

不同限制性内切核酸酶位点连接：由两种不同的限制性内切核酸酶切割的 DNA 片段，具有相同类型的黏性末端，即配对末端，也可以进行黏性末端连接。例如，*Mbo* I（▼GATC）和 *Bam* H I（G▼GATCC）切割 DNA 后均可产生 5′突出的 GATC 黏性末端，彼此可互相连接。

2. 平末端连接

DNA 连接酶可催化相同和不同限制性内切核酸酶切割的平末端之间的连接。原则上讲，限制酶切割 DNA 后产生的平末端也属配对末端，可彼此相互连接；如果目的序列和载体上没有相同的限制性内切核酸酶位点可供利用，用不同的限制性内切核酸酶切割后的黏性末端不能互补结合，则可用适当的酶将 DNA 突出的末端削平或补成平末端，再用 DNA 连接酶连接，但平末端连接要比黏性末端连接的效率低得多，若产生的黏性末端经特殊酶处理，使单链突出处被补齐或削平，变为平末端，也可实行平末端连接。

3. 同聚物加尾连接

同聚物加尾连接是人为在 DNA 两端做出能互补的核苷酸多聚物黏性末端，退火后能结合连接。利用同聚物序列，如 poly（A）与 poly（T）之间的退火作用完成连接。在末端转移酶（terminal transferase）作用下，在 DNA 片段末端制造出黏性末端，而后进行黏性末端连接。这是一种人工提高连接效率的方法，也属于黏性末端连接的一种特殊形式。

4. 人工接头连接

对平末端的基因 DNA 片段或载体 DNA，可先将磷酸化的接头（linker）或适当分子连到平末端，产生新的限制性内切核酸酶位点。再用识别新位点的限制性内切核酸酶切除接头的远端，产生黏性末端，再用连接酶连接。

四、重组 DNA 分子导入受体细胞

外源 DNA（含目的 DNA）与载体在体外连接成重组 DNA 分子（嵌合 DNA）后，需将其导入受体细胞（菌）。随着受体细胞（菌）的生长、增殖，重组 DNA 分子也复制、扩增，随后筛选出含目的 DNA 的重组体分子，这一过程即为无性繁殖系或克隆。在选择适当的受体细胞（菌）后，经特殊方法处理，使之成为感受态细胞（competent cell），即具备接受外源 DNA 的能力。根据重组 DNA 时所采用的载体性质不同，导入重组 DNA 分子的手段也不同：①转化（transformation），是指质粒 DNA 或以它为载体构建的重组 DNA 导入细菌宿主细胞，并使其获得新的表型；②转染（transfection），指以病毒、质粒为载体构建的重组子导入真核细胞的过程；③转导（transduction），指以噬菌体为载体，在细菌之间转移 DNA 的过程，有时也指在真核细胞之间通过逆转录病毒转移和获得细胞 DNA 的过程。④感染（infection）则是以噬菌体和真核细胞病毒为载体的重组 DNA 分子，在体外经过包装，成为具有感染能力的病毒或噬菌体颗粒，才能感染适当的细胞，并在细胞内扩增。

五、目的基因的筛选与鉴定

通过转化、转染或感染，重组体 DNA 分子被导入受体细胞，经适当涂布的培养基平板培养得到大量转化子菌落或转染噬菌斑。由于每一重组体只携带某一段外源基因，而转化或转染时每一受体菌又只能接受一个重组体分子，如何将众多的转化菌落或转染噬菌斑区分开来，并鉴定哪一菌落或噬菌斑所含重组 DNA 分子确实带有目的基因，这一过程即为筛选或鉴定。主要采用的方法：载体遗传标记法、菌落原位杂交法、限制性内切核酸酶酶切图谱法、DNA 序列测定法、外源基因表达产物法等。

1. 载体遗传标记法

载体遗传标记法的原理是利用载体 DNA 上携带的遗传标记基因筛选转化子或重组子。由于标记基因所对应的遗传表型与受体细胞是互补的，因此在培养基上施加合适的选择压力，即可保证转化子显现（长出菌落或噬菌斑），而非转化子隐去（不生长），这种方法称为正选择。经过一轮正选择，往往可使转化扩增物的筛选范围缩小成千上万倍。如果载体分子含有第二个标记基因，则可利用这个标记基因进行第二轮的正选择或是负选择，从而从众多转化子中筛选出重组子。

2. 菌落原位杂交法

在许多情况下，期望重组子与非重组子之间无法用遗传学方法区分，菌落原位杂交法（又称探针原位杂交法）则能从成千上万个重组子中迅速检测出期望重组子，其前提条件是必须拥有目的基因某一区域的探针序列。根据核酸杂交原理，探针序列特异性地杂交目的基因，并通过放射性核素或荧光基团进行定位检测。

3. 限制性内切核酸酶酶切图谱法

在外源 DNA 片段的大小以及限制性内切核酸酶图谱已知的情况下，对重组分子进行酶切鉴定，不仅能区分期望重组子与非重组子，有时还能初步确定期望重组子与非期望重组子。在经抗药性正选择后，从所有的转化子中快速抽提质粒 DNA，采用合适的限制性内切核酸酶消化，然后根据电泳图谱分析质粒大小，相对分子质量大于载体质粒

的为期望重组分子，最终利用载体上的已知酶切位点重组质粒插入片段的酶切图谱，并与已知数据进行比较，进而确定期望重组子。

4. DNA 序列测定法

通过次级克隆法去除大片段无关的 DNA 区域后，对含有目的基因的 DNA 片段进行序列测定与分析，以便最终获得目的基因的编码序列和基因的调控序列，精确界定基因的边界，这对目的基因的表达及其功能的研究具有重要意义。

第三节 真核细胞的转染

一、真核细胞转染技术

通常情况下，哺乳动物细胞摄取并表达外源基因的效率很低，这主要是由于真核细胞的脂质双层膜是阻止带电荷分子进入细胞的巨大障碍。目前人们已开发了数种转染技术来克服这个障碍。有了这些技术后，采用 DNA 或 RNA 转染的方法来研究培养细胞中的基因表达就成为生物学研究中的常规手段。

1. 常用转染方法

（1）DEAE-葡聚糖法

这是早在 1965 年出现的转染方法。带正电的 DEAE-葡聚糖可以结合带负电的 DNA 分子，使得 DNA 复合物结合在带负电的细胞表面。通过使用二甲基亚砜 DMSO 或甘油获得的渗透休克，也可能是细胞内吞作用使得 DNA 复合体进入细胞。DEAE-葡聚糖法仅限于瞬时转染，重复性好，转染时要去除血清。

（2）磷酸钙共沉淀转染

这种方法最早在 1973 年开始采用。具体方法：先将需要被导入的 DNA 溶解在钙盐溶液中，然后在不停地搅拌下逐滴加到磷酸盐溶液中，形成磷酸钙微结晶与 DNA 的共沉淀物。再将这种共沉淀物与受体细胞混合、保温，DNA 就可以进入细胞核内并整合到寄主染色体上。此法较易得到稳定转染，但转染原代细胞比较困难。这种方法多用于单层培养的细胞，也可用于悬浮培养的细胞。

（3）电穿孔法

通过短暂的高电场电脉冲处理细胞，沿细胞膜的电压差异会导致细胞膜的暂时穿孔。DNA 被认为是穿过孔扩散到细胞内的。电脉冲和场强的优化对于成功的转染非常重要，因为过高的场强和过长的电脉冲时间会不可逆地伤害细胞膜而裂解细胞。理论上说电穿孔法可用于各种细胞，且不需要另外采购特殊试剂，但需要昂贵的电转仪。此法每次转染需要更多的细胞和 DNA，因为细胞的死亡率高。每种细胞电转的条件都需要进行多次优化。

（4）脂质体法

中性脂质体是利用脂质膜包裹 DNA，借助脂质膜将 DNA 导入细胞内。带正电的阳离子脂质体则不同，带负电的 DNA 并没有预先包埋在脂质体中，而是自动结合到带正电的脂质体上，形成 DNA-阳离子脂质体复合物，从而吸附到带负电的细胞膜表面，经过内吞作用导入细胞。脂质体法始于 1987 年，此法的出现使得转染效率、转染的稳定性和可重复性大大提高。阳离子脂质体细胞毒性相对较高，对不同的细胞可能会干扰细胞的代谢。

（5）病毒介导的感染

感染需要将目的基因克隆到特定的病毒体系中，经过细胞的包装得到改造后的病毒，再进行感染。优点是转染效率特别高，尤其是对难以转染的原代细胞、活体细胞。

简而言之，现已建立并完善了数种不同的技术将各种分子，尤其是核酸输送进入真核细胞内，转染的目的是为了对真核细胞基因调节以及蛋白质的表达和功能等进行研究。

2. 转染技术和试剂的选择

并不是所有的转染技术都适用于所有类型的细胞，也没有一种转染试剂是可以普遍适用的。每种转染技术都会有其自身的优点和缺点，影响转染的因素包括：需要转染的细胞类型（对于不同的细胞系尤其如此）、需要被转染的分子（DNA、RNA、寡核苷酸、蛋白质）、对转染应用高通量的要求等。但无论在何种情况下，转染成功都取决于转染效率、低细胞毒性以及重复性这几个要素。为确保转染的高效性，需要选择一种可靠的转染技术及转染试剂，使之在细胞培养条件下能够发挥最佳效果。

3. 影响转染效率的因素

（1）细胞生理变化

为了研究细胞生理或用细胞进行目标验证，要注意确保转染过程本身不会改变研究目标的效应途径，即使有所改变的话，其改变程度也应尽量减小。如果所用的转染试剂对细胞有害，那么就很难区分所产生的效应到底是出自转染过程本身还是被转染的特定基因。正因为这个原因，一定要同时转染空载体或含有无关基因的载体作为空白对照。只有如此才能对目标基因所产生的效应做出正确评价。

（2）转染试剂的细胞毒性

转染试剂的细胞毒性可能是导致转染低效率的原因之一。因为转染试剂具有细胞毒性，只有少数细胞能够存活，从而导致蛋白质的产量很低。因为转染成功的细胞太少，那么那些未转染的细胞其生长速度大大超过转染成功的细胞，其结果同样也只能产生少量的目标蛋白。现在，为了获得蛋白质高通量表达，可以有一个更好的选择，即采用一种温和的、可以转染大多数细胞的转染试剂，可在转染后 3～7 天获得高产量的蛋白质，因为这种转染试剂几乎没有细胞毒性，可以转染大多数的细胞，并且每个能被转染的细胞都能获得高产量。

（3）其他因素

影响蛋白质高水平表达的其他关键因素包括细胞系（有些类型的细胞比其他细胞产量高）、质粒骨架（如增强子、启动子、转录调控元件）、目标蛋白（有些蛋白质很难获得高表达量）、目标蛋白的 cDNA 序列（如密码子的优化）、培养条件（营养物、代谢废物、转染抑制物）等。当这些因素都得到优化，并加入合适配比和数量的转染复合物（转染试剂-质粒）后，就可获得至少达到平均表达水平的稳定表达克隆。

二、瞬时转染和稳定转染

根据不同的实验目的，外源 DNA 导入哺乳细胞有两种类型：瞬时转染（transient transfection）和稳定转染（stable transfection）。瞬时转染是指外源基因进入受体细胞后，存在于游离的载体上，不整合到细胞的染色体上，在外源基因导入细胞 1～4 天后收获细胞进行分析；稳定转染需要外源基因整合到细胞的染色体上，从而得到稳定的转染细胞株。

第四节　克隆基因的表达

克隆基因的表达是基因工程的主要内容，包括外源基因（目的基因）的转录、翻译及蛋白质产物的加工等。

一、外源基因在原核细胞中的表达

大肠杆菌表达系统是最早开发的原核表达系统，是基因工程中应用较多、较为成熟的系统。

（一）原核生物基因表达的特点

原核生物在基因的组织、结构和表达调控机制等方面与真核生物有很多差异。与真核细胞相比，原核细胞的表达有以下特点：①原核生物只有一种 RNA 聚合酶（真核细胞有三种）识别原核细胞的启动子，催化所有 RNA 的合成。②原核生物的基因表达是以操纵子为单位的。操纵子是数个相关的结构基因及其调控区的结合，是一个基因表达的协同单位。调控区主要分为三部分：操纵基因、启动子及其他有调控功能的部位。③由于原核生物细胞无核膜，所以转录与翻译是偶联的，也是连续进行的。原核生物染色体 DNA 是裸露的环形 DNA，转录成 mRNA 后，可直接在胞质中与核糖体结合，翻译形成蛋白质。在翻译过程中，mRNA 可与一定数目的核糖体结合形成多核糖体。两个核糖体之间有一定长度的间隔，为裸露的 mRNA。每个核糖体可独立完成一条肽链的合成，即这种多核糖体可以同时在一条 mRNA 链上合成多条肽链，大大提高了翻译效率。④ 原核生物基因一般不含有内含子，在原核细胞中缺乏真核细胞的转录后加工系统，因此当克隆的含有内含子的真核生物基因在原核细胞中转录成 mRNA 前体后，其中内含子部分不能被切除。⑤原核生物基因的控制主要在转录水平，这种控制要比对基因产物的直接控制要慢。对 RNA 合成的控制有两种方式，一是起始控制（启动子控制），二是终止控制（衰减子控制）。⑥在大肠杆菌 mRNA 的核糖体结合位点上，含有一个转译起始密码子及同 16S 核糖体 RNA 3′端碱基互补的序列，即 SD 序列，而真核生物基因则缺乏此序列。

从上述特点可以看到，欲将外源基因在原核细胞中表达，必须满足以下条件：①通过表达载体将外源基因导入宿主菌，并指导宿主菌的酶系统合成外源蛋白；②外源基因不能带有间隔顺序（内含子），因而必须用 cDNA 或全化学合成基因，而不能用基因组 DNA；③必须利用原核细胞的强启动子和 SD 序列等调控元件控制外源基因的表达；④外源基因与表达载体连接后，必须形成正确的可读框；⑤利用宿主菌的调控系统，调节外源基因的表达，防止外源基因的表达产物对宿主菌产生毒害。

（二）基因表达的调控序列

如上所述，由于原核和真核细胞中基因表达的机制是不同的，因此必须详细了解基因表达过程中的各种调控因子，构建高效的表达载体，才能达到高效率、高水平表达外源基因的目的。对原核生物来讲，基因表达的调控序列主要涉及启动子、SD 序列、终止子、衰减子等序列。

1. 启动子

启动子是 DNA 链上一段能与 RNA 聚合酶结合并起始 RNA 合成的序列，它是基因表达不可缺少的重要调控序列。没有启动子，基因就不能转录。由于细菌 RNA 聚合酶不能识别真核基因的启动子，因此原核表达载体所用的启动子必须是原核启动子。

原核表达系统中通常使用的可调控的启动子有乳糖启动子（Lac）、色氨酸启动子（Trp）、乳糖和色氨酸的杂合启动子（Tac）、λ 噬菌体的左向启动子（IPL）、T7 噬菌体启动子等。

（1）Lac 启动子

Lac 启动子来自大肠杆菌的乳糖操纵子，是 DNA 分子上一段有方向的核苷酸序列，由阻遏蛋白基因（*lac* I）、启动基因（*P*）、操纵基因（*O*）和编码三个与乳糖利用有关的酶的基因结构所组成。Lac 启动子受分解代谢系统的正调控和阻抑物的负调控。正调控通过代谢激活蛋白（catabolite gene activation protein，CAP）因子和 cAMP 来激活启动子，促使转录进行。负调控则是由调节基因产生 Lac I 阻遏蛋白，该阻遏蛋白能与操纵基因结合阻止转录。乳糖及某些类似物，如 IPTG 可与阻遏蛋白形成复合物，使其构型改变，不能与 *O* 基因结合，从而解除这种阻遏，诱导转录发生。

（2）Trp 启动子

Trp 启动子可从大肠杆菌色氨酸操纵子中分离，其阻遏蛋白必须与色氨酸结合才有活性。当缺乏色氨酸时，该启动子开始转录。当色氨酸较丰富时，则停止转录。β-吲哚丙烯酸可竞争性抑制色氨酸与阻遏蛋白的结合，解除阻遏蛋白的活性，促使 Trp 启动子转录。

（3）Tac 启动子

Tac 启动子是一组由人工构建的 Lac 启动子和 Trp 启动子组合而成的杂合启动子，受 Lac 阻遏蛋白的负调节，它的启动能力比 Lac 启动子和 Trp 启动子都强。其中 Tac I 启动子是由 Trp 启动子的−35 区加上一个合成的 46 bp DNA 片段（包括 Pribnow 盒）和 *lac* 操纵基因构成，Tac 12 启动子是由 Trp 启动子−35 区和 Lac 启动子的−10 区，加上 Lac 操纵子中的操纵基因部分和 SD 序列融合而成。Tac 启动子受 IPTG 的诱导。

（4）IPL 启动子

IPL 启动子是来自 λ 噬菌体早期左向转录启动子，是一种活性比 Trp 启动子高 11 倍左右的强启动子。IPL 启动子受控于温度敏感的阻抑物 cIts857。在低温（30℃）时，cIts857 阻遏蛋白可阻遏 IPL 启动子转录。在高温（45℃）时，cIts857 蛋白失活，阻遏解除，促使 IPL 启动子转录。系统由于受 cIts857 作用，尤其适合于表达对大肠杆菌有毒的基因产物，缺点是温度转换不仅可诱导 IPL 启动子，也可诱导热休克基因，其中有一些热休克基因编码蛋白酶。如果用 λ 噬菌体 cI＋溶源菌，并用丝裂霉素 C 或萘啶酮酸进行诱导，可缓解这一矛盾。

（5）T7 噬菌体启动子

来自 T7 噬菌体的启动子，具有高度的特异性，只有 T7 RNA 聚合酶才能使其启动，故可以使克隆化基因独自得到表达。T7 RNA 聚合酶的效率比大肠杆菌 RNA 聚合酶的效率高 5 倍左右，它能使质粒沿模板连续转录几周，许多外源终止子都不能有效地终止它的转录，因此它可转录某些不能被大肠杆菌 RNA 聚合酶有效转录的序列。这个系统可以高效表达其他系统不能有效表达的基因。但要注意使用这种启动子时宿主中必须含有 T7 RNA 聚合酶。应用 T7 噬菌体表达系统需要两个条件：第一是具有 T7 噬菌体 RNA 聚合酶，它可以由感染的 λ 噬菌体或由插入大肠杆菌染色体上的一个基因拷贝产生；第二是在一个待表达基因上游带有 T7 噬菌体启动子。

2. SD 序列

1974 年 Shine 和 Dalgarno 首先发现，在 mRNA 上有核糖体的结合位点，它们是起始密码子 AUG 和一段位于 AUG 上游 3～10 bp 处的一段核苷酸序列，这段核苷酸序列有 3～9 个核苷酸组成，即称为 SD 序列。这段序列富含嘌呤核苷酸，刚好与 16S rRNA 3′端的富含嘧啶的序列互补，是核糖体 RNA 的识别与结合位点。以后将此序列命名为 Shine-Dalgarno 序列，简称 SD 序列。它与起始密码子 AUG 之间的距离是影响 mRNA 转录、翻译的重要因素之一，某些蛋白质与 SD 序列结合也会影响 mRNA 与核糖体的结合，从而影响蛋白质的翻译。另外，真核基因的第二个密码子必须紧接在 ATG 之后，才能产生一个完整的蛋白质。

3. 终止子

在一个基因的 3′端或是一个操纵子的 3′端往往有特定的核苷酸序列，且具有终止转录功能，这一序列称为转录终止子，简称终止子。转录终止过程包括：RNA 聚合酶停在 DNA 模板上不再前进，RNA 的延伸也停止在终止信号上，完成转录的 RNA 从 RNA 聚合酶上释放出来。对 RNA 聚合酶起强终止作用的终止子在结构上有一些共同的特点，即有一段富含 A/T 的区域和一段富含 G/C 的区域，G/C 富含区域又具有回文对称结构。这段终止子转录后形成的 RNA 具有茎环结构，并且有与 A/T 富含区对应的一串 U。转录终止的机制较为复杂，并且结论尚不统一。但在构建表达载体时，为了稳定载体系统，防止克隆的外源基因表达干扰载体的稳定性，一般都在多克隆位点的下游插入一段很强的 rrB 核糖体 RNA 的转录终止子。

4. 衰减子

衰减子（attenuator）是指在某些前导序列中带有控制蛋白质合成速率的调节区。在原核生物中，一条 mRNA 分子常常编码数种不同的多肽链。这种多顺反子 mRNA 的头一条多肽链合成的起始点，同 RNA 分子的 5′端间的距离可达数百个核苷酸。这段位于编码区之前的不转译的 mRNA 区段，叫做前导序列（leader）。此外，在 mRNA 的 3′端以及多顺反子 mRNA 中含有的长达数百个碱基的顺反子间序列（intercistranic-sequence），即间隔序列（spacer），也发现有不转译的序列。

（三）原核表达载体系统

克隆基因可以放在不同的宿主细胞中表达，常用的表达系统有原核表达系统和真核表达系统，即可用大肠杆菌、枯草芽孢杆菌、酵母、昆虫细胞、培养的哺乳类动物细胞。对于不同的表达系统，需要构建不同的表达载体。常用的原核表达载体系统有非融合型表达载体系统（如 pKK223-3）和融合蛋白表达载体系统（如 pGEX 和 pET 等）。

1. 原核表达载体的一般特性

大肠杆菌是当前采用最多的原核表达体系，用于大肠杆菌表达系统的载体应该符合下述标准：①含大肠杆菌适宜的选择标志；②具有能调控转录、产生大量 mRNA 的强启动子；③含适当的翻译控制序列和翻译起始点等；④含有合理设计的多接头克隆位点，以确保目的基因按一定方向与载体正确衔接及进行其表达产物的分离、纯化。

2. 非融合表达载体系统

在原核细胞中表达外源基因时，由于实验设计的不同，总的来说可表达融合和非融合蛋白。不与细菌的任何蛋白质或多肽融合在一起的表达蛋白称为非融合蛋白。非融合蛋白的优点在于它具有非常接近于真核细胞体内蛋白的结构，因此表达产物的生物学功能上也就更接近于生物体内天然蛋白质。非融合蛋白的最大缺点是容易被细菌蛋白酶所破坏。为了在原核生物细胞中表达出非融合蛋白，可将带有起始密码子 AUG 的真核基因插入到原核启动子和 SD 序列的下游，组成一个杂合的核糖体结合区，经转录、翻译，得到非融合蛋白。非融合表达载体 pKK223-3 是由 Brosius 等在哈佛大学的 Gilbert 实验室组建的。在大肠杆菌细胞中，它能极有效地高水平表达外源基因。它具有一个强的 Tac（Trp-Lac）启动子。这个启动子是由 Trp 启动子的−35 区、LacUV5 启动子的−10 区、操纵基因及 SD 序列组成。紧接 Tac 启动子的是一个取自 pUC8 的多位点接头，使之很容易把目的基因定位在启动子和 SD 序列后。在多位点下游的一段 DNA 序列中，还包含一个很强的核糖体 RNA 的转录终止子，目的是为了稳定载体系统。因为上游强的 Tac 启动子控制的转录必须由强终止子抑制，才不至于干扰与载体本身稳定性有关的基因表达。载体的其余部分由 pBR322 组成。在使用 pKK223-3 质粒时，应相应的使用一个 Lac I 宿主，如 JM 105。一个具有 pKK223-3 质粒类似结构的载体被用于表达 Lambda cI 基因时，经 IPTG 诱导产生的阻遏子蛋白占可溶性细胞抽提液中总蛋白的 18％～26％。由此可见，在需要获得大量的基因产物时，pKK223-3 确实是一种非常有用的工具。

3. 融合表达载体系统

融合蛋白的表达即表达的真核蛋白肽链 N 端含原核生物肽段。融合表达的方法是将真核基因插入启动子后能高效表达的原核基因下游，以生产融合蛋白的方式表达目的基因。这种融合蛋白具有真核细胞的抗原性，可以作为抗原，也能够抵御内源性蛋白酶的降解。

目前基因工程研究中较为理想的融合表达载体为硫氧还蛋白融合表达载体 pThio-His 系统。该系统具有如下优点：①利于目的蛋白的高水平稳定表达；②融合的硫氧还蛋白有助于可溶性表达产物的形成，如果产物为包涵体，则利于蛋白质的折叠复性；③带有标签，可用金属螯合层析进行纯化；④酶切可产生天然蛋白质产物。

表达融合蛋白应注意其可读框，其可读框应与融合的 DNA 片段的可读框一致，翻译时才不至于产生移码突变，如 pET、pGEX 等载体系列。pGEX 系统由 Pharmacia 公司构建，由三种载体 pGEX-1XT、pGEX-2T 和 pGEX-3X 以及一种用于纯化表达蛋白的亲和层析介质 Glutathione Sepharose 4B 组成。载体的组成成分基本上与其他表达载体相似，含有 tac 启动子及 lac 操纵基因、SD 序列、Lac I 阻遏蛋白基因等。这类载体与其他表达载体不同之处在于 SD 序列下游是谷胱甘肽巯基转移酶基因，而克隆的外源基因则与谷胱甘肽巯基转移酶基因相连。当进行基因表达时，表达产物为谷胱甘肽巯基转移酶和目的基因产物的融合体。这类载体有以下优点：①可诱导高效表达；②载体内含有 Lac I 阻遏蛋白基因；③表达的融合蛋白纯化方便；④ 使用血凝酶和 Xa 因子就可从表达的融合蛋白中切下所需的蛋白质和多肽；⑤用 EcoR I 限制酶从 λ gt11 载体中切割的基因可直接插入 pGEX-1/T 中。

4. 诱导表达

诱导表达将宿主菌的生长与外源基因的表达分成两个阶段，是减轻宿主细胞代谢负荷的最常用方法之一。一般采用温度诱导或药物诱导，如应用 Tac 启动子时，常用 F′ tac⁴ 的菌株或者将 lac I 基因克隆在表达质粒中。当宿主菌生长时，lac I 产生的阻抑物与 lac 操纵基因结合，阻碍了外源基因的转录及表达，此时，宿主菌大量生长。当加入诱导物（如 IPTG）时，阻遏蛋白不能与操纵基因结合，则外源基因大量转录并高效表达。有人认为，化学诱导比温度诱导更方便、有效，并且将相应的阻遏蛋白基因直接克隆到表达载体上，比应用含阻遏蛋白基因的菌株更为有效。

二、外源基因在哺乳动物细胞中的表达

与其他系统相比，哺乳动物细胞表达系统的优势在于能够指导蛋白质的正确折叠，提供复杂的 N 型糖基化和准确的 O 型糖基化等多种翻译后加工功能，因而表达产物在分子结构、理化特性和生物学功能方面最接近于天然的高等生物蛋白分子。

哺乳动物细胞也可用于外源基因表达并生产基因工程疫苗和多肽产品。哺乳动物细胞能识别和除去外源基因的内含子，剪接加工成成熟的 mRNA。另外，由哺乳动物细胞表达的蛋白质可正确进行翻译后加工，可使蛋白质产物具有较高的生物学活性。由哺乳动物细胞表达的蛋白质免疫原性好，为酵母型的 16～20 倍，哺乳动物细胞易被重组质粒转染，具有遗传稳定性和可重复性，经转染的哺乳动物细胞可将表达的产物分泌到培养基中，其提纯工艺简单、成本低。因此，在生物制药方面发展了许多哺乳动物细胞表达系统。

（一）哺乳动物细胞表达载体的调控元件

载体的选择取决于外源基因的导入方式和其调控元件是否有利于转录和翻译。真核生物基因高表达载体必须具有如下调控元件：①原核 DNA 序列；②启动子和增强子；③剪接信号；④终止信号和 poly（A）"加尾"信号。

1. 原核 DNA 序列

为了能在大肠杆菌中增殖，以便得到大量的重组 DNA，哺乳动物细胞表达载体一般带有的原核序列包括：一个能在大肠杆菌中自我复制的复制子，用于挑选重组 DNA 的细菌抗生素抗性基因（如氨苄青霉素抗性基因）以及便于把真核序列插入载体的少数单一限制性酶切位点。

2. 启动子

真核基因启动子是基因表达时与启动转录有关的顺式作用元件。通常位于基因转录起始位点的上游，含有启动子区和 TATA 框两种识别序列。TATA 框位于转录起始上游—25～—30 bp 处，是引导 RNA 聚合酶 II 在正确起始位点转录 RNA 所必需的 DNA 序列，即保证转录的精确起始。启动子的转录效率常常因细胞而异，因此需根据宿主细胞的类型选择不同的启动子，以便于真核基因的高效表达。

3. 增强子

增强子是提高启动子转录效率的一类顺式作用元件，由短的 DNA 序列组成。增强子的作用不具有方向性，可以在转录起始的 5′端或 3′端，其不论距离启动子和基因远近均可发挥作用，使同源启动子或异源启动子的转录水平提高近千倍。增强子具有组织特异性，在构建真核表达载体时，应根据重组基因的宿主细胞来选择启动子和增强子，以使其发挥作用。许多来自病毒的增强子具有广泛的宿主范围，尽管在不同类型的细胞中表现出差异，但在多种组织中都具有一定的活性。例如，SV40 的早期基因增强子在多种哺乳动物细胞中都具有活性，因此被广泛应用于载体中。

4. 剪切信号

真核基因由许多内含子和外显子组成，转录成前 mRNA，需经过剪切、拼接，除去内含子才能成为成熟的 mRNA。经研究发现，mRNA 拼接所需的基本序列位于内含子的 5′端或 3′端，在这些基本序列中至少最末端的两个碱基不变，内含子的 5′端总是 GU，3′端总是 AG。

5. 终止信号和多聚腺苷酸化信号

真核基因的确切终止信号和终止机制尚不清楚。但发现转录常终止于多聚腺苷酸化位点下游的一段长度为数百个核苷酸碱基的 DNA 区域内。准确而有效地进行多聚腺苷酸化需要两种序列：一是位于 poly（A）位点下游的 GC 富集区或 U 富集区；二是位于 poly（A）上游 11～30 个核苷酸处的一个由 6 个核苷酸组成的高度保守序列（5′-AAUAAA）。为了保证目的 mRNA 能有效的多聚腺苷酸化，在真核表达载体中必须包括多聚腺苷酸下游的一段序列。最常用的加 poly（A）信号是用 SV40 的一段长 7 bp 的 *Bam*H I-*Bcl* I 限制性片段，它含有早期和晚期转录单位的切割和多聚腺苷酸化信号，两套信号分别位于不同的 DNA 链上，作用方向相反，他们对各种 mRNA 的加工都同样有效。

（二）哺乳动物细胞基因转移的遗传性标记

通过真核基因表达载体，将外源基因导入哺乳动物细胞内进行表达，由于基因转移效率低，需要从众多细胞群体中检测出为数极少的转化细胞。因此，判断外源基因是否导入及筛选转化细胞就成为发展哺乳动物细胞表达系统的关键。目前，绝大多数哺乳动物细胞病毒载体，都是通过附加上选择标记基因进行筛选的。常用的标记基因有胸腺激酶基因（*tk*）、二氢叶酸还原酶基因（*dhfr*）、新霉素抗性基因（*neo*）、氯霉素乙酰基转移酶基因（*cat*）等。

(三) 常用哺乳动物细胞的表达载体系统

1. 表达载体的类型

哺乳动物细胞表达外源重组蛋白可利用质粒转染和病毒感染。根据进入宿主细胞的方式，可将表达载体分为质粒载体与病毒载体。

质粒载体借助于物理或化学作用导入细胞内。依据质粒在宿主细胞内是否具有自我复制能力，可将质粒载体分为整合型载体和附加体型载体两类。整合型载体无复制能力，需整合于宿主细胞染色体内方能稳定存在。附加体型载体则是在细胞内以染色体外可自我复制的附加体形式存在。整合型载体一般是随机整合入染色体，其外源基因的表达受插入位点的影响，同时还可能改变宿主细胞的生长特性。相比之下，附加体型载体不存在这方面的问题，但载体 DNA 在复制中容易发生突变或重排。附加体型载体在胞内的复制需要两种病毒成分：病毒 DNA 的复制起始点及复制相关蛋白质。

病毒载体是以病毒颗粒的方式，通过病毒包膜蛋白与宿主细胞膜的相互作用使外源基因进入细胞内。常用的病毒载体有腺病毒、腺相关病毒、逆转录病毒、semliki 森林病毒（sFv）载体等。另外，杆状病毒载体应用于哺乳动物细胞的表达在近几年颇受重视，这是因为它与其他病毒载体相比有特有优势，如可通过昆虫细胞大量制备病毒颗粒；可感染多种哺乳动物细胞，但在细胞内无复制能力，生物安全度高；可插入长达 38 kb 的外源基因等。

（1）质粒载体

pcDNA 表达载体系统是一种常用的非融合蛋白类型的高效真核表达载体。现以 pcDNA3.1（＋/－）为例说明其内部各个基因序列的功能。pcDNA3.1（＋/－）质粒载体以 pUC 质粒为基本骨架，它除了具有多克隆位点（MCS）、原核筛选标记（Amp）和原核复制起始位点（ori）外，还具有其特征的基因元件，如保证重组蛋白有效、高水平表达的人巨细胞病毒（CMV）早期启动子/增强子；可供重组质粒在体外转录/翻译检测实验的 T7 噬菌体 RNA 聚合酶结合的启动子，即它为 RNA 聚合酶的附着作用提供特异性的识别位点；使 mRNA 多腺嘌呤化和转录有效终止的牛生长激素多腺苷酸序列（BGH pA），这有利于具有毒性的抗原基因在此限制作用下有效发挥免疫功能；使 ZeocinTM 抑制基因有效的、高水平表达，并使 ZeocinTM 抑制基因在表达 SV40 大 T 抗原的细胞中复制 SV40 早期启动子和起始位点；可用于筛选大肠杆菌中的转化体和真核细胞中的转染子的 ZeocinTM 抑制基因等。

（2）病毒载体

当表达高等真核基因时，应用哺乳动物细胞表达体系比应用原核及昆虫细胞表达体系显得更为优越，这是因为原核及昆虫细胞表达体系不能对外源真核基因产物在转录及转译后进行有效的修饰。目前较广泛地应用于哺乳动物细胞表达体系的载体有猴病毒 40（simian virus40，SV40）、痘病毒（poxvirus）、腺病毒（adenovirus）、逆转录病毒（retrovirus）、腺病毒（adenovirus）等载体。下面仅简要介绍两种常用动物病毒载体。

逆转录病毒载体

逆转录病毒是从畜禽、鼠、爬行类及人体中分离到的一类种类繁多的动物病毒。由于这类病毒的一个重要特点是毒粒中含有依赖于 RNA 的 DNA 聚合酶即逆转录酶，故称逆转录病毒。逆转录病毒在现代分子生物学中具有非常重要的地位。这是因为逆转录病毒具有致瘤性，可引起白血病、淋巴瘤及恶性肿瘤，他们是用于研究癌变机制的重要模型。此外，逆转录酶的发现完善了"中心"法则，因此逆转录病毒基因组特殊的复制方式丰富了我们对基因的转录、复制及整合机制的认识。因此，逆转录病毒载体作为真核细胞基因工程载体与其他 DNA 动物病毒载体相比有其特殊的作用和功能。

所有逆转录病毒都有类似的基因组结构，重要特点是它具有二倍体基因，即有两个相同的单链 RNA 分子。两条 RNA 链均为正链，长 4～9 kb，不同毒株有较大差异。与真核 mRNA 一样，所有逆转录病毒基因组具有 5′甲基化帽子结构和 3′ poly（A）尾。两个 RNA 分子在 5′端非共价结合。逆转录病毒基因大致可分为两部分：①顺式作用元件，顺式作用元件即存在于病毒 RNA 中的信号，与病毒复制的蛋白质和其他因子反应有关。包括两端的长末端重复序列（LTR）、包装信号 ψ 位点等，是病毒复制所必需的调控区域；②反式作用因子，反式作用因子即病毒基因组所编码的蛋白质，是病毒 RNA 帽化、病毒进入细胞、病毒基因组逆转录以及病毒 DNA 整合入细胞 DNA 所必需。包括结构蛋白基因（gag）即编码病毒核心的结构蛋白，其序列相当保守；逆转录酶基因（pol）编码病毒所需的酶类，这种逆转录酶同时具 RnaseH 活性和依赖 RNA、DNA 聚合酶活性；糖蛋白基因（env）编码包膜糖蛋白，其起始位点在 pol 基因的 3′端，与 pol 基因有重叠，但它们的可读框不同，通常是由剪接后的亚基因组 RNA 转译而来。许多 RNA 肿瘤病毒还具有原癌基因（onc），onc 编码一种蛋白质，可使宿主细胞转化为癌细胞。

逆转录病毒 RNA 进入细胞后，由于它本身编码的逆转录酶作用，使其单链 RNA 逆转为双链 DNA，并整合到宿主细胞 DNA 中成为前病毒。前病毒是逆转录病毒增殖的正常途径和必由之路。这种整合以及子代病毒的形成对宿主细胞一般没有明显的影响，细胞被感染后仍然能正常生长，并不断地分泌病毒颗粒。前病毒在动物中可通过生殖细胞传递到下一代，病毒在诱导后（以感染形式从细胞中释放）往往只在异种动物中复制。带有三种对复制有重要作用的基因：gag、pol 和 env。它们的病毒属非缺陷型病毒。缺乏这三种基因的病毒属缺陷型病毒，要依靠非缺陷型的辅助病毒才能复制；逆转录病毒基因组中的反式作用因子可以除去，并以外源基因取代，这样所形成的重组病毒是复制缺陷型，只能产生病毒 RNA，不能编码病毒的结构蛋白。为了构建逆转录病毒载体，使重组病毒 RNA 能包装成病毒颗粒，人们构建了病毒包装细胞（viral packaging cell），它整合了除 ψ 位点外的整个辅助病毒基因组，由于缺 ψ 位点，该细胞能产生病毒蛋白，但本身不能包装成为病毒颗粒。病毒包装细胞仅能为重组病毒载体提供包装蛋白，使重组病毒形成有感染力的重组病毒颗粒，由此构建了逆转录病毒。

逆转录病毒基因组作为病毒载体，由其逆转录病毒特有的结构及生活周期所决定。①前病毒能以高效、低拷贝数及以细胞基因—LTR—病毒基因—LTR—细胞基因序列方向整合入宿主细胞 DNA 内；②整合的前病毒与宿主细胞保持稳定结合，无明显丢失或失活倾向；③5′端的 LTR 上有一个强大的启动子，提供的活性转录信号可使逆转录病毒基因或外源基因高效表达；④逆转录病毒对所感染的细胞无明显损害，对原有的基因表达也无干扰；⑤逆转录病毒能以高效率感染各种细胞，而无需其他 DNA 介导基因转移的多步骤操作；⑥一些逆转录病毒具广泛的宿主范围，还可通过改变逆转录病毒 *env* 基因产物而改变宿主范围；⑦前病毒 DNA 合成和整合所需的蛋白质可通过包装细胞系来提供。前病毒的表达通常不需要病毒编码的蛋白质，必须提供的病毒序列只是 LTR 和包装信号的顺式作用元件，一般总量少于 1 kb；⑧逆转录病毒所能容纳的基因组量范围有弹性，最大可达 10～15 kb 而无明显的最小限；⑨某些逆转录病毒基因（如 *env*）和某些癌基因（*onc*）的正常表达形式只剪接去初级转录物部分，而保留剩余部分作为基因组或其他基因（如 *gag* 基因）的 mRNA，因而可从同一启动子表达一个以上的导入基因；⑩与 *v-onc* 基因一样，可通过剪接，精确地除去插入基因内部的内含子，因而可把含内含子的真核细胞基因插入逆转录病毒中并连续传代，获得与外源基因完全互补的 DNA。

逆转录病毒载体系统就是利用逆转录病毒基因组作为载体，将目的基因带入其所感染的细胞内，因而载体感染细胞含有并表达全套 *gag*、*pol* 和 *env* 基因。一般可通过如下三条途径来实现。

1）加用非缺陷辅助病毒共感染　最初设计用于基因移植的病毒载体去除大部分病毒基因（包括可能有危险性的基因），代之以外源基因。经改变的逆转录病毒必须由辅助病毒制造侵入细胞所需的蛋白质外壳，因而要防止辅助病毒在体内释放出新一代病毒而使宿主细胞破裂，且要求缺陷病毒滴度尽可能高。

2）构建有复制能力（不依赖辅助病毒）的载体　由载体自身提供全套复制基因。这种结构类似于非缺陷 Rous 肉瘤病毒（RSV），迄今只构建 RSV 衍生的基因载体。在这种载体中除去大部分 *src* 基因，保留剪接受体位点，并插入限制性内切核酸酶裂解位点。这种载体的优点是病毒可快速播散到所有培养细胞或动物，缺点是插入片段小（约 3 kb），且不适用于哺乳动物细胞。

3）使用辅助细胞　这是依据重要包装信号在基因组非编码区内所提出包装缺陷基因组的一种战略，即产生含前病毒的辅助细胞系，含能指导 *gag*、*pol* 及 *env* 基因产物合成，但缺乏自身包装信号的前病毒。为使逆转录病毒包装细胞系尽可能少产生或不产生有复制能力的病毒，必须将病毒基因组作某些改变，即干扰顺式作用元件，保留反式作用因子的产生。因此，逆转录病毒基因转移系统由两种成分组成，第一种成分是由前病毒 DNA 构成的重组逆转录病毒载体，前病毒 DNA 分子的 *gag*、*pol* 和 *env* 基因被所转移的外源基因取代，但载体保留前病毒 DNA 的一些序列，即 5LTR、3LTR 和对该系统重要的包装信号，某些载体还保留部分 *gag* 序列。第二种成分是包装重组逆转录病毒载体的包装细胞，包装细胞含除去包装信号的前病毒 DNA。这两种成分的前病毒 DNA 分子有同源序列，即 5′LTR 和 3′LTR，某些载体还有 *gag* 序列的同源性。同源性越大，重组机会也越多。由于重组逆转录病毒分子只剩很少的病毒信息，不能表达完整的病毒蛋白，所以属复制缺陷型，仅在包装细胞系或辅助病毒的协助下才能包装成病毒颗粒。这种重组复制缺陷型病毒可直接作为载体运载外源基因，但还必须建立包装细胞系，以便在体外培养条件下产生载体病毒。含这种重组病毒分子的包装细胞系称为逆转录病毒产生细胞系，能高效、稳定地产生感染靶细胞的复制缺陷型病毒。载体中还可含选择标记或容易筛选的报告基因。目的基因可与逆转录病毒 LTR 的转录信号或与适当的内部启动子连接。载体可含 SV40 复制起始区，使携载体细胞与表达 SV40 T

抗原的 cos 细胞融合，从哺乳动物细胞中回收载体 DNA 顺序。

逆转录病毒载体的类型分为两类。①无辅助载体：这种载体不需要辅助病毒，由辅助细胞来获得感染性病毒。这种从辅助细胞产生出来的病毒称为无辅助病毒（helper free virus）。②穿梭载体：逆转录病毒载体的一个潜在应用是在真核细胞中选择特异基因或其变种，并再将其克隆至细菌质粒系统中供进一步分析。这种载体既可用于 DNA 克隆和表达研究，也有利于载体和旁侧序列的回收和分析，能实现动物和细菌细胞之间这种"穿梭"的载体称为穿梭载体（shuttle vector）。

逆转录病毒载体的特点：①基因转移效率高，病毒感染力达 100％；②逆转录病毒载体有广泛的宿主范围，不仅适用于单层细胞，而且适用于多种悬浮培养的淋巴细胞、前髓细胞及造血干细胞等多种骨髓来源的细胞，对迅速生长的细胞尤其有效；③重组病毒进入细胞以类似前病毒的形式进入 DNA，整合率较 DNA 转染以及 DNA 病毒感染高得多。

痘苗病毒载体

痘苗病毒（vaccinia virus，VCV）是人们熟悉的一种病毒，它被人们用作疫苗根除天花病，因此痘苗病毒是真核生物基因或外源性病毒基因表达的理想载体。利用痘苗病毒作为基因表达载体的优点：其一，基因组很大，故可用于插入的外源基因量也较大，病毒宿主范围广；其二，病毒在胞质中复制与增殖，无致癌性；其三，重组痘病毒可在人体接种作为活疫苗。

痘苗病毒的基因组十分庞大，为双链线状 DNA，长约 187 kb，其中有一个 28 kb 的区域，是病毒复制的非必需区，如 $Hind$Ⅲ A 片段中的 HA 基因，$Hind$Ⅲ J 片段中的胸腺嘧啶激酶基因（tk 基因），利用外源基因取代这些非必需区，不影响病毒的复制能力。但所插入的外源基因必须在病毒启动子下游，并且能够利用病毒启动子进行转录。由于插入到痘病毒载体的外源基因序列两侧具有与野生型病毒 DNA 相同的序列，因此在野生型病毒感染敏感细胞 1～2h 后，再转染这种重组载体，其重组载体中外源基因两侧的同源序列会与野生型病毒 DNA 发生同源重组，结果获得了带有外源基因的全痘苗 DNA，后者再经过包装形成重组痘苗病毒粒子，这种重组病毒粒子可以再感染敏感细胞，使外源基因在敏感细胞中得到表达。

用 DNA 重组技术将外源基因插入到痘苗病毒中，重组痘苗病毒仍然能正常复制。由于痘苗病毒易于培养，高度稳定，并能在大多数哺乳动物体内良好复制，因此，可以用它构建很好的载体。根据 Smith 等（1984）的报道，痘苗病毒载体基因工程通常分为三个步骤：①用痘苗病毒自身的带有启动子的胸腺嘧啶激酶基因（tk 基因）与质粒 pUC9 重组成 pGS20，并将外源基因嵌入到启动子下游成为嵌合载体，②将嵌合载体与野生型 VCV 合作感染人 tk 143 细胞，③tk 基因与细胞内同源序列发生等位交换，因而得到含有外源基因插入的重组病毒。由于痘苗病毒本身是一种高效活疫苗，我们可以利用它的 DNA 作为载体插入一种或几种外源基因，构建多价疫苗株，从而可以同时预防由几种致病微生物引起的传染病。迄今报道的应用痘苗病毒载体插入并表达的外源基因有乙型肝炎表面抗原基因、流感病毒血凝素基因、狂犬病毒表面糖蛋白基因，以及单纯疱疹病毒、EB 病毒、水疱性口炎病毒、伪狂犬病毒和疟原虫等有关抗原基因。用含乙肝表面抗原基因的重组痘苗病毒感染猩猩，一次接种可诱发抗体并可使其抵抗乙肝病毒的攻击；用含狂犬病毒表面糖蛋白基因的重组痘苗病毒给小鼠接种，可以使其抵抗强毒株狂犬病毒的攻击，这个重组株已经开始用于狗和绵羊等动物免疫。利用痘苗病毒作表达载体生产遗传工程活疫苗的潜力是很大的，今后很有可能出现能同时表达几种病毒的抗原的重组痘苗病毒，即能同时预防几种病毒病的多价活疫苗。可见，用痘苗病毒载体构建不同的基因工程疫苗，在病毒性疾病的预防中将发挥越来越重要的作用。

三、外源基因在其他细胞中的表达

1. 酵母表达系统

酵母是单细胞生物，也是重要的工业微生物。外源基因在酵母细胞内可较好地完成翻译后加工（糖基化、二硫键加工等），特别适合于某些蛋白质表达。酵母表达系统不仅具有大肠杆菌系统相同的简便、细胞生长迅速、能大规模发酵生产等优点，同时还兼顾到表达量及生物活性两方面的需要。该表达系统成本低，可以方便地进行扩增培养，并可利用发酵罐进行大规模的发酵，因此在生物制品工业中具有广阔的应用前景。

甲醇酵母基因表达系统是最近才发展起来的外源蛋白生产系统。目前该表达系统已成功地表达了多种异源蛋白，如白蛋白、肿瘤坏死因子、乙型肝炎病毒表面抗原等。

甲醇酵母的表达载体为整合型质粒，载体中含有与酵母染色体同源的序列，因而比较容易整合入酵母染色体中。大部分甲醇酵母的表达载体中都含有甲醇酵母醇氧化酶基因 1（$AOx1$），在该基因的启动子（PAOXI）作用下，外源基因得以表达。PAOXI 是一个强启动子，在以葡萄糖或甘油为碳源时，甲醇酵母中 $AOx1$ 基因的表达受到抑制，而在以甲醇为唯一碳源时 PAOXI 可被诱导激活，因而外源基因可在其控制下表达，将目的基因多拷贝整合入酵母染色体后可以提高外源蛋白的表达水平及产量。此外甲醇酵母的表达载体都为 $E.coli/Pichia\ pastoris$ 的穿梭载体，其中含有大肠杆菌复制起点和筛选标记，可在获得克隆后采用大肠杆菌细胞大量扩增。目前，将质粒载体转入酵母菌的方法主要有原生质体转化法、电击法及氯化锂法等。甲醇酵母一般先在含甘油的培养基中生长，培养至高浓度，再以甲醇为碳源，诱导表达外源蛋白，这样可以大大提高表达产量。利用甲醇酵母表达外源蛋白，其产量往往可达克级。与酿酒酵母相比其翻译后的加工更接近哺乳动物细胞，不会发生超糖基化。

利用 PAOXI 表达外源蛋白时，一般需很长时间才能达到峰值水平，而甲醇是高毒性、高危险性化工产品，在实验操作过程中存在不小的危害性，不宜用于食品等蛋白质生产，因此那些不需要甲醇诱导的启动子，如 GAP、FLD1、PEX8、YPTI 等受到青睐。利用三磷酸甘油醛脱氢酶（GAP）启动子代替 PAOXI，不需要甲醇诱导，培养过程中无需更换碳源，操作更为简便，可缩短外源蛋白到达峰值水平的时间。

酵母表达系统作为一种后起的外源蛋白表达系统，由于兼具原核以及真核表达系统的优点，正在基因工程领域中得到日益广泛的应用。

2. 昆虫细胞表达系统

昆虫细胞表达系统是以昆虫细胞作为表达的宿主细胞。真核基因在昆虫中表达时，外源基因表达产物不仅可利用信号肽分泌到细胞外，而且能有效地进行蛋白质翻译后的加工修饰，如糖基化、脂酰化和磷酸化等。所得到的表达产物的抗原性、酶活力等生物学活性与天然蛋白十分相似。

昆虫细胞表达系统具有易操作、安全等优点。已有几十种外源基因在昆虫细胞表达系统中获得成功的表达，是一种很有发展前途的真核表达系统。昆虫细胞的载体主要是昆虫杆状病毒载体。昆虫杆状病毒基因组较大（130 kb），因此可以插入大片段的 DNA（可达 100 kb），允许克隆多拷贝基因，从而提高表达效率。昆虫杆状病毒不感染脊椎动物，病毒启动子在哺乳动物细胞中无感染活性，因此用于表达癌基因或对人类有毒的蛋白质比较安全。

杆状病毒表达系统是目前应用最广的昆虫细胞表达系统，该系统通常采用苜蓿银纹夜蛾杆状病毒（AcNPV）作为表达载体。在 AcNPV 感染昆虫细胞的后期，核多角体基因可编码产生多角体蛋白，该蛋白质包裹病毒颗粒可形成包涵体。核多角体基因启动子具有极强的启动蛋白质表达的能力，故常被用来构建杆状病毒传递质粒。克隆入外源基因的传递质粒与野生型 AcNPV 共转染昆虫细胞后可发生同源重组，重组后多角体基因被破坏，因而在感染细胞中不能形成包涵体。利用这一特点可挑选出含重组杆状病毒的昆虫细胞。

昆虫细胞表达系统，特别是杆状病毒表达系统由于其操作安全，表达量高，目前与酵母表达系统一样被广泛应用于基因工程的各个领域。但是，此系统效率比较低，且载体构建时间长，一般需要 4～6 周，昆虫细胞表达系统不能表达带有完整 N 联聚糖的真核糖蛋白是其不足之处。

小　结

基因工程的基本特点是进行分子水平上基因操作和细胞水平上的蛋白质表达。基因工程的核心是基因克隆或重组 DNA 技术，该技术涉及"酶切、连接、转化、筛选" 4 个基本过程。这些过程中要使用多种工具酶，包括限制性内切核酸酶、DNA 连接酶和其他一些参与 DNA 合成与修饰的酶类。载体是能将分离或合成的基因导入细胞的 DNA 分子。DNA 重组分子在体外构建完成后，必须导入特定的受体细胞，使之无性繁殖并高效表达外源基因或直接改变其遗传性状。其重组 DNA 分子导入原核细胞的过程称为转化，导入哺乳动物细胞的过程称为转染。可通过瞬时转染和稳定转染将外源 DNA 导入哺乳动物细胞。此外，根据重组基因的大小与特性，选择合适的表达系统对克隆基因进行表达，以在体外表达目的蛋白。与原核表达系统相比，哺乳动物细胞与其他表达系统的优势在于能够指导蛋白质的正确折叠，提供表达蛋白翻译后的加工修饰，使表达蛋白在分子结构、理化特性和生物学功能方面更接近于天然的高等生物蛋白质分子，以用于表达蛋白的结构与功能研究。基因工程技术是现代生物技术的核心，目前在工业、农业和医疗中已经显示了巨大的应用前景。

（方　勤）

参 考 文 献

Alberts B, Johnson A, Lewis J et al. 2002. Molecular Biology of the Cell. 4th Ed. New York: Garland Science

Nicholl D S. 2002. An Introduction to Genetic Engineering. 2nd Ed. Cambridge University Press

Sambrook J, Fritsch E F, Maniatis T. 2001. Molecular Cloning, A Laboratory Manual. 3rd ed. New York : Cold
Spring Harbor

Chapter 6   Genetic Engineering and Its Application

Biotechnology has led the development of a new innovative research that combined basic and applied life sciences since 1970s. Genetic engineering, also known as recombinant DNA technology, means altering genes in a living organism to produce a genetically modified organism with a new genotype. The new advanced technology makes various kinds of genetic modification being possible: such as, inserting a foreign gene from one species into another, forming a transgenic organism; altering an existing gene so that its product is changed; or changing gene expression so that it is translated more often or not at all.

Section 1   Enzymes Used in Genetic Engineering

To manipulate DNA, there are many work need to be done in the processing of recombinant DNA, including cutting, ligation, synthesis, modification and so on. The tools that enable the manipulation to be performed are different enzymes, which are purified from a wide range of organisms.

1.1   Restriction Enzymes

Restriction enzymes are DNA-cutting enzymes found in bacteria. Because they cut within the molecule, they are often called restriction endonucleases. Due to their high specificity and ease of use, restriction enzymes are important tools in studies of DNA primary structure, recombinant DNA technology and other fields of molecular genetics and molecular biology.

1.1.1   Host Restriction Modification System and the Discovery of Restriction Enzymes

Bacteria are capable of modifying specific sequences within their genomes by methylation which prevents their own DNA from being recognized by the restriction enzymes encoded by their genomes. This process is termed modification and restriction. Although the phenomenon of host specificity was initially observed by Luria and Human in the early 1950s, it was nearly a decade later that Arber and Dussoix predicted its molecular basis. They proposed that host specificity was based on a two enzyme system: a restriction enzyme which recognizes specific DNA sequences and is able to cleave the foreign invading DNA upon entering the bacterial cell, and a modification enzyme (methylase) responsible for protecting host DNA against the action of its own restriction enzyme.

Restriction enzyme and modification methylase were thought to recognize the same nucleotide sequence and together form a restriction modification (R-M) system. In 1968, restriction modification enzymes *Eco*B and *Eco*K were isolated and classified as type I enzymes. Since they cleave DNA at random positions, they can not serve as tools to excise specific fragments. Two years later Smith and Wilcox isolated and characterized the first type II restriction enzyme, *Hind* II, that cleaved DNA in well-defined fragments. This discovery revolutionized research into gene structure and gene expression.

1.1.2   Nomenclature and Classification of Restriction Endonucleases

1.1.2.1   Nomenclature of restriction endonucleases

Restriction enzymes are named based on the organism in which they were discovered. For example, the enzyme *Hind* III was isolated from *Haemophilus influenzae*, strain Rd. The first three letters of the name are italicized because they abbreviate the genus and species names of the organism. The fourth letter typically comes from the bacterial strain designation. The Roman numerals are used to identify specific enzymes from bacteria that contain multiple restriction enzymes. Typically, the Roman numeral indicates the order in which restriction enzymes were discovered in a particular strain. *Eco*R I for another example (Table 6-1).

Table 6-1　Nomenclature of Restriction enzyme *Eco*R I

E	*Escherichia*	(genus)
co	*Coli*	(species)
R	RY13	(strain)
I	First identified	Order ID'd in bacterium

1. 1. 2. 2　Classification of restriction enzymes

There are three classes of restriction enzymes, named as types Ⅰ, Ⅱ, and Ⅲ. Neither type Ⅰ nor type Ⅲ restriction systems have found much application in recombinant DNA techniques.

Type I　Restriction modification systems

Type Ⅰ enzymes are complex, combination restriction and modification enzymes that cut DNA at random far from their recognition sequences. The cofactors for the enzyme are Mg^{2+} ions, ATP and S-adenosylmethioninc. Type Ⅰ can simultaneously hold two different sites on DNA creating a loop in nucleic acid. The enzyme consists of their types of subunit. The *Eco* B enzyme has the structure $R_2 M_2 S$. The R subunit is responsible for restriction and the M subunit for methylation. The enzyme's binding to DNA may be succeeded by either restriction or modification and this property is characterized by S subunit. These enzymes interact with an unmodified recognition sequence in double strand DNA and then attach to long DNA molecule. After travelling for distance between $1000\sim5000$ nucleotides the enzyme cleaves only one strand of the DNA at an apparently random site, and creates a gap of about 75 nucleotides in length. Acid soluble oligonucleotides removed from the gap are released. A second enzyme molecule is needed to cleave the remaining strand of DNA (Figure 6-1). Type I R-M enzymes are NOT sitepecific endonucleases and have not found a use in Genetic Engineering, which means they have been somewhat neglected over the years. However, they a very interesting multisubunit, multifunctional molecular motors with strong relationships to both recombination enzymes such as RecBC and to chromatin remodeling factors in eukaryotes. These enzymes were the first discovered of all of the restriction enzymes and pioneered studies of restriction endonucleases, which eventually led to the discovery of type Ⅱ restriction modification, their characterization as site-specific endonucleases and the development of genetic engineering.

Figure 6-1　Restriction enzyme type Ⅰ　(see page 173)

Restriction enzyme type Ⅲ

Type Ⅲ restriction systems have separate enzymes for restriction and methylation, but these enzymes share a common subunit. These enzymes cleave double stranded DNA at well defined sites. They require ATP, Mg^{2+} ions and have very partial requirement for S-adenosyl-methionine for restriction. They have intermediate properties between type Ⅰ and type Ⅱ. The systems recognize specific short DNA sequences and cleave a short distance, about $24\sim27$ bases, away from the recognition sequence to give specific double-stranded fragments with terminal 5'-phosphates. For the type Ⅲ enzymes *Eco*P Ⅰ, it is consisted of two subunits, the endonuclease subunit Res is controlled both at the translational level and by the associated Mod subunit (Figure 6-2).

Figure 6-2　Restriction enzyme type Ⅲ　(see page 174)

Restriction enzyme type Ⅱ

Type Ⅱ restriction enzymes, in contrast, are heavily used in recombinant DNA techniques. Type Ⅱ enzymes consist of single, separate proteins for restriction and modification. One enzyme recognizes and cuts DNA, the other enzyme recognizes and methylates the DNA. Type Ⅱ restriction enzymes cleave the DNA sequence at the same site at which they recognize it (Figure 6-3). The only exception are type IIs (shifted) restriction enzymes, which cleave DNA on one side of the recognition sequence, within twenty nucleotides of the recognition site. Type Ⅱ restriction enzymes discovered to date collectively recognizes over 200 different DNA sequences.

Figure 6-3　Restriction enzyme type Ⅱ　(see page 174)

Type Ⅱ restriction enzymes can cleave DNA in one of three possible ways (Table 6-2). In one case, these enzymes cleave both DNA strands in the middle of a recognition sequence, generating blunt ends. For example: (The notations 5′ and 3′ are used to indicate the orientation of a DNA molecule. The numbers 5 and 3 refer to specific carbon atoms in the deoxyribose sugar in DNA). These blunt ended fragments can be joined to any other DNA fragment with blunt ends, making these enzymes useful for certain types of DNA cloning experiments.

Table 6-2 The useful restriction enzymes and their recognition sequencest (see page 175)

Type Ⅱ restriction enzymes can also cleave DNA to leave a 3′ ("three prime") overhang. These 3′ overhanging ends can only join to another compatible 3′ overhanging end (that is, an end with the same sequence in the overhang). Finally, some type Ⅱ enzymes can generate 5′ overhanging DNA ends, which can only be joined to a compatible 5′ end. In the type Ⅱ restriction enzymes discovered to date, the recognition sequences range from 4~9 bp long. Cleavage will not occur unless the full length of the recognition sequence is encountered. Enzymes with a short recognition sequence cut DNA frequently; restriction enzymes with 8 or 9 bp sequences typically cut DNA very infrequently, because these longer sequences are less common in the target DNA.

1.1.3 Properties of Restriction Enzyme Type Ⅱ

1.1.3.1 Patterns of DNA Cutting by Restriction Enzymes

Restriction enzymes hydrolyze the backbone of DNA between deoxyribose and phosphate groups. This leaves a phosphate group on the 5′ ends and a hydroxyl on the 3′ ends of both strands. A few restriction enzymes will cleave single stranded DNA, although usually at low efficiency.

The restriction enzymes most used in molecular biology labs cut within their recognition sites and generate one of three different types of ends (Figure 6-4). ①5′ overhangs: The enzyme cuts asymmetrically within the recognition site such that a short single-stranded segment extends from the 5′ ends. BamH I cuts in this manner. ② 3′ overhangs: Again, we see asymmetrical cutting within the recognition site, but the result is a single-stranded overhang from the two 3′ ends. Kpn I cuts in this manner. ③ Blunts: Enzymes that cut at precisely opposite sites in the two strands of DNA generate blunt ends without overhangs. SmaI is an example of an enzyme that generates blunt ends.

Figure 6-4 Blunt and sticky ends generated by restriction enzymes (see page 176)

1.1.3.2 Cutting frequency and rate of restriction enzymes

One of the key parameters that affect the way in which we use these enzymes is the length of the recognition site, which will affect the frequency with which they cut DNA, and hence the average size of the fragments generated. For example, take the enzyme Sau 3A, which has a 4-base recognition site: GATC. If we assume, for the moment, that there is an equal proportion of all four bases, and also that the bases are randomly distributed, then at any position on the chromosome there is a one in four chance that it is a G. Then there is a one in four chance that the next base is an A; the chance of having the sequence GA is the product of the two or 1 in 4^2 (1 in 16). Extending the argument, the chance of the sequence GAT occurring is 1 in 4^3, and that of the 4-base sequence GATC is 1 in 4^4, or 1 in 256. So this enzyme (and any others with a 4-base recognition site) would cut DNA (on these assumptions) on average every 4^4 bases, and so would generate a set of fragments with an average size of 256 bases. The actual sizes of the fragments will be distributed quite widely on either side of this average value. A 6-base site such as that recognized by EcoR I (GAATTC) would occur every 4^6 bases (or 4096 bases). Thus, a 6-bp cutter will yield DNA fragments of average length about 4000bp.

1.1.3.3 Isocaudamer or isoschizomers

Although enzymes that generate cohesive ends are useful for gene cloning because of the efficiency of the ligation process, there is a limitation. One can only join together fragments with compatible ends. So two EcoR I fragments can be ligated to each other or two BamH I fragments can be joined, but you cannot ligate an EcoRI fragment directly to a BamH I fragment. However, there are circumstances in which compatible ends can be generated by different restriction enzymes. For example, the restriction enzymes

*Bam*H Ⅰ, *Bcl* Ⅰ, *Bgl* Ⅱ, *Sau*3A Ⅰ and *Xho* Ⅱ recognize different sequences, but they both generate the same sticky ends (with unpaired GATC sequences) and these can be joined. We call such enzymes that recognize identical sequences are isocaudamer.

In addition, different restriction enzymes can have the same recognition site—such enzymes are called isoschizomers; look at the recognition sites for *Sac* Ⅰ and *Sst* Ⅰ—they are identical. In some cases isoschizomers cut identically within their recognition site, but sometimes they do not. Isoschizomers often have different optimum reaction conditions, stabilities and costs, which may influence the decision of which to purchase. Another example is the pair of isoschizomers *Xma* Ⅰ and *Sma* Ⅰ. *Xma* Ⅰ cuts asymmetrically and produces sticky ends that can be ligated to other *Xma* I fragments, while *Sma* Ⅰ, asmentioned above, will generate blunt-ended fragments at the same site, allowing you to ligate the fragment to other blunt-ended DNA sequences.

1. 2 Other DNA Modifying Enzymes

Restriction enzymes as described above provide the cutting functions that are essential for the production of recombinant DNA molecules. Other enzymes used in genetic engineering, with the term used here include ligation, synthesis and alteration of DNA. Some of the most commonly used enzymes are described below (Table 6-3):

Table 6-3 Other DNA modifying enzymes (see page 176)

DNA ligase: The natural role of DNA ligase is to repair single-strand breaks (nicks) in the sugarphosphate backbone of a double-stranded DNA molecule, such as may occur through damage to DNA (or following the repair of such damage), as well as the joining of the short fragments produced as a consequence of replication of the "lagging strand" during DNA replication. The action of the ligase requires that the nick should expose a 3'-OH group and a 5'-phosphate.

DNA polymeraseⅠ: It has 5' exonuclease activity in addition to DNA synthesis. This bifunctional activity enables the enzyme to use nicks or gaps in double stranded DNA as a starting point for DNA synthesis. It is used with radioactive or biotinylated nucleotides to prepare labeled DNA of high specific activity.

Reverse transcriptase: Reverse transcriptase mediates the conversion of genetic information present in single stranded molecule of RNA into a double stranded molecule of DNA. It has several enzymatic activities such as: RNA-directed DNA polymerase, DNA-dependent DNA polymerase and RNase H activity.

Polynecleotide kinase: Polynecleotide kinase catalyses the transfer of the terminal phosphate group of ATP to the 5'-hydroxylated terminal of DNA or RNA. This enzyme is frequently used to end label nucleic acids with ^{32}P.

Terminal transferase: Terminal transferase is used to add homopolymer tails to the terminal 3'-hydroxyl group of a DNA molecule.

Alkaline phosphatases: Alkaline phosphatase catalyses the removal of the 5'-terminal phosphate residues from nucleic acids (RNA, DNA and ribo-and deoxyribonuncleotide triphoshate).

Section 2　Basic Process of Cloning a Gene

2. 1　Preparation of Interest Gene

2. 1. 1　Chemical Synthesis of Interest Gene DNA

The interest DNA synthesized by chemical method is based on DNA sequencing. The sequence of nucleotides or amino acids of a gene is known, one can synthesize the target DNA using chemical method. Genes such as human growth hormone releasing inhibitor, proinsulin, enkephalin and interferons are synthesized in this way. However, the application of this method is limited to synthesize relatively short nucleotides of some simple genes. The major advantage of this method is to synthesize new genes which do not exist in nature. So that it can benefit human beings by generating new biological features. In near future, this method will be improving and applied more widely.

2. 1. 2   Isolation of Interest Genomic DNA

The chromosomal DNA can be separated from tissues or cells and digested by restriction endonucleases to generate DNA fragments. These DNA fragments include the interest gene fragments. The interest DNA fragment is combined to an appropriate vector and then transformed into competent cells. Each recombinant DNA molecule can be amplified into multiple copies in bacterial cells. Different bacteria carry different recombinant DNA molecules which represent different chromosomal DNA fragments of an organism. Therefore the whole bacterial cells carrying all chromosome DNA fragments represent the entire genome. The collection of genomic DNA carried by the cloning vectors in bacteria is called genomic DNA library.

2. 1. 3   Systhesis of Complementary DNA (cDNA)

Complementary DNA of a given mRNA is inserted into a suitable vector and then is cloned into a host. Once inside the host cell, the cDNA in the hybrid will express the gene (s) it carries. Complementary DNAs (cDNAs) of the genome of an organism can be synthesized and cloned to generate a cDNA library. Thus, cDNA library is a collection of cDNA clones of an organism. The starting material for the synthesis of cDNA is the mRNA. Hence the resulting cDNAs will represent the mRNAs present in the cells from where these mRNAs were isolated. In the case of multicellular organisms there is specialization of individual cells.

2. 1. 4   Amplification of Interest Gene by Using Polymerase Chain Reaction (PCR)

The PCR is a powerful technique used to amplify DNA millions of fold, by repeated replication of a template, in a short period of time. The process utilizes sets of specific in vitro synthesized oligonucleotides to prime DNA synthesis. The design of the primers is dependent upon the sequences of the DNA that is desired to be analyzed. The technique is carried out through many cycles (usually 20~50) of melting the template at high temperature, allowing the primers to anneal to complimentary sequences within the template and then replicating the template with DNA polymerase. The process has been automated with the use of thermostable DNA polymerases isolated from bacteria that grow in thermal vents in the ocean or hot springs. During the first round of replication a single copy of DNA is converted to two copies and so on resulting in an exponential increase in the number of copies of the sequences targeted by the primers. After just 20 cycles a single copy of DNA is amplified over 2 000 000 fold.

2. 2   Choices of Cloning Vectors

In molecular biology, a vector is any vehicle used to transfer foreign genetic material into another cell or host cell. If it is used for reproducing the DNA fragment, it is called a cloning vector. If it is used for expressing certain gene in the DNA fragment, it is called an expression vector. There are three main clone vectors in genetic engineering, they are plasmid vector; phage and virus vector, and last, cosmid. These vectors should have following common features: ①the vector must be able to replicate in the hose cell. ② the vector must contain a selectable marker for identification and isolation of subpopulation of bacteria containing vector. ③ most commonly used vectors contain a short DNA sequence with many closely placed restriction enzyme cleavage sites, which is known as multicloning site (MCS) or polylinkers. In addition to meet the above conditions, the genetic engineering expression vector should also have other regulatory elements such as promoter, ribosomal binding site, enhancer, terminator.

2. 2. 1   Plasmid Vectors

Plasmids are double-stranded generally circular DNA sequences that are capable of automatically replicating in a host cell. Plasmid vectors minimalistically consist of an origin of replication that allows for semi-independent replication of the plasmid in the host and also the transgene insert. Modern plasmids generally have many more features, notably including a "multiple cloning site" which includes nucleotide overhangs for insertion of an insert, and multiple restriction enzyme consensus sites to either side of the insert. In the case of plasmids utilized as transcription vectors, incubating bacteria with plasmids generates hundreds or thousands of copies of the vector within the bacteria in hours, and the vectors can be extracted from the bacteria, and the multiple cloning sites can be restricted by restriction enzymes to excise the hundredfold or thousandfold amplified insert. These plasmid transcription vectors characteristically lack crucial sequences that code for polyadenylation sequences and translation termination sequences in translated mRNAs, making expression of transcription vectors impossible.

The plasmids most commonly used in recombinant DNA technology replicate in *E. coli*. Generally, these plasmids have been engineered to optimize their use as vectors in DNA cloning. Most plasmid vectors contain little more than the essential nucleotide sequences required for their use in DNA cloning: a replication origin, a drug-resistance gene, and a region in which exogenous DNA fragments can be inserted. Plasmids almost always exist and replicate independently of the chromosome of the cell in which they are found. Based upon the number of copies per cell, plasmids are classified into two types.

Stringent plasmids: These plasmids exist in small numbers, i. e. less than 100 copies/cell. Stringent plasmid is under the control of bacterial genome for replication and segregation. Generally, conjunctive plasmids are mostly stringent plasmids.

Relaxed plasmids: These plasmids exist in large numbers, i. e. , more than 100 copies/cell. Relaxed plasmid is not under the control of bacterial genome for replication and segregation. Generally, relaxed plasmids are of low molecular weight and most of them are of the non conjugative type.

The replication origin (ORI) is a specific DNA sequence of $50\sim100$ base pairs that must be present in a plasmid for it to replicate. Host-cell enzymes bind to ORI, initiating replication of the circular plasmid. Once DNA replication is initiated at ORI, it continues around the circular plasmid regardless of its nucleotide sequence. Thus any DNA sequence inserted into such a plasmid is replicated along with the rest of the plasmid DNA; this property is the basis of molecular DNA cloning.

At present, the most widely used vector is pcDNA and its derivatives series (Figure 6-5).

Figure 6-5　Map of pcDNA vector (see page 179)

2. 2. 2　Bacteriophages Vectors

The clone of single genes is usually best carried out using plasmid, since the insert will rarely be larger than about 2kb. But for cloning of larger pieces of DNA (e. g. during gene library construction), these plasmids are not suitable because large inserts increase the plasmid size, making the transformation inefficient. Large molecules can be injected in host bacterial cell by viral particles (bacteriophages). Commonly used bacteriophages are λ phage, M13, f1, fd and so on.

2. 2. 2. 1　λ phage vector

The DNA to be cloned is first inserted into the λ DNA, replacing a nonessential region. Then, by an in vitro assembly system (described below), the λ virion carrying the recombinant DNA can be formed. The λ genome is 49 kb in length which can carry up to 25 kb foreign DNA.

The extreme ends of the λ DNA are known as COS sites, each of them is single stranded, 12 nucleotides long. Because their sequences are complementary to each other, one end of λ DNA may be base pair with the other end of a different λ DNA, forming concatemers. The two ends of λ DNA may also bind together, forming a circular DNA. In the host cell, the λ DNA circularizes because ligase may seal the join of the COS sites. There are two kinds of λ vector. Which differ in size of DNA they accept, λ replacement vectors and λ insertion vectors. These are described below:

Replacement vectors: λ replacement vectors contain a restriction site for phage propagation in suitable bacterial host. Remaining part of λ genome is removed and is replaced by foreign DNA. Vector molecule containing target DNA and those without target DNA can be selected by infecting host. Recombinant phage in which the internal region is replaced by target DNA is viable and is mostly used for cloning eukaryotic DNA fragments.

Insertion vectors: The phage DNA is cleaved with a restriction enzyme that cuts it only once, and the target is inserted into this site. No phage DNA is removed therefore, much lower size of DNA can be inserted.

The capacity growth of λ phage is related with the size of the λ DNA packaging in the head protein. When the length of DNA is 78% shorter or 105% than wild-type, the activity of the phage dropped sharply. Therefore the λ vector DNA adding exogenous DNA should be 39~53 kb. Replacement vector DNA selected by the restriction enzyme digestion try to remove the middle segment, leaving only his left arm and his right arm to use exogenous DNA fragment connectivity, packaging them into a recombinant phage. This is because the left arm and right arm and middle have single-strand fragment of the sticky end which can recover to the original structure by the ligase, thereby affecting the exogenous DNA connected. Method to distinguish the plaque of recombinant phage and recovery phage is to use X-gal blue plaque test. The middle fragment of some phage vector encode acid β-galactosidase gene. When this phage infect the host cell with lac gene and grow in medium containing X-gal (5-chlorine-bromine-4-indole-3-β-D-galactosidase), semi-galactosidase acid will react to X-gal to produce the insoluble indigo dye. Therefore, the recovery phage containing the middle fragment form a blue plaque, while the plaque of recombinant phage connected with exogenous DNA is colorless and transparent (Figure 6-6).

Figure 6-6   The genome of λ phage (see page 180)

2.2.2.2   Filamentous phages as cloning vector

It includes Ff class of filamentous phages, including strains f1, fd and M 13, which infect E. coli cells displaying E. coli. These Ff virions are long and thin and contain a closed loop of single stranded DNA. Because the phages readily accept inserts of foreign DNA and they supply one strand of that DNA in all easily isolated form, vectors based on Ff phages have become standard choice. Following is a brief description of M13 Bacteriophage.

M13 is a filamentous bacteriophage of E. coli. It is 8701 nm long and 6 nm wide. It has a protein coat (capsid) made up of three kinds of capsomeres. These filamentous single-stranded DNA phages infect cells by adsorbing to and entering through F pili; thus, these phages only infect F+ or Hfr cells, not F-cells.

The DNA of wide type M13 bacteriophage is single-stranded and circular (Figure 6-7). It is 6407 bases long and has 10 genes that are closely packed. All these genes are essential for the replication of the phages. There is a segment of 507 nucleotide intergenic sequence which contains origin of replication. This intergenic region can be manipulated for cloning without disrupting the origin of replication. Hence the wild M13 phage has limited use in gene cloning experiments. The size of the phage particle is decided by the size of the phage DNA. Up to six times the normal length of M13 DNA can be packaged.

Figure 6-7   The genome of M13 phage (see page 187)

After reconstructing the interval sequence of M13 genome, it can be formed into a series of cloning vectors. M13 phage as a vector has the following important features: ① Infection and the release of the M13 will not kill the host bacteria, only lead to slow growth of the host; ② M13 DNA in the host bacteria may be both single-stranded and double-stranded. The infection or transformation method can insert M13 DNA into host bacteria; ③ The packaging of M13 has nothing to do with the DNA size. Even if the size of DNA is six times larger than the size of their own DNA, it still can be packaged. Because of these characteristics, M13 is the widely used in recombinant DNA. The main application of M13 phage vector is to generate single-strand DNA template for Sanger ddNTP chain termination method to sequence DNA.

2.2.3   Cosmid

In 1978 Collins and Hohn constructed of a new type of E. coli cloning vector, named cosmid (Keshi plasmid). It is composed with a normal plasmid and the cos sites of λ phage. For example, pBR322 and the cos site of the section of DNA in λ phage constitute cosmid pHC79 with a total length of 43Kb. In packaging, cos sites open and a sticky end of λ phage forms. pHC79 has tetracycline and ampicillin resistance due to the DNA of pBR322 (Figure 6-8).

Figure 6-8　Cosmid pHC 79（see page 182）

Cosmid is general of 4~6 kb. An important role of the *cos* site is to identify the capsid protein of phage. Any DNA molecule which includes *cos* site and is as long as equivalent to the length of phage genome, can be combined with the coat protein and packaged into similar λ phage particles. Therefore, foreign DNA insert into the cosmid can be greater than 40 kb. The recombinant cosmid can infect *E. coli* as the λ phage and replicate in the cell. If the host bacteria grow in the medium containing chloramphenicol, cosmid can be amplified to the host cell DNA in about 50 % of the total.

The problems for use of Cosmid: ①Vector is 1/10 of inserted fragment, so it can always be connected with itself, resulting in a reorganization of elements within a few cosmid linked together. But with alkaline phosphatase treatment, it can stop linking its own vector elements. ② Ranging from the size of exogenous fragments connect with each other and then insert into the vector, leading to the result that the fragments which were not adjacent to each other confused into a fragment. ③ Bacterial colony size much larger than phage plaque, so it will spend much time to screen for a DNA fragment containing the colony in the genetic library of cosmid.

2.3　DNA Recombination of Interest Gene and Its Vector

The next step of cloning a gene is to join the DNA fragment to the vector molecule, such as a plasmid or bacteriophage, that can be replicated by the host cell after transformation. The joining, or ligation, of DNA fragments is carried out by an enzyme known as DNA ligase. The natural role of DNA ligase is to repair single-strand breaks (nicks) in the sugar-phosphate backbone of a double-stranded DNA molecule, such as may occur through damage to DNA (or following the repair of such damage), as well as the joining of the short fragments produced as a consequence of replication of the "lagging strand" during DNA replication. The action of the ligase requires that the nick should expose a $3'$-OH group and a $5'$-phosphate. Some DNA ligases, such as T4 DNA ligase (encoded by the bacteriophage T4), are also capable of ligating blunt-ended fragments.

2.3.1　Cohesive end Ligation

The sticky ends created by restriction enzyme has the nick after a few base pairs in opposite strands. DNA ligase repairs these nicks and forms intact duplex. Some enzymes, such as *Pst* I, also form cohesive ends, but by cutting asymmetrically within the right-hand part of the recognition site; they thus generate sticky ends with unpaired single-strand sequence at $3'$ ends of fragment, and these ends are complementary to one another, so we can use DNA ligase to finally join the fragments together with a covalent bond to form recombinant molecule. So two *EcoR* I fragments can be ligated to each other, or two *BamH* I fragments can be joined, but you cannot ligate an *EcoR* I fragment directly to a *BamH* I fragment. However, there are circumstances in which compatible ends can be generated by different restriction enzymes. For example, the restriction enzymes *BamH* I and *Mbo* I recognize different sequences, but they both generate the same sticky ends (with unpaired GATC sequences) and these can be joined.

2.3.2　Blunt end Ligation

The *E. coli* DNA ligase can not catalyse ligation of blunt ends except under special reaction conditions of macromolecular crowding. Such blunt end ligation can be done with T4 DNA ligase. T4 DNA ligase can join the broken blunt ends of two DNA molecules respective of the sequences. The disadvantage of this technique is that any two broken ends may join including those belonging to same DNA molecule. However, blunt-ended ligation is a useful way of joining together DNA fragments, which have not been produced by the same restriction enzyme and therefore probably have incompatible sticky ends. These ends are removed prior to ligation using the enzyme S I nuclease, which digests single stranded DNA.

2.3.3　Ligation by Using Homo-Polylinker

By adding cohesive ends or linkers, blunt ended DNA fragment and vector can also be ligated. This technique makes use of producing and annealing complementary homopolymer sequences. An enzyme, terminal transferase (purified from calf-thymus), is used to synthesize homopolymeric extensions. Using precursor dGTP, poly dG is added at both the $3'$ cut ends of vectors, while poly dC is added to $5'$ cut end of foreign DNA. The vector and clone can then be joined by annealing the poly dG with poly dC tails and then ligating them using DNA ligase enzyme. The cloned sequence can be retrieved by using Pst I restriction enzyme. The method works out the chances of reannealing between the two cut ends of same DNA molecule or of two similar DNA molecules, thus is advantageous.

2.3.4　Ligation with Synthetic Oligoucleotides

They are self complementary synthetic oligoucleotides which contain restriction endonuclease sites. They may be used to clone DNA fragment into plasmids or other vectors. These blunt ended linkers are first phosphorylated using polynucleotide kinase and labeled or unlabelled ATP molecules. These are then ligated to the blunt ended DNA fragment and finally treated with restriction enzyme to generate sticky ends and ligated.

Blunt ligation is most usefully applied where blunt-ended fragments are joined via linker molecules (described later). The linker molecule can be ligated to both ends of foreign DNA to be cloned, and then treated with restriction enzyme to produce sticky ended fragments. The vector molecule is also cut with same restriction enzyme. Finally, the foreign DNA is ligated to vector via linker created sticky ends.

2.4　Introducing Recombinant Molecular into Host Cells

The DNA that we want to clone is inserted into a suitable vector, producing a recombinant molecule consisting of vector plus insert. This recombinant molecule will be replicated by the bacterial cell, so that all the cells descended from that initial transformant will contain a copy of this piece of recombinant DNA. This experiment rests on the natural ability of the *pneumococcus* to take up "naked" DNA from its surroundings. This ability is known as competence. There are four different ways to introduce recombinant into host cells. ① Transformation: bacterial transformation is the process by which bacterial cells take up naked DNA molecules. Bacteria which are able to uptake DNA are called "competent" and are made so by treatment with calcium chloride in the early log phase of growth. ② Transfection: transfection is the analogous to transformation except that the recipient is eukaryotes but not bacteria. Transfection also differs from transformation that transfection can be transient (no selection) but transformation always use the selection stress. ③ Transduction : transduction is the transfer of genetic material from one bacterium to another using a bacteriophage, or the process of transferring genetic material from one cell to another by a plasmid or bacteriophage. ④ Infection: *in vitro* assembled viruses or phages from recombinant DNA molecules generated by vectors from phage, cosmid or E viruses are able to infect proper host and get replicated in the cells.

2.5　Screening and Identification of Cloned Interest Gene

Previously, we describe how we can create a recombinant DNA. Next, we need to identify the clones carrying the target gene. There are four commonly used methods to screen and identify clones.

2.5.1　Plasmid Transformation and Antibiotic Selection

The process for the uptake of naked plasmid and bacteriophage DNA is the same; Plasmids used for the cloning and manipulation of DNA have been engineered to harbor the genes for antibiotic resistance. Thus, if the bacterial transformation is plated onto media containing ampicillin, only bacteria which possess the plasmid DNA will have the ability to metabolize ampicillin and form colonies. In this way, bacterial cells containing plasmid DNA are selected.

2.5.2　Colony *in situ* Hybridization

As mentioned above, several thousands of recombinants need to be plated in each culture dish in order to keep the total number of plates reasonably low. This means that you are unlikely to be able to pick your positive recombinants pure and uncontaminated by their neighbors the first time around. Instead, you would lift a small agar plug containing the positive and neighboring recombinants and elute them into a buffer. The eluted mixture would then be replated onto a new plate for rescreening with the same probe or antibody. This serves two purposes. First, it is almost inevitable that some of the clones are false positives, caused by non-specific binding of probe to the membrane. Indeed the likelihood of this is sufficiently high that it is common practice to use two duplicate filters for hybridization; only those clones that give a positive signal on both filters are regarded as worth following up. Second, for those that are true positives, the secondary screening will allow you to pick a recombinant that is free from contaminating neighbors.

2.5.3　Restriction Digests and Agarose Gel Electrophoresis

More accurate and more reliable estimates of the size of the insert require some degree of plasmid purification so that the plasmid can be digested with a restriction endonuclease.

The purified plasmid can then be digested with an appropriate restriction enzyme to generate linear fragments that can be accurately sized on an agarose gel. A common procedure is to use the same enzyme that was involved in the cloning step. For example, if you have inserted an *Eco*R I fragment into a unique *Eco*R I site on the cloning vector, then digestion with *Eco*R I will yield two fragments: one corresponding to the linearized vector, and the second being the inserted fragment. Using a standard marker containing a set of linear DNA fragments of known size enable the gel to be calibrated so that the size of the insert can be determined (within the limits of accuracy of the system).

2.5.4　DNA Sequencing

Restriction enzyme mapping and Southern blotting help the gross organization of a piece of DNA, and allow the approximate position of sequences of interest to be determined. The next step in analysis is normally to determine the nucleotide sequence.

Section 3　Transfection of Eukaryotic Cells

This section covers general information on transfection techniques and considerations for transfection efficiency and optimization. In addition, we discuss the various transfection agents as well as general protocols for transfection and specific examples using transfection reagents.

3.1　Techniques Used in Eukaryotic Cell Transfection

3.1.1　Methods for Transfection

Essentially, transfection is a method that neutralizes or obviates the issue of introducing negatively charged molecules (e. g. , phosphate backbones of DNA and RNA) into cells with a negatively charged membrane. Chemicals like calcium phosphate and DEAE-dextran or cationic lipid-based reagents coat the DNA, neutralizing or even creating an overall positive charge to the molecule. This makes it easier for the DNA: transfection reagent complex to cross the membrane, especially for lipids that have a "fusogenic" component, which enhances fusion with the lipid bilayer. Physical methods like microinjection or electroporation simply punch through the membrane and introduce the DNA directly into the cytoplasm. Each of these transfection technologies is discussed in the following sections.

3.1.1.1　DEAE-dextran

One of the first chemical reagents used for transfer of nucleic acids into cultured mammalian cells was DEAE-dextran. DEAE-dextran is a cationic polymer that tightly associates with negatively charged nucleic acids. An excess of positive charge, contributed by the polymer in the DNA: polymer complex, allows the complex to come into closer association with the negatively charged cell membrane. Uptake of the complex is presumably by endocytosis. This method is successful for delivery of nucleic acids into cells for transient expression; that is, for short-term expression studies of a few days in duration. However, this technique is not generally useful for stable or long-term transfection studies that rely upon integration of the transferred DNA into the chromosome. Stable transfection requires several weeks for selection, cloning and characterization of the transformed cells.

3. 1. 1. 2　Calcium phosphate

Calcium phosphate co-precipitation became a popular transfection technique following the systematic examination of this method in the early 1970s. The authors examined the performance of various cations and the effect of cationic concentration, phosphate concentration and pH on transfection. Calcium phosphate co-precipitation is widely used because the components are easily available and inexpensive, the protocol is easy-to-use, and many different types of cultured cells can be transfected. The protocol involves mixing DNA with calcium chloride, adding this in a controlled manner to a buffered saline/phosphate solution and allowing the mixture to incubate at room temperature. The controlled mixing generates a precipitate that is dispersed onto the cultured cells. The precipitate is taken up by the cells via endocytosis or phagocytosis. Calcium phosphate transfection is routinely used for both transient and stable transfection of a variety of cell types. In addition, calcium phosphate also appears to provide protection against intracellular and serum nucleases.

3. 1. 1. 3　Electroporation

Electroporation was first reported for gene transfer studies into mouse cells. This technique is often used for cell types such as plant protoplasts that are difficult to transfect by other methods. The mechanism for entry into the cell is based upon perturbing the cell membrane by an electrical pulse, which forms transient pores that allow the passage of nucleic acids into the cell. The technique requires fine-tuning and optimization for duration and strength of the pulse for each type of cell used. In addition, electroporation often requires more cells than chemical methods because of substantial cell death, and extensive optimization is often required to delicately balance transfection efficiency against cell viability. More modern instrumentation allows nucleic acid delivery to the nucleus and has been successful for transfer of DNA and RNA to primary and stem cells.

Another physical method of gene delivery is biolistic particle delivery, also known as particle bombardment. This method relies upon high velocity delivery of nucleic acids on microprojectiles to recipient cells by membrane penetration. This method has been successfully employed to deliver nucleic acid to cultured cells as well as to cells in vivo. Biolistic particle delivery is relatively costly for many research applications, but the technology can also be used for genetic vaccination and agricultural applications.

3. 1. 1. 4　Cationic lipids

The advance in liposomal vehicles was the development of synthetic cationic lipids by Felgner and colleagues. The cationic head group of the lipid compound associates with negatively charged phosphates on the nucleic acid. Liposome-mediated delivery offers advantages such as relatively high efficiency of gene transfer, ability to transfect certain cell types that are resistant to calcium phosphate or DEAE-dextran, in vitro and in vivo applications, successful delivery of DNA of all sizes from oligonucleotides to yeast artificial chromosomes, delivery of RNA, and delivery of protein. Cells transfected by liposome techniques can be used for transient expression studies and long-term experiments that rely upon integration of the DNA into the chromosome or episomal maintenance. Unlike the DEAE-dextran or calcium phosphate chemical methods, liposome-mediated nucleic acid delivery can be used for in vivo transfer of DNA and RNA to animals and humans.

A lipid with overall net positive charge at physiological pH is the most common synthetic lipid component of liposomes developed for gene delivery. Often the cationic lipid is mixed with a neutral lipid such as L-dioleoyl phosphatidylethanolamine (DOPE). The cationic portion of the lipid molecule associates with the negatively charged nucleic acids, resulting in compaction of the nucleic acid in a liposome/nucleic acid complex, presumably from electrostatic interactions between the negatively charged nucleic acid and the positively charged head group of the synthetic lipid. For cultured cells, an overall net positive charge of the liposome/nucleic acid complex generally results in higher transfer efficiencies, presumably because this allows closer association of the complex with the negatively charged cell membrane. Entry of the liposome complex into the cell may occur by the processes of endocytosis or fusion with the plasma membrane via the lipid moieties of the liposome. Following cellular internalization, the complexes appear in the endosomes and later in the nucleus. It is unclear how the nucleic acids are released from the endosomes and lysosomes and traverse the nuclear membrane. DOPE is considered a "fusogenic" lipid, and its role may be to release these complexes from the endosomes as well as to facilitate fusion of the outer cell membrane with the liposome/nucleic acid complexes. While DNA will need to enter the nucleus, the cytoplasm is the site of action for RNA, protein or antisense oligonucleotides delivered via the liposomes.

3.1.1.5　Viral methods

While transfection has been used successfully for gene transfer, the use of viruses as vectors has been explored as an alternative method for delivery of foreign genes into cells and as a possible in vivo option. Adenoviral vectors are useful for gene transfer due to a number of key features: ① they rapidly infect a broad range of human cells and can achieve high levels of gene transfer compared to other available vectors; ② adenoviral vectors can accommodate relatively large segments of DNA (up to 7.5kb) and transduce these transgenes in nonproliferating cells; and ③ adenoviral vectors are relatively easy to manipulate using recombinant DNA techniques. Other vectors of interest include adeno-associated virus, herpes simplex virus, retroviruses and a subset of the retrovirus family, lentiviruses. Lentiviruses (e.g., HIV-1) are of particular interest because they have been well-studied, can infect quiescent cells, and can integrate into the host cell genome to allow stable, long-term expression of the transgene.

As with all viral transfer methods, there are drawbacks. For adenoviral vectors, packaging capacity is low, and production is labor-intensive. With retroviral vectors, there is the potential for activation of latent disease and, if there are replication competent viruses present, activation of endogenous retroviruses and limited expression of the transgene.

3.1.2　Reagent Selection

With so many different methods of gene transfer, how to choose the right transfection reagent or technique for transfection? Any time a new element, like a new cell line, the optimal conditions for transfection will need to be determined. This may involve choosing a new transfection reagent.

3.1.3　Factors Influencing Transfection Efficiency

With any transfection reagent or method, cell health, degree of confluency, number of passages, contamination, and DNA quality and quantity are important parameters that can greatly influence transfection efficiency. Note that with any transfection reagent or method used, some cell death will occur.

Cell Health: Cells should be grown in medium appropriate for the cell line and supplemented with serum or growth factors as needed for viability. Contaminated cells and media (e.g., yeast or mycoplasma) should never be used for transfection. If the cells have been compromised in any way, discard them and reseed from a frozen, uncontaminated stock. Make sure the medium is fresh if any components are unstable. Medium lacking necessary factors can negatively affect cell growth. Be sure the 37℃ incubator is supplied with CO_2 at the correct percentage (usually $5\%\sim10\%$) and kept at 100% relative humidity.

Confluency: As a general guideline, transfect cells at $40\%\sim80\%$ confluency. Too few cells will cause the culture to grow poorly without cell-to-cell contact. Too many cells results in contact inhibition, making cells resistant to uptake of foreign DNA. Actively dividing cells take up introduced DNA better than quiescent cells.

Number of Passages: Keep the number of passages low (less than 50). In addition, the number of passages for cells used in a variety of experiments should be consistent. Cell characteristics can change over time with immortalized cell lines, and cells may not respond to the same transfection conditions after repeated passages, resulting in poor expression.

DNA Quality and Quantity: Plasmid DNA for transfections should be free of protein, RNA, chemical and microbial contamination. Suspend ethanol-precipitated DNA in sterile water or TE buffer to a final concentration of $0.2\sim1\text{mg} \cdot \text{ml}^{-1}$. The optimal amount of DNA to use in the transfection will vary widely depending upon the type of DNA, transfection reagent, the target cell line and number of cells.

3.2　Transient and Stable Transfection

The process of introducing nucleic acids into eukaryotic cells by nonviral methods is defined as "transfection". Using various chemical, lipid or physical methods, this gene transfer technology is a powerful tool for studying gene function in the context of a cell.

3.2.1 Transient Expression Versus Stable Transfection

Another parameter to consider is the time frame of the experiment you wish to conduct. Is it short-term or long-term? For instance, determining which of the promoter deletion constructs still function as a promoter can be accomplished with a transient transfection experiment, while establishing stable expression of an exogenously introduced gene construct will require a longer term experiment.

3.2.2 Transient Expression

Cells are typically harvested 24~72 hours post-transfection for studies designed to analyze transient expression of the transfected genes. The optimal time interval depends upon the cell type, research goals and the specific characteristics of expression for the transferred gene. Analysis of gene products may require isolation of RNA or protein for enzymatic activity assays or immunoassays.

3.2.3 Stable Transfection

The goal of stable, long-term transfection is to isolate and propagate individual clones containing transfected DNA that has integrated into the cellular genome. Distinguishing nontransfected cells from those that have taken up the exogenous DNA involves selective screening. This screening can be accomplished by drug selection when an appropriate drug resistance marker is included in the transfected DNA. Alternatively, morphological transformation can be used as a selectable trait in certain cases.

Section 4　Expression of Cloned Gene

High-level production of recombinant proteins as a prerequisite for subsequent purification has become a standard technique. Important applications of recombinant proteins are: ①immunization, ② biochemical studies, ③ three-dimensional analysis of the protein, and ④ biotechnological and therapeutic use. Production of recombinant proteins involves cloning of the appropriate gene into an expression vector under the control of an inducible promoter. But efficient expression of the recombinant gene depends on a variety of factors such as optimal expression signals (both at the level of transcription and translation), correct protein folding and cell growth characteristics. Display of recombinant proteins on the bacterial surface has many potential biotechnological applications and requires further knowledge on targeting motifs present on carrier proteins usually used as fusion partners. In addition, the selection of a particular expression system requires a cost breakdown in terms of design, process and other economic considerations.

4.1　Prokaryotic Expression System (*E. coli* Expression System)

(*E. coli*) is one of the most widely used hosts for the production of heterologous proteins and its genetics are far better characterized than those of any other microorganisms. Recent progress in the fundamental understanding of transcription, translation, and protein folding in *E. coli*, together with serendipitous discoveries and the availability of improved genetic tools are making this bacterium more valuable than ever for the expression of complex eukaryotic proteins.

4.1.1 Characteristics of Prokaryotes Expression

Gene expression in bacteria, as in all cells, involves: ① Production of messenger RNA by copying of the DNA template by RNA polymerase. ② Translation of the message into protein by the protein synthesis machinery. In bacteria, these two steps are closely linked. Transcription starts when RNA polymerase binds to the promoter region of the gene, and proceeds to copy the DNA sequence into mRNA. As the 5′ end of the transcript (i. e., the mRNA) is made, ribosomes immediately attach to it, and start translation, even before the entire message is synthesized. The ribosomes move along the message, translating the mRNA, as it is being made. Behind the ribosomes, the 5′ end of the message is rapidly degraded. The entire process could be completed in a matter of a couple of minutes.

The important features of this process are: ① RNA polymerase binding to the promoter region of the gene, followed by transcription to produce a messenger RNA. When the polymerase reaches a terminator sequence, transcription stops, and the mRNA is released from the DNA template. This mRNA contains a 5′ flanking sequence, the actual coding sequence that will be translated, and a 3′ flanking sequence. Because translation does not usually begin at the 5′ end of the message, the mRNAs also carry signals that define the beginning and end of each encoded protein. ② Initiation of protein synthesis (translation): The signal that indicates the start site of protein synthesis is the AUG, or start codon, which specifies methionine. This site is recognized by the initiator tRNA, which carries a formylmethionine residue (this tRNA is different from the one that inserts methionines within the protein sequence). Another important signal on the mRNA is a purine-rich region about 10 nucleotides upstream of the AUG. This sequence, called a Shine-Dalgarno sequence, contains a sequence complementary to the 3′ end of 16S ribosomal RNA, which is part of the 30S ribosomal subunit. Because of this, this sequence is also called the Ribosome Binding Site, or RBS. The initiator tRNA will bring a formylmethionine residue to the AUG nearest to the RBS, to initiate translation. ③ Termination of protein synthesis: The end of the protein coding sequence is indicated by the presence of a stop codon (UAA, UGA, or UAG). This signal is recognized by release factors, which detach the completed protein from the ribosome.

DNA sequences involved in translation: Due to the complexity of the process the determinants of protein synthesis initiation have been difficult to decipher. It became clear that the wide range of efficiencies in translation of different mRNAs is predominantly due to the structure at the 5′ end of each mRNA species. Therefore, no universal sequence for the efficient initiation of translation has been devised. The translation initiation region comprises four different sequences: ① the Shine-Dalgarno sequence, ② the start codon, ③ the spacer region between the Shine-Dalgarno sequence and the start codon, ④ sometimes translational enhancers.

4.1.2 Regulation Systems in *E. coli*

Tight expression of transcription of recombinant genes is often desirable or necessary since leaky expression can be detrimental or even lethal to cell growth. Regulated gene expression requires an inducible or repressible system, and therefore, all expression systems are based on controllable promoters. Promoters allowing constitutive expression turned out not to be adequate for the production of recombinant proteins due to two main reasons: First, they do not allow the production of toxic proteins and second, even non-toxic proteins produced at physiological concentrations can be deleterious to the cells when produced at higher levels. One prominent example are integral membrane proteins which, when overproduced, cause jamming of the inner membrane leading to cell death. Four regulatable promoter systems are widely used, where three are based on the repressors (such as LacI, TrpR and phage λ cI) and the fourth on a phage RNA polymerase.

The *lac* system consists of the promoter/operator region preceding the *lac* operon and the LacI repressor encoded by the *lacI* gene. In the absence of an inducer, the Lac repressor binds to its operator situated immediately downstream from the promoter as a homotetramer. The wild type *lac* promoter sequence contains one deviation in the 35 and two in the 10 box, and the spacer region encompasses 18 nucleotides if compared to the consensus sequence. One of the many promoter mutations isolated has been termed *lacUV5*. If its DNA sequence is compared to that of the wild type promoter, it becomes apparent that two nucleotides have been exchanged resulting in the consensus 10 box. The promoter strength of *lacUV5* has increased 2.5 fold, and mutations increasing the promoter strength are called promoter up mutations in general. The promoter of the *trp* operon exhibits the consensus 35 box and the optimal spacer length, but three deviations within the 10 box. Based on the *lacUV5* and the *trp* promoters, an artificial promoter was constructed exhibiting the consensus sequence of σ70 dependent promoters and termed Ptac (from *trp* and *lac*). How are the LacI and TrpR repressors inactivated to initiate expression of the recombinant genes? In the case of the Plac, the PlacUV5 and the Ptac promoters, the repressor is inactivated by addition of isopropyl-β-D-thiogalactopyranoside (IPTG). This compound binds to the active LacI repressor and causes dissociation from its operator. IPTG has two advantages over lactose: First, its uptake is not dependent on the Lac permease (it diffuses through the inner membrane) and second, it cannot be cleaved by β-galactosidase preventing turn-off of transcription. The *lacI* gene is either part of the expression plasmid or it is present within the chromosome. Since the wild-type level of the LacI repressor is not sufficient to repress expression of the recombinant gene in the absence of IPTG, two derivates have been isolated resulting in an increase in the amount of repressor based on promoter-up mutations called *lacIq* and *lacIq1*. Expression systems based on the *trp* system make use of synthetic media with a defined tryptophan concentration. The concentration is chosen in such a way that the system becomes self-inducible when the tryptophan concentration within the cells falls below a threshold level. Additionally, 3-β-indole-acrylic acid can be added which inactivates the TrpR repressor and inhibits charging of tRNA^{trp} by tryptophanyl-tRNA synthetase. The third system makes use of the bacteriophage λ repressor cI. This repressor is synthesized from the λ prophage and prevents expression of all the lytic genes by interacting with two operators termed OL and OR. These two operators overlap with two strong promoters, PL and PR, respectively, and as long as the cI repressor is bound to its two operators, binding of RNA polymerase is prevented. Expression vectors carry the cI repressor gene and either PLOL or PROR. How can the λ expression system be induced? The wild-type cI repressor protein can be inactivated by UV-irradiation or treatment of the cells by mitomycin C. A more convenient way is the application of a temperature-sensitive version of the cI repressor called *cI857*. Therefore, *E. coli* cells carrying a λ-based expression system are grown to mid-exponential phase at low temperature and then transferred to high temperature to induce expression of the recombinant gene. The most widely applied expression system makes use of the phage T7 RNA polymerase which recognizes only promoters found on the T7 DNA, and not promoters present on the *E. coli* chromosome. Therefore, the expression vectors contain one of the T7 promoters (normally the promoter present in front of gene 10) to which the recombinant gene will be fused. The gene coding for the T7 RNA polymerase is either present on the expression vector itself or on a second compatible plasmid or integrated into the *E. coli* chromosome. In all three cases, the gene is fused to an inducible promoter allowing its transcription and translation during the expression phase. The T7 RNA polymerase offers three advantages over the *E. coli* enzyme: First, it consists of only one subunit, second it exerts a higher processivity, and third it is insensitive towards rifampicin. The latter characteristic can be used especially to enhance the amount of recombinant protein by adding this antibiotic about 10 min after induction of the gene coding for the T7 RNA polymerase. During that time, enough polymerase has been synthesized to allow high-level expression of the recombinant gene, and inhibition of the *E. coli* enzyme prevents further expression of all the other genes present on both the plasmid and the chromosome. Since all promoter systems are leaky, low-level expression of the gene coding for T7 RNA polymerase may be deleterious to the cell in those cases where the recombinant gene codes for a toxic protein. These polymerase molecules present during the growth phase can be inhibited by expressing the T7 encoded gene for lysozyme. This enzyme is a bifunctional protein that cuts a bond in the cell wall of *E. coli* and selectively inhibits the T7 RNA polymerase by binding to it, a feedback mechanism that ensures a controlled burst of transcription during T7 infection.

4.1.3 General Strategy for Gene Expression in *E. coli*

The basic approach used to express all foreign genes in *E. coli* begins with insertion of the gene into an expression vector, usually a plasmid. This vector generally contains several elements: ① Sequences encoding a selectable marker that assure maintenance of the vector in the cell. ② A controllable transcriptional promoter (e. g. , *lac*, *trp* or tac) which, upon induction, can produce large amounts of mRNA from the cloned gene. ③ Translational control sequences, such as an appropriately positioned ribosome-binding site and initiator ATG. ④ A polylinker to simplify the insertion of the gene in the correct orientation within the vector. Once constructed, the expression vector containing the gene to be expressed is introduced into an appropriate *E. coli* strain by transformation. Following are common used systems in *E. coli* expression.

Vectors for non-fusion protein expression system: plasmid pKK223-3 is for over-expression of proteins under the control of the strong tac promotor in prokaryotes. The Expression vector pKK233-2 is designed to provide transcriptional and translational signals to cloned genes for high-level expression in

E. coli. It contain following 4 characteristics: ① Tac promoter is Strong, and is a regulatable promoter that is IPTG inducible . ② Ribosome-binding site and ATG start codon optimally spaced (8 nucleotides apart). ③ Transcription terminators. ④ Also has restriction enzyme cloning site and a selectable marker.

Vectors for expression fusion proteins: For fusion expression, the two most commonly used tags are glutathione S-transferase (GST tag) and 6×histidine residues (His× 6 tag). As for the selection of host and vectors, the decision to use either a GST or a His tag must be made according to the needs of the specific application.

The pET System is the most powerful system yet developed for the cloning and expression of recombinant proteins in *E. coli*. Based on the T7 promoter-driven system, the pET System has been greatly expanded and now includes over 35 vector types, 11 different *E. coli* host strains and many other companion products designed for efficient detection and purification of target proteins.

Target genes are cloned in pET plasmids under control of strong bacteriophage T7 transcription and translation signals; expression is induced by providing a source of T7 RNA polymerase in the host cell. T7 RNA polymerase is so selective and active that almost all of the cell's resources are converted to target gene expression.

Choosing a pET vector for expression usually involves a combination of factors. Consider the following three primary factors:

- The application intended for the expressed protein.
- Specific information known about the expressed protein.
- Cloning strategy.

For fusion expression, various fusion tags available. A number of pET vectors carry several of the fusion tags in tandem as 5′ fusion partners. In addition, many vectors enable expression of fusion proteins carrying a different peptide tag on each end by allowing in-frame read-through of the target gene sequence. Using vectors with protease cleavage sites (thrombin, Factor Xa, enterokinase) between the 5′ tag and the target sequence enables optional removal of one or more tags following purification.

4. 2　Mammalian Expression System

The production of proteins in mammalian cells is an important tool in numerous scientific and commercial areas. For example, the proteins expressed in and purified from mammalian cell system are routinely needed for life science research and development. In the field of biomedicine, proteins for human therapy, vaccination or diagnostic applications are typically produced in mammalian cells. Gene cloning, protein engineering, biochemical and biophysical characterization of proteins also require the use of gene expression in mammalian cells. Other applications in widespread use involve screening of libraries of chemical compounds in drug discovery and the development of cell-based biosensors.

4. 2. 1　Regulatory Elements in Mammalian Expression Vectors

Expression vectors require not only transcription but translation of the vector's insert, thus requiring more components than simpler transcription-only vectors. The basic regulatory elements used in mammalian expression vectors are prokaryote DNA sequence, promoter and enhancer; untranslated regions (UTRs), and polyadenylation and transcription termination.

4. 2. 1. 1　Prokaryote DNA sequence

Mammalian expression vectors contain either the pUC ori (high copy number) or pBR322 ori (low copy number). However, the vectors with pBR322 ori have been discontinued and all of currently available mammalian expression vectors contain the pUC ori.

4. 2. 1. 2　Promoter and enhancer

Promoter is the necessary component for all vectors used to drive transcription of the vector's transgene. CMV and SV40 promoters are commonly used for high-level expression promoters in mammalian expression. Enhancers are genetic elements that, although not absolutely necessary for expression, act to increase the expression of a gene from a promoter through interactions with different transcription factors. An enhancer can be located upstream, downstream, or even overlapping the promoter, as long as it is in *cis* with the promoter. The effect of a promoter/enhancer combination on expression of a given gene depends upon the cell line.

4.2.1.3　Untranslated regions (UTRs)

UTRs contain specific characteristics that may impede transcription or translation, and thus the shortest UTRs or none at all are encoded for in optimal expression vectors. Kozak surveyed 699 Eukaryotic 5′ untranslated regions and revealed that most are 50~150 bp in length. However, in some cases, these 5′ untranslated regions are longer due to the presence of an intron which is subsequently removed before translation initiation. The fact that only 4 of these 699 Eukaryotic transcripts do not have 5′ untranslated regions indicates that these 5′ untranslated regions may be needed for proper translation initiation. Since translation initiation in eukaryotes is not well characterized, it is recommended that the 5′ untranslated region be kept to 50~100 bp in length and that mRNA secondary structure formation be minimized in this region.

Untranslated sequences (including introns, 3′ and 5′ untranslated regions) do not, as a general rule, improve expression. However, if there is evidence to suggest that certain gene-specific sequences improve the efficiency of ectopic expression of specific genes, then it may be worthwhile to include those sequences during cloning. However, in nearly all cases, no untranslated sequences are necessary for high level expression from current vectors.

4.2.1.4　Polyadenylation and transcription termination

Polyadenylation tail creates a polyadenylation tail at the end of the transcribed pre-mRNA that protects the mRNA from exonucleases and ensures transcriptional and translational termination (stabilizes mRNA production).

4.2.2　Other Features

Genetic markers: Genetic markers for viral vectors allow for confirmation that the vector has integrated with the host genomic DNA.

Antibiotic resistance: Vectors with antibiotic-resistance open reading frames allow for identification of which cells have uptaken the vector through antibiotic selection.

Epitope: Vectors containing a sequence for a specific epitope that is incorporated into the expressed protein allow for antibody identification of cells expressing the vector.

β-galactosidase: Vector's multiple cloning site contains sequence for β-galactosidase, an enzyme that digests galactose, to either side of the region intended for an insert. If the insert has not successfully ligated into the vector, cells expressing the empty vector will generate β-galactosidase and digest galactose. However, cells that express a vector with a transgene will have the coding sequence for β-galactosidase and be unable to digest galactose, and a subsequent color dye for galactose (X-gal) subsequently identifies cells expressing a vector with an insert, although it is unknown whether the insert is the intended one.

Targeting sequence: Expression vectors may include encoding for a targeting sequence in the finished protein that directs the expressed protein to a specific organelle in the cell.

Selection markers
- **Neomycin:** Reliable and convenient selection using widely used selection agent, Geneticin.
- **Zeocin:** Efficient and high potent selection.
- **Hygromycin:** Different mode of action than Geneticin or Zeocin, hence useful for dual selection.

4.2.3　Plasmid Vectors System

The following pcDNA3.1 vectors are designed for high level expression of recombinant proteins in mammalian cells. These vectors offer the following features: ①High level constitutive transcription from mammalian enhancer promoter sequences plus transcription termination and polyadenlyation signals. ② Prokaryotic sequences to permit growth and maintenance in *E. coli* as well as isolation of single stranded DNA. ③ Versatile multiple cloning sites to permit unidirectional or bidirectional cloning. ④ Neomycin, Blasticidin, Hygromycin or Zeocin resistance genes for selection of stable cell lines.

pcDNA3. 1 (+/−) are 5. 4 kb vectors derived from pcDNA3 and designed for high-level stable and transient expression in mammalian hosts. High-level stable and non-replicative transient expression can be carried out in most mammalian cells. The vectors contain the following elements: ①Human cytomegalovirus immediate-early (CMV) promoter for high-level expression in a wide range of mammalian cells. ② Multiple cloning sites in the forward (+) and reverse (−) orientations to facilitate cloning. ③ Neomycin resistance gene for selection of stable cell lines. ④ Episomal replication in cells lines that are latently infected with SV40 or that express the SV40 large T antigen (e. g. COS-1, COS-7).

4. 2. 4　Viral Vectors

Viral vectors are generally genetically-engineered viruses carrying modified viral DNA or RNA that has been rendered noninfectious, but still contain viral promoters and also the transgene, thus allowing for translation of the transgene through a viral promoter. However, because viral vectors frequently are lacking infectious sequences, they require helper viruses or packaging lines for large-scale transfection. Viral vectors are often designed for permanent incorporation of the insert into the host genome, and thus leave distinct genetic markers in the host genome after incorporating the transgene. For example, retroviruses leave a characteristic retroviral integration pattern after insertion that is detectable and indicates that the viral vector has incorporated into the host genome.

4. 2. 4. 1　Retroviral vectors

Basics of the retrovirus virion and infection: Retrovirus virions contain a protein capsid that is lipid encapsulated. Virions range in diameter from 80~130 nm. The viral genome is encased within the capsid along with the proteins integrase and reverse transcriptase. The genome consists of two identical positive (sense) single-stranded RNA molecules ranging in size from 3. 5~10 kb. Following cellular entry, the reverse transcriptase synthesizes viral DNA using the viral RNA as its template. The cellular machinery then synthesizes the complementary DNA which is then circularized and inserted into the host genome. Following insertion, the viral genome is transcribed and viral replication is completed. The majority of retroviruses are oncogenic although the degree to which they cause tumors varies from class to class.

The retroviral genome: The retroviral genome consists of little more than the genes essential for viral replication. The prototype and simplest genome to describe is that of the **moloney murine leukemia virus (MMLV)** in contrast to the highly complex genomes of the HTLV and HIV retroviruses. The genome can be divided into three transcriptional units: **gag, pol** and **env**. The **gag** region encodes genes which comprise the capsid proteins; the **pol** region encodes the reverse transcriptase and integrase proteins; and the **env** region encodes the proteins needed for receptor recognition and envelope anchoring. An important feature of the retroviral genome is the **long terminal repeat (LTR)** regions found at each of the gene. The LTR plays an important role in initiating viral DNA synthesis and its integration as well as regulating transcription of the viral genes.

MMLV vectors To date, these vectors have been used more than any other gene transfer vehicle. They are produced simply by replacing the viral genes required for replication with the desired genes to be transferred. Thus, the genome in retroviral vectors will contain an LTR at each end with the desired gene or genes in between. The most commonly used system for generating retroviral vectors consists of two parts, the retroviral DNA vector and the packaging cell line.

Retroviral DNA vectors are plasmid DNAs which contain two retroviral LTRs in the region internal to these LTRs for insertion of the desired gene. The gene of interest is cloned into the multicloning site following the simian virus SV40 promoter (SV). As one can imagine, these plasmid DNAs can be manipulated to meet a variety of needs allowing for multiple applications and the design of very elegant vectors.

Packaging cell lines provide all the viral proteins required for capsid production and the virion maturation of the vector. These packaging cell lines have been made so that they contain the gag, pol and env genes. Early packaging cell lines contained replication competent retroviral genomes and a single recombination event between this genome and the retroviral DNA vector could result in the production of a wild type virus. This led to the term "helper virus contamination" and to this date, with all viral vectors systems, it is important to insure that the vector be free of helper virus. Current packaging systems require that three homologous recombination events occur for any wild type virus production. Since this is an extremely rare event, current cell lines are considered to be helper free, although it is still wise to test for contamination of any wild type virus. There have been a wide variety of packaging cell lines produced as well as retroviral DNA vectors, and most are commercially available. For the MMLV vectors it is the packaging cell line which determines if the vector is ecotropic, xenotropic or amphotropic. Dependent upon the target cells it is important to ensure that the correct packaging cell line be chosen.

HIV-Based Vectors　Human immunodeficiency viruses (HIV) are the most recently discovered members of the retrovirus family and have led to the new classification of lentivirus. Like other members of the retroviral family, the HIV genome contains the *gag*, *pol* and *env* genes. In addition, several other nonstructural proteins which serve regulatory functions are contained within its genome.

HIV-based vectors are recent developments in the field of gene therapy and focus on treatment of AIDS. Vector production is similar to that of MMLV vectors, and HIV vectors possess all the same advantages and disadvantages as their MMLV counterparts. An exciting application of HIV vectors is to use HIV vectors to target genes selectively into HIV-containing cells. The goal is that genes delivered by the HIV vector would allow selective killing of any cell previously or subsequently infected with HIV. An example of how this might be accomplished focuses on the *tat* and *rev* genes found in the HIV genome. These genes are also found only in cells infected with HIV. By using a *tat* and *rev* induced promoter to drive the expression of a toxin gene, any cell infected with HIV should be selectively killed upon expression of the toxin gene. Of course this system does require that a cell become infected with HIV but once infected the cell would terminate as would the HIV virus. A major concern in using HIV vectors is the fact that there is a strong possibility for genetic recombination between infectious HIV and the HIV vector itself. Such an event would result in the HIV vector acting as an infectious HIV particle. One possible method to overcome this risk would be the inclusion of a suicide gene into the HIV vector genome. The suicide gene would make cells infected with the HIV vector genome sensitive to a drug that would poison only those cells containing the HIV vector genome. An example of a suicide gene is the Herpes simplex virus thymidine kinase gene, and ganciclovir is the drug which would yield selective toxicity.

4.2.4.2　Vaccinia Vectors

Vaccinia virus has been utilized as a means for immunization against smallpox and a variety of other infectious agents. Vaccinia virus is a member of the pox virus family. They are large brick shaped virions measuring $300 \sim 450$ by $170 \sim 270$ nm. They are enveloped and contain an extremely large genome of $130 \sim 200$ kb. This large genome enables large genes to be inserted into vaccinia-based vector. Vaccinia vectors can infect a large variety of cell types, but a primary concern is their safety. As with smallpox vaccination, the adverse reaction rate for administration of vaccinia vectors is 1 in 50 000 doses. Obviously, vaccinia vectors for gene therapy applications are limited to individuals not previously vaccinated against smallpox or are immune compromised. Recombinant vector production is similar to the other vector systems that we had discussed requiring a viral vector DNA and a packaging cell line. These vectors offer the potential to develop a large variety of gene therapy based vaccinations. Currently these applications are not as cost-effective as the more traditional vaccination protocols but this system might allow for the vaccination against diseases presently lacking a vaccine. The system is also being examined in the treatment of HIV, although currently little attention is given to vaccinia-based systems in the field of gene therapy.

4.3　Other Expression Systems

4.3.1　Yeast Expression System

Yeast is a eukaryotic organism and has some advantages and disadvantages over *E. coli*. One of the major advantages is that yeast cultures can be grown to very high densities, which makes them especially useful for the production of isotope labeled protein. The two most used yeast strains are *Saccharomyces cerevisiae* and the methylotrophic yeast *Pichia pastoris*.

Various yeast species have proven to be extremely useful for expression and analysis of eukaryotic proteins. These yeast strains have been genetically well characterized and are known to perform many posttranslational modifications. These single-celled eukaryotic organisms grow quickly in defined medium, are easier and less expensive to work with than insect or mammalian cells, and are easily adapted to fermentation. Yeast expression systems are ideally suited for large-scale production of recombinant eukaryotic proteins.

4.3.2　Insect Cells Expression System

Insect cells are a higher eukaryotic system than yeast and are able to carry out more complex post-translational modifications than the other two systems. They also have the best machinery for the folding of mammalian proteins and, therefore, give you the best chance of obtaining soluble protein when you want to express a protein of mammalian origin. The most commonly used vector system for recombinant protein expression in insect is baculovirus, although baculoviral also can be used for gene transfer and expression in mammalian cells.

Baculovirus-assisted insect cell expression is optimal for glycosylated protein expression in a cost-effective manner. There are many advantages to using baculovirus for heterologous gene expression. Heterologous cDNA is expressed well. Proper transcriptional processing of genes with introns occurs but is expressed less efficiently. As with other eukaryotic expression systems, baculovirus expression of heterologous genes permits folding, post-translational modification and oligomerization in manners that are often identical to those that occur in mammalian cells. The insect cytoplasmic environment allows proper folding and S-S bond formation, unlike the reducing environment of the $E.$ $coli$ cytoplasm. Post-translational processing identical to that of mammalian cells has been reported for many proteins. These include proper proteolysis, N-glycosylation and O-glycosylation, acylation, amidation, carboxymethylation, phosphorylation, and prenylation. Proteins may be secreted from cells or targeted to different subcellular locations. Single polypeptide, dimeric and trimeric proteins have been expressed in baculoviruses. Finally, expression of heterologous proteins is under the control of the strong polyhedrin promoter, allowing levels of expression of up to 30% of the total cell protein. The benefits of protein expression with baculovirus can be summarized as: ① Eukaryotic post-translational modification. ② Proper protein folding and function. ③ High expression levels. ④ Easy scale up with high-density suspension culture. ⑤ Safety.

Baculoviruses infect primarily insects with a narrow host range. $Autographa$ $californica$, the most commonly used baculovirus for protein expression, infects only 2 lepidopteran (moth) families in nature. Although these viruses may enter other cells types (perhaps by phagocytosis), they are not infectious in them. For example, nucleocapsid proteins are not removed in most human cells. In human hepatic cell lines that do remove these proteins, the virus fails to replicate and express proteins due to the absence of insect transcription factors. Thus, working with baculoviruses is considered safe for humans and contamination of mammalian cell lines in shared biosafety hoods is not a problem.

Despite these potential advantages, particular patterns of post-translational processing and expression must be empirically determined for each construct. Differences in proteins expressed by mammalian and baculovirus infected insect cells have been described and overcome in some cases. For example, inefficient secretion from insect cells may be circumvented by the addition of insect secretion signals (e. g. honeybee melittin sequence). Improperly folded proteins and proteins that occur as intracellular aggregates may be due to expression late in the infection cycle. In such cases, harvesting cells at earlier times after infection may help. Low levels of expression can often be increased with optimization of time of expression and multiplicity of infection. The complete analysis of carbohydrate structures has been reported for a limited number of glycoproteins. Potential N-linked glycosylation sites are often either fully glycosylated or not glycosylated at all, as opposed to expression of various glycoforms that may occur in mammalian cells. Species-specific or tissue-specific modifications are unlikely to occur.

Summary

The term genetic engineering is also name as recombinant DNA technology or gene manipulation, which is consisted of restriction enzymes digestion, ligation, transformation and screening a recombinant gene. The essence of the technology is to manipulate DNA in molecular level and express protein in $vitro$

at host cell level. There are many enzymes used in making recombinant DNA, which are restriction enzymes, DNA ligase and some of other related DNA enzymes involved in DNA synthesis and modification. When a single recombinant DNA molecule, composed of a vector plus an inserted DNA fragment, is introduced into a host cell, the inserted DNA is reproduced along with the vector, producing large numbers of recombinant DNA molecules that include the fragment of DNA originally linked to the vector. Two types of vectors are most commonly used in molecular cloning: *E. coli* plasmid vectors and bacteriophage λ vectors. Transformation is defined as recombinant DNA molecules introduced into *E. coli* host cells, while transfection is to mammalian host cells. Transient and stable transfection are commonly used to introduce recombinant DNA into mammalian cells for different research purposes. In addition, to choose a proper system for recombinant DNA expression is also very important in order to achieve a right over-expressed interest protein *in vitro*. Comparing prokaryotes expression, the advantages of mammalian or other eukaryotic expression system are more excellent for protein folding and post-modification. Production of proteins expressed in mammalian cells or other cells is an important tool in numerous scientific and commercial areas. Great perspectives have been taken in the field of biomedicine, industry/manufacturing, and agriculture through recombinant DNA techniques.

(Qin Fang)

第七章 细胞信号转导

生物体中，单细胞生物可直接对外界环境变化做出反应，而多细胞组成的有机体，细胞已经分化并具有特殊的结构与功能，大多数细胞不与外界直接接触，因此多细胞生物对外界的刺激（包括物理、化学因素），需要细胞间复杂的信号传递系统来传递，从而协调、控制体内每个细胞的新陈代谢和行为，以保证整体生命活动的正常进行。细胞间的相互识别、联络和相互作用称细胞通讯（cell communication）。信号转导是指（signal transduction；也译作信号传导等）是指细胞外的信息，经过一系列的生化反应之后，活化了细胞内部的信息，进而使细胞产生一系列反应的过程。如果机体细胞间不能准确有效地进行通讯和信息转导，机体就可能出现代谢紊乱、细胞癌变甚至死亡。细胞信号转导的异常与人类许多常见疾病相关，其中包括肿瘤、内分泌代谢性疾病、感染性疾病、心血管疾病以及某些精神疾病等。因此，掌握细胞信号通讯与转导机制，将有助于深入探讨疾病的发生、发展以及防治机制。

第一节 信 息 分 子

人体细胞之间的信号转导可通过相邻细胞的直接接触来实现，但更重要的则是通过细胞分泌各种化学物质来调节自身和其他细胞的代谢和功能。这些具有调节细胞生命活动的化学物质称为信息分子（signal molecule）。细胞间的信息传递是跨膜的信号转导。

一、细胞间信息分子

细胞间信息分子是由细胞分泌的调节靶细胞生命活动的化学物质。目前已知的细胞间信息物质的化学本质为蛋白质和肽类（如生长因子、细胞因子、胰岛素等）、氨基酸及其衍生物（如甘氨酸、甲状腺素、肾上腺素等）、类固醇激素（如糖皮质激素、性激素等）、脂酸衍生物（前列腺素）和气体（如一氧化氮、一氧化碳）等。根据细胞分泌信息分子的方式又可将细胞间信息分子分为神经递质、内分泌激素、局部化学介质和气体信号分子四大类。

（一）神经递质

神经递质又称为突触分泌信号（synaptic signal）。神经递质是神经系统细胞间通讯的化学信号分子，由神经元突触前膜释放，如乙酰胆碱和去甲肾上腺素等，其作用时间较短。

（二）内分泌激素

内分泌激素又称为内分泌信号（endocrine signal），是由特殊分化的内分泌细胞释放的化学信号分子，它们通过血液循环到达靶细胞，经过受体介导而对靶细胞产生作用，大多数对靶细胞的作用时间较长。

按照内分泌激素的化学组成可分为含氮类激素和类固醇激素。含氮类激素包括氨基酸衍生物（如肾上腺素、甲状腺素）、小肽类（促甲状腺素、促甲状腺素释放因子、胰高血糖素、促乳素）、蛋白质类（胰岛素、生长激素）和糖蛋白类（促黄体激素、促甲状腺激素）。类固醇激素包括各种性激素、皮质醇、醛固酮等。

按激素受体的性质，可将激素分为胞内受体激素和胞膜受体激素。胞内受体激素包括甲状腺素、类固醇激素，它们的受体在胞质或胞核中。胞内受体激素很容易通过细胞膜进入靶细胞内与相应的受体结合。除甲状腺素外，其他的含氮激素均为胞膜受体激素，它们均为水溶性的，很难直接通过细胞膜的脂质双分子层进入细胞，必须与靶细胞表面的受体结合才能引发细胞的应答反应。

（三）局部化学介质

局部化学介质又称为旁分泌信号（paracrine signal）。此类信息物质分泌到细胞外液后，进入血液的数量极少，绝大多数通过扩散作用作用于附近的靶细胞，但又不像神经递质那样由专一的突触结构释放。由此可见，局部化学介质既不同于激素，又不同于神经递质。体内的局部化学介质包括组胺、花生四烯酸（AA）及其代谢产物〔前列腺素（PG）、血栓素（TX）和白三烯〕、生长因子、细胞生长抑素等。局部化学介质也必须与细胞膜受体结合才能引发细胞的应答反应，但花生四烯酸的某些代谢产物，如LTC$_4$能以受体非依赖方式，在生理和病理情况下调节心脏兴奋性。除生长因子外，局部化学介质的作用时间均较短。

（四）气体信号分子

一氧化氮（NO）是一种结构简单、半衰期短、化学性质活泼的气体信号分子。NO合酶（NO synthase，NOS）通过氧化L-精氨酸的胍基来产生NO。NOS可分为组成型（constitutive NO synthase，cNOS）和诱生型（inducible NO synthase，iNOS）两类，前者又分为内皮型（endothelial NO synthase，eNOS）和神经型（neuronal NO synthase，nNOS）。cNOS主要存在于上皮、脑和血小板，在生理状态下，只生成少量NO。乙酰胆碱、缓激肽、组胺、内皮素等均能刺激血管内皮细胞，导致其质膜上Ca^{2+}通道开放，细胞内Ca^{2+}浓度升高，Ca^{2+}与钙调蛋白（calmodulin，CaM）结合而激活cNOS，增加NO的合成。α-干扰素（α-interferon）、肿瘤坏死因子-β（tumour necrosis factor-β）、脂多糖（lipopolysaccharide，LPS）、白细胞介素-1（interferon-1，IL-1）等细胞因子可以激活核因子-κB（nuclear factor-κB，NF-κB）和干扰素调节因子-1（IFN regulatory factor-1，IRF-1）及丝裂原激活蛋白激酶（mitogen-activated protein kinase，MAPK）等，从而上调iNOS活性而引起NO合成增加。

除NO外，最近又发现一种与NO相似的气体信息分子——一氧化碳（CO）。CO可在血红素单加氧酶氧化血红素（heme）的过程中产生。CO具有与NO类似的功能。

除了上述4种主要的细胞间信号分子外，还有一些信号分子能与分泌细胞自身的受体结合而起调节作用，称为自分泌信号（autocrine signal），如一些癌蛋白。而有些细胞间信息物质可在不同的个体间传递信息，如昆虫的性激素。

二、细胞内信息分子

在细胞内传递特定调控信号的化学物质称为细胞内信息分子（intracellular signal molecule）。细胞内信息物质主要包括：第二信使、第三信使、信号转导蛋白或酶等。

通常将在细胞内传递信息的小分子化学物质称为第二信使。第二信使主要包括：环核苷酸类，如 cAMP 和 cGMP；脂类衍生物，如二酰甘油（DAG）、神经酰胺（ceramide，Cer）、花生四烯酸；无机物，如 Ca^{2+}、NO；糖类衍生物，如 1，4，5-三磷酸肌醇（inositol 1,4,5-triphosphate，IP_3）等。需要指出的是，一方面，花生四烯酸及其代谢产物可作为一些激素、递质和生物活性因子的第二信使调节细胞的功能；另一方面，细胞受刺激后，细胞内产生的花生四烯酸及其代谢产物可释放至细胞外，作为第一信使以自分泌或旁分泌方式发挥作用。细胞膜上或细胞内能够传递特定信号的蛋白质或酶分子，常与其他蛋白质或酶构成复合体以传递信息。主要包括：G 蛋白及其调节蛋白，如 GEF、GAP 等；信号连接蛋白，如 SOS、GRB2 等；具有酪氨酸激酶活性的胰岛素受体底物-1/2（IRS-1/2）；各种蛋白激酶和蛋白磷酸酶等，见图 7-1。

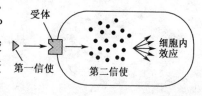

图 7-1　第二信使

负责细胞核内外信息传递的物质称为第三信使，是一类可与靶基因特异序列结合的核蛋白，能调节基因的转录，因此又称为 DNA 结合蛋白，发挥着转录因子或转录调节因子的作用，如立早基因（immediate-early gene）编码的蛋白质常作为第三信使参与基因调控、细胞增殖与分化以及肿瘤的形成等。立早基因多数为细胞原癌基因（如 c-fos、AP_1/c-jun 等）。

细胞内信号分子在传递信号时绝大部分通过酶促级联反应进行。它们最终通过改变细胞内有关酶的活性、开启或关闭细胞膜离子通道及细胞核内基因的转录等，达到调节细胞代谢和控制细胞生长、繁殖和分化的功能。所有信息物质在完成信息传递后，必须立即灭活。通常细胞通过酶促降解、代谢转化或细胞摄取等方式灭活信息物质。一些细胞信号分子影响细胞内代谢的途径见表 7-1。

表 7-1　信号分子影响细胞功能的途径

	信息物质	受体	引起细胞内的变化
神经递质	乙酰胆碱、谷氨酸、γ-氨基丁酸	质膜受体	影响离子通道开闭
生长因子	类胰岛素生长因子-1、表皮生长因子、血小板衍生生长因子	质膜受体	引起酶蛋白和功能蛋白的磷酸化和脱磷酸化,改变细胞的代谢和基因表达
激素	蛋白质、多肽及氨基酸衍生物类激素	质膜受体	同上
	类固醇类激素、甲状腺素	胞内受体	调节转录
维生素	维生素 A、维生素 D	胞内受体	同上

三、细胞间信号转导的作用方式

1）内分泌信号途径（endocrine signaling pathway）是内分泌细胞产生的激素通过血液循环作用于生物体其他远端部位的靶细胞，见图 7-2。

2）旁分泌信号途径（paracrine signaling pathway）是由细胞分泌的信号分子只是作为局部的介导物，作用于邻近的靶细胞，见图 7-3。

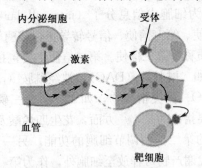

图 7-2　内分泌信号途径

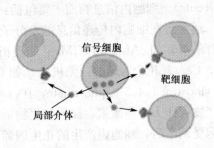

图 7-3　旁分泌信号途径

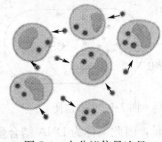

图 7-4　自分泌信号途径

3）自分泌信号途径（autocrine signaling pathway）是联系同一种细胞，即信号的靶细胞就是产生信号的细胞本身，见图 7-4。

信号转导的基本原理在于受体的活性形式能够激发细胞内酶的活性。这种内移胞质信号比胞外的原始信号（如配体）要强很多，能直接活化一系列蛋白质或者伴随着细胞内小分子数量的增加。

第二节　受　体

受体（receptor）是指存在于靶细胞膜上或细胞内能特异识别与结合生物活性分子（配体），进而引起靶细胞生物学效应的分子。绝大部分受体为蛋白质，少数为糖脂。能与受体呈特异性结合的生物活性分子称配体（ligand）。细胞间信号分子就是一类最常见的配体。除此之外，某些药物、维生素和毒物也可作为配体发挥生物学作用。

受体在细胞信号转导过程中起着极为重要的作用。其中，位于胞质和细胞核中的受体称为胞内受体，它们大多数为 DNA 结合蛋白；存在于细胞质膜上的受体则称为膜受体，它们绝大部分是镶嵌蛋白。

一、受体的分类、结构及功能

（一）膜受体

这类受体是细胞膜上的结构成分，一般是糖蛋白、脂蛋白或糖脂蛋白。当配体与受体结合后，往往引起细胞膜结构和功能的改变，导致细胞内某种化学物质的浓度改变，由此触发一系列的化学和生理变化。图 7-5 所示为多肽及蛋白质类激素、儿茶酚胺类激素、前列腺素以及细胞因子通过这类受体进行的跨膜信号传递。

膜受体主要分如下几类：

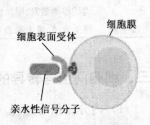

图 7-5　细胞表面膜受体

1) 环状受体 即配体依赖型离子通道（ion-channel-linked receptor）。受体本身是离子通道的组成部分，当神经递质与这类受体结合后，离子通道打开或关闭，从而改变膜对某种离子的通透性。这类受体主要在神经冲动的快速传递中起作用。此类受体包括烟碱样乙酰胆碱受体（N-AchR）(图 7-6)、A 型-氨基丁酸受体（GABAAR）、谷氨酸受体、甘氨酸受体及 5-羟色胺受体（5-HTR）等。

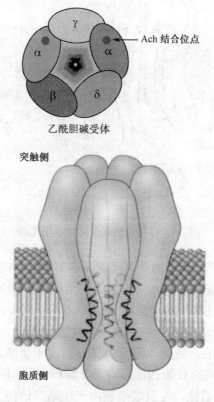

图 7-6 烟碱样乙酰胆碱受体的分子结构

2) G 蛋白偶联受体（G-protein coupled receptor，GPCR） 它又称 7 个跨膜螺旋受体或蛇型受体。此类受体通常由单一的多肽链或均一的亚基组成，其肽链可分为细胞外区、跨膜区、细胞内区三个区。受体的 N 端位于细胞外区，而 C 端位于细胞内区；中段形成 7 个跨膜的 α 螺旋结构和 3 个细胞外环与 3 个细胞内环。每个 α 螺旋结构分别由 20～25 个疏水氨基酸组成。受体的疏水螺旋区的一级结构是高度同源的，亲水环的一级结构有较大的变异。在第五及第六跨膜 α 螺旋结构之间的细胞内环部分（第三内环区），是与鸟苷酸结合蛋白（guanylate binding protein，G 蛋白）偶联的区域（图 7-7）。大多数常见的激素受体和慢反应神经递质受体是属于 G 蛋白偶联型受体。该类受体对多种激素和神经递质作出应答。配体包括生物胺、感觉刺激（如光和气体等）、脂类衍生物、肽类等。GPCR 是研究得最为广泛和透彻的一类受体。它们组成不同功能的超大家族。目前已知的 GPCR 已达 1000 多种，而且数量还在增加。

GPCR 是糖蛋白，不同的受体有不同的糖基化模式，它们经常发生在受体的 N 端。GPCR 有一些保守的半胱氨酸残基，其中一些半胱氨酸残基对维持受体的结构起到关键

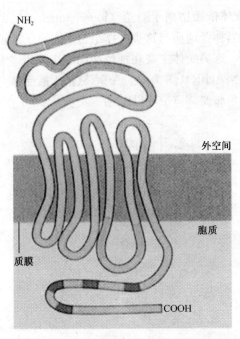

图 7-7 G 蛋白偶联受体分子模式图

作用。在胞外的第二环和第三环有两个高度保守的半胱氨酸残基，参与形成连接第二环和第三环的二硫键，维持蛋白质胞外结构域的正常构象。许多 GPCR 的 C 端也存在一个高度保守的 Cys 残基。在肾上腺素能 α 受体（adrenergic α receptor，α-AR）、肾上腺素能 β 受体（adrenergic β receptor，β-AR）和视紫质受体（rhodopsin receptor）中，此 Cys 残基是被棕榈酰化的，使受体的胞内部分锚定于质膜，从而稳定受体胞内部分的三级结构。受体通过不同的 G 蛋白而影响腺苷酸环化酶（adenylate cyclase，AC）或磷脂酶 C（lipase C）等的活性，再引起细胞内产生第二信使。这类受体的信号转导可总结为激素→受体→G 蛋白→酶→第二信使→蛋白激酶→酶或功能蛋白→生物学效应。这类受体分布极广，主要参与细胞物质代谢的调节和基因转录的调控。

G 蛋白是一类和 GTP 或 GDP 结合的、位于细胞膜胞液面的外周蛋白，由三个亚基组成。它们是 α 亚基（45kDa）、β 亚基（35kDa）和 γ 亚基（7kDa）（图 7-8）。其中，α 亚基可与 GTP 或 GDP 结合，并具有 GTPase 活性。G 蛋白通过 β、γ 亚基的异戊二烯化的基团或 α 亚基的豆蔻酰化的基团锚定于细胞膜。G 蛋白有两种构象，一种以 αβγ 三聚体存在并与 GDP 结合（图 7-9），为非活化型；另一种是 α 亚基与 GTP 结合并导致 βγ 二聚体的脱落，此型为活化型。

G 蛋白有许多种，常见的有激动型 G 蛋白（stimulatory G protein，Gs）、抑制型 G 蛋白（inhibitory G protein，Gi）和磷脂酶 C 型 G 蛋白（PI-PLC G protein，Gp）。不同的 G 蛋白能特异地将受体和与之相适应的效应酶偶联起来。各种 G 蛋白的 α 亚基均有一个可被霍乱毒素或百日咳毒素进行 ATP-核糖基化修饰部位，并能改变 G 蛋白的功能。霍乱毒素能引起 α_s 的 ADP 核糖基化，使 α_s 丧失 GTPase 活性，因而不能把 GTP 水解为 GDP，所以 G 蛋白处于持续激活状态。百日咳毒素通过促使 α_i 的 ADP 核糖基

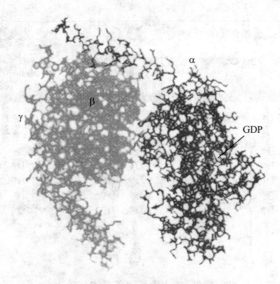

图 7-8　G 蛋白异三聚体的分子结构

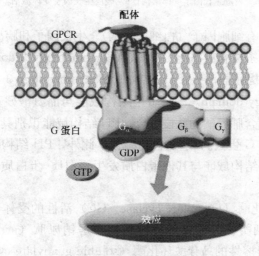

图 7-9　G 蛋白的活化

化并阻止了 α_i 被激活而使腺苷酸环化酶不可逆激活。

3）酶联受体（enzyme-linked receptor）　它又称单个跨膜 α 螺旋受体。这类受体主要有酪氨酸蛋白激酶受体型和非酪氨酸蛋白激酶受体型。前者为催化型受体（catalytic receptor）（如胰岛素受体和表皮生长因子受体等），它们与配体结合后即有酪氨酸蛋白激酶活性，即可导致受体自身磷酸化，又可催化底物蛋白的特定酪氨酸残基磷酸化（图 7-10）；后者（如生长激素受体、干扰素受体）与配体结合后，可与酪氨酸蛋白激酶偶联而表现出酶活性。这类受体全部为糖蛋白且只有一个跨膜螺旋结构。催化型受体跨膜区 22～26 个氨基酸残基构成一个 α 螺旋，高度疏水，胞外区一般有 500～850 个氨基酸残基，有的含与免疫球蛋白（Ig）同源的结构，有的富含半胱氨酸区段，该区为配体结合部位。细胞内为近膜区和功能区。酪氨酸蛋白激酶功能区位于 C 端，包括 ATP结合和底物结合两个功能区。该型受体与细胞的增殖、分化、分裂及癌变有关。能与这

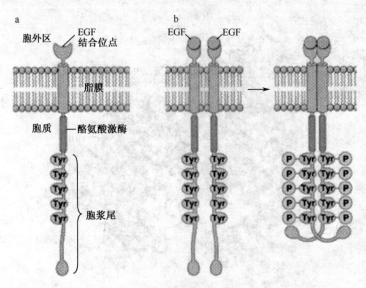

图 7-10　表皮生长因子受体及激活

a. 表皮生长因子受体结构；b. 表皮生长因子受体激活

类受体结合的配体主要有细胞因子（白细胞介素）、生长因子和胰岛素等。

该类受体的下游分子常含有 SH_2 结构域（Scr homology 2 domain，该结构域与原癌基因 *scr* 编码的 2 结构域同源）、SH_3 结构域（Scr homology 3 domain）和 PH 结构域（pleckstrin homology domain）等。SH_2 结构域能与酪氨酸残基磷酸化的多肽链结合；SH_3 结构域能与富含脯氨酸的肽段结合；PH 结构域能识别具有磷酸化的丝氨酸和苏氨酸的短肽，并能与 G 蛋白的 βγ 复合物结合，此外，PH 结构域还能与带电的磷脂结合。由此可见，这些结构域能与其他蛋白质发生蛋白质-蛋白质相互作用，参与细胞间的信号转导。

4）具有鸟苷酸环化酶（guanylate cyclase，GC）活性的受体　该类受体分为膜受体和胞液可溶性受体两类。膜受体的配体包括心房钠尿肽（arrionatriuretic peptide，ANP）和鸟苷蛋白。可溶性的鸟苷酸环化酶（soluble guanylate cyclase，SGC）的配体为 NO 和 CO。

GC 膜受体由同源的三聚体或四聚体组成。每一条亚基包括 N 端的胞外受体结构域、跨膜区域、膜内的蛋白激酶样结构域和 C 端的鸟苷酸环化酶催化结构域（图7-11）。单个跨膜结构域和胞内近膜区为一长度为 37 个氨基酸残基的片段，有蛋白激酶样结构域激酶活性，目前尚不知其功能。每条亚基通过胞外受体结构域间的氢键连接成三聚体或四聚体。GC 是一个高度磷酸化的酶。受体与配体结合后，GC 的活性大为提高。随后迅速去磷酸化使 GC 活性复原。

胞液可溶性受体是由 α、β 两个亚基组成的杂二聚体，分子质量分别为 76 kDa 和 80 kDa。每个亚基具有一个鸟苷酸环化酶催化结构域和血红素结合结构域。当杂二聚体解聚后，酶活性丧失。酶活性依赖于 Mn^{2+}。在脑、肺、肝及肾等组织中大部分具鸟苷酸环化酶活性的受体是胞液可溶性受体，而在心血管组织细胞、小肠、精子及视网膜杆状细胞则大多数为膜受体。

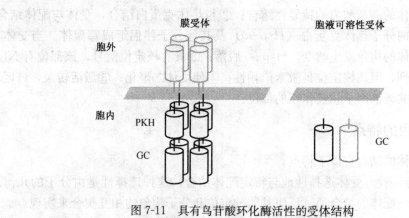

图 7-11　具有鸟苷酸环化酶活性的受体结构

PKH. 激酶样结构域；GC. 鸟苷酸环化酶结构域

（二）胞内受体

胞内受体（intracellular receptor）位于细胞质或细胞核内，通常为单体蛋白。胞内受体多为反式作用因子，某些激素进入细胞后，能与特异性的胞内受体结合形成活性复合物，作用于染色体 DNA，调节基因表达，从而影响细胞的物质代谢和生理活动。能与该型受体结合的信息物质有类固醇、甲状腺激素和维 A 酸等。胞内受体通常为400～1000 个氨基酸残基组成的单体蛋白，包括 4 个区域（图 7-12）。

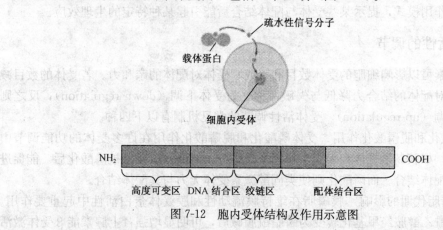

图 7-12　胞内受体结构及作用示意图

1）高度可变区　位于 N 端，长度不一，氨基酸残基有 20～600 个。具有一个非激素依赖性的组成型转录激活功能（activation function）区。该区还是多数核受体抗体的结合部位。

2）DNA 结合区（DNA-binding domain）　位于受体分子的中部，主要包含 66～68 个氨基酸残基组成的核心结构和后续的 C 端延伸组成。核心结构含两个锌指模型（模体），它能顺 DNA 螺旋旋转并与之结合。

3）铰链区　除部分甾体激素受体外，多数胞内受体主要定位于核内。核受体中有与 SV40 大 T 抗原核定位信号（nuclear localization signal，NLS）相似的氨基酸序列。核受体在胞质中合成后，NLS 相似序列能引导核受体进入细胞核。

4）配体结合区　位于 C 端，其作用包括①与配体结合，该区域的某些氨基酸残基参与受体和配体的高亲和力的特异性结合；②与热休克蛋白结合，受体与配体结合前，一分子受体、两分子热休克蛋白（Hsp 90）及其他分子伴侣组成寡聚体。当受体与配体结合后，受体的构象发生改变，Hsp 90 脱落；③具有核定位信号，该部位有 NLS 相似的氨基酸序列，但该核定位具激素依赖性；④使受体二聚化；⑤激活转录，该区域还为与其他转录共激活因子相互作用的部位。

二、受体作用的特点

受体与配体的结合有以下特点。

1）高度专一性　受体选择性地与特定配体结合，这种选择性是由分子的几何形状决定的。受体与配体的结合通过反应基团的定位和分子构象的相互契合来实现。

2）高度亲和力　无论是膜受体还是胞内受体，他们与配体间的亲和力都极强。体内信息物质的浓度非常低，通常 $\leqslant 10^{-8}\,mol \cdot L^{-1}$，但却具有显著的生物学效应，足见两者间的亲和力之高。

3）可饱和性　受体-配体结合曲线（Scatchard 曲线）为矩形双曲线。增加配体浓度，可使受体饱和。

4）可逆性　受体-配体以非共价键结合，当生物效应发生后，配体即与受体解离。受体可恢复到原来的状态，并再次被利用，而配体则常被立即灭活。

5）特定的作用模式　受体在细胞内的分布，从数量到种类，均有组织特异性，并出现特定的作用模式，提示某些受体与配体结合后能引起某种特定的生理效应。

三、受体活性的调节

许多因素可以影响细胞的受体数目和（或）受体对配体的亲和力。若受体的数目减少和（或）对配体的结合力降低与失敏，称之为受体下调（down-regulation），反之则称为受体上调（up-regulation）。受体活性调节的常见机制有以下四种。

1）磷酸化和脱磷酸化作用　受体磷酸化和脱磷酸化作用在许多受体的功能调节中起重要作用，如胰岛素受体和表皮生长因子受体分子的酪氨酸残基被磷酸化后，能促进受体与相应配体结合。而磷酸化则使类固醇激素受体无力与其配体结合。

2）膜磷脂代谢的影响　膜磷脂在维持膜流动性和膜受体蛋白活性中起重要作用。质膜的磷脂酰乙醇胺经甲基化转变为磷脂酰胆碱后，可明显增强肾上腺素能 β 受体激活腺苷酸环化酶的能力。

3）酶促水解作用　有些膜受体可通过内化（internalization）方式被溶酶体降解。

4）G 蛋白的调节　G 蛋白可在多种活化受体与腺苷酸环化酶之间起偶联作用，当一受体系统被激活而使 cAMP 水平升高时，就会降低同一细胞受体对配体的亲和力。

第三节　细胞信号转导途径

细胞信号转导（cellular signal transduction）是指信号分子通过与靶细胞受体（膜受体或胞内受体）结合，引起受体的构象和（或）酶活性改变，即可将信号传递给细胞内的各种信号转导分子，通过一系列的级联反应，最后产生特定的细胞生物学效应。由细胞内若干信号转导分子所构成的级联反应系统被称为细胞信号转导途径。同一信号转导途径可由不同的信号分子通过不同的机制激活，同一信号分子也可通过若干不同的途径进行信号转导，且绝大多数信号转导途径之间存在广泛的信号串话（crosstalk），从而使各个信号转导途径在细胞间或细胞内形成一个复杂的信号转导网络。

信号转导一般包括以下步骤：特定的细胞释放信息物质→信息物质经过扩散或血循环到达靶细胞→与靶细胞的受体特异性结合→受体对信号进行转换并启动细胞内信使系统→靶细胞产生生物学效应。

一、膜受体介导的信号转导途径

膜受体介导的信号转导途径的共同特征，是通过细胞外的信号分子与靶细胞表面受体的特异结合来触发细胞内的信号转导过程，信号分子本身并不进入细胞内。在这种信号转导机制中，常将在细胞外传递特异性信号的信号分子称为第一信使，而将在细胞内传递特异性信号的小分子物质（如 cAMP、cGMP、Ca^{2+}、DAG、IP_3）以及 TPK 等称为第二信使。

（一）环核苷酸信号转导途径

此类信号转导途径的共同特征是以小分子的环核苷酸（cAMP 或 cGMP）作为第二信使，通过细胞内环核苷酸浓度的改变进行信号转导。

1. cAMP 信号转导系统

这是一条经典的途径，信号分子通常与 G 蛋白偶联型受体结合而激活此途径。构成 cAMP 信号转导途径的级联反应为信号分子→膜受体→G 蛋白→AC→cAMP→PKA→效应蛋白（酶）→细胞生物学效应。

1）受体通过调节腺苷酸环化酶来控制 cAMP 浓度　作为一种细胞内信使，cAMP 浓度的变化相当快，在细胞对激素的反应中，几秒钟内 cAMP 的浓度变化达 5 倍以上。这种快速反应的机制是通过两种酶实现的，即腺苷酸环化酶和 cAMP 磷酸二酯酶。腺苷酸环化酶的底物是 ATP，产物是 cAMP，此酶是一种细胞膜结合蛋白。cAMP 磷酸二酯酶能快速水解 cAMP，产生 5'-AMP。细胞外信号主要通过改变腺苷酸环化酶的活性而不是磷酸二酯酶的活性来控制 cAMP 的水平。不同的激素和靶细胞膜上的相应受体结合后，通过不同的途径影响细胞内 cAMP 浓度。有些通过 Gs 蛋白激活腺苷酸环化酶，从而升高细胞内 cAMP 浓度，如促甲状腺素、促肾上腺皮质激素、促黄体生成素、肾上腺素、甲状旁腺素、胰高血糖素、血管升压素等；有些通过 Gi 蛋白抑制腺苷酸环化酶，能降低细胞内 cAMP 浓度。α2-肾上腺能受体与 Gi 蛋白偶联，β-肾上腺能受体与 Gs 蛋白偶联，因此肾上腺素和受体结合后通过与不同类型的 G 蛋白（Gi 或 Gs），刺激或抑制腺苷酸环化酶，从而控制细胞内 cAMP 浓度。

2）G 蛋白偶联受体介导的信号转导　在 G 蛋白介导的信号转导中，一方面，G 蛋白可以通过 GTP 酶水解 GTP 为 GDP，重新形成不具活性的三聚体 G 蛋白，使得 G 蛋白的信号传递及时终止，有利于 G 蛋白接收下一次信号。另一方面，当信号分子长期存在时，一类特定的 G 蛋白偶联受体激酶（G-protein coupled receptor kinase，GRK）使得 G 蛋白偶联受体 C 端的多个丝氨酸残基发生磷酸化，从而受体与 G 蛋白解偶联；

同时捕获蛋白（arrestin）可以识别并结合磷酸化的受体，阻断受体与 G 蛋白之间的相互作用。

3) cAMP 依赖的蛋白激酶介导的 cAMP 效应　动物细胞中，cAMP 主要通过激活 cAMP 依赖的蛋白激酶 A（cAMP-dependent protein kinase A or protein kinase A，PKA）发挥其生物效应。PKA 催化 ATP 分子末端磷酸基团转移到选择性靶蛋白的特异丝氨酸残基或苏氨酸残基上，被共价磷酸化修饰的氨基酸残基进而调控该靶蛋白的活性。无活性状态的 PKA 具有两个相同的催化亚基和两个相同的调节亚基，调节亚基能结合 cAMP。当 cAMP 和调节亚基结合后，该亚基的构象发生变化，使调节亚基从酶分子上解离下来，释放出的催化亚基则被激活，催化底物蛋白分子的磷酸化。肾上腺素与骨骼肌细胞膜上的 β-肾上腺能受体结合后，通过 Gs 蛋白使细胞内腺苷酸环化酶激活，cAMP 浓度升高，激活 PKA。PKA 活化后在 ATP 的存在下可磷酸化许多蛋白质，调节细胞物质代谢和基因表达。例如，在能分泌生长激素释放抑制激素（somatostatin 或 GHRIH）的细胞中（下丘脑和胰腺 δ 细胞），cAMP 能使编码该激素的基因开放。这类基因的调控区有一短序列的顺式作用元件，称为 cAMP 反应元件（cAMP response element，CRE），能识别 CRE 的转录因子称为 CRE 结合蛋白，简称 CREB。CREB 被 PKA 磷酸化并与 CRE 结合后，就能促进有关基因的转录。

cAMP 浓度的增加往往是短暂的，多为一过性升高，随后迅速被特异的磷酸二酯酶降解而灭活。

2. cGMP 信号转导系统

该信号转导途径以 GC 催化 GTP 生成第二信使 cGMP 为特征，即通过胞质中 cGMP 浓度的改变来完成信号转导过程。构成 cGMP 信号转导途径的级联反应为信号分子→膜受体/GC→cGMP→PKG→底物蛋白（酶）→细胞生物学效应。

（1）GC 与 cGMP 的生成

GC 广泛存在于动物各种组织细胞中，按其存在的亚细胞部位及分子结构的不同而分为两类。一类是具有受体作用的跨膜蛋白，分子结构与催化型受体类似，其细胞外区带有与特异信号分子结合的结构域，而细胞内区则带有 GC 结构域，因此也被称为膜结合性 GC。心房钠尿肽、脑钠肽及 C 类钠肽等信号分子可特异地与膜结合性 GC 结合而激活其酶活性，导致胞质 cGMP 浓度升高。另一类为胞质中可溶性 GC，分子结构为 α 及 β 两个亚基构成的二聚体。可溶性 GC 可被气体信号分子 NO 特异地激活，使胞质中的 cGMP 浓度升高。

两类 GC 的组织细胞分布有所不同，膜结合性 GC 主要分布于心血管组织、小肠、精子和视网膜杆状细胞，而可溶性 GC 则主要分布于脑、肝、肾、肺等组织中，这种分布可导致不同组织细胞对同一信号产生不同的反应。

（2）PKG 及其生物学效应

cGMP 的生物学效应几乎都是通过激活蛋白激酶 G（protein kinase G，PKG）来实现的，PKG 也是一种 Ser/Thr 蛋白激酶，可催化特异底物蛋白（酶）的 Ser/Thr 残基的磷酸化修饰而使其活性或其他生物学效应发生改变（图 7-13）。

已知 PKG 有两种同工酶，Ⅰ型酶为均一的二聚体，而Ⅱ型酶则为单体，每个亚基或单体都有 2 个 cGMP 的结合位点。PKG 的高度选择性底物包括 cGMP，磷酸二酯酶

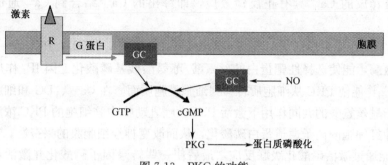

图 7-13　PKG 的功能

（cGMP-PDE）、PKA 的调节亚基、磷酸化酶激酶 α 亚基、脂肪细胞激素敏感脂肪酶、受磷蛋白（phospholamban）、钙泵（Ca^{2+}-ATP 酶）、Ca^{2+} 通道蛋白等。已知 PKG 可引起血管平滑肌细胞质膜和肌浆网膜受磷蛋白磷酸化，使 Ca^{2+} 与 Ca^{2+} 泵的亲和力增高，导致质膜和肌浆网 Ca^{2+} 泵的活性升高，摄取 Ca^{2+} 泵增多，引起胞质 Ca^{2+} 浓度降低而致平滑肌舒张。心房钠尿肽、NO 及硝基扩血管药正是通过 cGMP 信号转导途径激活 PKG 而致血管平滑肌舒张。

（二）Ca^{2+} 信号转导系统

Ca^{2+} 作为细胞信号在许多细胞反应中发挥作用，如细胞增殖、分泌、肌肉收缩和细胞骨架的重排等。胞质内 Ca^{2+} 浓度很低，小于 $10^{-7}\,mol \cdot L^{-1}$，远远低于细胞外液的 Ca^{2+} 浓度。细胞的内质网、线粒体、肌浆网是细胞内 Ca^{2+} 的储存库。许多信号分子引起细胞外液 Ca^{2+} 内流或亚细胞器中 Ca^{2+} 释放，使得胞质内 Ca^{2+} 迅速升高，调节各种生命活动。Ca^{2+} 信号有两条途径：一条途径在神经细胞中存在，细胞膜去极化（depolarization）导致 Ca^{2+} 流入神经末梢，启动神经递质分泌，有关这方面内容在生理学中有详细介绍；另一条途径是细胞外信号与 G 蛋白偶联受体结合后，将信号转导至内质网，使内质网内的 Ca^{2+} 释放到细胞质，引起胞质 Ca^{2+} 浓度升高，而参与信号转导反应。

1. 通过 G 蛋白偶联受体激活肌醇磷脂信号途径

肌醇磷脂（inositol phospholipid）位于细胞膜磷脂双分子层的内层。与信号转导有关的肌醇磷脂是磷脂酰肌醇（phosphatidylinositol，PI）的磷酸化衍生物：PI 一磷酸（PIP）、PI 二磷酸（IP_2）、PI 三磷酸（IP_3）。细胞外信号分子结合并激活受体后，使特定蛋白（Gq）激活，活化 Gq 激活与细胞膜附着的磷脂酶 C-β，然后磷脂酶 C-β 使 IP_2 裂解，产生两种分子：IP_3 和二酰甘油，两者都在信号转导中起重要作用。通过肌醇磷脂信号途径作用的细胞外信号有激素，如血管升压素（vasopressin）；有神经递质，如乙酰胆碱（作用于胰腺和平滑肌）；有抗原（作用于肥大细胞）；有凝血酶（作用于血小板）等。

2. IP_3 和 DG 的作用

由 IP_2 水解产生的 IP_3 是水溶性的小分子物质，离开细胞膜后能在细胞质内很快地扩散，IP_3 与内质网膜上的特异 Ca^{2+} 通道结合后，就能使内质网腔里的 Ca^{2+} 释放到细胞质，而且释放的 Ca^{2+} 具有正反馈效应，即释出的 Ca^{2+} 结合到 Ca^{2+} 通道，再促进 Ca^{2+} 释放。

DG 的重要作用是激活蛋白激酶 C（protein kinase C，PKC），PKC 是一类 Ca^{2+} 依赖的蛋白激酶，能使选择性靶蛋白的丝氨酸/苏氨酸残基磷酸化。因 IP_3 作用而升高的细胞质内 Ca^{2+} 能使 PKC 从细胞质转移到细胞膜胞质面。在 Ca^{2+}、DG 和细胞膜磷脂成分中的磷脂酰丝氨酸的共同作用下激活 PKC。哺乳动物中脑细胞的 PKC 浓度最高，其作用是使神经细胞的离子通道蛋白磷酸化，从而改变神经细胞膜的兴奋性。在许多细胞中，PKC 能通过激活磷酸化级联反应，最后使一些转录因子磷酸化并激活，从而调控相关基因的表达。

3. 钙调蛋白的作用

钙调蛋白（calmodulin）是一种特异的 Ca^{2+} 结合蛋白，几乎在所有的真核细胞中都存在。作为细胞内 Ca^{2+} 受体，钙调蛋白介导多种由 Ca^{2+} 调节的生物过程。钙调蛋白的一级结构高度保守，只有一条多肽链，约含 150 个氨基酸残基，有 4 个高亲和力钙结合部位，与 Ca^{2+} 结合后构象会发生改变。Ca^{2+} 通过别构作用激活钙调蛋白。Ca^{2+}-钙调蛋白复合物的作用是能与多种靶蛋白结合并改变靶蛋白的活性。这些靶蛋白包括各种酶和细胞膜上的转运蛋白，如细胞膜上的 Ca^{2+}-ATP 酶（使胞质内 Ca^{2+} 泵出细胞）。但是 Ca^{2+}-钙调蛋白的效应主要是通过 Ca^{2+}-钙调蛋白依赖的蛋白激酶（如 CaM 激酶，CaMPK）的介导来实现的。CaM 激酶也是通过使靶蛋白上特异的丝氨酸或苏氨酸产生磷酸化而激活靶蛋白。CaM 激酶具有较广的特异性，这也说明这类酶在动物细胞中介导 Ca^{2+} 的多种作用。

4. Ca^{2+} 途径和 cAMP 途径的相互作用

Ca^{2+} 细胞内信号途径和 cAMP 细胞内信号途径虽然是两条独立的途径，但相互之间也有作用。第一，细胞内 Ca^{2+} 水平和 cAMP 水平能相互影响，如直接与 cAMP 水平有关的腺苷酸环化酶和磷酸二酯酶都受到 Ca^{2+}-钙调蛋白复合物的调节。蛋白激酶 A（PKA）能磷酸化一些 Ca^{2+} 通道和 Ca^{2+} 泵，使它们改变活性，例如，PKA 能磷酸化内质网上的 IP_3 受体，启动或抑制 IP_3 引起的 Ca^{2+} 释放。第二，直接受 Ca^{2+} 和 cAMP 调控的酶之间能相互影响，如有些 CaM 激酶能被 PKA 磷酸化而改变活性。第三，这些酶能共同对一些靶分子产生相互作用，在这种情况下，是 PKA 和 CaM 激酶分别使一些蛋白质的不同部位磷酸化。

（三）酶偶联受体介导的信号转导系统

酶偶联受体和 G 蛋白偶联受体一样也是一类跨膜蛋白，与细胞外信号分子结合的结构域在细胞膜外，细胞内的胞质结构域本身即具有酶活性，或直接与其他酶相关联。已知有 5 种类型酶偶联受体：①受体鸟苷酸环化酶（receptor guanylyl cyclase）；②受体酪氨酸激酶（receptor tyrosine kinase）；③酪氨酸激酶相关受体（tyrosine-kinase associated receptor）；④受体酪氨酸磷酸酶（receptor tyrosine phosphatase）；⑤受体丝氨酸/苏氨酸激酶（receptor serine/threonine kinase）。本章只介绍前三种酶偶联受体介导的信号转导系统。

1. 受体鸟苷酸环化酶信号转导系统

这类受体与细胞外信号分子结合后，能催化细胞质内 cGMP 的生成，因该跨膜受体的胞质结构域具有鸟苷酸环化酶活性，催化 GTP 生成 cGMP，cGMP 再激活 cGMP 依赖的蛋白激酶（cGMP dependent protein kinase，G 激酶），G 激酶能使靶蛋白上的丝氨酸或苏氨酸残基磷酸化，激活靶蛋白。在此信号转导系统中，cGMP 是细胞内信号分子。与 cAMP 信号不同之处：在 cAMP 信号途径中联系受体与环化酶的是 G 蛋白，而 cGMP 是酶联受体的一种，受体本身就是酶，因此，在 cGMP 信号途径中，联系受体与环化酶的是 cGMP，此联系通过受体本身。但在某些细胞中，如视觉细胞，cGMP 的生成也通过 G 蛋白。通过受体鸟苷酸环化酶途径的细胞外信号有心房钠尿肽等。

2. 受体酪氨酸激酶信号转导系统

1）受体酪氨酸激酶　第一个被确认具有酪氨酸特异的蛋白激酶活性的受体是表皮生长因子（epidermal growth factor，EGF）受体。EGF 受体只有一条肽链，约有 1200 个氨基酸残基。当 EGF 与 EGF 受体结合后，受体的细胞质酪氨酸激酶结构域即被激活，激活的酪氨酸激酶能选择性地使受体蛋白本身的酪氨酸残基或其他靶蛋白上的酪氨酸残基磷酸化。现已发现，大多数生长因子和分化因子的受体都属这一类受体，这些受体都可以通过自身磷酸化（auto-phosphorylation）来启动细胞内信号的级联反应。

2）受体酪氨酸激酶信号转导系统中的其他成分　具有 SH 结构域的蛋白质不是指含有 SH 基团（巯基）的蛋白质，而是指最初在 Src（src 癌基因编码）蛋白中发现的一段序列，SH 是 Src 同源性（Src homology）的缩写。已发现有许多种含有 SH 结构域的蛋白质，如 GTP 酶激活蛋白（GTPase-activating protein，GAP）、磷脂酶 C-γ（PLC-γ 作用与 PLC-β 相同）、类 Src 非受体型蛋白酪氨酸激酶（Src-like nonreceptor protein tyrosine kinase）和 IRS-1 等。这些蛋白质都具有两种 SH 结构域——SH2 和 SH3。SH2 能识别磷酸化的酪氨酸残基，使含有 SH2 的蛋白质与激活的受体酪氨酸激酶结合。SH3 的作用是与细胞内其他蛋白质结合。在具有 SH2 和 SH3 的蛋白质中有些是酶蛋白，如 GAP、PLC-γ 等，有的只是作为一种"连接器"，如生长因子受体结合蛋白 α(growth factor receptor bound protein2，GRB2)，它的作用就是作为连接受体酪氨酸激酶和其他蛋白质的桥梁。当受体酪氨酸激酶被激活后，通过 GRB2 的作用，使调节分子 SOS 蛋白（son of sevenless protein）活化；活化的 SOS 使 Ras 蛋白激活（由原癌基因 ras 编码，性质类似于 G 蛋白），Ras 在通过受体酪氨酸激酶介导的信号转导中发挥中心作用，是一种关键的成分，这种信号转导控制细胞的生长和分化。Ras 的突变使其失去信号转导作用，能引起细胞的癌变。

3. 酪氨酸激酶相关受体信号转导系统

它是酪氨酸激酶相关受体信号转导系统（JAK-STAT）中一个比较典型的例子。这是一种比较简单的信号系统，只有三种成分：受体、JAK（janus kinases）激酶及信号转导子和转录激动子（signal transductor and activator of transcription，STAT）。JAK-STAT 信号转导过程：细胞因子或生长因子 ＋受体→受体二聚化→JAK →STAT→基因转录活性改变 → 细胞生理功能改变。

1）酪氨酸激酶相关受体　这类受体包括多种细胞因子（cytokine）的受体，如干扰素受体、白细胞介素-2 受体等。这类受体本身不具有内在的激酶活性，但是细胞外信号分子与之结合后，能形成二聚体，受体二聚体能与 JAK 激酶结合，并激活 JAK 激酶。

2）JAK 激酶　JAK 激酶是一族分子，有多种成员，每种成员能特异地和相应的细胞因子受体结合。JAK 激酶原来称为 Janus 激酶（Janus 意为罗马神话中的天门神，也称两面神，专司守护门和万物的始末面神），因为这种分子上有两个激酶结构域。JAK 激酶属酪氨酸激酶，主要的底物是 STAT。

3）STAT　STAT 是一类转录因子，目前已知至少有 7 种 STAT，每种 STAT 分别由相应的 JAK 激酶激活。STAT 的磷酸化导致 STAT 复合体的形成，形成复合体的基础是分别存在于两个 STAT 上的 SH2 结构域和磷酸化的酪氨酸残基之间的相互作用。STAT 复合体从细胞质转移到细胞核内，和相应的顺式作用元件结合，调控靶基因的表达。JAK-STAT 信号转导途径见图 7-14。

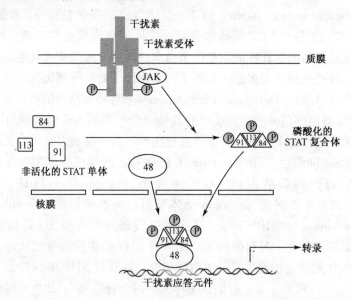

图 7-14　干扰素诱导 JAK-STAT 复合体核内转移及调节基因转录机制

二、细胞内受体介导的信号转导系统

小分子脂溶性（疏水性）的细胞外信号分子有类固醇激素、甲状腺素、维 A 酸类、维生素 D 等。这类信号分子虽然在结构上差异很大，但是信号转导的机制是相同的。这些脂溶性分子都能通过细胞膜直接扩散进入细胞质或细胞核，与细胞内受体蛋白结合，并激活受体，最后通过受体去调控基因的表达，因此这类细胞内受体是一类反式作用因子。这类受体称为细胞内受体超家族（intracellular receptor superfamily）或类固醇激素受体超家族（steroid-hormone receptor superfamily）。

（一）细胞内受体的结构域

这类受体都有两个结构域（DNA 结合结构域、激素结合结构域）和一个可变区。不同受体的 DNA 结合结构域显示较高的一级结构同源性，激素结合结构域的同源性低一些，可变区的一级结构则不具同源性。可变区含有激活结构域，DNA 结合结构域具有 4 个半胱氨酸残基的锌指结构。细胞内受体通过 DNA 结合结构域与相应靶基因上的反应元件结合。

（二）脂溶性细胞外信号分子调控基因表达的机制

以糖皮质激素作用机制为例（图 7-15）。糖皮质激素透过细胞膜进入细胞后与糖皮质激素受体结合，受体被激活，激活的受体进入细胞核识别结合糖皮质激素反应元件（glucocorticoid response element，GRE 位于增强子内），当激素-受体复合物与增强子结合后就激活启动子，启动转录。其他脂溶性激素的作用机制基本相同。

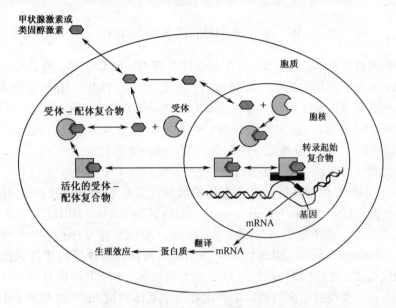

图 7-15　糖皮质激素作用机制的示意图

第四节　可控性蛋白降解与信号转导

在机体发育时许多活动都是受到调控的，这些调控活动通过不同的信号转导来实现，除了前面几节提到的信号途径以外，还有可控性蛋白水解相关的信号途径，如 Wnt、Hedgehog、Notch、NF-κB 等，这些信号途径往往影响相邻细胞的分化，称为侧向信号发放（lateral signaling）。

一、Notch 信号途径

Notch 基因最早发现于果蝇，部分功能缺失导致翅缘缺刻（notch）（图 7-16）。在胚胎发育中，当上皮组织的前体细胞中分化出神经元细胞后，其细胞表面 Notch 配体 Delta 与相邻细胞膜上的 Notch 结合，启动信号途径，防止其他细胞发生同样的分化，这种现象叫作侧向抑制（lateral inhibition）。*Notch* 基因位点突变的半合子或纯合子果蝇在胚胎期死亡，在死亡的胚胎中，神经组织取代了上皮组织从而使神经组织异常丰富。

普通翅　　　　　　　　　　　　Notch 缺陷翅

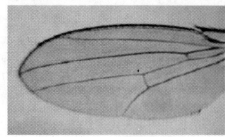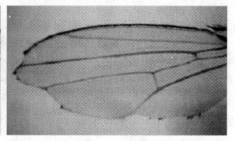

图 7-16　Notch 缺陷引起果蝇翅缘缺刻

Notch 信号途径由 Notch、Notch 配体（DSL 蛋白）和 CSL（一类 DNA 结合蛋白）等组成。Notch 及其配体均为单次跨膜蛋白，当配体（如 Delta）和相邻细胞的 Notch 结合后，Notch 被蛋白酶体切割，释放出具有核定位信号的胞内区 Notch（intracellular domain of Notch，ICN），进入细胞核与 CSL 结合，调节基因表达。可概括：Delta→Notch→酶切→ICN→进入细胞核→CSL-ICN 复合体→基因转录。

Notch 为分子质量约 300 kDa 的蛋白质，果蝇只有 1 个 Notch 基因，人类有 4 个（Notch1～4）。Notch 的胞外区是结合配体的区域，具有不同数量的 EGF 样重复序列（EGF-R）和 3 个 LNR（Lin/Notch repeat）。胞内区由 RAM（RBP-J kappa associated molecular）结构域、6 个锚蛋白（cdc10/ankyrin，ANK）重复序列、2 个核定位信号（Nuclear localization signal，NLS 它为一段信号序列，可引导蛋白质进入细胞核）和 PEST 结构域组成。RAM 结构域是与 CSL 结合的区域，PEST 结构域与 Notch 的降解有关。Notch 蛋白要经过三次切割，第一次在高尔基体内被 furin 切割为 2 个片段，转运到细胞膜形成异二聚体。当配体结合到胞外区，Notch 蛋白又发生两次断裂，先是被肿瘤坏死因子-α-转化酶（TNF-α-converting enzyme，TACE）切割，然后被 γ-促分泌酶（γ-secretase）切割，后者需要早老蛋白（presenilin，PS）参与。酶切以后释放 ICN，进入细胞核发挥生物学作用。

Notch 配体在果蝇中为 Delta 和 Serrate，线虫为 Lag-2，取首字母，Notch 的配体又被称为 DSL 蛋白（在哺乳动物中叫作 Jagged），都是单次跨膜糖蛋白，其胞外区含有数量不等的 EGF 样重复区，N 端有一个结合 Notch 配体必需的 DSL 基序。

CSL 为转录因子，在哺乳动物中叫作 CBF1，在果蝇中叫作 Suppressor of Hairless，在线虫中叫作 Lag-1。CSL 能识别并结合特定的 DNA 序列（GTGGGAA），这个序列位于 Notch 诱导基因的启动子上。ICN 不存在时，CSL 为转录抑制因子。当结合 ICN 时，CSL 能诱导相关基因的表达。Notch 信号的靶基因多为碱性螺旋-环-螺旋类转录因子（basic helix-loop-helix，bHLH），它们又调节其他与细胞分化直接相关基因的转录，如哺乳动物中的 HES（hairy/ enhancer of split）、果蝇中的 E（spl）（enhancer of split）及非洲爪蟾中的 XHey-1 等。Notch 信号途径示意图见图 7-17。

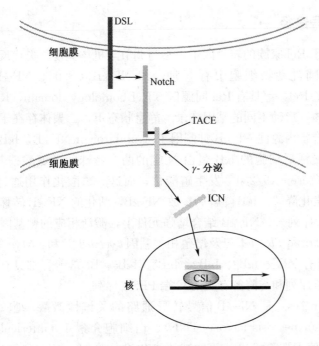

图 7-17 Notch 信号途径

二、Wnt 信号途径

Wnt 是一类分泌型糖蛋白，通过自分泌或旁分泌发挥作用。在小鼠中，肿瘤病毒整合到 Wnt 之后导致乳腺癌，命名为 Int1，它与果蝇的无翅基因（wingless，wg）有高度同源性。Wnt 信号途径能引起胞内 β-连锁蛋白（β-catenin）积累。Wnt 的受体是卷曲蛋白（frizzled，Frz），为 7 次跨膜蛋白，结构类似于 G 蛋白偶联型受体，Frz 胞外 N 端具有富含半胱氨酸的结构域（cysteine rich domain，CRD），能与 Wnt 结合。Frz 作用于胞质内的蓬乱蛋白（dishevelled，Dsh 或 Dvl），Dsh 能切断 β-连锁蛋白的降解途径，从而使 β-连锁蛋白在细胞质中积累，并进入细胞核，与 T 细胞因子/淋巴增强因子（T cell factor / lymphoid enhancer factor，TCF/LEF）相互作用，调节靶基因的表达。

三、Hedgehog 信号途径

Hedgehog 是一种共价结合胆固醇的分泌性蛋白，在动物发育中起重要作用。果蝇的该基因突变导致幼虫体表出现许多刺突，形似刺猬，故名 Hedgehog。脊椎动物中至少有三个基因编码 Hedgehog 蛋白，即 *Shh*（sonic hedgehog）、*Ihh*（Indian hedgehog）和 *Dhh*（desert hedgehog），其中 *Shh* 是根据电子游戏中的角色命名的，后两者是用刺猬的两个种命名的。

两个跨膜蛋白 Patched（Ptc）和 Smoothened（Smo）介导 Hedgehog 信号向胞内传递。Ptc 是 12 次跨膜蛋白，能与 Hedgehog 结合；Smo 为 7 次跨膜蛋白，与 G 蛋白偶联型受体同源。在无 Hedgehog 的情况下，Ptc 抑制 Smo。当 Hedgehog 与 Ptc 结合时，则解除了 Ptc 对 Smo 的抑制作用，引发下游事件。

四、NF-κB 信号途径

NF-κB 是属于 Rel 家族的转录因子，参与调节与机体免疫、炎症反应、细胞分化有关的基因转录。哺乳动物细胞中有 5 种，即 NF-κB1（P50）、NF-κB2（P52）、RelA（P65）、RelB 和 C-Rel，都具有 Rel 同源区（Rel homology domain，RHD），能形成同二聚体或异二聚体，启动不同的基因转录。静息状态下，二聚体存在于细胞质中，并且以无活性形式络合于内源性 NF-κB 抑制因子（inhibitor of NF-κB，IκB）上。在 IκB 激酶的作用下，经激活后细胞的 IκB 蛋白 N 端的两个特殊的丝氨酸残基［IκBα（serine 32/36）或 IκBβ（serine 19/23）］发生磷酸化，而这一磷酸化作用通过 26S 蛋白酶体，促进 IκB 发生泛素化降解。IκB 降解后释放 NF-κB，活化的 NF-κB 转移进入细胞核，结合在特异的 DNA 序列——NF-κB 结合增强元件上，激活相应的靶基因，如抗凋亡相关基因 *Bcl-2*、*cFLIP* 和 *TRAF1* 以及增殖相关基因 *cyclin D1* 和 *c-Myc* 等的转录。

IκB 家族成员有 IκBα、IκBβ、IκBγ、IκBδ、IκBε、Bcl-3 等，都具有与 Rel 蛋白相互作用的锚蛋白重复序列和与降解有关的 C 端 PEST 序列。

IKK（IκB kinases）是 NF-κB 信号转导通路的关键性激酶。胞外信号，如肿瘤坏死因子 α（tumor necrosis factor，TNF）、白细胞介素-1（interleukin-1，IL-1）等可以激活 IKK，使 IκB 磷酸化，随后被泛素化途径降解。NF-κB 信号转导途径见图 7-18。

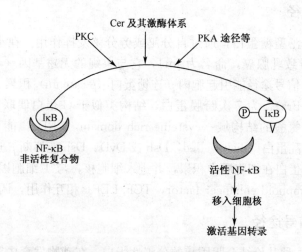

图 7-18　NF-κB 的激活过程示意图

第五节　信号转导途径的一般特性

尽管信号转导途径错综复杂，但众多的信号转导途径也有一些共同的特性。

1）信号转导分子存在的暂时性　许多信号蛋白的半衰期都很短，如 Fos 只有 2h，*c-fos* 基因表达在刺激后 2h 就停止；*junB*、*erg-1* 的表达在刺激后 14h 停止；*c-jun* 则为 6h。尽管如此，这些基因产物的作用时间却是很长的，如 *c-fos* 诱导的与 AP-1 结合的活性可以持续达 6h 以上。

2）信号转导分子活性的可逆性变化　被激活的各种信号转导分子在完成任务后又恢复钝化状态，准备接受下一次的刺激。它们不会总处在兴奋状态。例如，激酶的磷酸化与去磷酸化，就有磷酸酪氨酸膦酸酯酶在调节着。

3）信号转导分子激活机制的类同性　Fos的激活要其丝氨酸和苏氨酸的磷酸化；JAK激活要其酪氨酸磷酸化。它们在传递信息后又都要去磷酸化。可见，磷酸化和去磷酸化是绝大多数信号分子可逆激活的共同机制。

4）信号转导途径的连贯性　信号转导途径上的各个反应相互衔接，形成一个级联反应过程，有序地依次进行，直至完成。其间，任何步骤的中断或者出错，都将给细胞甚至机体带来严重后果。

5）作用的一过性与效果的永久性的有机统一　一条信号转导途径的总的作用时间一般不会很长。例如，编码转录因子的原癌基因的诱导只有几到几十分钟，许多功能基因的被诱导过程也是以小时计算的。尽管时间相对不长，但是，刺激经由信号转导途径所造成细胞增殖、分裂、分化、成熟、恶变、转化，或者自我凋亡等效果却往往是不可逆的。由此可见，信号转导过程一定受到严格的调节控制。

6）网络化　细胞内的各种信号转导途径是相互联系的，形成一张遍布整个细胞的信号转导途径网络。在这张网中，各条途径相互沟通，相互串联，相互影响，相互制约，相互协调，相互作用。这样，细胞才能够对各种刺激作出迅速而准确的响应，才能应环境的变化而变化。

7）专一性　鉴于各条信号转导途径有共同的信号转导分子，鉴于细胞内存在有信号转导途径网络，那么，为什么不同的刺激能够产生特殊的细胞响应呢？这说明，信号转导有专一性。有赖于此，细胞能够对不同的刺激作出不同的反应。

第六节　信号转导与疾病

信号转导在生命现象中有非常重要的作用，它的失误会造成疾病，甚至日常生活中常见的疾病和问题也与信号转导有关，如酒精中毒。研究表明，急性酒精刺激时，大鼠小脑中cAMP含量和蛋白激酶A（PKA）的活性比对照增加了80%，说明急性酒精摄入时腺苷酸环化酶信号转导途径被激活。但是，经常性、慢性地摄入酒精时，这些分子的活性和含量都没有变化。还有，内源性NO在心肌局部缺血和再灌流损伤的病理过程中起作用。这个作用是通过局部缺血心肌中的跨膜信号转导途径实现的，具体说是cGMP介导的。细胞因子在机体中起着非常重要的作用，这些作用也都是通过信号转导发挥的。不仅如此，意识方面的信号也会产生物质性的结果。例如，经常处于忧郁与悲伤状态的人容易患癌症。也就是说，精神状态与机体的免疫力高低有关。

因此，研究和设计以信号转导途径为靶点的药物和疾病治疗方法，就成为临床医学和药物产业的新领域。近几年来，这两方面都有很大发展。其中，有的比较成熟，针对性强，疗效也比较好；有的药物已经成为产品，在临床上广为使用；有的虽然很不成熟，却表现出很大的应用潜力和前景。

小　结

信号转导级联反应系统介导了刺激的识别及处理过程。这些分子环路系统识别、放大和整合不同的外界信号，从而产生一系列酶活性、基因表达或离子通道活性的变化。通过对信号转导的理解，有助于我们有效地治疗疾病，甚至构建人工器官（理论上）。

膜受体传递着细胞内外环境的信息。一些非极性的信号分子，如雌激素以及其他类固醇激素能通过细胞膜而渗透扩散进入细胞。这些分子一旦进入细胞内即可与 DNA 直接作用的蛋白质相结合而调控基因的转录。

第二信使通过受体-配体复合物来传递信号。这些被称为第二信使的小分子的浓度变化构成分子信息传递环节中重要的一步。一些特别重要的第二信使有 cAMP、cGMP、Ca^{2+}、肌醇 1,4,5-三磷酸（三磷酸肌醇）和二酰甘油（DAG）。

蛋白质的磷酸化是一种普遍的传递信号方式。许多第二信使就是通过活化一些蛋白激酶来产生反应的。这些酶将一些磷酸化基团从 ATP 转移到蛋白质的丝氨酸、苏氨酸及酪氨酸残基上。

任何信号转导都最终将进入终止状态。蛋白质磷酸酶参与了信号转导的终结机制。从信号转导起始到其产生其他细胞的影响效应后，信号转导进程终究要终止。无法终止该信号转导，细胞就会失去对新信号的反应。而且，无法终止的信号转导可导致细胞陷入生长增殖无法控制的状态，最后导致肿瘤的产生。

（姚　军）

参 考 文 献

Bodine P V. 2008. Wnt signaling control of bone cell apoptosis. Cell Res, 18 (2)：248～253

David L. Nelson，Michael M. Cox. 2005. Lehninger Principles of Biochemistry. 4th ed. New York：W H Freeman and Company

Luttrell L M. 2006. Transmembrane signaling by G protein-coupled receptors. Methods Mol Biol，332：3～49

Luttrell. L M. 2008. Reviews in molecular biology and biotechnology：transmembrane signaling by G protein-coupled receptors. Mol Biotechnol，39 (3)：239～264

Sahlgren C，Lendahl U. 2006. Notch signaling and its integration with other signaling mechanisms. Regen Med，1(2)：195～205

Spiegelberg B D, Hamm H E. 2007. Roles of G-protein-coupled receptor signaling in cancer biology and gene transcription. Curr Opin Genet Dev, 17 (1)：40～44

Weaver R F. 2005. Molecular Biology. 3rd ed. New York：The McGraw Hill Companies

Chapter 7　Signal Transduction

　　Cells respond to their environment by reorganising their structure, regulating the activity of proteins and altering patterns of gene expression. The stimulus for such responses is termed a signal, and may be a small molecule, a macromolecule or a physical agent, such as light. Signals interact with the responding cell through molecules termed receptors.

　　In biology, signal transduction refers to any process by which a cell converts one kind of signal or stimulus into another. Most processes of signal transduction involve ordered sequences of biochemical reactions inside the cell, which are carried out by enzymes, activated by second messengers, resulting in a signal transduction pathway. Such processes are usually rapid, lasting on the order of milliseconds in the case of ion flux, or minutes for the activation of protein-mediated and lipid-mediated kinase cascades, but some can take hours, and even days (as is the case with gene expression), to complete. The number of proteins and other molecules participating in the events involving signal transduction increases as the process emanates from the initial stimulus, resulting in a "signal cascade," beginning with a relatively small stimulus that elicits a large response. This is referred to as amplification of the signal.

Section 1　Signal Molecules

　　A **signal molecule** is a chemical involved in transmitting information between cells. Such molecules are released from the cell sending the signal, cross over the gap between cells, and interact with receptors in another cell, triggering a response in that cell.

1. 1　Extracellular Signaling Molecules

　　Most signal transduction involves the binding of extracellular signaling molecules (or ligands) to cell-surface receptors that face outward from the plasma membrane and trigger events inside the cell. Also, intracellular signaling cascades can be triggered through cell-substratum interactions, as in the case of integrins, which bind ligands found within the extracellular matrix. Steroids represent another example of extracellular signaling molecules that may cross the plasma membrane due to their lipophilic or hydrophobic nature. Many, but not all, steroids have receptors within the cytoplasm, and usually act by stimulating the binding of their receptors to the promoter region of steroid-responsive genes. Within multicellular organisms, there is a diverse number of small molecules and polypeptides that serve to coordinate a cell's individual biological activity within the context of the organism as a whole. These molecules have been functionally classified as neurotransmitter, hormone and chemokine.

　　a) Neurotransmitters are chemicals that are used to relay, amplify and modulate signals between a neuron and another cell. Neurotransmitters are packaged into vesicles that cluster beneath the membrane on the presynaptic side of a synapse, and released into the synaptic cleft, where they bind to receptors located in the membrane on the postsynaptic side of the synapse. Release of neurotransmitters is most commonly driven by arrival of an action potential at the synapse, but may also be driven by graded electrical potentials. Also, there is often a low level of "baseline" release even in the absence of electrical stimulation.

　　b) Hormones are chemicals released by cells that affect cells in other parts of the body. Only a small amount of hormone is required to alter cell metabolism. It is also a chemical messenger that transports a signal from one cell to another. All multicellular organisms produce hormones; plant hormones are also called phytohormones. Hormones in animals are often transported in the blood. Cells respond to a hormone when they express a specific receptor for that hormone. The hormone binds to the receptor protein, resulting in the activation of a signal transduction mechanism that ultimately leads to cell type-specific responses.

　　Endocrine hormone molecules are secreted (released) directly into the bloodstream, while exocrine hormones (or ectohormones) are secreted directly into a duct, and from the duct they either flow into the bloodstream or they flow from cell to cell by diffusion in a process known as paracrine signalling.

c) Chemokines are a family of small cytokines, or proteins secreted by cells. Proteins are classified as chemokines according to shared structural characteristics such as small size (they are all approximately $8\sim10$ kDa in size), and the presence of four cysteine residues in conserved locations that are key to forming their 3-dimensional shape. Their name is derived from their ability to induce directed chemotaxis in nearby responsive cells; they are chemotactic cytokines. However, these proteins have historically been known under several other names including the SIS family of cytokines, SIG family of cytokines, SCY family of cytokines, Platelet factor 4 superfamily or intercrines. Some chemokines are considered pro-inflammatory and can be induced during an immune response to promote cells of the immune system to a site of infection, while others are considered homeostatic and are involved in controlling the migration of cells during normal processes of tissue maintenance or development. Chemokines are found in all vertebrates, some viruses and some bacteria, but none have been described for other invertebrates. These proteins exert their biological effects by interacting with G protein-linked transmembrane receptors called chemokine receptors, which are selectively found on the surfaces of their target cells.

1. 2   Intracellular Signaling Molecules

The secondary messenger is the small molecular material that transmits messages in cells. The secondary messengers include Ca^{2+}, DAG, IP_3, Cer, cAMP, cGMP, arachidonic acid and its metabolite etc. The principle of signal transduction is that the active form of a receptor triggers a catalytic activity in the cytosol. The amplitude of the cytosolic signal is much greater than the original extracellular signal (the ligand). The cytosolic signal may take the form of directly activating a series of proteins or it may be accomplished by increasing the quantity of a small molecule inside the cell. A molecule produced in response to transduction of an extracellular signal is called a second messenger (by contrast with the first messenger, which was the extracellular ligand) (Figure 7-1).

Figure 7-1   The second messenger (see page 223)

The third messenger is the material responsible for information transmission inside and outside cell nucleus, is also called the DNA binding protein. The third messenger is one kind of nuclear protein that may bind with the special sequence of target genes and can regulate gene transcription. The encoding protein of immediate-early gene is an example of the third messenger (Table 7-1).

Table 7-1   Intercellular signaling molecule that affect cell functions

	Signaling substances	Receptors	Intracellular changes
Neurotransmitter	Acetylcholine, glutamate, γ-aminobutyric acid	plasma membrane receptor	Effects on open and close of ion channel
Growth factor	insulin-like growth factor I, epidermal growth factor, platelet derived growth factor	plasma membrane receptor	phosphorylation and dephosphorylation of apoenzyme and functional proteins, changes of metabolism and gene expression
Hormone	Hormones of protein, peptide and amino acid derivate; steroid hormones; thyroxine	intracellular receptor	
Vitamin	Vitamin A, Vitamin D	intracellular receptor	the same as above

1. 3   Modes of Cell-Cell Signaling

a) Endocrine signaling is a type of cell-cell signaling in which endocrine cells secrete hormones that are carried through the circulation to distant target cells (Figure 7-2).

Figure 7-2　The endocrine signaling pathway（see page 224）

b) Paracrine signaling is local cell-cell signaling in which a molecule released by one cell acts on a neighboring target cell（Figure 7-3）.

Figure 7-3　The paracrine signaling pathway（see page 224）

c) Autocrine signaling is a type of cell signaling in which a cell produces a signaling molecule to which it or the same kind of cells also respond（Figure 7-4）.

Figure 7-4　The autocrine signaling pathway（see page 224）

Section 2　Receptors

Receptors are proteins or glycolipids located in the plasma membrane or inside the target cells, which can specifically recognize biological active molecules（signal molecules）and transmit signals into the cell. Extracellular signaling molecules bind to either cell-surface receptor or intracellular receptors. The biological active molecules that can recognize and bind to receptors are called ligands.

2. 1　Transmembrane Receptors

These receptors span the membrane（Figure 7-5）. The signal activates a channel, an enzyme, or a G-protein cascade.

Figure 7-5　The transmembrane receptors（see page 224）

2. 1. 1　Ion-channel-linked Receptors Open an Ion Channel in Response to the Signal Molecule

Ion channels are water-soluble pores in the cell membranes whose activity is controlled by opening and closing, thus allowing or preventing the movement of ions and other small molecules in or out of the cell. Ion channels are closed as a default state but open in response to a specific signal. Some are ligand-gated, i. e. they open in response to the binding of a particular ligand (e. g. glutamate and γ-aminobutyric acid (GABA) receptors). Others, particularly those found in neurons, are voltage-gated, i. e. they open in response to the electrical changes associated with an action potential. Finally, there are second-messenger-gated ion channels, which respond to second messengers in the cell, e. g. calcium ions, cyclic nucleotides and lipids. Ion channels may allow the passive movement of ions along an electrochemical gradient, or may actively pump ions against such a gradient, a process which requires ATP hydrolysis（Figure 7-6）.

Figure 7-6　The structure of nicotinic acetylcholine receptor（see page 225）

The transmembrane region of ion channels comprise several amphipathic helices (q. v.) which pack together so that hydrophobic residues surround a hydrophilic core. Some channels are single, large proteins with a number of similar domains. Others are multisubunit proteins. Neurotransmitter-gated ion channels have been particularly well characterized. They comprise five similar subunits, each of which possesses a four-helix transmembrane domain. One of the four helices is amphipathic, and the pore of the channel is lined by the amphipathic helices from each subunit. The opening and closing of ion channels is mediated by conformational changes stimulated by the gating mechanism (i. e. binding of a ligand or voltage changes). The particular arrangement of charged residues at the entrance to the pore, and lining it, control the ion selectivity of the channels.

2. 1. 2　G-protein Linked Receptors Activate an Intracellular G-protein That in Turn Activates Intracellular Enzymes

The G-protein coupled receptors (GPCRs) are single polypeptides whose central, hydrophobic region forms a seven-span transmembrane domain. The external, N-terminal region is often the ligand-binding domain, although some ligands (e.g. adrenaline) bind within the membrane domain. The internal, C terminal region is associated with a trimeric guanine nucleotide binding protein, or G-protein. The genes for several hundred GPCRs have been cloned and characterized. The receptors demonstrate remarkable conservation considering the diversity of their ligands, which include small molecules such as adrenaline and serotonin, peptides such as glucagon, the neurokinins and the opioids, and large glycoproteins such as follicle-stimulating hormone. Further GPCRs are stimulated by odorants and light. GPCRs are also widely distributed as well as the hundreds of vertebrate receptors identified, GPCRs are responsible for the transduction of yeast mating-type signals, and the response to cAMP in Dictyostelium discoideum. Two putative GPCRs have also been identified in viral genomes, although their functions are unknown. A membrane protein with similar structure, but no sequence homology, is also found in the halophilic bacterium Halobacterium hulobium (Figure 7-7).

Figure 7-7 The schematic diagiam of G-protein linked receptor (see page 226)

G-proteins are heterotrimeric complexes (comprising α-subunit, β-subunit and γ- subunit) which associate with GDP in their inactive form (Figure 7-8). Activation of the receptor by ligand binding stimulates an interaction which increases the rate at which GDP dissociates from the G-protein. Dissociation allows GTP to replace GDP because it is more abundant than GDP in the cell, and this causes the α-subunit to separate from the $\beta\gamma$ dimer. G-proteins have intrinsic GTPase activity, and slowly hydrolyze their cognate GTP thus inactivating themselves, whereupon they reassemble into trimers (Figure 7-9).

Figure 7-8 The structure of heterotrimeric G-proteins (see page 227)

Figure 7-9 The activation of G-protein (see page 227)

More than 20 different types of α-subunit and also multiple types of the β-subunit and γ-subunit have been identified in mammals: potentially hundreds of combinatorial trimers can form, some of which can be coexpressed in a particular cell type. The α-subunit is often the primary activator of downstream effector molecules, and four major families of G-proteins have been identified whose α-subunit interact with different types of effector and have different effects on the availability of intracellular second messengers. The $\beta\gamma$ dimers also mediate downstream responses, e.g. the STE1 and STE2 proteins of S. cerevisiae activate a kinase which links G-protein signaling into the MAP kinase pathway. The precise response to G-protein activation is governed both by the particular α-subunit, β-subunit and γ-subtypes and the particular isoforms of downstream targets. For example, the various isoforms of adenylate cyclase show differential responses to Gs (stimulatory G protein), and Gi (inhibitory G protein) regulation.

Cholera toxin (CT) interferes with the function of a key G-protein in the small intestine by deactivating its intrinsic GTPase activity. The consequence is uncontrolled activation of the signal transduction pathway which produces the second messenger $3', 5'$-cyclic AMP leading to a life-threatening diarrhea following the activation of ion and water secretion.

2. 1. 3   Enzyme-linked Receptors

The receptors may divide into two kinds. One kind has "the enzyme" activity, including the receptors of peptide growth factors, such as EGF, PDGF, CSF. The other kind does not have "the enzyme" activity, but they can bind with non-receptor tyrosine kinase, such as cytokines receptor superfamily. The common characteristics of the receptors are single-pass membrane protein usually; and actived by dimerization upon binding with ligand.

Receptor tyrosine kinases (RTKs) are single membrane spanning receptors with intrinsic, ligand-activated protein tyrosine kinase activity on their cytoplasmic domains. Over 50 RTKs have been characterized. They are divided into 14 major families based on structural motifs found in the extracellular domain. The tyrosine kinase domain is strongly conserved across all families, although in the PDGFR, FGFR and VEGFR subfamilies the domain is divided by an internal sequence which binds to other signaling molecules.

In the absence of the ligand, RTKs are monomeric and have no kinase activity. Although not all receptor activation mechanisms have been investigated, activation is probably brought about in each case by oligomerization, which stimulates kinase activity and autotransphosphorylation (the phosphorylation of one receptor monomer by another). Many RTKs are active as dimers, dimerization being brought about by ligand binding, often because the ligands themselves are dimeric and can bind two receptors simultaneously (e. g. PDGF). The insulin receptor family is anomalous in this respect as the receptors are constitutively dimeric, the two subunits being held together by disulfide bonds. In this case, ligand binding is likely to induce a conformational change facilitating activation of the kinase domain.

Once the receptor is activated, RTK signaling can be initiated in two ways: ① By phosphorylating downstream targets; ② By recruiting signaling complexes including proteins which specifically recognize phosphotyrosine residues. Several domains of other proteins interact specifically with phosphotyrosine, including the SH2 domain and the PTB domain The insulin receptor family is again slightly different from other RTKs in that many, if not all, of its downstream reactions are mediated by a small protein, insulin receptor substrate-1 (IRS-1), which is phosphorylated by the insulin receptor tyrosine kinase. Major signaling pathways initiated by RTK activation include the Ras-Raf-MAP kinase pathway, and the phospholipase C-γ-activated second-messenger system.

Receptors with associated tyrosine kinase activity. A number of receptors phosphorylate target proteins upon activation but possess no intrinsic enzyme activity. Such receptors, which include many cytokine receptors and the receptors which process antigens in the immune system, recruit and activate cytoplasmic protein tyrosine kinases upon ligand binding.

There are two major families of cytokine receptors. Class I receptors include most hematopoietic and immune system cytokine receptors and three ubiquitous receptors: gp130, βc and γc. The distantly related class II receptors include the interferon receptors (Figure 7-10) and the interleukin 10 receptor. Receptors of each class are identified by conserved motifs in both the extracellular and intracellular domains of the molecule.

Figure 7-10 The interferon induced signaling transduction (see page 228)

Ligand binding induces receptor dimerization which juxtaposes the intracellular domains of two monomers. Many class I receptors function as homodimers, whereas others form heterodimers with the ubiquitous receptors p130, βc and γc. Heterodimerization is nonpromiscuous, so the hematopoietic cytokine receptors can be divided into three functional subfamilies depending upon which of the ubiquitous molecules each interacts with. Within each subfamily there is a degree of functional redundancy, reflecting the role of the common receptor monomer. Class II receptors are formed from multiple subunits, many of which are uncharacterized. However, activation of the class II receptors also involves oligomerization (Figure 7-11).

Figure 7-11 The structure and activation of the EGF receptor (see page 229)

Other membrane receptors. The three receptor types discussed above represent nearly all known receptor molecules involved in signal transduction in animal cells. Some signals, however, are transmitted through receptors with distinct signaling mechanisms. The atrial natriuretic peptides, for instance, bind receptors with intrinsic guanylate cyclase activity. These increase the intracellular levels of cGMP and activate protein kinase G (PKG). A number of receptors with protein tyrosine phosphatase activity have also been described, including the leukocyte common antigen CD45 receptor. The tumor necrosis factor receptor family signal using death domains which bind to cytoplasmic proteins with similar domains. For other receptors well characterized at the genetic level, the signal transduction mechanisms remain far from clear (e. g. Notch and Delta which control lateral inhibition in the nervous system) (Figure 7-12).

Figure 7-12 The structure of guanylate cyclase receptor (see page 229)

2. 2 Cytoplasmic and Nuclear Receptors

Most receptors for extracellular signals are on the cell surface because the signaling molecules are hydrophilic and unable to penetrate the plasma membrane. Other molecules can cross the plasma membrane directly and interact with cytoplasmic or nuclear receptors, and these often initiate simple signal transduction pathways. Examples of such signals include the steroid and thyroid hormones and retinoic acid, which are lipid-soluble, and the gas nitric oxide (q. v. receptor guanylate cyclases) (Figure 7-13).

Figure 7-13　The structure and function of intracellular receptor (see page 233)

2.3　Defining a Receptor

- Specificity—A receptor must be able to distinguish between often closely—related signals.
- High affinity—Signals are often present in low concentrations. The effective receptors can often detect nmol $\cdot$ L^{-1} ～pmol $\cdot$ L^{-1} concentrations.
- Saturability—A cell has a finite number of receptors and, thus there is a limit to the number of ligand molecules a cell can bind.
- Reversibility—Ligand-receptor association is not covalent. When the ligand concentration drops, the complex can dissociate.

2.4　Two Basic Forms of Signaling are Recognized

An amphiphilic chemical messenger diffuses across the plasma membrane. Inside the cell the messenger molecules interact with specific receptors (binding proteins). In the ligand-bound state these receptors regulate the transcription of specific messenger RNAs and thereby control the expression of key enzymes, receptors or transporters.

The chemical messenger is unable to cross the plasma membrane because of low lipid solubility. The messenger binds to specific plasma membrane receptors (integral membrane proteins). In the ligand-bound state the receptors activate either an intrinsic ion channel (ionotropic receptors) or a defined sequence of enzymes (metabotropic receptors).

Section 3　Cellular Signaling Pathway

Since the ultimate targets of the signal transduction pathway are usually not found at the plasma membrane, there must be further intracellular signal transduction. This is often mediated by a cascade of sequential enzyme activation: one signaling component phosphorylates and activates a second component, which performs the same process on a third, and so on until the ultimate target is reached. The signaling process may also be inhibitory.

3.1　cAMP Signaling

The discovery of the physiological roles of cyclic AMP (cyclic $3'$, $5'$ adenosine monophosphate) in the 1960s, can be considered as the initial foray into the then unknown field of signal transduction. As the years have passed, our understanding of the importance of cAMP as a central second messenger in cells has continued to grow. The primary cAMP target is protein kinase A (PKA).

Over the years, cyclic AMP has been shown to regulate a diverse variety of processes that include: transport across the plasma membrane, assembly/disassembly of microtubules, glucose metabolism, fatty acid formation, protein synthesis, DNA synthesis and gene regulation.

Cyclic AMP is made from ATP by the action of adenylyl cyclase (also called adenyl cyclase and adenylate cyclase). The key to cAMP's function is the cyclic phosphate group. Cyclic AMP is subsequently broken down by cAMP phosphodiesterase to $5'$AMP which does not have the cyclic phosphate group and does not regulate cellular processes to any great extent. This cycle of synthesis and degradation are very precise regulation of events controlled by cAMP. Here's a diagram of the basic reaction.

A ligand binds to a receptor causing a conformational change that allows the G protein to associate with it. This interaction leads to a change in the conformation of the G protein as subunit causing GTP to replace GDP which leads to the dissociation of the GTP-Gαs from the G β/γ subunits. The Gα and Gγ subunits are linked to the membrane by lipid groups. The "s" on the Gα subunit indicates it stimulates cAMP synthesis by adenylyl cyclase. In keeping with this, the GTP-Gαs subunit binds to adenylyl cyclase (AC). This activates AC leading to the synthesis of cAMP from ATP.

Protein kinase A is a primary target of cAMP in a diverse number of tissues and cell types. In its inactive state two regulatory subunits keep two substrate binding (catalytic) subunits in an inactive state. A pseudosubstrate domain in the regulatory subunits binds to the catalytic domain of the catalytic subunits. Binding of cAMP to the regulatory subunits changes the conformation of the pseudosubstrate domains resulting it the disassociation of the now active PKA catalytic subunits. The active PKA can now phosphorylate target proteins.

In addition to the two targets mentioned above, PKA has many other substrate proteins which it can phosphorylate. These include protein phosphatase-1 (regulation of glucose metabolism), heart muscle troponin (contraction), myosin light chain kinase (muscle contraction), phosphofructokinase (anaerobic metabolism) and CREB (transcription factor) (Figure 7-14).

Figure 7-14 The functions of PKG (see page 236)

3.2 Calcium Signaling

Increases in the concentration of calcium in the cytosol provides a signal that can initiate muscle contraction, vision and other signaling pathways. The response depends on the cell type. In muscle, a transient rise in the cytosolic calcium levels (from opening calcium channels in the sarcoplasmic reticulum) causes contraction. This signaling in contraction is a direct consequence of electrical activation of the voltage-gated channel.

Calcium signaling is also involved in the response to growth factors. Normally, cells maintain low calcium levels in the cytosol. A low cytosolic calcium level is maintained by pumps that use ATP hydrolysis to move Ca^{2+} out of the cytosol. Ca^{2+} concentration in the cytosol increases by activating a calcium channel that lets Ca^{2+} flow back into the cytosol.

Calcium signaling can be activated directly by regulating Ca^{2+} channels. However, there is an indirect way to cause cytosolic calcium to rise. The inositol phosphate signaling pathway can also activate calcium signaling in response to a number of hormones and effectors. Phosphatidyl inositol (PI) is a phospholipid found on the plasma membrane. A kinase phosphorylates the head group leading to inositol bisphosphate (IP_2). When an extracellular signal activates phospholipase C (PLC), it cleaves the PIP_2 into IP_3 (inositol triphosphate) and diacylglycerol (DAG). IP_3 activates a Ca^{2+} channel in the ER, and Ca^{2+} comes rushing out. The increased calcium causes protein kinase C (PKC) to bind Ca^{2+}, move to the plasma membrane, and combine with DAG. The C kinase is responsible for activating the final effector—generally activating transcription through transcription factors.

Calcium can also directly affect signaling through another pathway. The major calcium binding protein in the cell is calmodulin (CAM). CAM is not an enzyme, but it will activate some enzymes when it binds to them. CAM binds to its target enzymes only when calcium is bound so it propagates the calcium signal. This can be used to integrate signals in glycogen breakdown, but CAM can also activate a specific cellular kinase (serine/threonine), called CAM kinase (calmodulin-dependent kinase). This kinase then carries the signal.

3.3 MAP Kinase Signaling

MAP kinase (mitogen-activated protein kinase, also known as Erk, extracellular signal regulated kinase) is a serine/threonine protein kinase activated by many growth factors. The pathway to MAP kinase activation involves Ras and Raf, with Raf phosphorylating a kinase upstream of MAP kinase, termed MAP kinase kinase (MAPKK) or Mek (MAPK/Erk kinase). MAP kinase requires phosphorylation of tyrosine and threonine residues to become activated, and Mek is an example of a dual specificity kinase, with the combined activities of both a tyrosine kinase and a serine/threonine kinase.

The MAP kinase pathway channels mitogenic signals (signals driving cell proliferation) from the cell surface to the nucleus, and many components of the pathway are therefore oncogenic when inappropriately activated. The activation of MAP kinase causes it to be translocated to the nucleus, where it phosphorylates and activates a number of transcriptional regulators including Elk-1, C/EBP-β and c-Myc. MAP kinase also phosphorylates a further kinase called Rsk which translocates to the nucleus and may activate other transcription regulators, such as the serum response factor (SW).

The MAP kinase pathway is highly conserved in eukaryotes. As well as mediating an important growth response in mammals, pathways with homologous components are found in *Drosophila* and *C. elegans* (where they control cellular differentiation) and in yeast, where the function is environmental monitoring. In vertebrates, there is considerable redundancy of signaling components in the MAP kinase pathway, with multiple genes encoding the adaptors, Ras activating and inhibiting proteins, Ras and Raf, Mek and MAP kinase itself. These components may have varying substrate specificities and some are cell type specific or developmentally regulated, allowing them to play specific roles in signal transduction.

3.4　Notch Signaling Pathway

Notch is one of the fundamental signaling pathways that regulate metazoan development and adult tissue homeostasis. The Notch mutant was initially described in *Drosophila*, based on its dominant wing-notching phenotype (Figure 7-15). The study of the embryonic lethal phenotype caused by complete lack of Notch function and its complex allelic series and genetic interactions brought Notch to the forefront, so that in the mid-1980s the *Drosophila* Notch gene product was identified.

Figure 7-15　Notch in the wings of *Drosophila* (see page 237)

This pathway is found throughout the animal kingdom. It differs from many of the other signaling pathways discussed here in that the ligands as well as their receptors are transmembrane proteins embedded in the plasma membrane of cells. Thus, signaling in this pathway requires direct cell to cell contact.

Notch proteins are single-pass transmembrane glycoproteins. They are encoded by four genes in vertebrates. However, the first notch gene was discovered in *Drosophila* where its mutation produced notches in the wings. Their ligands are also single-pass transmembrane proteins. There are many of them and often several versions within a family (such as the serrate and delta protein families).

The Notch signaling pathway is evolutionarily conserved and the basic molecular players in this pathway are ligands (Delta and Jagged), Notch receptors, and the transcription factors. Notch is a transmembrane heterodimeric receptor and there are four distinct members (Notch1 to Notch4) in humans and rodents. In a physiologic condition, binding of the Notch ligand to its receptor initiates Notch signaling by releasing the intracellular domain of the Notch receptor (Notch-I C) through a cascade of proteolytic cleavages by both α-secretase (also called tumor necrosis factor-α-converting enzyme) and γ-secretase. The released intracellular Notch-I C then translocates into the nucleus where it modulates gene expression primarily by binding to a ubiquitous transcription factor, CBF1, suppressor of hairless, Lag-1 (CSL). This binding recruits transcription activators to the CSL complex and converts it from a transcriptional repressor into an activator, which turns on several downstream effectors (Figure 7-16).

Figure 7-16　Notch signaling pathway (see page 238)

It would appear that proper development of virtually all organs (brain, pancreas, GI tract, heart, blood vessels, mammary glands—to name a few) depends on notch signaling. Notch signaling appears to be a mechanism by which one cell tells an adjacent cell which path of differentiation to take (or not take).

The physiologic functions of Notch signaling are multifaceted, including maintenance of stem cells, specification of cell fate, and regulation of differentiation in development as well as in oncogenesis. Defects in notch signaling have been implicated in some cancers, e. g. melanoma.

3.5　Wnt Signaling

Wnt proteins are cysteine-rich glycoproteins that interact with coreceptors low-density lipoprotein receptor-related protein (LRP) 5/6 and Frizzled, a seven-span transmembrane protein. Some Wnts (Wnt1, Wnt3, Wnt3a) stimulate β-catenin signaling by triggering the release of β-catenin from its destruction complex.

Frizzled receptors, like GPCRs, are transmembrane proteins that wind 7 times back and forth through the plasma membrane. Their ligand-binding site is exposed outside the surface of the cell. Their effector site extends into the cytosol. Their ligands are Wnt proteins. These get their name from two of the first to be discovered, proteins encoded by wingless (wg) in *Drosophila* and its homolog Int-1 in mice.

The roles of β-catenin molecules connect actin filaments to the cadherins that make up adherens junctions that bind cells together. Any excess β-catenin is quickly destroyed by a multiprotein degradation complex of which one component is the protein encoded by the APC tumor suppressor gene.

Phosphorylated β-catenin can have ubiquitin molecules attached to prepare it for destruction in proteasomes. But undegraded β-catenin takes on a second function: it becomes a potent transcription factor. The binding of a Wnt ligand to Frizzled (done with the aid of cofactors) activates Frizzled. This, in turn, activates a cytosolic protein called Dishevelled. Activated Dishevelled inhibits the β-catenin degradation complex so β-catenin escapes destruction by proteasomes and is free to enter the nucleus where it binds to the promoters and/or enhancers of its target genes. Wnt-controlled gene expression plays many roles in embryonic development and regeneration as well as regulating activities in the adult body.

3.6   Hedgehog Signaling Pathway

The hedgehog (Hh) family of secreted proteins regulates many developmental processes in both vertebrates and invertebrates. The Hh gene was first identified in *Drosophila* because of its role in embryonic segment polarity and was later shown to act in other aspects of *Drosophila* development, such as patterning of the imaginal discs. Soon after the molecular identification of the *Drosophila* hh gene, which showed that it encodes an unusual secreted protein, vertebrate homologs of Hh were identified in chick and mouse, and were implicated in patterning of the limb and the neural tube.

Mammals have three hedgehog genes encoding three different secreted hedgehog proteins (Hh). However, hedgehog was first identified in *Drosophila*, and the bristly phenotype produced by mutations in the gene gave rise to the name. Patched (Ptc) receptor is a 12-pass transmembrane protein embedded in the plasma membrane.

In mammals, when there is no hedgehog protein present, the patched receptors bind a second transmembrane protein called smoothened (Smo). However, when Hh protein binds to patched, the Smo protein separates from Ptc enabling Smo to activate a zinc-finger transcription factor designated GLI. GLI migrates into the nucleus when it activates a variety of target genes. Hedgehog signaling plays many important developmental roles in the animal kingdom. For example, wing development in *Drosophila* development of the brain, GI tract, fingers and toes in mammals. Mutations or other sorts of regulatory errors in the hedgehog pathway are associated with a number of birth defects as well as some cancers. Basal-cell carcinoma, the most common skin cancer (and, in fact, the most common of all cancers in much of the world), usually reveals mutations causing extra-high hedgehog or suppressed patched activity (both leading to elevated GLI activity).

3.7   NF-κB Signaling Pathway

NF-κB transcription factors play an important role in the regulation of immune response, embryo and cell lineage development, cell apoptosis, inflammation, cell cycle progression, oncogenesis, viral replication, and various autoimmune diseases. The activation of NF-κB is thought to be part of a stress response as it is activated by a variety of stimuli that include growth factors, cytokines, lymphokines, UV, pharmacological agents, and other stresses. In its inactive form NF-κB is sequestered in the cytoplasm, bound by members of the IκB family of inhibitor proteins. The various stimuli that activate NF-κB cause phosphorylation of IκB, which is followed by its ubiquitination and subsequent degradation. IκB proteins are phosphorylated by a IκB kinase complex consisting of IKKalpha, IKKbeta, and IKKgamma. This phosphorylation results in the exposure of the nuclear localization signals (NLS) on the NF-κB subunits and the subsequent translocation of the molecule to the nucleus. In the nucleus, NF-κB binds with a consensus sequence (5'-GGGACTTTCC-3') of various genes, activating their transcription.

A total of five NF-κB subunits that form dimers have been identified in mammalian cell; RelA (p65), p50, RelB, c-Rel, and p52. The most common and best characterized form of NF-κB is the p65-p50 heterdimer. Each dimer combination exhibits differences in DNA binding affinity and transactivation potential.

NF-κB often exists as a heterodimer of p50 and p65 subunits and is sequestered in the cytoplasm as an inactive complex bound to an endogenous inhibitor IκB (13, 14). Following cellular stimulation, IκB proteins are phosphorylated by IκB kinase. The phosphorylation of IκB promotes its ubiquitination and degradation through the 26S proteasome). Degradation of IκB protein liberates NF-κB, allowing it to translocate to the nucleus and bind to DNA to activate transcription of responsive genes involved in anti-apoptosis (Bcl-2, cFLIP, and TRAF1) and proliferation (cyclin D1 and c-Myc) (Figure 7-17).

Figure 7-17   NF-κB signaling pathway (see page 239)

3.8   Signal Transduction Through Cytoplasmic and Nuclear Receptors

The steroid hormone receptor superfamily includes receptors for steroid and thyroid hormones, and vitamins A and D and their derivatives, including the important developmental molecule retinoic acid. Interaction with the ligand causes a conformational change (transformation) which stimulates DNA binding activity. Some receptors are located in the cytoplasm (e. g. the glucocorticoid receptor) (Figure 7-18), and transformation allows nuclear translocation. Others are already present in the nucleus. The transformed receptors are transcription factors and interact with DNA through a highly conserved zinc-binding domain. They can be divided into several families based on the architecture of the recognition site. The activity of some receptors can be regulated by phosphorylation as well as ligand-binding (e. g. the progesterone and estrogen receptors), allowing integration with other signaling pathways.

Figure 7-18   The mechanism of glucocorticoid receptor (see page 240)

Section 4   Features of Signal-Transducing Systems

4.1   Specificity

Signal molecule fits binding site on its complementary receptor;
other signals do not fit. Signal transductions are remarkably specific and exquisitely sensitive. Specificity is achieved by precise molecular complementarity between the signal and receptor molecules, mediated by the same kinds of weak (noncovalent) forces that mediate enzyme-substrate and antigen-antibody interactions.

Multicellular organisms have an additional level of specificity, because the receptors for a given signal, or the intracellular targets of a given signal pathway, are present only in certain cell types.

4.2   Amplification

When enzymes activate enzymes, the number of affected molecules increases geometrically in an enzyme cascade. Amplification by enzyme cascades results when an enzyme associated with a signal receptor is activated and, in turn, catalyzes the activation of many molecules of a second enzyme, each of which activates many molecules of a third enzyme, and so on. Such cascades can produce amplifications of several orders of magnitude within milliseconds.

4.3   Desensitization/Adaptation

Receptor activation triggers a feedback circuit that shuts off the receptor or removes it from the cell surface. The sensitivity of receptor systems is subject to modification. When a signal is present continuously, desensitization of the receptor system occurs; when the stimulus falls below a certain threshold, the system again becomes sensitive. Think of what happens to your visual transduction system when you walk from bright sunlight into a darkened room or from darkness into the light.

4.4   Integration

A final noteworthy feature of signal-transducing systems is integration the ability of the system to receive multiple signals and produce a unified response appropriate to the needs of the cell or organism. Different signaling pathways converse with each other at several levels, generating a wealth of interactions that maintain homeostasis in the cell and the organism. When two signals have opposite effects on a metabolic characteristic such as the concentration of a second messenger X, or the membrane potential Vm, the regulatory outcome results from the integrated input from both receptors.

4.5 Receptor-Ligand Promiscuity

Where ligands activate multiple receptors or many ligands activate the same receptor, or where receptors comprise multichain oligomers with distinct signaling specificities.

4.6 Divergence

Where the stimulus of a receptor activates two parallel pathways. An example is the activation of the Ras-Raf-MAP kinase pathway and phospholipase C-γ (which increases levels of the second messengers IP_3 and DAG) by RTK activity. This is mediated by different signaling molecules possessing domains, in this case the SH2 domain, which can interact with the activated receptor.

4.7 Cross-Talk

Where one pathway branches off and interacts with another. All the major signaling pathways in the cell use protein kinases and phosphatases. There is a high level of interaction between them, e. g. protein kinase A, which is activated by G-protein mediated increases in cAMP levels, inactivates Ras. Conversely, protein kinase C, which is calcium-dependent, stimulates Ras. The major second-messenger systems of the cell are also interdependent: IP_3 induces calcium transport from the ER and DAG cooperates with calcium to activate protein kinase C. The calcium-dependent molecule CaM regulates the activities of PI (3) kinase and adenylate cyclase, which control the levels of IP_3 and cAMP, respectively. cAMP can activate ion channels and hence influence the levels of Ca^{2+} in the cytoplasm.

With these myriad interconnections superimposed upon a regulatory network of broad specificity kinases and phosphatases, it is a wonder that any signaling specificity is maintained at all; the mechanisms of signaling specificity are a major topic of current research. However, cells synthesize only a subset of the many signaling molecules that have been described, so that individual cells can respond only to certain signals and can respond through pathways restricted by the particular components and active in the cell. Another mechanism which could be used to regulate signal response is the global regulation of gene expression. Thus activated transcription factors in the nucleus would find different arrays of genes available to them through selective epigenetic silencing.

Section 5　Signal-Transduction and Diseases

In light of their complexity, it comes as no surprise that signal-transduction pathways occasionally fail, leading to pathological or disease states. Cancer, a set of diseases characterized by uncontrolled or inappropriate cell growth, is strongly associated with defects in signal-transduction proteins. Indeed, the study of cancer, particularly cancer caused by certain viruses, has contributed greatly to our understanding of signal-transduction proteins and pathways.

Impaired GTPase activity in a regulatory protein also can lead to cancer. Indeed, Ras is one of the genes most commonly mutated in human tumors. Mammalian cells contain three 21 kDa Ras proteins (H-Ras, K-Ras, and N-Ras) that cycle between GTP and GDP forms. The most common mutations in tumors lead to a loss of the ability to hydrolyze GTP. Thus, the Ras protein is trapped in the "on" position and continues to stimulate cell growth.

We consider here some pathologies of the G-protein dependent signal pathways. Let us first consider the mechanism of action of the cholera toxin, secreted by the intestinal bacterium Vibrio cholera. Cholera is an acute diarrheal disease that can be life threatening. It causes voluminous secretion of electrolytes and fluids from the intestines of infected persons. The cholera toxin, choleragen, is a protein composed of two functional units—a B subunit that binds to G_{M1} gangliosides of the intestinal epithelium and a catalytic A subunit that enters the cell. The A subunit catalyzes the covalent modification of a $G_{\alpha s}$ protein: the α-subunit is modified by the attachment of an ADP-ribose to an arginine residue. This modification stabilizes the GTP-bound form of $G_{\alpha s}$, trapping the molecule in the active conformation. The active G protein, in turn, continuously activates protein kinase A. PKA opens a chloride channel and inhibits the Na^+-H^+ exchanger by phosphorylation. The net result of the phosphorylation of these channels is an excessive loss of NaCl and the loss of large amounts of water into the intestine. Patients suffering from cholera for $4\sim6$ days may pass as much as twice their body weight in fluid. Treatment consists of rehydration with a glucose-electrolyte solution.

Summary

Signal-transduction cascades mediate the sensing and processing of stimuli. These molecular circuits detect, amplify, and integrate diverse external signals to generate responses such as changes in enzyme activity, gene expression, or ion channel activity. By understanding cell signaling, diseases can be treated effectively and, theoretically, artificial tissues could be built.

Membrane receptors transfer information from the environment to the cell's interior. A few nonpolar signal molecules such as estrogens and other steroid hormones are able to diffuse through the cell membranes and, hence, enter the cell. Once inside the cell, these molecules can bind to proteins that interact directly with DNA and modulate gene transcription.

Second messengers relay information from the receptor-ligand complex. Changes in the concentration of small molecules, called second messengers, constitute the next step in the molecular information circuit. Particularly important second messengers include cyclic AMP and cyclic GMP, calcium ion, inositol 1,4,5-trisphosphate (IP_3), and diacylglycerol (DAG).

Protein phosphorylation is a common means of information transfer. Many second messengers elicit responses by activating protein kinases. These enzymes transfer phosphoryl groups from ATP to specific serine, threonine and tyrosine residues in proteins.

Protein phosphatases are one mechanism for the termination of a signaling process. After a signaling process has been initiated and the information has been transduced to affect other cellular processes, the signaling processes must be terminated. Without such termination, cells lose their responsiveness to new signals. Moreover, signaling processes that fail to be terminated properly may lead to uncontrolled cell growth and the possibility of cancer.

(Jun Yao)

第八章 细胞周期与细胞凋亡

第一节 细胞周期及调控

细胞周期（cell cycle）是指连续分裂的细胞从一次有丝分裂结束到下一次有丝分裂完成所经历的整个序贯过程。在这一过程中，细胞的遗传物质经过复制平均分配到2个子细胞中。细胞周期中每一事件都是有规律、精确地发生，并且在时间与空间上受到严格调控。

一、细胞周期时相

细胞周期大致可分4期：G_1 期，DNA 合成前期，细胞开始生长；S 期，DNA 合成期；G_2 期，DNA 合成后期；M 期，有丝分裂期（图 8-1）。

除了上述 4 个时期，G_1 期的细胞在进入 DNA 复制之前，可以进入静止期，即 G_0 期。实际上，位于 G_0 期的非增殖细胞是机体的主体。而特定环境中的生长因子和有丝分裂原则推动细胞由 G_0 期进入增殖状态。哺乳动物细胞进入细胞周期依赖于细胞外信号，即促增殖信号和促分化信号的平衡。

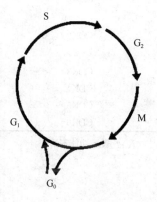

图 8-1 细胞周期

在每一个细胞周期，两个关键事件必须协调：DNA 复制和有丝分裂，使每一个子代细胞得到相同的 DNA 拷贝。

二、细胞周期时相的调控因子及作用机制

细胞周期的运转是一个有序的基因调控过程（$G_1 \rightarrow S \rightarrow G_2 \rightarrow M$），这是与细胞周期相关的基因在时间和空间上进行有序表达的结果，不能正常完成一个周期的细胞将死亡或发生癌变。2001 年，诺贝尔生理及医学奖授予了在这一领域有突出贡献的科学家利兰·哈特韦尔（Leland Hartwell），他发现和研究了细胞分裂周期基因（cell division cycle gene，*cdc*）；保罗·纳斯（Paul Nurse）和蒂姆·亨特（Tim Hunt），他们分别发现了调节细胞周期的关键分子周期蛋白依赖性激酶（cyclin-dependent kinase，CDK）以及周期蛋白（cyclin）。三位先驱的成果，构筑了细胞周期调控机制的框架，并迅速推动了细胞周期、肿瘤发生与抑制的机制等方面的研究。

（一）CDK、Cyclin 和 CKI

哺乳类细胞周期的运行主要受 CDK 的调控。这是一组丝/苏氨酸蛋白激酶，它的周期性激活与失活是推动细胞周期运行的主要因素。CDK 必须结合 Cyclin 的调节亚基。有活性的全酶由 1 分子催化亚基（CDK）和 1 分子调节亚基（Cyclin）组成，称为

CDK-Cyclin 复合体。Cyclin 和 CDK 形成的二聚体复合物为细胞周期调节的核心，但在细胞周期的不同时期形成 CDK-Cyclin 复合体是不同的。近年来的研究显示，这一机制还涉及 CDK 抑制因子（CDK inhibitor，CKI），CKI 可以和 CDK 或 CDK-Cyclin 复合体结合并抑制其活性，从而对细胞周期进展起负调控作用。Cyclin、CDK、CKI 多因素相互协同作用，并通过 pRb 途径和 p53 途径共同影响细胞周期进程。

目前，已发现的 CDK 家族主要有 CDK1、CDK2、CDK3、CDK4、CDK5、CDK6 和 CDK7；Cyclin 可分为 A、B、C、D、E，其中 D 又分为 D1、D2 和 D3；各 CDK 只能与相应的 Cyclin 结合（表8-1，图 8-2）。

表 8-1　哺乳动物的各细胞周期时相中存在的 CDK 和 Cyclin

CDK	Cyclin	细胞周期时相
CDK1	A(432 个氨基酸)	G_2
	B(433 个氨基酸)	G_2，M
CDK2	A	S
	E(395 个氨基酸)	晚 G_1
CDK3	?	?
CDK4	D(295 个氨基酸)	早 G_1
CDK5	（存在于神经元中，相当于体细胞中的 CDK1）	
CDK6	D	早 G_1

注：? 表示功能不清楚。

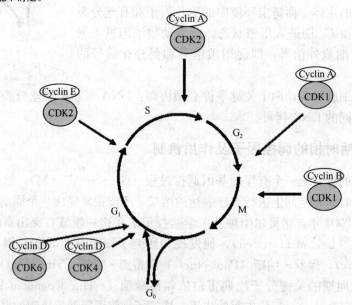

图 8-2　细胞周期中 Cyclin 和 CDK 的表达

（二）CDK-Cyclin 复合体的活性调节

在细胞周期各时相中，特异的 CDK 含量基本恒定，CDK-Cyclin 复合体的活性高低决定于下列几种因素：①相应 Cyclin 水平的高低；②CDK-Cyclin 复合体的磷酸化水平；③CKI 含量的高低。至于 Cyclin 和 CKI 含量的变化，决定于二者的合成和分解速率。

1. Cyclin 的合成与降解

Cyclin 基因表现出周期性地表达变化，当处于静止期（G_0）的哺乳动物细胞受生长因子等刺激后，首先出现 Cyclin D，并同时诱导 CDK4 及 CDK6 的表达，结合成相应的 CDK4-Cyclin D 及 CDK6-Cyclin D 复合物，其主要底物是 Rb 及相关蛋白质，促使细胞通过细胞周期 G_1 的限制点（restriction point，R）。在 G_1 中/晚期，Cyclin E 表达，并与 CDK2 结合成 CDK2-Cyclin E，使 DNA 复制的启动因子磷酸化而被激活，促进越过 G_1/S 期控制点（checkpoint）进入 S 期；Cyclin A 在 Cyclin E 出现之后也被表达，并与 CDK2 形成复合物，在 S 期内，以 CDK2-Cyclin A 的作用为主。在 S 期晚期和 G_2 期，Cyclin B 表达。G_2/M 期时则表达 CDK1，以 CDK1-Cyclin B 及 CDK1-Cyclin A 复合物为主。

泛素（ubiquitin）化降解系统是细胞周期组分降解的主要途径。两种结构和功能相似的复合体 SCF（Skp-Cullin-F-box）和 APC（anaphase-promoting complex）负责介导细胞周期组分的泛素化降解。SCF 作用在 G_1 期末期、S 期以及 G_2 期早期；而 APC 在 G_2 期末活化并介导周期进入分裂期。目标组分泛素化降解的靶向性是通过一些因子与复合体相互作用而实现的。例如，SCF 需要含有 F-盒子基序的蛋白质家族，而 APC 则需要含有 WD-40 重复序列的特异性因子。在 Cyclin 的近 N 端有一区段与其降解有关，称降解盒。当 CDK 活性全盛时，其蛋白激酶可使一种识别蛋白磷酸化而激活，该识别蛋白专门识别和结合 Cyclin 的降解盒，随后 Cyclin 通过泛素化途径而被降解。或者当 CDK-Cyclin 复合体活性达到高峰时，除了磷酸化激活有关靶分子外，亦可催化复合体自身的 Cyclin 磷酸化，促使后者被泛素化降解。随着 Cyclin 的降解，CDK 也就失去活性，因而 Cyclin 的降解是 CDK 活性增高后的一个自我调节过程。另外，还有磷酸酶可专门使磷酸化的识别蛋白脱磷酸化，以终止其降解活动。在细胞周期顺序推进时，各时相 Cyclin 的降解和下一时相特征性 Cyclin 的积聚合成都交替地进行着。

2. CDK-Cyclin 复合体的磷酸化/脱磷酸化

CDK 和 Cyclin 均受到磷酸化的调节。而且 CDK-Cyclin 复合体形成之后仅具有部分活性，只有当 CDK 被磷酸化后，CDK-Cyclin 复合体才完全活化。使 CDK 磷酸化的激酶是 CAK（CDK activating kinase）。CAK 的活性在细胞周期中稳定，但在部分细胞由 G_0 期到 S 期的转变过程中被诱导表达。CDK 相关的磷酸酶 KAP（CDK-associated protein phosphatase）可拮抗 CDK 被 CAK 磷酸化的作用。KAP 的作用对象是 Cyclin 降解后形成的 CDK 单体。

磷酸化也可以负性调节 CDK 活性。这种抑制性的磷酸化出现在所有 CDK 的 N 端。即便是存在 CAK 介导的激活性磷酸化，这些位点的磷酸化仍会抑制 CDK 活性。

3. CKI

CKI 是 CDK 抑制分子，可阻止细胞通过限制点，具有抑癌基因的活性。但与抑癌基因 $p53$ 不同，CKI 的作用方式是直接与 CDK 或 Cyclin-CDK 复合物结合，调节细胞周期进程。目前已发现 7 种 CKI，分为 INK4（inhibitor of CDK4）和 CIP/KIP（CIP：CDK-interacting protein；KIP：kinase inhibiting protein）两大家族。INK4 家族，又称 p16 家族，包括 p15、p16、p18、p19，它们同 CDK4 和 CDK6 结合，能够特异性抑制 CDK4-CyclinD、CDK6-CyclinD1 的活性。CIP/KIP 家族，又称 p21 家族，包括 p21、 p27、p57 等，能广泛抑制 CDK-Cyclin 的作用。抗增殖、促分化和促细胞生存的信号可以诱导 CKI 的表达。

（三）控制点的概念及调控

在细胞周期中有一个限制点和两个控制点，对正常细胞周期的运行具有重要的调控作用。控制点是确保真核细胞准确分裂的调控机制。在细胞周期中，后一个事件的发生依赖于前一个事件的完成。例如，有丝分裂只有在 DNA 复制完成后才能进行。这种依赖性的调控机制就称为"控制点"。现在认为，控制点在细胞周期中广泛存在，但主要的是 G_1/S 控制点和 G_2/M 控制点。限制点位于 G_1 时相的中间区。

1. G_1/S 控制点

G_1/S 控制点位于 G_1 期晚期，控制 G_1 期到 S 期的转换。如果环境中缺乏细胞增殖信号，大多数细胞在此时停滞并进入静止期 G_0 期。哺乳动物细胞在 G_0 期受生长因子等分裂原刺激后，CDK4 和 CDK6 与 Cyclin D1 升高，形成的 CDK4,6-Cyclin D1 复合体经磷酸化修饰呈现激酶活性，在 G_1 期中期达到高峰，越过限制点。在 G_1 期晚期还出现 CDK2-Cyclin E 复合体的蛋白激酶。由这两种激酶磷酸化 Rb，释放出转录因子 eE2F，促使许多与 DNA 复制有关的基因表达，因此，细胞越过 G_1/S 控制点而进入 S 期（图 8-3）。

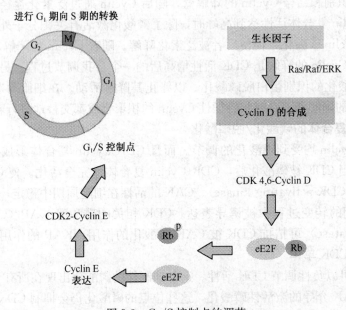

图 8-3　G_1/S 控制点的调节

G_1/S 控制点主要受 p16 的调控。p16 抑制 CDK4/6 使之不能与 Cyclin D1 结合，CDK4 活性的抑制可使细胞停滞于 G_0/G_1 期。当有分裂刺激信号或癌基因促进 Cyclin D1 表达上调时，则可克服 p16 的阻碍作用。一旦 CDK4/6 与 Cyclin D1 结合，则磷酸化抑癌基因产物 Rb，后者失去了对转录因子 eE2F 的抑制作用，eE2F 进而促进 Cyclin E 的表达，并与 CDK2 结合而使细胞进入到 S 期。

2. G₂/M 控制点

G_2/M 控制点位于 G_2 期晚期，控制 M 期的启动。G_2 期到 M 期转换的标志是 CDK2-Cyclin A 复合体的分解和 CDK1-Cyclin B 的形成。在 G_2 期内主要是由 CDK1-Cyclin B 复合体呈现的功能，因为它的功能直接与细胞成熟进行有丝分裂有关，故将 CDK1-Cyclin B 复合体特称为 MPF（maturation promoting factor）。

当哺乳动物细胞处于 G_2 期进入 M 期的控制点时，MPF 被 CAK 磷酸化 T161 位点而激活，MPF 磷酸化下游相应的蛋白质，促使细胞进入有丝分裂。MPF 受 WEE1 激酶和 Cdc25（cell division cyclin 25）磷酸酶的双重调节，WEE1 使 MPF 的 T14、Y15 位点磷酸化从而导致 MPF 失活，Cdc25 则可以水解 MPF 的 T14、Y15 磷酸基团而使其活化。然而，在有丝分裂前，DNA 往往被损伤，为了防止这种损伤传递到子代细胞，细胞会通过使磷酸酶 Cdc25 失活而使细胞周期在控制点暂时停滞。此时细胞进行损伤修复，等待细胞外刺激信号以及必需的生长因子、激素或营养素等。如果细胞损伤不能修复，控制点信号会活化致细胞死亡的通路。因此，由于基因突变、染色体损伤或非整倍体等导致的控制点缺陷，可导致肿瘤发生。

三、细胞周期与衰老和疾病

细胞周期与衰老以及肿瘤、AIDS 等多种人类疾病相关。

衰老细胞中，一些与细胞周期有关的蛋白质的表达多数降低，如 CDK2、Cyclin A 和 Cyclin B 的表达均降低；Cyclin E 虽然表达不受影响，但不易被磷酸化而激活，使 CDK2-Cyclin E 的活性降低，不能使 Rb 磷酸化，则 E2F 仍与 Rb 结合而不能发挥其转录作用，细胞仍不能进入 S 期。

肿瘤和癌症的主要原因是细胞周期失调后导致的细胞无限制增殖。从分子水平看，是由于基因突变致使细胞周期的促进因子不恰当的活化，和（或）抑制因子失活，造成细胞周期调节失控的结果。其中，破坏 R 点的正常控制、使细胞同期调控系统总得到"增殖"的指令，是肿瘤细胞"常耍的手腕"。

第二节　细胞凋亡

细胞凋亡（apoptosis），也称程序性细胞死亡（programmed cell death，PCD），是一个受基因调控的细胞自主的有序的死亡过程。与细胞坏死（necrosis）不同，凋亡涉及一系列基因的激活、表达，是一个主动的、耗能的过程，使得机体清除不需要的或者功能失常的细胞，更好地适应生存环境。

凋亡的异常活化将导致许多疾病，如 AIDS、神经退行性疾病和脑卒中等。相反，凋亡不足也是许多疾病的原因，如肿瘤、自身免疫病和病毒感染。

一、细胞凋亡的特征

形态学特征：凋亡中的细胞表现出的主要形态学变化，包括细胞皱缩和染色体聚集等。基因组断裂，细胞会皱缩，部分胞体会崩解为凋亡小体（apoptotic body）。染色体紧缩呈月牙状，聚集在核膜周围。不像细胞坏死，往往是细胞膨胀、破裂并将内容物释放而导致炎症。凋亡的细胞成分严格地包裹在细胞膜内（胞膜完整）并被吞噬细胞清除，不会导致不必要的炎症。

生化特征：从生物化学角度上看，凋亡细胞的这些形态学特征与磷脂酰丝氨酸转移到细胞膜外侧面、细胞骨架结构的破坏和限制性内切核酸酶的活化有关。细胞凋亡的形态学、生物化学和分子水平的变化见表 8-2。这些特征也被用于细胞凋亡的检测。

表 8-2　细胞凋亡的特征

形态学特征	生化特征
胞膜完整,外形发泡状	胞质内 Ca^{2+} 浓度升高
胞质浓缩,细胞器紧聚	细胞内活性氧增多
染色体紧缩呈月牙状,凝聚在核膜周围	质膜通透性变大
形成膜包的凋亡小体,被附近的细胞或巨噬细胞吞噬消化清除	DNA 内切核酸酶活性被激活升高,双链 DNA 在核小体之间切断形成 185～250 bp 的倍数的有序片段
细胞凋亡可发生在正常细胞群中的单个细胞	Ⅱ型谷氨酰胺转移酶和需钙蛋白酶活性升高

二、细胞凋亡的影响因素

在正常生理情况下，细胞凋亡与细胞增殖一样，是自稳平衡进行的，但亦可受许多诱导因素激发而加强，或受阻抑因素而减弱。细胞凋亡相关因素分诱导性因素和抑制性因素两大类。

1. 凋亡诱导因素

激素和生长因子失衡、理化因素、免疫性因素和微生物等。

2. 凋亡抑制因素

某些激素（ACTH、睾酮）或细胞因子（IL-2，神经生长因子）等的去除；某些二价金属阳离子，如 Zn^{2+}；药物，如苯巴比妥、半胱氨酸蛋白酶抑制剂；病毒，如 EB 病毒、牛痘病毒 CrmA 等以及中性氨基酸，都具有抑制细胞凋亡的作用。

三、参与细胞凋亡的主要分子

细胞凋亡过程中有多种生物分子参与，包括半胱天冬蛋白酶（caspase）、Bcl-2 家族、转录因子（NF-κB、c-Jun、c-Fos）、线粒体通透性转变孔（mitochondrial permeability transition pore，MPTP）以及 Ca^{2+} 等。细胞凋亡相关基因多达数十种，根据功能的不同可将其分为三类：抑制凋亡基因（*EIB*、*IAP*、*Bcl-2*）、促进凋亡基因（*Fas*、*Bax*、*ICE*、*p53*）和双向调控基因（*c-myc*、*Bcl-x*）。

1. Caspase 家族

半胱天冬蛋白酶（cysteinyl aspartic acid-specific protease，Caspase）是一类进化上保守的天冬氨酸残基特异性的半胱氨酸蛋白酶家族，在细胞凋亡过程中 Caspase 是引起细胞形态学和生物学改变的关键执行者。

所有的 Caspase 以酶原的形式合成，以无活性的 Procaspase 形式存在，只有在一定形式下或在上游活性的 Caspase 作用下，才能被分解为有活性的分子。活性的 Caspase 均是由两个大亚基（17～20 kDa）和两个小亚基（10～12 kDa）组成的异源四聚体。激活的 Caspase 可以水解一些参与细胞调节、细胞信号转导、DNA 修复、组织平衡和细胞存活等过程的重要蛋白质，从而使细胞表现为凋亡特有的形态学及生化特征。目前属于这类蛋白质的家族成员约有 14 种，大部分位于胞质中，少量位于线粒体等部位。按功能可分为 3 组：①起始分子：包括 Caspase 2、8、9 和 10。②效应分子：包括 Caspase 3、6、7。起始 Caspase 分子活化效应分子，再作用于细胞内的底物分子，包括限制性内切核酸酶、细胞骨架以及细胞周期蛋白等。③其他：包括 Caspase 1、4、5、11、12、13 和 14。

2. Bcl-2 家族

Bcl-2 家族是最先被注意到与细胞凋亡有密切关系的基因家族之一。Bcl-2 最初发现于 B-淋巴细胞瘤/白血病-2（B cell lymphoma/leukemia-2，Bcl-2）而得名，其分子结构的 N 端含有 4 个 BH 结构域（BcL-2 homologous domain），其 C 端具有质膜锚定作用，借以定位于线粒体膜表面，维护线粒体膜的通透性。目前，在哺乳类细胞中发现的 Bcl-2 家族有 24 个成员。根据它们的结构和功能的不同分为两个亚族：其中以 Bcl-2、Bcl-XL 为代表的 8 个成员形成一个具有抗凋亡作用的 Bcl-2 亚族；而其他以 Bax、Bak、Bad、Bid 等为代表的 16 个成员形成另一个促凋亡作用的 Bcl-2 亚族。Bcl-2 基因家族成员通常以二聚体的形式发挥作用，Bcl-2/Bax 和 Bcl-2/Bcl-XL 均可抑制细胞凋亡，而 Bad/Bax、Bax/Bax 和 Bcl-2/Bcl-XS 则可促进细胞凋亡。抗凋亡和促凋亡二者相互制约，维持平衡状态，平衡一旦打破，就会引起疾病的发生。

3. Fas

自杀相关因子（factor associated suicide，Fas）是广泛表达于正常细胞和肿瘤细胞表面的 I 型受体，是 TNF 受体家族中发现与凋亡有关的第一个成员，其胞质段含有死亡结构域（death domain，DD）及阻抑结构域（suppressive domain，SD）。在 Fas 未结合其配体 FasL 前，该 SD 上结合 Fas 结合磷酸酶-1（fas-associated phosphatase-1，FAP-1），使 Fas 呈抑制状态。Fas 广泛表达于多种细胞上。Fas 通过与其配体 FasL 结合而引起凋亡。

4. *p53*

p53 基因是一种抑癌基因，其基因表达产物的分子质量大小约为 53 kDa，是一种 DNA 转录调节蛋白。p53 蛋白在维持细胞正常生长、抑制恶性增殖中发挥重要作用，其作用机制是多方面的，其中之一是参与细胞周期调控和诱导细胞凋亡。当细胞的 DNA 损伤时，p53 使细胞周期停滞于 G_1 期，即 p53 表达增加使细胞中止增殖以便为损伤的 DNA 进行修复争取时间；如果 DNA 受损严重无法修复，则 p53 蛋白持续增高，导致细胞凋亡。若 p53 蛋白发生突变，则不能抑制细胞增殖和诱导细胞凋亡。

四、细胞凋亡途径

细胞凋亡受两种信号途径的控制，一种为外源性途径，另一种为内源性途径。外源性途径由细胞外配体，如 FasL 或 TRAIL 与相应受体结合而引起；内源性途径由刺激因素导致细胞器（特别是线粒体和内质网）功能受损而引起。

（一）死亡受体介导的细胞外凋亡途径

细胞外凋亡途径由细胞外小分子配体与受体结合而引发。配体包括 FasL（Fas ligand）、肿瘤坏死因子 α（tumor necrosis factor α，TNFα）和肿瘤坏死因子相关的诱导细胞凋亡的配体（TNF related apoptosis inducing ligand，TRAIL）。有 5 种死亡受体：TNF 受体 1、Fas、TRAMP 受体 1、TRAIL 受体 1、TRAIL 受体 2。

死亡受体介导凋亡途径很相似。配体结合诱导受体寡聚化，随后募集衔接分子通过死亡结构域与受体结合。衔接分子与邻近的 Caspase 结合，形成死亡诱导信号复合体（death inducing signaling complex，DISC）。活化的 Caspase 能够直接活化下游效应分子的级联反应，继而作用于细胞内的底物，最后导致细胞死亡。以 Fas 和 FasL 为例，过程如下。

Fas 的配体 FasL 主要表达于 T 淋巴细胞和肿瘤细胞，是细胞膜表面的 II 型受体。Fas 和 FasL 都是以同源三聚体形成存在。Fas 和配体 FasL 结合后，诱导 Fas 胞质段 C 端的 SD 脱离 FAP-1，同时诱导其 DD 结合 Fas 缔合蛋白（Fas associated protein with DD，FADD），FADD 再以其 N 端的死亡效应子结构域（death effector domain，DED）结合 Procaspase 8（或 10），形成一个由 Fas-FADD-Procaspase 8（或 10）三种分子组成的复合体，称为诱导死亡的信号复合体（death inducing signaling complex，DISC），其中 Procaspase 8（或 10）就可自身催化形成活性的 Caspase 8（或 10），接下去进行级联反应，激活下游的靶分子 Procaspases（包括 Procaspase 3，6，7）。活性的 Caspase 8（或 10）尚可催化 Bid（Bcl-2 家族的促凋亡分子）。Bid 的 N 端 1～60 位氨基酸是 Bid 的抑制区段，N 端第 60 位和第 61 位氨基酸之间的肽键断裂，释放出活化的 C 端部分（61～195 位氨基酸）转位到线粒体膜，降低其跨膜压，引起线粒体内细胞色素 c（cytochrome c，Cyt c）和 Procaspase 2，3，7，9 等死亡因子释放出来，在胞质内凋亡蛋白酶激活因子-1（apoptotic protease activating factor-1，Apaf-1）的参与下形成凋亡复合体（apoptosome），其中 Procaspase 9 自身激活，由此扩大 Fas 介导的死亡信号转导效应（图 8-4）。

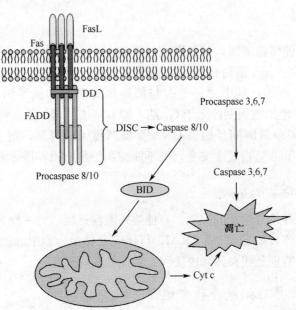

图 8-4　Fas /FasL 介导的凋亡途径

（二）线粒体介导的细胞内凋亡途径

线粒体凋亡途径由一系列的细胞内应激因素活化，如 Bax、氧化剂、钙超载、活化的 Caspase 和神经酰胺等。线粒体介导的凋亡途径如下：凋亡信号使 Cyt c 从膜间腔释放进入胞质，继而与 Apaf-1、Procaspase 9 结合形成凋亡复合物，在 ATP 参与下，Caspase 9 活化。Caspase 9 随后活化效应分子 Caspase 3，6，7，从而导致凋亡（图8-5）。

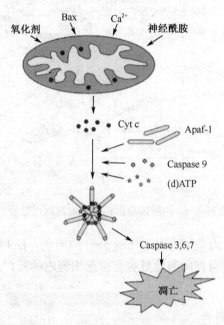

图 8-5　线粒体介导的细胞凋亡途径

除了 Cyt c 之外，还有其他的促凋亡分子从线粒体释放出来，如凋亡诱导因子（apoptosis-inducing factor，AIF）可以促进 DNA 片段化，SMAC（second mitochondria-derived activator of caspase）是线粒体内第二个激活 Caspase 的因子。

（三）内质网介导的细胞内凋亡途径

以上死亡受体活化（外源性途径）和线粒体损伤途径（内源性途径）是细胞内两条经典的凋亡途径。内质网（endoplasmic reticulum，ER）应激启动的凋亡途径是近年才发现的一种新的凋亡途径。内质网调节蛋白质合成、折叠和运输，对应激做出反应，调节胞内 Ca^{2+} 水平，是介导内源性凋亡途径的第二个主要场所。内质网的功能受到各种内外因素的影响，称为内质网应激。内质网通过未折叠蛋白反应、内质网超载反应、内质网有关的降解等方式来应对应激刺激。如果内质网损伤过重而内质网应激反应不能使其继续存在，则诱导凋亡。

严重且持续的内质网应激通过诱导 CHOP（CCAAT enhancer binding protein-homologous protein，也称为 CEBPE）、活化 Caspase 12 和 JNK 途径以及通过 Bax/Bak 调节细胞内 Ca^{2+} 水平而引起凋亡效应。内质网介导的细胞凋亡途径见图 8-6。

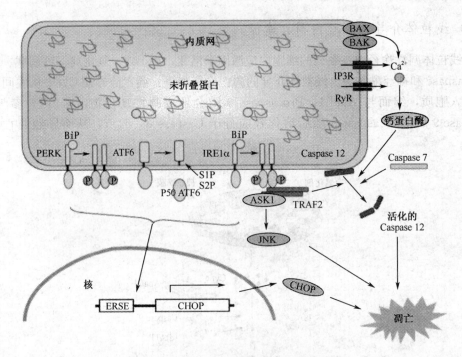

图 8-6　内质网应激诱导的细胞凋亡途径

除了直接介导凋亡，内质网还参与 Fas 介导的凋亡，DNA 损伤和癌基因表达引起的 p53 依赖的途径，内质网钙库的动员也会引起细胞内的死亡途径，以及增强线粒体对凋亡信号的敏感性。

五、病毒感染与细胞凋亡

病毒感染与复制常常导致细胞凋亡，这种效应往往造成感染性疾病的病变。但病毒也会采取各种措施抑制凋亡，以延长被感染细胞的寿命，使病毒的复制、传播和生存达到最大化。

（一）病毒感染导致细胞凋亡

许多细胞因病毒感染而死亡，虽然随后子代病毒的释放也减少了。外源性凋亡信号可以在病毒颗粒穿过细胞膜时被引发。例如，人免疫缺陷病毒（HIV）、牛痘病毒、甲型肝炎病毒（HAV）等与细胞表面的结合就是导致凋亡的。病毒颗粒跨膜时与 TNF 受体相关蛋白 CAR1 结合而使得 TNF 受体介导的死亡信号活化。某些病毒表面蛋白与该受体的结合也足以诱导凋亡。病毒，如 HBV 感染介导细胞表达 Fas/FasL 或上调 Fas、FasL 的表达，促进细胞凋亡。某些病毒感染使 TNF α 产生增加。

病毒感染也可以通过细胞内成分导致内源性凋亡途径，如流感病毒可以激活双链 RNA 活化蛋白激酶 R（PKR），呼吸道合胞病毒则可以引起内质网应激反应，一些病毒可以诱导一些核内的感受器（如 p53），或是引起线粒体损伤。

一些病毒基因产物可与细胞周期关键分子结合而刺激宿主细胞 DNA 复制，同时可稳定 p53 蛋白使细胞内 p53 含量增高。

病毒感染可激活细胞毒性 T 细胞（CTL）介导细胞凋亡。CTL 主要通过两种机制杀死感染细胞：一种是分泌穿孔素使细胞坏死；另一种则是由 CTL 介导靶细胞凋亡。CTL 可通过上调 Fas/FasL 基因表达，激活酸性鞘磷脂酶，产生神经酰胺，神经酰胺进一步激活丝/苏蛋白激酶，通过信号转导激活细胞凋亡通路。CTL 通过分泌因子诱导凋亡。

（二）病毒感染抑制细胞凋亡

通过抑制凋亡来维持宿主细胞生存，这是许多病毒和一些细菌类的病原体在进化中产生的重要的生存策略。其机制包括：编码 Bcl-2 类似物或促进 Bcl-2 表达；编码蛋白酶抑制剂以灭活 Caspase；干扰线粒体 Cyt c 的释放；病毒基因编码蛋白调节 p53 介导的细胞凋亡是多种病毒共同的途径。

六、细胞凋亡与疾病

随着对人类疾病的深入了解，认识到细胞凋亡是机体维持细胞群体数量稳态的重要手段。细胞凋亡失调（凋亡不足和（或）凋亡过度）可成为某些疾病的重要发病机制。

（一）细胞凋亡过度引发的疾病

细胞凋亡导致细胞消失是许多疾病的中心环节，举例如下。

1）获得性免疫缺陷综合征（AIDS）的特征是 $CD4^+$ T 细胞的减少。CD4 是 T 淋巴细胞（免疫系统的重要细胞）表面上的一种糖蛋白。该受体蛋白能结合主要分子完成许多免疫系统的理想任务。令人遗憾的是，它也是 HIV 病毒蛋白 gp120 的受体。病毒蛋白 gp120 均能与 CD4 抗原以高亲和力结合，这种结合没有同时与 MHCⅡ类作用，因而引起 $CD4^+$ T 细胞凋亡，导致免疫系统的崩溃。另外，HIV 感染的 T 细胞对 FasL 和 TRAIL 途径诱导细胞凋亡特别敏感。

2）许多神经元退行性疾病，如阿尔茨海默症、帕金森病、亨廷顿病等，已经发现凋亡过度确实导致了这些疾病的发生。

3）在脑和心脏急性梗塞区域附近的细胞死亡往往是以凋亡的方式进行的。

4）丙型肝炎的肝细胞高表达 Fas，因而易于被细胞毒性 T 细胞破坏。

5）患桥本（Hashimoto）甲状腺炎的甲状腺细胞表达 Fas 和 FasL，在人白细胞介素-1（IL-1）作用下可以互相残杀，因为 IL-1 能诱导 Fas 表达。

6）肿瘤细胞高表达 FasL 于细胞膜或者分泌 FasL，可以与浸润 T 细胞上的 Fas 结合而摧毁 T 细胞，使肿瘤细胞免受抗肿瘤的淋巴细胞作用。

（二）凋亡不足引发的疾病

细胞数目过度增加是肿瘤的特点之一，这是由于细胞过度增殖和细胞凋亡不足引起。抑癌基因 *p53* 和 *Rb* 的缺失或突变（见于 50％的肿瘤），或 Bcl-2 表达上调或 Bax 下调可以导致细胞凋亡不足。例如，低分化滤泡的非霍奇奎淋巴瘤（Hodgkin）就是由于 *Bcl-2* 基因染色体易位而导致其过度表达所引起的。

也有很多报道表明自身免疫疾病与凋亡的关系：凋亡不足或者无法清除凋亡细胞。

在凋亡小体中存在自身抗原，因此凋亡细胞对抗原呈递、天然免疫活化以及调节巨噬细胞的细胞因子释放很重要。由于 Fas 表达缺陷，引起自身反应性 T 细胞阴性选择的凋亡功能丧失，才形成自身免疫性疾病（如系统性红斑狼疮）。

小　　结

细胞周期是指连续分裂的细胞从一次有丝分裂结束到下一次有丝分裂完成所经历的整个序贯过程。大致可分 G_1 期（DNA 合成前期）；S 期（DNA 合成期）；G_2 期（DNA 合成后期）；M 期（有丝分裂期）。G_1 期的细胞在进入 S 期前，可以进入 G_0 期（静止期）。哺乳动物细胞依赖于细胞外的生长因子和有丝分裂原推动细胞由 G_0 期进入细胞周期。

哺乳类细胞周期的运行主要受周期蛋白依赖性激酶和周期蛋白形成的 CDK-Cyclin 复合体的调控。不同细胞周期形成的 CDK-Cyclin 复合体是不同的，它的周期性激活与失活是推动细胞周期运行的主要因素，这一机制还涉及 CDK 抑制因子（CKI）对细胞周期的负调控作用。在细胞周期各时相中，特异的 CDK 含量基本恒定，CDK-Cyclin 复合体的活性高低决定于下列几种因素：①相应 Cyclin 水平的高低；②CDK-Cyclin 复合体的磷酸化调节；③CKI 含量的高低。在细胞周期顺序推进时，各时相 Cyclin 的降解和下一时相特征性 Cyclin 的积聚合成，交替地进行。泛素化降解系统是细胞周期组分降解的主要途径。

正是由于细胞周期各时相中，尤其是运行至两个主要调控点，即 G_1/S 控制点和 G_2/M 控制点时，特异性 CDK-Cyclin 复合物的顺序形成、激活和失活，成为推进细胞周期运行的关键。细胞周期必须依序进行，若某一时相中 DNA 受损，在其未修复前将阻止细胞周期继续运行。

对细胞周期运行及其调控的分子机制的研究有助于阐明衰老以及肿瘤、AIDS 等病理生理现象发生发展的机制。

细胞凋亡，也称程序性细胞死亡，是一个受基因调控的细胞自主的有序的死亡过程。凋亡涉及一系列基因的激活和表达，是主动的、耗能的过程，使得机体清除不需要的或者功能失常的细胞，更好地适应生存环境。细胞凋亡在形态学、生物化学和分子水平具有其特有的特征，这些特征也被用于细胞凋亡的检测。

细胞凋亡过程中有多种生物分子参与，根据功能的不同可将其分为三类：抑制凋亡基因、促进凋亡基因和双向调控基因。Caspase 是一类进化上保守的天冬氨酸残基特异性的半胱氨酸蛋白酶家族，是细胞凋亡过程中引起细胞形态学和生物学改变的关键执行者。所有的 Caspase 以无活性的 Procaspase 形式存在，在一定形式下被激活。

细胞凋亡受两种信号途径的控制，一种为内源性途径；另一种为外源性途径。细胞外凋亡途径由细胞外小分子配体（FasL、TNFα 和 TRAIL）与死亡受体（TNF 受体、Fas、TRAMP 和 TRAIL 受体）结合而引发。通过衔接分子与邻近的 Caspase 结合，形成死亡诱导信号复合体（DISC）。活化的 Caspase 能够直接活化下游效应分子的级联反应，继而作用于细胞内的底物，最后导致细胞凋亡。

线粒体凋亡途径由一系列的细胞内应激因素引起 Cyt c、AIF、SMAC 等从线粒体释放进入胞质，继而与 Apaf-1、Procaspase 9 结合形成凋亡复合物，在 ATP 参与下，活化的 Caspase 9 随后活化效应分子 Caspase 3、6、7，从而导致凋亡。

内质网应激启动的凋亡途径是近年才发现的一种新的凋亡途径。内质网通过未折叠蛋白反应、内质网超载反应、内质网有关的降解等方式来应对应激刺激。严重且持续的内质网应激通过诱导 CHOP、活化 Caspase 12 和 JNK 途径，以及通过 Bax/Bak 调节细胞内 Ca^{2+} 水平而引起凋亡效应。

病毒感染与复制常常导致细胞凋亡，这种效应往往造成感染性疾病的病变。但病毒也会采取各种措施抑制凋亡，以延长被感染细胞的寿命，使病毒的复制、传播和生存达到最大化。凋亡的异常活化将导致许多疾病，如 AIDS、神经退行性疾病和脑卒中等。相反，凋亡不足也是许多疾病的原因，如肿瘤、自身免疫病和病毒感染。

（孙 军 袁 萍）

参 考 文 献

邓耀祖，屈伸. 2002. 医学细胞分子生物学. 北京：科学出版社

刘景生. 1999. 细胞信息与调控. 北京：北京医科大学中国协和医科大学联合出版社

沈大棱，吴超群. 2006. Cell Biology 细胞生物学双语教材. 上海：复旦大学出版社

anchez I S', Dynlacht B D. 2005. New insights into cyclins, CDKs, and cell cycle control. Seminars in Cell & Developmental Biology, 16：311~321

Darnell J E, Lodish H, Baltimore D. 2004. Molecular Cell Biology. 5th ed. New York：Freeman and Company

Lockshin R A, Zakeri Z. 2007. Cell death in health and disease. J Cell Mol Med, 11(6)：1214~1224

Obayaa J, Sedivyb J M. 2002. Regulation of cyclin-Cdk activity in mammalian cells, CMLS, Cell. Mol Life Sci. 59, 126~142

Syeed S A, Vohra H, Gupta A et al. 2001. Apoptosis：Molecular machinery. Current Science, 80 (3)：349~360

Chapter 8　Cell Cycle and Apoptosis

Section 1　General Information and Control of Cell Cycle

The cell cycle means the whole process from the end of one mitosis to the end of the next mitosis in a continuous divided cell. During a cell cycle, DNA will be replicated and distributed into the two daughter cells equally. Every event in the cell cycle takes place regularly and correctly, and is tightly controlled.

1.1　The Phases of Cell Cycle

One standard cell cycle is divided into 4 phases: ①G_1 phase the gap before DNA replication. Cell begins germination. ② S phase DNA synthetic phase. DNA is synthesized. ③ G_2 phase the gap after DNA replication. ④M phase mitosis phase (Figure 8-1).

Figure 8-1　The four phases of cell cycle (see page 255)

Beside these four phases, before commitment to DNA replication, cells in G_1 can enter a resting state called G_0. Actually, Cells in G_0 account for the major part of the non-growing, non-proliferating cells in the human body. The availability of growth factors and mitogens in the immediate environment promote the cell enter into the proliferation state from G_0. In mammalian cells, the re-entry into the cell cycle depends on extracellular signals, namely the balance between mitogenic stimuli (growth factors such as FGF and EGF, thrombin, serum) and differentiating factors (cytokines, TGF-β, contact inhibition, IL-1, TNF-α and cellular stresses such as DNA damaging agents and UV radiation).

During each cell cycle, two key events need to be coordinated: DNA replication by which the genome is partitioned into two identical copies, and mitosis by which one copy is inherited by each of the daughter cells. The mechanisms that control these important tasks are highly conserved in evolution.

1.2　The Regulatory Factors of Cell Cycle

The running of cell cycle is an ordered gene-regulated process, which is the result of ordered expression of cell cycle related genes in time and space. Those cells which can not finish a complete cycle will enter into apoptosis or be transformed. In 2001, the Nobel Prize in Physiology or Medicine was shared by several scientists because of their outstanding work in this field, Leland Hartwell, who discovered and researched the cell division cycle genes; Paul Nurse and Timothy Hunt, who discovered the key regulator of cell cycle, Cyclin-dependent kinases (CDKs) or Cyclins, respectively.

1.2.1　CDKs, Cyclins and CKIs

The cell cycle is majorly controlled by CDKs. CDKs are proline-directed serine/threonine protein kinases. The periodic activation and inactivation of CDKs is the key factor that promotes the operation of cell cycle. However without their Cyclin partners, the CDKs are blind and inactive. Cyclin binding provides the CDK with targeting domains important for substrate selection and subcellular localization, which in turn determine the biological specificity. Recently, CDK inhibitors (CKIs) are found which also regulate the cell cycle as a negative control by binding CDKs or CDK-Cyclin complexes to inhibit their activity.

A variety of CDK: Cyclin complexes are formed during distinct phases of the cell cycle. Up to date, the described CDKs family includes, CDK1, CDK2, CDK3, CDK4, CDK5, CDK6, and CDK7. The Cyclins are divided into 2 groups according to G_1 phase and M phase, Cyclin D, Cyclin E in Cyclin G_1 phase, and Cyclin B in M phase (Table 8-1 and Figure 8-2).

Table 8-1 CDKs and Cyclins exist in different cell cycle phases of mammalian cells

CDKs	Cyclins	cell cycle phases
CDK1	A (432aa)	G_2
	B (433aa)	G_2, M
CDK2	A	S
	E (395aa)	Late G_1
CDK3	ND*	ND*
CDK4	D (295aa)	Early G_1
CDK5	Exist in neuron, corresponding with CDK1 in somatic cell	
CDK6	D	Early G_1

*: not determined.

Figure 8-2 Expression patterns of Cyclins and CDKs activities during the cell cycle (see page 256)

1. 2. 2 The Activity Regulation of CDK-Cyclin Complex

In general, the levels of CDKs are relatively constant throughout the cell cycle, whilst the levels of the Cyclins vary substantially. The activity of CDK-Cyclin complex is decided by the following factors: ①The corresponding level of Cyclins. ② The phosphorylation state of CDK-Cyclin complexes. ③The level of CKIs.

1. 2. 2. 1 The synthesis and degradation of Cyclins

As a general rule, CDK genes are constitutively expressed and CDKs are relatively stable, whereas Cyclin genes show periodic patterns of expression and Cyclins are subject of regulated degradation.

As cells emerge from quiescence in response to mitogenic stimuli, the synthesis of Cyclin D is induced firstly, then CDK4 and CDK6 are induced. Cyclin D assembles with CDK4 and CDK6 to form CDK4-Cyclin D, or CDK6-Cyclin D complexes, whose major substrates are the retinoblastoma protein (Rb) and the related proteins p107 and p130 (Rb pocket proteins), and promotes the cell cycle passing the restriction point of G_1 phase. In mid/late G_1, Cyclin E is induced and forms complexes with CDK2. CDK2-Cyclin E complexes and the resulting kinase activity is required for S phase entry and the initiation of DNA replication. Cyclin A is expressed soon after Cyclin E at the G_1/S boundary and also forms complexes with CDK2, to a lesser extent, with CDK1. The activity of CDK2-Cyclin A is required for S phase transition and control of DNA replication. Cyclin B1 associates with CDK1 and is expressed in late S and G_2 phases; however, CDK1-Cyclin B complexes remain inactive until late G_2 when their activation is required for entry into mitosis. The complexes CDK1-CyclinB and CDK1-CyclinA are important for G_2/M phase.

The ubiquitin-mediated proteasome system is the main pathway employed for the degradation of cell cycle components. Two structurally and functionally similar complexes, the Skp-Cullin-F-box (SCF) complex and the anaphase-promoting complex (APC), target specific cell cycle components for ubiquitination at discrete points in the cell cycle. The SCF complex is employed at the end of G_1, through S and into early G_2 phase, whereas the APC becomes active at the end of G_2 and mediates the transition through mitosis. Target specifity of ubiquitination is provided by a large number of factors that interact with the complexes. In the case of SCF, these factors are comprised by a family of proteins containing an F-box sequence motif. In the case of the APC, recent data indicate the existence of WD-40 repeats in the specificity factors. A fragment close to N terminal of Cyclins, called degradation box, is related with its degradation. The activated CDK can phosphorylate a recognization protein, which specifically recognizes and binds the degradation box of Cyclin, that makes the degradation of Cyclin by ubiquitin-mediated pathway. The Cyclin in a CDK-Cyclin complex also can be phosphorylated and then degraded. With the degradation of Cyclin, CDK lose its activity. So the degradation of Cyclin is regarded as a self-regulation mechanism of CDK. During the ordered promotion of cell cycle, the degradation of Cyclins of last phase and the synthesis of Cyclins of next phase take place alternately.

1. 2. 2. 2 Phosphorylation and dephosphorylation of CDK-Cyclin complexes

Both Cyclins and CDKs are subject to post-translational regulation by phosphorylation. The assembly of a CDK with its corresponding Cyclin yields only a partially active complex, full activity being achieved after phosphorylation of the CDK on a conserved threonine residue proximal to the ATP-binding cleft (Thr172 in CDK4/6, Thr160 in CDK2, and Thr161 in CDK1). Cyclin-CDK binding seems to precede the activating phosphorylation. The kinase responsible for catalyzing the activating threonine phosphorylation has been designated CAK, for CDK-activating kinase. CAK activity appears to be constant throughout the cell cycle, but is induced in some cell types during the transition from G_0 into S. The phosphorylation of CDKs by CAK is antagonized by the action of a specific phosphatase known as KAP. KAP is believed to act on monomeric CDKs that are the result of Cyclin degradation.

Phosphorylation of CDKs can also negatively regulate their kinase activity. The inhibitory phosphorylations occur near the N termini of all CDKs, specifically on Tyr15 of CDK2 and CDK1 and on Tyr17 of CDK4 and CDK6. Phosphorylation of these sites results in the inhibition of CDK activity even in the presence of the CAK-catalyzed activating phosphorylation. Wee1 and Myt1 have been identified as the kinases responsible for the phosphorylation of the inhibitory sites on CDK2 and CDK1. The activities of Wee1 and Myt1 are regulated by both phosphorylation and subcellular localization in a cell cycle dependent manner. The inhibitory phosphorylations of CDKs are removed by the action of the Cdc25 family of protein phosphatases. Interestingly, the Cdc25 proteins are themselves substrates of Cyclin/CDKs, and their phosphorylation stimulates the phosphatase activity.

1. 2. 2. 3 CKIs

CDKs are negatively regulated by a group of functionally related proteins called CDK inhibitors (CKIs). These CKIs fall into two families: the INK4 (inhibitor of CDK4) inhibitors and the Cip/Kip (Cip: CDK-interacting protein, Kip: kinase inhibiting protein) inhibitors. There are four known INK4 family members: $p16^{INK4A}$, $P15^{INK4B}$, $p19^{INK4D}$, and $p18^{INK4C}$, and three known Cip/Kip family members: $p21^{Waf1/Cip1}$, $p27^{Kip1}$, and $p57^{Kip2}$. The INK4 family specifically inhibits CDK4 and CDK6 activity during the G_1 phase of the cell cycle, while the Cip/Kip family can inhibit CDK activity during all phases of the cell cycle.

CKI expression is induced by anti-proliferative signals, differentiation and survival factors.

1. 2. 3　The Definition of Checkpoint and Mechanisms of Cell Cycle Regulation

Cell cycle checkpoints are control mechanisms that ensure the fidelity of cell division in eukaryotic cells. These checkpoints verify whether the processes at each phase of the cell cycle have been accurately completed before progression into the next phase. Checkpoints are now accepted to exist at every single point in the cell cycle. The most important two checkpoints are G_1/S checkpoint and G_2/M checkpoint.

1. 2. 3. 1　G_1/S checkpoint

The first checkpoint is located at the end of the cell cycle's G_1 phase, just before entry into S phase, making the key decision of whether the cell should divide, delay division, or enter a resting stage. Most cells stop at this stage and enter a resting state called G_0. The G_1 checkpoint is where eukaryotes typically arrest the cell cycle if environmental conditions make cell division impossible or if the cell passes into G_0 for an extended period. In animal cells, the G_1 phase checkpoint is called the restriction point, and in yeast cells it is called the start point.

The restriction point is mainly controlled by action of the CKI-p16. This protein inhibits the CDK4/6 and ensures that it can no longer interact with Cyclin D1 to cause the cell cycle progression. In mitogenic signaling induced or oncogenic induced Cyclin D expression, this checkpoint is overcome because the increased expression of Cyclin D allows its interaction with CDK4/6 by competing for binding. Once active CDK4/6-Cyclin D complexes form, they phosphorylate the tumour suppressor retinoblastoma (Rb), this relieves the inhibition of the transcription factor E2F. E2F is then able to cause expression of Cyclin E, which then interacts with CDK2 to allow for G_1/S phase transition. The regulation of G_1/S checkpoint was described in Figure 8-3.

Figure 8-3　The regulation of G_1/S checkpoint by growth factors (see page 258)

1.2.3.2 G_2/M checkpoint

The second checkpoint is located at the end of G_2 phase, triggering the start of the M phase (mitosis). In order for this checkpoint to be passed, the cell has to check a number of factors to ensure that the cell is ready for mitosis. If this checkpoint is passed, the cell initiates the many molecular processes that signal the beginning of mitosis. The G_2/M transition is marked by the degradation of CDK2-Cyclin A complex and the expression of Cyclin B that activates the CDK1 kinase.

In higher eukaryotes, completion of the G_2/M transition requires two Cyclins, A2 and B1, that pair with CDK1. The CDKs associated with this checkpoint are activated by phosphorylation by maturation promoting factor (or mitosis promoting factor, MPF). The MPF activates the CDK in response to environmental conditions being right for the cell and allows the cell to begin DNA replication. The molecular nature of this checkpoint involves an activating phosphatase, known as Cdc25, which under favourable conditions removes the inhibitory phosphates present within the MPF complex. However, DNA is frequently damaged prior to mitosis, and to prevent transmission of this damage to daughter cells, the cell cycle is arrested via inactivation of the Cdc25 phosphatase (via phosphorylation with other protein kinases).

Cells can arrest at cell cycle checkpoints temporarily to allow for: ① Cellular damage to be repaired. ② The dissipation of an exogenous cellular stress signal. ③ Availability of essential growth factors, hormones, or nutrients. Checkpoint signaling may also result in activation of pathways leading to programmed cell death if cellular damage cannot be properly repaired. Defects in cell cycle checkpoints can result in gene mutations, chromosome damage, and aneuploidy, all of which can contribute to tumorigenesis.

1.3 The Association of Cell Cycle with Senescence and Diseases

The cell cycle is associated with senescence and multiple diseases like tumor and AIDS.

In senescent cells, the expression of many proteins involved in cell cycle is decreased, such as CDK2, Cyclin A, and B. The expression of some other proteins, like Cyclin E, is not influenced. However Cyclin E is not easily phosphorylated and actived, which in turn results in the low activity of CDK2-Cyclin E complex and prevention of cell proceeding into S phase.

The major reason of tumorigenesis is that the cell cycle is out of control, leading to the unlimited proliferation of cells. The molecular mechanisms underling tumorigenesis are that abnormal activation of cell cycle promoting factors and inactivation of inhibitors due to genetic mutations. Disturbing the normal control of R point, always giving the order of proliferation to the cell cycle regulation system, are the general tricks of tumor cells.

Section 2 Apoptosis

Apoptosis, or programmed cell death, is a highly regulated process that allows a cell to self-degrade in order for the body to eliminate unwanted or dysfunctional cells. The word "apoptosis" comes from the ancient Greek apo'pto'sis, meaning 'the falling off of petals from a flower or of leaves from a tree in autumn'. The name was first introduced by John Kerr in 1972 and refers to the morphological feature of formation of "apoptotic bodies" from a cell. Apoptosis describes the process by which cells are "silently" removed under normal conditions when they reach the end of their life span, are damaged, or superfluous.

Apoptosis is an active, energy-dependent process. Apoptosis is the major physiological mechanism of cell removal and general tissue phenomenon necessary for development and homeostasis: elimination of redundant cells during embryogenesis, cell atrophy upon endocrine withdrawal or loss of essential growth factors or cytokines, tissue remodeling and repair, and removal of cells that have sustained genetic damage. Normal apoptotic cell removal and cell replacement in tissue remodeling is estimated at some 1×10^{11} cells per day—equivalent to the turnover of an adult's total body weight every $18 \sim 24$ months.

Although apoptosis is important for the normal development and health of an animal, its aberrant activation may contribute to a number of diseases, for example, AIDS, neurogenerative disorders, and ischemic injury. In contrast, impaired apoptosis may be a significant factor in the etiology of such diseases as cancer, autoimmune disorders, and viral infections.

2.1　Characteristics of Apoptosis

During apoptosis, the genome of the cell will fracture, the cell will shrink and part of the cell will disintegrate into smaller apoptotic bodies. Unlike necrosis, where the cell dies by swelling and bursting its content in the area, and causes an inflammatory response, apoptosis is a very clean and controlled process where the content of the cell is kept strictly within the cell membrane as it is degraded. The apoptotic cell will be phagocytized by macrophages before the cell's contents have a chance to leak into the neighborhood. Therefore, apoptosis can prevent unnecessary inflammatory response.

Cells undergoing apoptosis show a sequence of cardinal morphological features including ruffling, membrane blebbing, cellular shrinkage and condensation of chromatin. Biochemically, these alterations are associated with the translocation of phosphatidylserine to the outer leaflet of the plasma membrane, disruption of the cytoskeletal architecture, the activation of an endonuclease which cleaves genomic DNA into multiple internucleosomal fragments. Finally, the cell shrinks and then fragments into a cluster of membrane-enclosed "apoptotic bodies" that are rapidly ingested by adjacent macrophages or other neighboring phagocytic cells.

The characteristic pattern of morphological, biochemical and molecular changes, are described in Table 8-2. These properties are always used for detection of cell apoptosis.

Table 8-2　The characteristic pattern of apoptosis cell

Morphological Changes	Functional/Biochemical Changes
Cell shrinkage	Free calcium ion rise
Cell shape change	Bcl-2/Bax interaction
Condensation of cytoplasm	Cell dehydration
Nuclear envelope changes	Loss of mitochondrial membrane potential
Nuclear fragmentation	Proteolysis
Loss of cell surface structures	Phosphatidylserine externalisation
Apoptotic bodies	Lamin B proteolysis
Cell detachment	DNA denaturation
Phagocytosis of remains	$50\sim300$kb cleavage
	Intranucleosomal cleavage
	Protein cross-linking

2.2　Factors Influencing Apoptosis

In normal conditions, the rate of cell growth and cell death is balanced. There are many factors to promote or inhibit the apoptosis. The related factors of apoptosis are divided into inducing factors and inhibiting factors.

2.2.1　Inducing Factors of Apoptosis

These factors include the hormones and growth factors, physical or chemical factors, immunological factors and microorganisms, etc. For example, glucocorticoid, glutamate and transforming growth factor-β (TGF-β) can induce the apoptosis of lymphocytes, neurons, or hepatic cells respectively.

2.2.2　Inhibiting Factors of Apoptosis

These factors include the removal of some hormones (ACTH and andrusol) or cytokines (IL-2 and neurogrowth factors), some bivalent metallic ions (Zn^{2+}), drugs, virus (EB virus and cowpox virus) and neutral amino acid.

2. 3　Molecules Involved in Apoptosis

Many molecules are involved in the process of apoptosis, such as caspase (cysteinyl aspartic acid-specific protease), Bcl-2 family, NF-κB, c-Jun, c-Fos, MPTP (mitochondrial permeability transition pore), and Ca^{2+}.

According to their function, the apoptosis related genes are divided into three classes, apoptosis inhibiting genes, such as *EIB*, *IAP*, *Bcl-2*; apoptosis promoting genes, such as *Fas*, *Bax*, *ICE*, *p53*, and bidirectional regulating genes, such as *c-myc*, *Bcl-x*.

2. 3. 1　Caspases Family

Caspases, Cys-dependent Asp specific proteases, are key executors in apoptosis. Proteolysis by caspases is restricted and limited, peptide cleavage occurs only after specific aspartic acid residues (thus c-asp-ases), producing disassembly of the protein—not general proteolysis.

All caspases are synthesized as zymogens that need to be activated by proteolytic cleavage. The active enzyme is composed of a heterotetrameric complex, caspase activity results in: cleavage of cytoskeletal proteins, disruption of the nuclear membrane, disruption of cell-cell contact, and the freeing of the DNA nuclease (CAD, caspase-activated deoxyribonuclease) from its associated protein inhibitor (ICAD) to allow DNA fragmentation. The restricted proteolysis results not in cellular lysis but in membrane-bound sealed apoptotic bodies. These irreversible proteolytic events are responsible for the morphological changes characteristic of apoptotic cells.

There are 14 mammalian caspases identified to date. Most of them locate in cytosol. Some locate in other places, such as mitochondria. Functionally, caspases are classified as:

a) The initiator　including Caspase 2, Caspase8, Caspase9 and Caspase10. This group is characterized by long prodomains (up to 90 amino acids) containing either DED domains (Caspase 8 and Caspase 10) or a caspase recruitment domain (CARD) (Caspase 2 and Caspase 9; CED 3);

b) The effector　including Caspase 3, Caspase 6 and Caspase 7, which have short prodomains. Initiator caspases cleave and activate effector caspases. Then effector caspases function on cellular substrates, such as cytoskeletal architectures, the endonucleases, and proteins involved in cell cycle.

c) Other　the remaining caspases whose main role lies in cytokine maturation rather than apoptosis. This group includes Caspase 1, Caspase 4, Caspase 5, Caspase 11, Caspase 12, Caspase 13 and Caspase 14.

2. 3. 2　Bcl-2 Family

A major downstream regulation of the apoptotic death signal resides with the *bcl-2/bax* gene family. Bcl-2 (B-cell lymphoma-2) family proteins are typically small (20~30 kDa) and are defined according to the presence of one or more homology domains (BH domains), which are typically α helices. More than 20 different members of this family have been recognized: some (including Bcl-2 and Bcl-XL) are apoptosis-inhibitory proteins, others (such as Bax, Bad, and Bid) are promoters of apoptosis. Areas of commonality of structure allow these proteins to homo and hetero-dimerize. The dimmers of Bcl-2/Bax and Bcl-2/Bcl-XL inhibit apoptosis, the dimmers of Bad/Bax, Bax/Bax and Bcl-2/Bcl-XS promote apoptosis. The balances of these members maintain the equilibrium state, once the balance is broken, apoptosis is induced.

2. 3. 3　Fas

factor associated suicide (Fas) was the first member of the TNF receptor superfamily described in terms of its function in apoptosis. It is expressed in a wide range of cell types: lymphoid cells, hepatocytes, some tumor cells, as well as in lung and myocardium. Fas causes apoptosis by binding with Fas ligand. Fas ligand (FasL) is a member of the TNF family; its expression is more restricted: it is found on cytotoxic T cells and natural killer cells. FasL, by binding and cross-linking to the Fas receptor, sets the apoptotic process in motion. This mechanism plays a significant role in: removal of activated T cells at the end of an immune response, deletion of virus-infected target cells, killing of tumor cells, and destruction of cells in numerous pathological states.

2.3.4 p53

The p53 is a regulator of DNA transcription. It binds directly to DNA, recognizes DNA damage (single-strand or double-strand breaks), and mediates at least two important cellular events: it can induce cell cycle arrest in G1 or it can promote apoptosis. If cellular damage is "considered" reparable, p53-induced cell cycle arrest allows time for DNA repair. With more extensive damage, to prevent the cell with an impaired DNA sequence from proliferating as a defective or malignant clone, p53 moves the cell into the apoptotic pathway through the Bax/Bcl-2 pathway. In the event of DNA damage, p53 gene induction is accompanied by increased synthesis and phosphorylation of p53. As a result, p53 protein activity may increase a hundred fold.

2.4 Apoptosis Pathways

Apoptosis can be triggered mainly through either the extrinsic pathway or the intrinsic pathway. The extrinsic pathway is triggered by extracellular ligands, such as FasL and TRAIL, binding in a conventional manner to the extracellular domains of transmembrane receptors. The intrinsic apoptosis pathway is initiated after cell stressors perturbing intracellular organelle function, specially mitochondria and endoplasmic reticulum (ER). Some other new discovered apoptosis pathways are also simply introduced.

2.4.1 The Extrinsic Apoptosis Pathway Mediated by Death Receptors

The extrinsic pathway is triggered by binding of small molecular weight ligands to one or more membrane-expressed death receptors. Ligands for these receptors include TNFα, Fas ligand, and TNF receptor apoptosis-inducing ligand (TRAIL). Death receptors are characterized by an intracellular region, called the death domain, an intracellular region of about 80 amino acids, which is required for the transmission of the cytotoxic signal. Five different such death receptors are known including tumor necrosis factor (TNF) receptor-1, Fas (CD95/APO-1), TNF-receptor-related apoptosis-mediated protein (TRAMP) and TNF-related apoptosis-inducing ligand (TRAIL) receptor-1 and receptor-2.

The signaling pathways by which these receptors induce apoptosis are rather similar. Ligand binding induces receptor oligomerization, followed by the recruitment of an adaptor protein to the death domain through homophilic interaction. Following self-association, the DD of the receptors recruits other DD-containing proteins which then serve as adaptors in the signaling cascades. Then the adaptor protein binds a proximal caspase, and forms the death inducing signal complex (DISC). The activated caspase is capable of directly activating effector caspases and then degrade cellular targets.

In the case of Fas/FasL system, the binding of the FasL to Fas on the target cell will trigger multiple receptors to aggregate together. The aggregation of these receptors recruits an adaptor protein known as Fas-associated death domain protein (FADD) on the cytoplasmic side of the receptors. FADD, in turn, recruits Caspase 8, an initiator protein, to form the death-inducing signal complex (DISC). Through the recruitment of Caspase 8 to DISC, Caspase 8 will be activated and it is now able to directly activate Caspase 3, an effector protein, to initiate degradation of the cell. Active Caspase 8 can also cleave BID protein to tBID, which acts as a signal on the membrane of mitochondria to facilitate the release of Cyt c in the intrinsic pathway (Figure 8-4).

Figure 8-4 Fas /FasL mediated apoptosis (see page 262)

2.4.2 The Intrinsic Apoptosis Pathway Mediated by Mitochondria

The mitochondrial pathway is activated by a variety of intracellular stresses, including Bax, oxidants, Ca^{2+} overload, active caspases, and perhaps ceramide. The general manner of mitochondrial apoptosis pathway is described as the following. The apoptotic signal leads to the release of Cyt c from the mitochondrial intermembrane space into the cytosol, where it binds to the apoptotic protease activating factor-1 (Apaf-1), a mammalian CED-4 homologue. Binding of Cyt c to Apaf-1 triggers the formation of the apoptosome. It contains seven Apaf-1, seven Cyt c, seven (d) ATP and seven procaspase 9 molecules. Procaspase 9 is the initiator caspase of the apoptosome. The apoptosome-bound procaspase 9 is activated and then activate an effector caspase (e. g. , caspase 3), which then can cleave the cellular substrates needed for the orchestration of apoptosis.

Upon activation of the intrinsic pathway, a range of proapoptotic molecules in addition to Cyt c is released from the mitochondria. Besides the release of Cyt c from the intermembrane space, the intermembrane content released also contains apoptosis inducing factor (AIF) to facilitate DNA fragmentation, and SMAC (second mitochondria-derived activator of caspase) /Diablo proteins to inhibit the inhibitor of apoptosis (IAP) (Figure 8-5).

Figure 8-5 The intrinsic apoptosis pathway mediated by mitochondria (see page 263)

2. 4. 3 The Intrinsic Apoptosis Pathway Mediated by Endoplasmic Reticulum

The endoplasmic reticulum (ER) regulates protein synthesis and folding, protein trafficking, responses to stress, and intracellular Ca^{2+} levels, and is the second major site for intrinsic apoptosis pathway initiation. Functions of the ER are affected by various intracellular and extracellular stimuli, so called ER stress, which includes inhibition of glycosylation, reduction of disulfide bonds, calcium depletion from the ER lumen, impairment of protein transport to the Golgi, and expression of mutated proteins in the ER. The ER responds by triggering specific signaling pathways including the unfolded protein response (UPR), the ER-overload response (EOR) and the ER-associated degradation (ERAD), to survive the ER stress. If the damage of ER is too great and the ER stress response can not make it survived, an apoptosis response is elicited.

In the case of the unfolded protein response (UPR), mechanisms of apoptosis are described in Figure 8-6. Severe and prolonged ER stress triggers the apoptosis signal through induction of CHOP (CCAAT enhancer binding protein-homologous protein), activation of caspase 12 and ASK1-JNK pathway, and the regulation of the ER Ca^{2+} homeostasis by Bax/Bak.

Figure 8-6 Pathways of ER stress-induced apoptosis (see page 264)

In addition to propagating death-inducing stress signals itself, the ER also contributes in a fundamental way to Fas-mediated apoptosis and to p53-dependent pathways resulting from DNA damage and oncogene expression. Mobilization of ER calcium stores can initiate the activation of cytoplasmic death pathways as well as sensitize mitochondria to direct proapoptotic stimuli.

2. 5 Viral Infection and Apoptosis

Viral infection and replication commonly result in apoptosis and this effect may be responsible for much of the pathology associated with infectious disease. Viruses also have adopted diverse strategies for inhibiting apoptosis and, thereby, prolonging the life of the infected cell, such that virus replication, spread and persistence is maximised.

2. 5. 1 Viral Infection Causes Apoptosis

Many cells undergo apoptosis in response to viral infection, even though with a consequent reduction in the release of progeny virus.

Extrinsic apoptosis signals can be initiated by adsorption of viral particles during the entry process at the plasma membrane. Attachment of noninfectious, conformationally intact HIV, vaccinia virus, and Herpes simplex virus (HSV) particles to the surface of cells is proapoptotic. Receptor mediated death signaling also appears to be triggered during entry of virus which attaches to cells by means of a TNF receptor-related protein, CAR1. Engagement of this receptor by the viral envelope proteins is sufficient to initiate apoptosis. Some viruses can induce or increase the expression of Fas/FasL, TNFα, TRAIL to promote apoptosis.

Viral infection may also stimulate intrinsic cell death pathways mediated by intracellular triggers, such as ① the double stranded RNA-activated protein kinase R (PKR) that is activated by diverse viruses including influenza A virus; ② the endoplasmic reticulum stress response that is activated by, for example, respiratory syncytial virus; ③ nuclear sensors such as p53 that detect genotoxic damage or inappropriate transcriptional activation that are induced by a number of viruses; ④ events linked to mitochondria. For example, expression of adenovirus protein E1A promotes viral replication but also triggers host cell apoptosis by activating the tumor suppressor protein p53.

Viral infection also cause target cell apoptosis by activation of CTL. In addition of releasing perforin to initiate cell necrosis, CTL also trigger cell apoptosis by increase expression of Fas/FasL, producing ceramide.

2.5.2　Viral Infection Inhibits Apoptosis

In addition to conferring a selective advantage to the virus, the capacity to prevent apoptosis has an essential role in the transformation of the host cell by oncogenic viruses.

An imperative in the evolutionary process of survival for viruses and some bacterial pathogens is the preservation of the host cell. It is to be expected then that these organisms will inhibit apoptosis. Examples of this interference or abrogation of the apoptotic process include organisms that: Ⓐ encode a protein similar to Bcl-2 (adenovirus), Ⓑpromote expression of Bcl-2 (Epstein-Barr virus), Ⓒ encode a protease inhibitor that inactivates Procaspase-1 and 8 (cowpox), Ⓓ interfere with mitochondrial Cyt c release into the cytosol (chlamydia), Ⓔ encode proteins which can regulate p53-mediated apoptosis.

2.6　Apoptosis and Disease

A deep look into the molecular pathogenesis of human diseases has revealed an apoptotic component that either contributes to disease progression or is responsible for the process in a growing number of pathological conditions. While apoptosis is an essential building block of physiological events, it seems to have an equally important function in pathological events.

2.6.1　Disease Mediated by Increased Apoptosis

Disappearance of cells by apoptosis is central to a number of diseases. Followings are some examples.

a) AIDS The hallmark of AIDS (acquired immunodeficiency syndrome) is the decline in the number of the patient's $CD4^+$ T cells. HTV can induce cell death other than T cells, including neurons and haemopoietic progenitors in the bone marrow. These cells may die in the absence of intracellular infection by virus. Both virus and the viral protein gp120 have a high affinity for CD4 antigen. Binding of virus or its gp120 protein to CD4 helper cells, without the simultaneous engagement of the MHC class II complex, triggers apoptosis. All T cells, both infected and uninfected, express Fas. Expression of a HIV gene (called Nef) in a HIV-infected cell causes the cell to express high levels of FasL at its surface, while preventing an interaction with its own Fas from causing it to self-destruct. However, when the infected T cell encounters an uninfected one (e. g. in a lymph node), the interaction of FasL with Fas on the uninfected cell kills it by apoptosis.

b) In many neurodegenerative diseases, such as Alzheimer's disease, Parkinson's disease, Huntington's disease, an apoptotic component that significantly contributes to the pathogenesis of the disease has been found.

c) The death of cells in proximity to an area of acute infarction in the brain or in the heart is in the manner of apoptosis.

d) The overexpression of Fas in hepatocytes infected with hepatitis C renders these cells vulnerable to destruction by cytotoxic T lymphocytes.

e) Thyroid cells in Hashimoto's thyroiditis express both Fas and Fas ligand, as a result, they may destroy one another under the stimulus of IL-1, which induces an increase in Fas expression.

f) Tumor cells that constitutively express FasL on their membrane or secrete FasL, may escape cytotoxic T cell destruction. Because FasL activates the T-cell Fas pathway, destroying T cells that have infiltrated the tumor, thus eliminating these anti-tumor cytotoxic lymphocytes.

2.6.2　Disease Mediated by Impaired Apoptosis

An inappropriate increase in cell number is one of the hallmarks of tumor. It may be due to either increased proliferation or decreased cell death. The decreased apoptosis may result from an absent or mutated form of the tumor suppressor gene *p53* or *Rb* present (in 50% and 40% of human tumors, respectively), from increased expression of Bcl-2, or a decrease in Bax. For example, low-grade follicular non-Hodgkin's lymphoma is characterized not by a rapidly proliferating tumor cell population but by the gradual accretion of slow-growing lymphoma cells, which is caused with deregulation and overexpression of Bcl-2 occurring with the translocation of the *bcl-2* gene from chromosome 18 into the Ig locus on chromosome 14.

Several reports suggest a correlation between apoptosis and autoimmunity through an impairment of apoptosis, or an ineffective removal of apoptotic cells. Autoantigens are found within apoptotic bodies and that apoptotic cells are critical in the presentation of antigens, activation of innate immunity and regulation of macrophage cytokine secretion. Apoptotic bodies have been also described as B cell autoantigens. Autoimmunity, like SLE, also caused by deficiency of Fas expression which makes apoptosis needed for the negative selection of auto-reactive T cell lost.

Summary

The cell cycle means the whole process from the end of one mitosis to the end of the next mitosis in a continuous divided cell. One standard cell cycle is divided into 4 phases: G_1 phase, S phase, G_2 phase and M phase. Before commitment to DNA replication, cells in G_1 can enter a resting state called G_0. In mammalian cells, the re-entry into the cell cycle depends on extracellular signals, growth factors and mitogens.

The cell cycle is majorly controlled by Cyclin-dependent kinases (CDKs). CDKs are proline-directed serine/threonine protein kinases. The periodic activation and inactivation of CDKs is the key factor that promotes the operation of cell cycle. Cyclin binding provides the CDK with targeting domains important for substrate selection and subcellular localization, which in turn determine the biological specificity. Recently, CDK inhibitors (CKIs) are found which also regulate the cell cycle as a negative control by binding CDKs or CDK-Cyclin complexes to inhibit their activity.

In general, the levels of CDKs are relatively constant throughout the cell cycle, whilst the levels of the Cyclins vary substantially. The activity of CDK-Cyclin complex is decided by the following factors: ① the corresponding level of Cyclins; ② the phosphorylation state of CDK-Cyclin complexes; ③ the level of CKIs. During the ordered promotion of cell cycle, the degradation of Cyclins of last phase and the synthesis of Cyclins of next phase take place alternately. The ubiquitin-mediated proteasome system is the main pathway employed for the degradation of cell cycle components.

Cell cycle checkpoints are control mechanisms that ensure the fidelity of cell division in eukaryotic cells. These checkpoints verify whether the processes at each phase of the cell cycle have been accurately completed before progression into the next phase. The most important two checkpoints are G_1/S checkpoint and G_2/M checkpoint. Cells can arrest at cell cycle checkpoints temporarily to allow for cellular damage to be repaired.

The cell cycle is associated with senescence and multiple diseases like tumor and AIDS.

Apoptosis, or programmed cell death, is a highly regulated process that allows a cell to self-degrade in order for the body to eliminate unwanted or dysfunctional cells. Apoptosis is an active, energy-dependent process. Apoptosis is the major physiological mechanism of cell removal and general tissue phenomenon necessary for development and homeostasis. During apoptosis, there are special characteristic pattern changes in morphology, biochemistry and molecular level. These properties are always used for detection of cell apoptosis.

Many molecules are involved in the process of apoptosis. According to their function, the apoptosis related genes are divided into three classes, apoptosis inhibiting genes, apoptosis promoting genes and bidirectional regulating genes. Caspases, Cys-dependent Asp specific proteases, are key executors in apoptosis. All caspases are synthesized as zymogens that need to be activated by proteolytic cleavage. Proteolysis by caspases is restricted and limited, peptide cleavage occurs only after specific aspartic acid residues, producing disassembly of the protein—not general proteolysis.

Apoptosis can be triggered mainly through either the extrinsic pathway or the intrinsic pathway. The extrinsic pathway is triggered by binding of small molecular weight ligands (FasL, TNFα and TRAIL) to one or more membrane-expressed death receptors (TNF receptor, Fas, TRAMP and TRAIL receptor). Following self-association, the receptors recruits adaptor proteins. Then the adaptor proteins bind proximal caspases, and form the death inducing signal complex (DISC). The activated caspase is capable of directly activating effector caspases in the signaling cascades and to initiate degradation of the cell.

The mitochondrial pathway is activated by a variety of intracellular stresses. The apoptotic signal leads to the release of Cyt c, AIF and SMAC from the mitochondrial intermembrane space into the cytosol, where it binds to the Apaf 1 and Procaspase 9 to form the apoptosome. The apoptosome-bound Procaspase 9 is activated and then activate an effector caspase (e. g. Caspase 3), which then can cleave the cellular substrates needed for the orchestration of apoptosis.

The ER responds by triggering specific signaling pathways including the unfolded protein response (UPR), the ER-overload response (EOR) and the ER-associated degradation (ERAD), to survive the ER stress. If the damage of ER is too great and the ER stress response can not make it survived, an apoptosis response is elicited. Severe and prolonged ER stress triggers the apoptosis signal through induction of CHOPCCAAT enhancer binding protein-homologous protein), activation of caspase 12 and ASK1-JNK pathway, and the regulation of the ER Ca^{2+} homeostasis by Bax/Bak.

A deep look into the molecular pathogenesis of human diseases has revealed an apoptotic component that either contributes to disease progression or is responsible for the process in a growing number of pathological conditions. While apoptosis is an essential building block of physiological events, it seems to have an equally important function in pathological events. Disappearance of cells by apoptosis is central to a number of diseases, such as AIDS, neurodegenerative diseases and apoplexy. On the contrary, an impairment of apoptosis, or an ineffective removal of apoptotic cells is a reason of many diseases, such as tumor, autoimmunity disease and viral infection.

(Jun Sun   Ping Yuan)

第九章 肿瘤分子生物学

进入 21 世纪，恶性肿瘤仍然是危害人类生命健康的重要疾病。肿瘤研究是医学研究中最复杂、最困难的课题，涉及多学科的知识和理论，其中包括病理学、细胞生物学、胚胎发育学、生物化学、遗传学、肿瘤免疫学、肿瘤病毒学、肿瘤药理学等。20世纪 70 年代起对基因、蛋白质的研究取得了重大突破，尤其是基因克隆技术及其他相关技术的迅速发展，肿瘤研究在以上学科研究的基础上发展成为独特的肿瘤分子生物学，从此肿瘤研究得到了迅猛发展。

第一节 癌 基 因

癌基因（oncogene）可定义为一大类促进细胞分裂并有潜在致癌或促癌作用的基因群。它的发现可追溯到动物致癌病毒的研究。

一、病毒癌基因

早在 1911 年，Rous 首先发现鸡肉瘤病毒（RSV）能使鸡胚成纤维细胞在培养中转化，也能在接种给鸡后诱发肉瘤。1970 年 Temin 和 Batimore 证实 RSV 是一种逆转录病毒（retrovirus）；20 世纪 70 年代 H. Varmus 和 Bishop 研究小组发现 RSV 中的致瘤基因是 src，它能导致培养细胞转化和呈恶性表型，也能在动物中引起肿瘤。这种基因被称为病毒癌基因（virus oncogene，V-onc）；V-src 是第一个被鉴定出的病毒癌基因，以后从许多动物中分离出致癌的逆转录病毒，并鉴定出 30 多种 V-onc。V-onc 不仅与致癌密切相关，而且与正常细胞的某些 DNA 序列同源，从而进一步有了细胞癌基因的概念。

二、细胞癌基因

1. 细胞癌基因与原癌基因

V-onc 不参与病毒复制周期，它们是从哪儿来？又是怎样整合到病毒基因组去的呢？此项研究最终导致人类癌症中细胞癌基因（cellular oncogene，C-onc）的发现。Varmus 和 Bishop 用 src 的 cDNA 和其他基因组 DNA 杂交，发现 src 的同源物普遍存在于动物细胞（如鸡、鸭、果蝇）中。原来 src 编码一种胞质酪氨酸激酶，参与细胞增殖相关的信号转导，是细胞的正常组分。由于 RSV 等逆转录病毒的基因组是整合在宿主基因组上复制的，会将宿主的某些基因复制到了自身的基因组中，被这样的病毒感染的细胞，src 拷贝增多了，引起细胞过度增殖。为了区别起见，将存在于正常细胞中的癌基因序列称为细胞癌基因（C-onc），由于 C-onc 在正常细胞中以非激活的形式存在，故又称为原癌基因（proto-onc）。从结构上看 C-onc 或 proto-onc 是间断的，存在内含子，这是真核基因的特点。而 V-onc 是连续的，基因跨度较小。

2. 细胞癌基因的分类及功能

C-onc 或 proto-onc 在生物进化中是高度保守，已发现其表达的蛋白质产物分布于细胞膜、细胞质及细胞核中。目前已分离到 200 多种细胞癌基因。根据癌基因蛋白的功能，可将这些癌基因分为以下几类：①生长因子类的癌基因，包括 *sis*、*fgf-5*、*hst*、*int-1* 和 *int-2* 等。Waterfield 等发现 *v-sis* 的表达产物 P28 与血小板源生长因子（PDGF）的 B 链高度同源。Chiu 等在分析了 *v-sis* 在细胞内的对应物 *c-sis* 的核苷酸序列后，证实 *c-sis* 是 PDGF 的结构基因。已知 PDGF 能促进平滑肌细胞等多种细胞分裂增殖。Yoshida 等还发现 *c-hst* 的编码产物与成纤维细胞生长因子（FGF）高度同源。而已知 FGF 可促进多种细胞增殖。②生长因子受体类的癌基因，包括 *yes*、*fgr*、*erbB*、*kit*、*met*、*ros*、*fms*、*trk* 和 *neu* 等。已发现 *erb* 的产物在结构上与表皮生长因子受体（EGFR）高度同源；且有 TRK 活性；*c-fms* 的产物与集落刺激因子同源；*c-mas* 则编码血管紧张素受体，而 *c-Kit* 则编码肥大细胞生长因子受体。③酪氨酸蛋白激酶（非受体）类的癌基因，包括 *src*、*fyn*、*syn*、*lyn*、*slk*、*sea*、*lck*、*hck*、*ros*、*abl*、*fps*、*fes*、*yes*、*fgr*、*tkl*、*ret*、*ptc* 和 *rel* 等编码的蛋白质产物，有的与细胞内膜共价相连，有的游离于细胞质之中。Src 族激酶的结构在很大程度上具有相似性，有相同的结构域和调节机制。他们含有一个或者多个 N 端的酰基化位点（膜定位所需的），一个独特的结构域、SH3 结构域、SH2 结构域、催化结构域和 C 端 Tyr529（抑制位点）。*src* 族在生长因子诱导的胞内反应中起着重要作用，包括增殖、存活、黏连和迁移。许多直接的证据证明 *c-src* 与人类癌症密切相关，如乳腺癌、肝癌和结肠癌等。在恶性肿瘤细胞中 *c-src* 大量聚集在核周，而在正常的细胞内 *c-src* 相对均匀的散布于细胞质内。④丝氨酸/苏氨酸蛋白激酶类的癌基因，包括 *col*、*mos*、*raf/mil* 以及 *pim-1* 等，这些癌基因的表达产物不仅是 cAMP 和 PIP2 途径的核心成分，而且与细胞信号转导密切相关。⑤G 蛋白类的癌基因，包括 *gsp*、*gip*、*bcl-2*、*Ha-ras*、*ki-ras* 和 *N-ras* 等，G 蛋白与 GTP 结合可介导多种信号。研究表明，*ras* 基因家族编码的产物 P21 与 G 蛋白 α 亚基同源，也有 GTP 结合活性。Lorry 等发现，P21 与 GTP 结合后，可激活磷脂酶 C，从而促进 PI 代谢，产生细胞增殖的第二信使 IP_3 与 DG，参与细胞增殖的调节。⑥核转录因子蛋白类的癌基因，包括 *evi-1*、*ets-1/2*、*gil-1*、*hox2.4*、*lyl-1*、*maf*、*pbx*、*ski*、*spi-1*、*vav*、*myc*、*myb*、*c-jun*、*c-fos*、*erbA*、*rel* 等。它们具有下列共同特点：①当受到细胞外生长或分化因子刺激时能快速、瞬间地被诱导转录。②其 mRNA 半衰期短、不稳定。这是由于在 3′端非翻译区域存在 AT 丰富的不稳定顺序。③其编码的多数蛋白质半衰期短（20～90min）。它们均为翻译后修饰，通常为丝氨酸磷脂化作用。④核内癌蛋白类的癌基因，具有亮氨酸拉链（leucine zipper，LZ）、锌指（zine finger，ZF）、同源区（homeo domain，HD）或螺旋-环-螺旋（helix-loop-helix，HLH）结构，能与 DNA 特异顺序结合。已证明 *fos*、*myc* 及 *myb* 的产物均定位于细胞核内。这表明它们在有关细胞增殖及分化的核内过程中起重要作用。通常在不分裂的静止细胞中其不表达或表面很低，而在进入细胞周期时，*myc*、*fos* 可持续表达，且 *fos* 产物可与 *jun* 的产物形成异二聚体，并与 DNA 有关部位结合后，调节基因转录。这类 C-onc 通过表达产物与 DNA 结合，促进与细胞增殖有关的基因开放，产生细胞增殖效应。

三、癌基因的激活机制

　　proto-onc 是细胞基因组的正常成分，在正常情况下，不表达或只是有限制地表达，处于非激活状态，不具有致癌性；但在一些特定因素作用下，其结构或调控发生异常，使之被"激活"才会具有致癌活性。癌基因活化与肿瘤的发生有密切关系。目前有关癌基因的激活机制有以下 7 种假说。

1. 基因突变

　　1982 年，美国麻省理工学院的 Weinberg、国立癌症研究所的 Barbacid 和哥伦比亚大学的 Wigler 几乎同时发现，人膀胱癌细胞的恶性转化活性是和癌基因的一个点突变直接相关的。他们发现人膀胱癌 EJ 株中有一段 6.6 kb 的 *BamH* I 片段，具有非常强的恶性转化活性，且和大鼠肉瘤病毒的癌基因 *V-Ha-ras* 是同源的，称为 *EJ-ras*，相应的细胞原癌基因是 *c-Ha-ras*。*EJ-ras* 和 *c-Ha-ras* 之间的结构比较表明，*c-Ha-ras* 基因的第一外显子中第 12 号密码子的第二个核苷酸，由 G 突变成了 T。这个碱基替代突变使 *c-Ha-ras* 编码的 P21 蛋白的第 12 位氨基酸，由甘氨酸变成了缬氨酸。

　　1984 年，Santos 等用限制性内切核酸酶的酶谱分析技术，对一例肺鳞状上皮细胞癌患者的癌细胞和多种非癌组织细胞中的 *Ki-ras* 基因进行了系统的结构分析。证实只有癌细胞中的 *Ki-ras* 基因的第 12 号密码子由原癌基因的 GGT 突变成了 CGT，并由此获得了转化活性，而同一患者的非癌组织细胞中的 *Ki-ras* 都没有发生突变。这项研究为我们提供了体细胞突变致癌的第一例临床实证，同时也为肿瘤的分子诊断开创了先例。迄今为止，已在膀胱癌、乳腺癌、结肠癌、肾癌、肝癌、肺癌、胰腺癌和胃癌等实体瘤中发现了 *ras* 基因的突变。另外还发现 *ras* 基因突变和急性淋巴细胞白血病、B 细胞淋巴瘤、Burkitt's 淋巴瘤，以及急性和慢性细胞白血病都有关系。常见的 *ras* 突变发生于密码 12、13 和 61。人和其他哺乳动物有三种 *ras* 基因，即 *Ha-ras*、*Ki-ras* 和 *N-ras*，其编码的蛋白质 P21ras 能通过和 GTP 或 GDP 的结合来调节细胞的生长与分化，一旦 *ras* 基因突变导致 P21 结构改变和功能异常，导致正常 *ras*、GTPase 活性的丧失，突变的 *ras* 仍然和 GTP 的激活蛋白（GAP）形成复合物，但其 GTPase 的活性不能被 GAP 刺激，因此引起 ras-GTP 突变体半衰期延长，使细胞生长发育失控，则有可能引起恶性肿瘤。

2. 基因插入

　　带有病毒致癌基因的病毒感染宿主细胞后，其 DNA 随机地整合到宿主基因中，如果病毒的 DNA 恰好位于原癌基因附近，癌毒基因启动子便激活原癌基因，转化为癌基因，由此引起细胞癌变。例如，宫颈癌中发现人乳头瘤病毒（特别是 16 型、18 型）的 DNA 序列，在宫颈细胞中与宿主发生整合。又如，带有长末端重复顺序（LTR）的禽白血病前病毒插入到鸡 *c-myc* 第一外显子内，LTR 含有促进因子（promoter element），后者与 RNA 聚合酶 II 转录起始有关，这种插入重排就使 *c-myc* 激活而诱发 B 细胞淋巴瘤。

3. 染色体易位

　　在 Burkitt 淋巴瘤中，有三种常见的染色体易位，都涉及第 8 号染色体长臂 2 区 4 带的原癌基因 *c-myc*：①8-14 易位 t（8；14）（q 24；q32），将 *c-myc* 移至 14 号染色体的免疫球蛋白基因 IgH 附近，约有 90％的患者出现这类畸变；②8-2 易位 t（8；2）（q 24；p12），将 *c-myc* 移至 2 号染色体的免疫球蛋白基因 IgK 附近；③8-22 易位 t（8；22）（q24；q11），将 *c-myc* 移至 22 号染色体的免疫球蛋白基因 Ig 附近。这三种易位可能造成的共同后果是在 B 细胞分化后期不再表达的原癌基因 *c-myc*，置于免疫球蛋白基因的表达增强子的调控之下，使 *c-myc* 基因的表达增强。*c-myc* 编码的核内蛋白能通过控制转录来激活某些与细胞增殖有关的基因，使细胞获得在体外条件下无限生长的能力，所以 c-myc 被称为"永生化基因"。*c-myc* 基因的活化往往会使细胞处于癌前病变状态。

1960 年 Nowell 和 Hungerford 发现了慢性粒细胞白血病的细胞遗传学标记——费城染色体。1973 年 Rowley 证实费城染色体起源于 9 号染色体和 22 号染色体之间的交互易位，t（9；22）（q34；q11）。近年来的研究表明，这个易位使 9 号染色体上的原癌基因 *c-abl* 和 22 号染色体上的断裂点集丛区（breakpoint cluster region，bcr）融合成一体。易位形成的费城染色体上的 *bcr/c-abl* 融合基因编码一个分子质量为 210 kDa 的融合蛋白 P210$^{\text{bcr-abl}}$。值得注意的是 P210 和 P145 虽都具有酪氨酸激酶活性，但是 P210 多了自我催化性，它能以自身为底物来催化磷酸化反应。这种由于酶的底物特异性的改变而引起的结构变化，也许与融合蛋白逃避正常调控，变更生长信号传递的渠道或频率，最终造成细胞恶性转化是密切相关的。

4. 基因扩增

癌基因可通过一定的机制在原来的染色体上复制出许多拷贝，为转录提供超量的模板，随着转录水平的增加，癌基因产物剧增，从而增加了转化蛋白的剂量，促使细胞的恶性转化。1982 年 Collins 和 Dalla Favera 等首次报道了 *c-myc* 基因的扩增，发现 HL-60 细胞株和早幼粒细胞白血病患者的白血病细胞中 *c-myc* 扩增 8～32 倍，当达到 30 倍时，染色体上出现染色体均染区（HSR）和双微粒染色体（DMS），随后在结肠癌、乳腺癌、子宫颈癌、食管癌等中均发现 *c-myc* 癌基因扩增。

5. 基因偶联

细胞癌变是两种以上癌基因协同作用的结果。例如，利用分子杂交实验证明仅用人的癌基因（*c-ras*）转染大鼠胚胎的成纤维细胞，没有得到转化灶，不会发生癌变。如果再加第二种癌基因（*c-myc*），结果就不一样，能得到转化灶，在裸鼠皮下产生一个大而稳定的肿瘤。不同癌基因之间进行相互作用，使细胞癌变。此过程称为癌基因偶联或者叫基因协作。因为癌基因偶联后，使本来不表达或表达不利的癌基因受到强烈的激活，导致癌变。

6. DNA 低甲基化

DNA 甲基化是基因转录的重要调控方法，它的状态维持与基因转录活性密切相关。DNA 转录非常活跃的区域，通常甲基化程度较低，肿瘤组织 DNA 具有普遍的低甲基化特点，这种现象早在癌前病变就出现，并随着肿瘤进展，低甲基化程度呈进行性下降。它可能通过影响原癌基因的表达促进肿瘤的发生发展，人类 *c-myc* 基因全长有 17 个 CCGG 位点，外显子 1、外显子 2 CCGG 位点比较密集，难以对每个 CCGG 位点的甲基化状态进行研究，但外显子 3 只有一个 CCGG 的位点，而且 *c-myc* 表达高的组织细胞中易发生脱甲基化。O-htsuki 等发现人类骨髓瘤细胞的 *c-myc* 外显子 3 CCGG 序列与正常人扁桃体 B 淋巴细胞相比甲基化程度显著偏低，而且这种低甲基化与 *c-myc* 在骨髓瘤细胞中高表达有关。Sharrard 等用 *c-myc* 基因亚片段作为探针，分别对该基因各段的 CCGG 位点甲基化程度进行检测，发现 *c-myc* 5′端和外显子 1、外显子 2 的甲基化水平在正常结肠黏膜、腺瘤和腺癌都一致，而外显子 3 在腺瘤和腺癌是明显低甲基化，在淋巴瘤、肝癌、肺癌等也发生 *c-myc* 等外显子 3 CCGG 位点的低甲基化，而这些肿瘤组织 *c-myc* 蛋白常常是增高的。因而癌基因的低甲基化在肿瘤的发生发展中起着重要作用。

7. 微卫星异常

微卫星（microsatellite）又称短的串联重复序列（short tandem repeat，STR）、简单重复顺序（simple sequences repeat，SSR），是由 1～6 个核苷酸组成的重复单元串联排列而成的 DNA 序列，广泛分布于原核、真核生物基因组中。微卫星等位基因的差异主要是由于重复单元串联重复次数不同。肿瘤中微卫星异常主要表现为微卫星不稳定（microsatellite instability，MSI）和微卫星的杂合性缺失（loss of heterozygosity，LOH）。

肿瘤的 MSI 是指肿瘤组织与其相应非肿瘤组织相比，其结构性等位基因的大小发生改变，即微卫星重复单元的增加或丢失。在扩增片段长度多态性分析中，表现为肿瘤出现非肿瘤组织没有的额外带，或肿瘤的一个等位基因与非肿瘤组织相应等位基因相同，而另一个等位基因的位置发生改变。MSI 的机制尚未完全清楚，目前普遍接受的观点是 DNA 错配修复缺陷是其主要原因。

肿瘤中 LOH 比微卫星不稳定更常见。表现为与非肿瘤组织相比，肿瘤的一个等位基因消失。微卫星的缺失是相应区域 DNA 片段缺失的标志。由于许多微卫星都与基因紧密连锁甚至位于基因内部，微卫星缺失常表示其连锁基因的缺失。

微卫星异常的肿瘤发生机制：微卫星可能参与基因表达调控、基因复排与变异、维持基因组稳定等重要生命活动，微卫星的不稳定，势必会造成细胞生命活动的紊乱，有可能发生肿瘤。另外，MSI 是错配修复缺陷的标志，而错配修复缺陷除造成 MSI 外，还可能使癌基因、抑癌基因及一些对细胞生长起调控作用的基因的复制错误得不到修复，从而细胞生长失控，最终发生肿瘤。Branch P 等证明错配修复缺陷能增强细胞对烷化剂的耐受，在受到烷化剂的致命损伤后，错配修复缺陷细胞不会像正常细胞一样停顿于 G_2 期进行修复或发生死亡，而是继续复制，使大量错误信息在细胞内不断积聚，最终导致癌变。

四、近年发现与人类肿瘤相关的癌基因

1. 与乳腺癌相关的癌基因

HOXA1（Hox genes A1）基因是人乳腺上皮细胞的癌基因。人生长激素 hGH 增强 *HOXA1* 的表达和转录活性，而 *hGH* 基因与乳腺上皮细胞病理增殖有关。*HOXA1* 在乳腺癌细胞中过表达能促进细胞增殖，这是由于 *HOXA1* 增强了 *bcl-2* 的转录水平，而 Bcl-2 蛋白与抑制细胞凋亡有关。另外，*HOXA1* 的过表达能抵抗细胞凋亡作用。

DCD（dermcidin）是乳腺癌细胞中的一个新的生长和存活因子。DCD 编码的蛋白质是分泌蛋白，在 10％的侵袭性乳腺癌中表达，过表达增强细胞生长和存活，减少对

血清的依赖。其在乳腺肿瘤形成中扮演重要角色。

2. 与膀胱癌相关的癌基因

CDC 91L1（cell division cycle 91-like 1）基因也被称为 U 类磷脂酰肌醇聚糖基因（phosphatidylinositol glycan class U，PIG-U），是磷脂酰肌醇聚糖锚定通路中转酰胺基酶复合物的组成单位。CDC 91L1 的表达受到染色体易位的影响。在 1/3 的膀胱癌细胞株和原发性肿瘤中 CDC 91L1 基因扩增且过表达导致 NIH3T3 细胞恶性转化；CDC 91L1 的过表达也导致尿激酶受体增加，进而增强膀胱癌细胞中的 STAT3 磷酸化。Guo 等的研究表明，CDC91L1 是膀胱癌的一个相关癌基因。

3. 与多种癌相关的癌基因

KIF14（kinesin family member 14，也称 KIF14）基因定位于染色体 1q。利用 RT-PCR 检测发现，KIF14 在各种肿瘤中都过量表达。在 20/22 的视网膜成神经细胞瘤中，KIF14 mRNA 的表达量是正常视网膜成神经细胞中的 100～1000 位；肿瘤细胞株中的表达比肿瘤组织表达水平高。经免疫印迹发现，视网膜成神经细胞瘤和乳腺癌细胞中 KIF14 蛋白过量。在 4/4 乳腺癌细胞中 KIF14 的表达量是正常乳腺组织的 31～92 倍；在 5/5 成神经管细胞瘤细胞中 KIF14 的表达量是胎儿大脑的 22～79 倍；在 10/22 原发性肺癌中 KIF14 过表达则会降低存活率。因此，KIF14 是一个潜在的癌基因。

综上所述，由于癌基因是作为引起恶性肿瘤的逆转录病毒的部分而被发现和命名的。而实际上，癌基因是涉及细胞重要功能的一些正常基因的突变型，所以原癌基因的正常功能被描述为产生癌基因的概念是不正确的，然而由于这一名称已被广泛应用，再要改变为时已晚。

第二节　抑癌基因

癌基因的发现，激发人们以更大热情研究与癌发生相关的其他基因。不久，人们发现了存在于细胞内的另一类基因——抑癌基因（又称抗癌基因）。

一、抑癌基因的概念及发现

抑癌基因（tumor suppressor gene）是一大类可抑制细胞生长并能潜在抑制癌变作用的基因群。它们对细胞的生长发育和终末分化起负调控作用。

20 世纪 60 年代，有人将癌细胞与同种正常成纤维细胞融合，所获杂种细胞的后代只要保留某些正常亲本染色体时就可表现为正常表型，但是随着染色体的丢失又可重新出现恶变细胞。这一现象表明，正常染色体内可能存在某些抑制肿瘤发生的基因，它们的丢失、突变或失活，使激活的癌基因发挥作用而致癌。1986 年视网膜母细胞瘤（retinoblastoma，Rb）抑癌基因 Rb 被克隆和完成全序列测定，抑癌基因的存在得以证实。

抑癌基因的研究起步晚，目前已阐明的抑癌基因不多，作用机制也尚未清楚认识，但可以肯定抑癌基因和癌基因一样，也是一类细胞的正常基因，它除了抑癌作用外也必然有其他重要生理功能。

二、比较明确的抑癌基因

抑癌基因应符合三个基本条件：①在该癌的相应正常组织中必须有正常的表达；②在该种恶性肿瘤中该基因理应有所改变，包括点突变、DNA 片段或全基因的缺失或表达缺陷；③导入该基因缺陷的恶性肿瘤细胞中将部分或全部抑制其恶性表型。目前已知的一些抑癌基因及抑癌候选基因有以下几种。

1. *Rb* 基因

第一个抑癌基因模型来自视网膜细胞癌基因的研究。视网膜母细胞瘤（retinoblastoma）是一种儿童期发病的恶性家族性肿瘤。Benedict 通过对视网膜母细胞瘤患者的正常体细胞和肿瘤细胞作比较，发现癌细胞的一条 13 号染色体丢失，从而引起视网膜母细胞癌变。通过 7 年的研究，他们最终证实视网膜细胞瘤来自正常亲本的正常等位基因，即视网膜细胞 13 号染色体上的 2 个等位基因发生丢失，从而导致癌变。与此同时，Cavenee 等通过研究也证明体细胞中的杂合性在癌细胞中丢失，因而推断 *Rb* 基因能抑制肿瘤形成，是"肿瘤抑制因子"，或名"抗癌基因"。1987 年 Lec 等通过两项研究确认，*Rb* 基因确是能抑制视网膜细胞瘤发生的抗癌基因。*Rb* 基因位于 13q14，DNA 全长 100 kb 以上，mRNA 约 4.7 kb，表达产物为核磷酸蛋白（P110RB），含 928 个氨基酸，属反式作用因子。*Rb* 的表达及磷酸化状态与细胞周期密切相关：即在 G_1/S 表达增高，并从有功能的非磷酸化状态转为无功能的磷酸化状态，前者有阻断细胞进入 S 期的功能。非磷酸化的 pRB 可与 SV_{40} 大 T 抗原、腺病毒 E1A 及 HPVE7 蛋白结合而失活；pRB 亦可阻断 *c-fos* 的表达及抑制 *c-myc* 的表达；而 TGF-β 可阻断 P110RB 在 G_1 期的磷酸化。*Rb* 在视网膜母细胞瘤中存在基因缺失、点突变、mRNA 及蛋白质产物表达下降等，一些骨肉瘤、小细胞肺癌、乳腺癌及膀胱癌中亦有 *Rb* 缺陷。

2. *p53* 基因

p53 基因位于染色体 17p13.1，产物亦为核磷酸蛋白及反式作用因子，含 375 个氨基酸，其中 N 端 73 个氨基酸为调控活性区。*p53* 的表达及磷酸化状态也与细胞周期相关。与其他抑癌基因不同的是野生型属抗癌基因，但其突变型可出现两种情况：一是显性阴性作用（dominant negative effect），即突变型 p53 蛋白能使同时表达的野生型 p53 蛋白失活；二是显性致癌作用（dominant oncogenic effect），即某些 *p53* 突变体在丢失功能的同时具有致癌作用而成为癌基因。起源于不同组织的人类肿瘤，如结肠癌、肺癌、乳腺癌、肝癌、食管癌、白血病等几乎都有 *p53* 基因的缺失或点突变（点突变主要集中在进化保守的第 5~8 外显子），总的缺陷率为 10%~70%。*p53* 的变异常见于癌及癌前期损伤，可能是散发及原位癌发生的重要原因。突变 *p53* 有以下特性：①能激活 *c-myc* 的启动子进而调控内源性 *c-myc* 基因；②外显子 1 的 3′ 端是其调控 *c-myc* 的功能区；③反应区是定向和定位的，无增强子功能；④功能作用需要 C 端来参与，而野生型 *p53* 不需要。针对上述特性选择性抑制突变 *p53* 的功能，可减少癌变的发生。*p53* 等位基因的全部缺失，不是癌转化的必要条件，仅降低 *p53* 表达水平就能导致癌变。人胃癌中已发现 *p53* 和 *erbB-2* 基因的重排和扩增。研究表明，p53 蛋白能诱导一些基因的转录，这些基因的上游调节区含有 *p53* 特异的 DNA 序列，如肌肉肌苷激酶（*MCK*）基因、*GADD45*（growth arrest DNA damage enducible）基因及 *mdm-2*（mu-

rine double minute）基因等即属此类基因。p53 蛋白也可作用于其他生长调控有关的基因（如 *hsp70*）和抑癌基因（如 *Rb*、*p53* 自身等）。另外，*p53* 对含有 TATA 启动子序列的许多基因有负调节作用。但 *p53* 对细胞其他基因转录的调控受 *p53* 本身的改变（如点突变、基因缺失）、*p53* 翻译后的改变（如磷酸化状态）及 *p53* 与其他细胞和病毒蛋白结合等因素影响。这些蛋白质包括 mdm-2、SV_{40} 大 T 抗原、腺病毒 E1B 及 HPV E6 蛋白等，其中 SV_{40} 大 T 抗原、腺病毒 E1B 及 HPV E6 蛋白与 p53 蛋白结合后抑制了 p53 蛋白对其他基因的转录调控或促进 *p53* 降解而使其失活；而位于 12q12-13 的 *mdm-2* 癌基因其产物（可能亦是一种转录因子）可与 p53 蛋白结合而抑制后者的转录激活活性，已知在约 30% 的人软组织肉瘤中存在此基因扩增及产物过度表达。另外，*p53* 还能与 WT1 结合而影响其活性。

另外，*p53* 可能与基因组的"可塑性"（plasticity）有关，*p53* 与 *GADD45* 及 *AT*（ataxia telangiectasia）等基因一起控制 DNA 的修复，体细胞接触放射线后 *p53* 半衰期延长，稳定性增高，使细胞阻断于 G_1 期以修复 DNA 损伤，从而降低和避免了基因组的突变。另外，接触化疗药物或放射线时 *p53* 的失活阻止细胞进入凋亡，从而使细胞处于分裂状态而增高突变及恶性转化的概率。再者，突变的 *p53* 可使癌细胞对放疗、化疗产生抗性而导致转移。

3. *p16* 基因

p16 基因是 1993 年 Serrano 和 Beach 等在用酵母双杂交筛选系统研究与 CDK 作用的蛋白质时，首先确定的一个编码人类周期素依赖性激酶 4（CDK4）抑制因子的基因，又称多肿瘤抑制基因（MTS1）。它是人们发现的第一个直接作用于细胞周期而抑制细胞分裂的抑癌基因，在多种肿瘤细胞中广泛而高频率变异。人的 *p16* 基因定位于 9p2.1，在酵母人工染色体中含有该基因克隆的 *EcoR* 片段长度大约为 23 kb，由 3 个外显子和 2 个内含子组成，外显子序列编码的蛋白质产物的分子质量为 16 kDa，即 P16 蛋白，由 4 个锚蛋白重复序列组成。P16 蛋白在正常细胞中起负反馈作用，通过抑制 CDK4 蛋白与 Cyclin D_1 的结合，阻断 pRB 蛋白磷酸化，并通过 E2F 的不断分离，使细胞滞留在 G_1 期，导致细胞生长抑制。

4. *p15* 基因

p15 基因又称为 *MTS2* 基因，该基因定位于 9p21 染色体上，其产物和 P16 有 95% 的同源性。P15 蛋白含 137 个氨基酸，能与 CDK4 和 CDK6 结合，抑制 CDK4-Cyclin D_1 和 CDK6-Cyclin D_1 的结合，从而抑制细胞增殖。TGF-β 可显著诱导 *p15* 转录，提示 *p15* 使细胞停止在 G_1 期的生长抑制作用可能通过接触抑制，作为细胞生长抑制信号的效应因子而起作用。在肿瘤的形成过程中，*p16*、*p15* 在细胞分裂增殖的调节中可互相发挥作用。在乳腺癌和其他肿瘤中 *p16* 缺失比 *p15* 缺失更常见，*p16* 缺失不一定总伴有 *p15* 缺失，而 *p15* 缺失总伴有 *p16* 缺失，由此说明，*p16* 基因可能是 9p21 染色体异常的真正靶点。*p15* 与 *p16* 在乳腺癌中的作用基本相同。

5. *p21* 基因和 *p27* 基因

p21 基因定位于 6p21.2，编码 164 个氨基酸，C 端有一 PCNA 结合区，为核内蛋白。*p27* 基因是 p21 家族成员，定位于 12p13，可读框长 591 bp，是高度保守的细胞周期素依赖性激酶抑制因子，对细胞周期起负调控作用。*p27* 的表达水平与 Cyclin D1 和

Cyclin E 的水平明显相关。但免疫组化对 *p27* 表达检测显示，某些肿瘤中 *p27* 表达相对增高，可能是因为癌细胞对 p27^{kip1} 的抑制有了抗性造成的。但提高 p27^{kip1} 的表达水平或模仿 p27^{kip1} 活性的治疗策略仍在癌症治疗中有重要意义。

6. *APC* 基因和 *DCC* 基因

APC(adenomatous polyposis coli) 基因为腺瘤样结肠息肉的易感基因，定位于 5q21，由 15 个外显子和 14 个内含子组成，编码 2843 个氨基酸，分子质量为 300 kDa。*APC* 与结肠腺瘤样息肉综合征、散发性结肠癌、肺癌等肿瘤均有关。*APC* 在正常细胞中促使 β 连环蛋白的降解，但在 *APC* 突变癌细胞中缺失该能力，使核内有大量 β 连环蛋白聚集，*APC* 的这种功能对肿瘤抑制作用非常关键。

DCC(deleted in colorectal carcinoma gene) 定位于染色体 18q21，含有 29 个外显子，表达产物为 190 kDa 的跨膜磷蛋白。*DCC* 基因异常有等位基因丢失，5′端同源性丧失，内含子及外显子点突变，微卫星的杂合性缺失（LOH）和 DCC 低表达等。其产物位于细胞表面，属免疫球蛋白超家族成员，与神经细胞黏附因子同源。表明其于细胞-细胞外基质具有相互作用，其功能的丢失可能导致细胞间接触、黏附能力下降，从而提高癌细胞的转移能力。

7. *FHIT*

FHIT(fragile histidine triad) 基因定位于 3p14.2。FHIT 的灭活可能发生在肿瘤 G$_1$ 期的早期阶段，并且与 G$_2$、G$_3$ 期肿瘤发展密切相关。根据该基因的缺失仅见于白血病而非正常造血细胞，推测 FHIT 与白血病发生相关。此外，FHIT 蛋白的缺失在弥散性小叶乳腺癌发生中亦起重要作用。有证据表明 3p14.2-21.3 的杂合性缺失在所有非典型类癌瘤中广泛存在，并且伴有 FHIT 蛋白的表达减少，暗示了它们都有 FHIT 的缺失。

8. *NF-1* 和 *NF-2* 基因

NF-1 (neurofibro matosis type 1) 基因定位于 17q11，基因全长 60 kb，产物含 2485 个氨基酸，位于胞质，其中 350 个氨基酸与 ras GAP 及酵母 IRA 同源，并有 GTPase 活化蛋白（GAP）的活性，但仅对 ras P21 专一。NF1 在神经纤维瘤及部分肉瘤中存在基因丢失、插入突变、终止密码形成、mRNA 表达下降及 LOH 等。

NF-2 位于染色体 22q，产物位于胞质，可能与细胞骨架及膜的连接有关，在脑膜瘤和神经鞘瘤中有基因丢失及突变。

9. 肾母细胞肿瘤抑制基因

肾母细胞肿瘤抑制基因（Wilm's tumor gene1，*WT1*）定位于染色体 11p13，DNA 全长约 50 kb，有 10 个外显子。*WT1* 基因在正常肝组织胚胎期和成熟期非正常肝组织中表达，而在成熟期正常肝组织中不表达。蛋白质产物含 4 个锌指结构，可能亦为核内反式作用因子。在肾母细胞瘤中存在等位基因突变，部分缺失或部分伴有 11p15 的 LOH。

10. *MMAC*

MMAC (mutated in multiple advanced cancer) 又称 *PTEN* (phosphatase and tension homology deleted donchromosome ten) 是定位于 10q23.3 的抑癌基因。该基因的变异及表达水平变化对成胶质细胞瘤的生长有重要影响，与患者的康复密切相关。

MMAC/PTEN 导入细胞能抑制 Akt 转导的信号和细胞的生长，暗示了其缺失会促进细胞的增生不死性。此外有报道证实 *PTEN* 在许多肿瘤中是因为体细胞突变而失活。

11. 候选肿瘤抑制基因 *DPC4/SMAD4*

DPC4 是近年来一种新发现的肿瘤抑制基因，近年来国内外对其研究非常活跃，主要在结直肠癌、乳腺癌、胃癌、胰腺癌等疾病的发生、发展中起作用。基因定位于人染色体 18q21，有个 11 个外显子和 10 个内含子，具有 2680 bp 的转录单位，编码 552 个氨基酸。SMAD4 蛋白是 DPC4 的表达产物，主要作用是参与转化生长因子-β（TGF-β）超家族的信号转导，SMAD4 蛋白为 TGF-β 信号转导的细胞内信使，它将信号从细胞膜传至细胞核，从而影响细胞内信号转导。青少年家族性息肉病患者基因的突变进一步证实 *DPC4* 是一种肿瘤抑制基因。

12. *caveolin-1*

基因位于人染色体 7q31.1。Caveolin-1 蛋白是分子质量为 22 kDa 的整合膜蛋白，广泛存在于多种类型的细胞中，甚至在中枢神经、胎盘组织中都有表达。Caveolin-1 与多种信号分子的共同序列相互作用。但在发生肿瘤时，会出现 Caveolin-1 的表达异常。*caveolin-1* 在肿瘤细胞中表达及其作用机制尚不甚明确，它是一种直接参与细胞生长和增殖的负调控基因，通过抑制 CyclinD1 的表达和 ERK 的活化、负性调控 EGF、FGF、PDGF 等生长因子、抑制 Src 酪氨酸激酶的磷酸化而阻止细胞周期于 G_1 期，从而抑制细胞增殖、分裂。*caveolin-1* 可能是一个抑癌基因。将反义 *caveolin-1* 注入 NIH3T3 细胞后，细胞呈现出不锚定的独立生长特性，并在免疫缺失鼠体内形成肿瘤，显示过激的 p42/44MAP 激酶级联反应。而 *caveolin-1* 表达的上调能够介导接触抑制，同时对 p42/44MAP 激酶级联反应进行负调控，抑制肿瘤的生长。

三、其他新近报道的抑癌基因

1. 与腺癌相关的抑癌基因

ST18（乳腺癌抑癌基因 18）定位于染色体 8q11，是乳腺的肿瘤抑制基因，编码一个锌指结构的 DNA 结合蛋白，具有潜在的转录调节功能。在包括乳腺上皮细胞在内的大多数正常组织中，*ST18* 都能表达，但表达水平较低。在乳腺癌细胞系和主要的原发性乳腺肿瘤中，*ST18* mRNA 显著减少。在 80％的乳腺癌样品和主要的乳腺癌细胞系中 *ST18* 基因启动子 160 bp 区域发生超甲基化。*ST18* 在 MCF7 细胞中异位表达可抑制细胞集落的形成，在小鼠中表达可阻止肿瘤的形成。

LOT1（lost-on-transformation 1）基因定位于染色体 6q24-25，是一个锌指结构的核转录因子，与抑制细胞增殖有关，常在乳腺癌细胞、卵巢癌细胞中发生基因沉默。体外实验表明，*LOT1* 启动子甲基化显著降低其启动萤光素酶的转录。体内实验表明，在乳腺癌和卵巢癌细胞中，LOT1 的 5′CpG 岛是甲基化区；在这些细胞中，这一区域的甲基化水平为 31％～99％。

2. 与胰腺癌相关的抑癌基因

WWOX（ww domain-containing oxidoreductase）基因定位于染色体16q23.3-24.1，跨越不稳定区域 FRA16D。在 Kuroki 等的研究中，*WWOX* 在 9 种胰腺癌细胞的 4 种中表达改变；在 2 种细胞 22％中和 2 种样品 43％中检测到 *WWOX* 启动子的超甲基化，

利用 RT-PCR 和 Western 印迹检测到在所有 3 个细胞株中 *WWOX* 表达水平都降低；研究表明，*WWOX* 通过影响细胞凋亡，进而在胰脏癌发生中扮演重要角色。

3. 与鼻咽癌相关的抑癌基因

BLU（也称 as ZMYND10，zinc finger，MYND-type containing 10）基因定位于染色体 3q21，是鼻咽癌的候选抑癌基因。*BLU* 的异位表达阻止了癌细胞集落的形成。*BLU* 启动子在环境压力下激活，受 E2F 调节。*BLU* 启动子和第一个外显子定位在 CpG 岛。在包括鼻咽在内的正常上呼吸道组织中，*BLU* 都有较高的表达，但是在 7 种鼻咽癌细胞株中 *BLU* 表达较低。*BLU* 在肿瘤中的甲基化水平经检测为 7/29.24%。因遗传或后天因素，24/28.83% 鼻咽肿瘤中的 *BLU* 基因断裂，因此 *BLU* 是一个压力应激基因。

4. 与多种癌相关的抑癌基因

DLC-1（deleted in liver cancer 1）基因定于染色体 8p21-22，在大多数人类癌症，如乳腺癌、肝癌、结肠癌和前列腺癌中 *DLC-1* 基因都有所改变，包括基因组缺失或启动子甲基化。*DLC-1* 对乳腺癌和肝癌细胞生长具有阻断作用。Yuan 等检测了具有高缺失频率的染色体 8p21-22 区域的非小细胞肺癌（NSCLC）中 *DLC-1* 基因的变化，发现在 95%（20/21）的原发性 NSCLC 和 58%（11/19）的 NSCLC 细胞中，*DLC-1* mRNA 的表达水平显著降低。去甲基化试剂 5-偶氮-2-脱氧胞苷可重新激活 *DLC-1* 下调的 82%（9/11）NSCLC 细胞株 *DLC-1* mRNA 的表达，说明 *DLC-1* 的转录沉默主要与 DNA 甲基化有关，而不是基因组缺失。将 *DLC-1* 转入 *DLC-1* 阴性的细胞株中，导致细胞增殖被阻断，减少了细胞集落形成。这表明 *DLC-1* 在 NSCLC 中是一个新的抑癌基因。

RBSP3（RB1serine phosphatase from human chromosome 3）基因定位于染色体 3p21.3，有 2 个剪切形式 RBSP3 A/B。序列分析表明，*RBSP3* 的 C 端有磷酸化酶区域，可以控制 RNA 酶 II 的转录。该基因在 12 种肿瘤细胞株的 11 种和 8 种肿瘤活组织检查的 3 种中表达水平显著降低。*RBSP3* 在体外的表达可抑制肿瘤细胞生长，在 SCID 鼠体内可抑制肿瘤形成，通过 ppRB 去磷酸化可使细胞周期停滞在 G_1/S 期。

RASSF1A（RAS association family1 gene）基因定位于染色体 3p21.3，由 Dammann 等在 2000 年分离得到。*RASSF1A* 基因的失活主要是由甲基化和等位基因丢失造成的。RASSF1A 蛋白与人类 DNA 修复蛋白 XPA 相互作用，其 C 端与小鼠 Ras 的效应子 Norel 和 Maxp1 具有 55% 的同源性，但 N 端没有很高的同源性。研究表明，*RASSF1A* 与 Ras 以 GIP 领带的形式相结合，调节 Ras，这表明 *RASSF1A* 与 Ras 以 GTP 依赖的形式相结合，RASSF1A 蛋白参与 DNA 修复体系或是 Ras 通路。另有报道，33/48 个肝外胆管癌样品 68.75% *RASSF1A* 未表达，说明 *RASSF1A* 在肝外胆管癌的发生中占有重要地位。在 58.33%（28/48）的样品和细胞株中 *RASSF1A* 的 CpG 全部或部分甲基化，使 *RASSF1A* 转录沉默。利用 LOH 分析检测了染色体 3p21.3 的等位基因，*RASSF1A* 基因定位在小随体 D3S4604，LOH 在 D3S4604 的频率为 68.75%，远高于普通癌细胞。

四、抑癌基因在细胞周期调控中的作用

细胞周期的发现是 20 世纪 50 年代细胞学的重大发现，它的发现标志着仅以注重细

胞形态观察的细胞学向形态和功能并重的细胞生物学发展。其中最重要的是：①阐明细胞分裂的驱动和调控机制；②发现并确定细胞内的监控系统。一些抑癌基因（如 *Rb*、*p16*、*p53* 等）参与调控监控系统的功能及检查细胞周期各期的状态。

1. *p53* 基因与细胞周期调控

p53 基因的产物是一个分子质量约为 53 kDa 的蛋白质（p53），其氨基酸序列与某些转录因子类似，参与细胞周期和细胞增殖的调节。在正常细胞中，p53 蛋白控制着处于长停滞状态的静止期细胞从 G_0 期到 G_1 期转变，并在 G_1 期的生长限制性位点控制着细胞进入细胞周期后的增殖。野生型 p53 蛋白也是 G_1 期 DNA 损伤的检查点，像"分子警察"监视着细胞基因组的完整性。如果 DNA 遭到破坏，p53 蛋白能通过某些机制使 DNA 得以修复，如果修复失败，则 p53 蛋白能够通过"细胞程序性死亡"而引发细胞自杀，阻止具有癌变倾向基因突变的细胞产生，从而起到抑癌作用。突变型 p53 则使细胞无限生长，与肿瘤的发生密切相关。野生型 p53 蛋白功能：①生化功能，与特定序列 DNA 结合，激活含有 p53 反应元件的基因启动子的转录活性，抑制不含 p53 反应元件的基因启动子的转录活性，促进单链 DNA 的退化，抑制解链酶的活性，抑制 DNA 复制。②生物学功能，诱导细胞生长停滞后 G_1 期，抑制肿瘤细胞生长，保护遗传稳定性，DNA 损伤后诱导细胞程序性死亡。

p53 能与 DNA 特异结合，而且有一个酸性激活区，可以推测 p53 能激活其邻近基因表达，如果目的基因上游带有 p53 结合位点的特异 DNA 序列并与 p53 表达载体共转染，则目的基因得到高表达。这种激活也可以是间接的，即通过野生型 p53 大量增加而影响细胞内各种基因表达。事实上，p53 能影响一些基因表达水平，而它们中很少含有 p53 结合位点，此外 p53 可以抑制 SV40T 抗原对聚合酶 α 的结合，从而抑制 SV40 DNA 复制。

2. *Rb* 基因在细胞周期调控中的作用

Rb 基因的蛋白质产物的分子质量为 105 kDa（pRB），它是存在于核内的蛋白质，有磷酸化与非磷酸化两种形式。pRB 的磷酸化作用随着细胞周期发生改变，在 G_1 早期以去磷酸化活性形式存在的 pRB 不但可作用于 E2F 等转录因子，阻断 mRNA 转录，抑制 DNA 合成，还可激活转录因子 ATF2，促进细胞分裂抑制因子的转录，从而负调控细胞生长，防止肿瘤发生。在 G_1 中期、S 期以及 G_2/M 期，通过磷酸化作用，pRB 失去活性，而 pRB 的磷酸化通过 CDK 4-Cyclin D 或 CDK 6-Cyclin D 激酶来实现。

与 pRB 磷酸化有关的蛋白质是 Cyclin。D 型细胞周期蛋白包括 D1、D2、和 D3 三个成员，已证明三者可与 CDK4 或 CDK6 形成复合物。细胞周期蛋白 D1 的异位表达可诱导 pRB 的磷酸化，并可加速细胞通过 G_1 期的进程。细胞周期蛋白 E 也与 pRB 的磷酸化有关，当它异位合成时，可诱导 pRB 的磷酸化，在 G_1 后期，当 pRB 的磷酸化加强时，细胞周期蛋白 E 的水平明显增加。pRB 还可能是细胞周期蛋白 A 依赖激酶的生理底物。已发现在细胞周期蛋白阻遏的启动基因控制下，合成细胞周期蛋白 A 的细胞表现出 pRB 的磷酸化并提前进入 S 期。另一个重要的证据是，细胞周期蛋白 A-CDK2 复合物可稳定地与转录因子 E2F-1 结合并可使 Dp-1（differentiation-regulated transcription factor protein-1）磷酸化，从而导致 E2F 从 DNA 上释放并抑制靶基因的表达。此外，在 G_2 期和 M 期，细胞周期蛋白 B-CDK 2 可能也参加了 pRB 的磷酸化。此方面的研究证据虽然还不多，但至少已清楚了在体外 pRB 是细胞周期蛋白 B-cdc 2 激酶复合物作用的底物。

3. *p16* 基因是一个重要的细胞周期调节物

p16 基因是一个重要的细胞周期调节物。体外实验表明，其表达产物 p16^{IMK4} 能与 CDK4 或 CDK6 特异性结合。Parry 等发现在 *Rb* 阴性细胞中，由于 *p16* 高水平表达，导致 CDK4-Cyclin D 和 CDK 6-Cyclin D 激酶的缺乏，p16^{INK4} 能与周期蛋白 D1 竞争与 CDK4 或 CDK6 结合。同样，在神经胶质瘤细胞系中，*CDK4* 基因的扩增也表现出与 *p16* 缺失相同的效用，即 *p16* 与 *CDK4* 基因作用参与拮抗。可见 p16^{INK4} 主要通过与 CDK4 或 CDK6 结合，阻碍其与周期蛋白 D1 形成复合物，使 CDK4-Cyclin D1 和 Cyclin D1-CDK 6 激酶失活，导致一系列底物去磷酸化，其中主要是使 pRB 去磷酸化，保持活化形式，从而使起始调控点下游事件不能发生，细胞周期停滞在 G_1/S 期，细胞增殖受到抑制，呈现抑癌基因作用。反之，当 p16^{INK4} 由于 *p16* 突变而不能正常表达，不能竞争 CDK4 和 CDK6 与 Cyclin D1 的结合，将会导致细胞异常增殖。研究表明，在 *Rb* 基因失活的细胞系中存在高水平的 *p16* 表达，反之，在 *Rb* 基因高活性的细胞系中 *p16* 的表达很低或不表达，即 *p16* 的表达与 *Rb* 活性呈负相关。pRB 也能调节 *p16* 的表达，这种调节通过对 *p16* 转录活性的抑制来实现（图 9-1）。

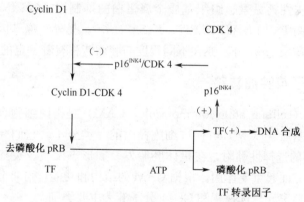

图 9-1　*p16* 基因在细胞周期调控中的作用
（＋）促进；（－）抑制

由图 9-1 可见，p16^{INK4} 与 pRB 相互调节，构成一个环路，保证细胞正常地从 G_1 期进入 S 期。一旦这个环路被打破，将导致细胞异常增殖向癌变发展，或者死亡。

总之，*p16* 基因的产物 p16^{INK4} 直接抑制 *CDK4*（或 *CDK6*），对细胞周期调控与肿瘤发生起重要作用，其作用机制还有待进一步深入研究。

4. *p27* 基因是细胞周期的负调控因子

p27 是近年发现的抑癌基因，*p27* 基因在及其产物对于细胞的生长及其肿瘤发生发展和转移有着极其重要的作用。P27 位于细胞核内，*p27* 定位在人体染色体 12p12.0～12p31.1 交接处。

p27 主要通过抑制 CDK 来抑制细胞周期。*p27* 对 CDK 的抑制主要有 6 个方面：①通过其 C 端抑制 cdk2-Thr 160 的磷酸化，从而抑制 CDK2 的前活性状态复合物的激活过程，使 CDK 处于非活性状态对细胞周期实现负调控。②通过直接与细胞周期蛋白-

CDK 结合抑制已激活的复合物的激酶活性。③通过阻断 pRB 蛋白的磷酸化形式控制细胞周期的进程。④$p27$ 可作为启动子诱导细胞凋亡。⑤$p27$ 对细胞增殖的启动和放大起决定作用。⑥$p27$ 具有诱导未成熟的细胞分化的作用。以上表明 $p27$ 的积聚部分参与了决定前体细胞何时停止增殖，启动分化的内在监控机制，同时又部分地参与在启动分化时使细胞周期停滞的机制。p27 作为细胞周期的负调控因子，具有阻止细胞通过 G_1/S 期转换的关卡作用，从而抑制细胞的增殖，使细胞有机会修复损伤的 DNA 和 DNA 复制过程中产生的错误。$p27$ 的异常在人类肿瘤中属罕见事件，提示 $p27$ 表达可能与肿瘤的发生发展密切相关。

第三节　肿瘤转移的分子机制

肿瘤转移（cancer metastasis）是恶性肿瘤最重要的生物学特征之一，是肿瘤临床治疗的难题，也是影响疗效和预后乃至危及肿瘤患者生命的主要原因。肿瘤转移指癌细胞脱离原发部位向远端器官的扩散，形成与原发肿瘤性质相同的继发肿瘤。肿瘤转移是极其复杂的多步骤的序贯过程，是众多因子相互协调以及共同作用的结果。其中，促进肿瘤转移的因子主要有黏附分子与受体、基质降解酶、细胞运动因子等。而抑制肿瘤转移的因子主要有上皮性钙黏素、组织基质金属蛋白酶抑制因子（TIMP）、$nm23$ 基因、MHC 基因及其他基因（如 $KAIPCD82$、$KiSS21$ 等）。另外，癌基因与抑癌基因（如 c-ras、c-myc、erb-$B2$、$p53$ 等）亦在不同程度与肿瘤转移有着一定的关联。

一、细胞黏附分子与肿瘤转移

细胞黏附分子（cellular adhesion molecule，CAM）是指由细胞合成并组装于细胞表面或分泌至胞外基质（ECM）可促进细胞黏附的一类分子。它们介导细胞之间或细胞与（ECM）之间的选择性黏附，在机体胚胎发育、形态发生、炎症反应、凝血和维持组织结构完整等方面起重要作用。根据 CAM 的结构和功能，可将其分为钙黏素、整合素、免疫球蛋白超家族、选择素和 CD44 分子五大主要类别。

1. 钙黏素功能的丧失及上皮间质的转变

钙黏素（cadherin，CD）是分子质量为 120 kDa 的跨膜糖蛋白，其家族成员众多，各类 CD 中，E-cadherin（E-CD）与肿瘤关系最密切。E-CD 及其连环蛋白复合体在上皮细胞间黏附和组织形态的维持上起了至关重要的作用，其组成主要包括 E-CD，α-连环蛋白、β-连环蛋白、γ-连环蛋白以及 p129CTN。E-CD 是上皮型钙黏素，主要存在于人和动物的上皮细胞，它是一种介导同型细胞间黏附的钙依赖跨膜糖蛋白，是维持上皮细胞形态和结构完整性和极性的重要分子，E-CD 包括细胞外区、跨膜区及细胞质区。细胞外区的功能最重要，有 5 个重复的结构域（Ec1～5），第 6～11 外显子能与钙特异结合，相邻细胞间 E-CD 的胞外域可相互结合形成拉链样结构，从而发挥黏附功能。连环蛋白位于细胞质内，是一类细胞质蛋白，与 E-CD 的细胞内结构结合形成复合体，α-连环蛋白再与肌动蛋白结合，使 α-连环蛋白锚着于肌动蛋白骨架上，该复合体中任何一种成分异常都会影响到 α-连环蛋白介导的细胞黏附功能。

上皮细胞间质转化（epithelial-mesenchymal transition，EMT）是多细胞生物胚胎

发生中的基础过程，它是以上皮细胞极性丧失及其间质特性的获得为主要特征。EMT与肿瘤的原位侵袭和远处转移有着密切的关系，被认为是晚期癌症的关键所在。E-CD不能启动转录，表达被抑制就会诱导 EMT 的发生，E-CD 因此被认为是转移抑制基因。E-CD 的下调常发生在转录水平，E-CD 启动子常常被特异性转录抑制因子所抑制，包括 Snail1、Snail2、SPIP1、δEF1、Twist 和 E12/E47 及启动子的甲基化，有些抑制因子已被发现在肝癌和乳腺癌浸润前有特异性的表达，这些抑制因子的表达被某些信号转导途径调节，包括标准的 Wnt 信号转导途径，TGF-β、FGF、EGF、Stat3 和细胞核因子 κ-B（NFκ-B）。E-CD 的缺失破坏了相邻细胞之间的黏附作用，使迁移细胞脱离上皮层，E-cadherin 与细胞骨架相互作用的丧失、Rho 家族 GTPase 的活化将促使肿瘤细胞的迁移和浸润。

2. 整合素

整合素（integrin）是广泛存在于动植物细胞表面的一类细胞黏附分子，由 α、β 亚单位形成的跨膜异二聚体，目前发现有 19 种 α 亚基和 8 种 β 亚基，相互结合形成 24 组不同的整合素家族。整合素在细胞迁移过程中将细胞外基质（EMC）与细胞内骨架肌动蛋白连接在一起，不同 α、β 亚单位形成不同的整合素，其通过配体细胞外基质，如纤维结合蛋白、玻璃黏连蛋白、层黏连蛋白、胶原蛋白与胞内骨架蛋白、肌动蛋白结合，重要的是通过束缚 EMC 成分激活整合素从而诱导细胞内信号转导途径，调节细胞增殖、存活、极性、运动和分化。整合素的功能是通过与相应配体结合介导细胞与基底膜、细胞与细胞的黏附。然而在肿瘤发生过程中，癌基因、抑癌基因的突变可使非黏附肿瘤细胞在细胞基质黏附因子缺乏的情况下存活和增殖。某些整合素信号转导途径有利于癌细胞的存活和增殖。在黑色素瘤、恶性胶质瘤、前列腺癌、乳腺癌及卵巢癌都有整合素 $\alpha_V\beta_3$、$\alpha_2\beta_1$、$\alpha_4\beta_1$ 和 $\alpha_6\beta_1$ 整合素的过度表达，肿瘤细胞的迁移和转移通过整合素介导黏附斑和肌动蛋白依赖的收缩完成的，此外 EMC 是肿瘤侵袭的屏障，它的溶解会促进肿瘤的侵袭转移，活化的整合素可以和基质金属蛋白酶 MMP-9 结合并正向调控 MMP-9 的表达，MMP-9 的高表达可直接破坏和降解 EMC，从而是肿瘤细胞向周围组织迁移和浸润。

3. 免疫球蛋白超家族

IgSF 的黏附分子是一类位于细胞表面与免疫球蛋白结构相似的跨膜蛋白质，其分子结构与免疫球蛋白有很高的同源性，由一个或多个免疫球蛋白同源单位组成，中间由二硫键相连。IgSF 的多数成员与细胞之间的识别有关。

NCAM（neural cell adhesion molecule，CD56）和 L1（neural cell adhesion molecule L1 CD171，L1CAM）在肿瘤细胞的浸润和转移扩散中起着重要的作用，两者属于免疫球蛋白超家族（immunoglobulin super family，IgSF），并在中枢神经系统的发育过程中起着关键性的作用，它们不仅介导静止神经元的黏附，而且尤为重要的是涉及神经突的生长轴突的引导及神经细胞的迁移。此外还在细胞与细胞、细胞与 ECM 的黏附机制中起一定的作用，同时活化信号受体、诱导细胞内信号的级联反应。

NCAM 在恶性肿瘤的进展中起着重要的作用，成熟 NCAM 120 kDa 亚型向胚胎 140 kDa 和 180 kDa 的转换以及 NCAM 表达的普遍下调与某些癌症的不良预后相关，在胰 β 细胞致癌转基因鼠肿瘤模型中，NCAM 的缺失导致淋巴结转移的形成，此过程由淋巴血管生成因子 VEGF-C 和 VEGF-D 的过度表达及其上调周围淋巴血管生成所介导，在神经元中具有与此相似的功能，NCAM 也可通过束缚 FGFR 和诱导不同的信号转导途径，从而导致 β_1 整合素介导的细胞基质的黏附和神经突的生长，调节肿瘤细胞基质的黏附。

L1 在神经元中与 NCAM 的功能相似，在神经胶质瘤、黑色素瘤、肺癌、前列腺癌、肾癌、卵巢癌和子宫内膜癌浸润前可特异性表达，并且它的表达常与转移有关。与 NCAM 相似，L1 在体外已证实可增强肿瘤细胞的迁移和浸润，L1 启动细胞基质黏附及依靠 ECM 蛋白的肿瘤细胞的迁移，如纤维结合蛋白和层黏连蛋白可通过整合素识别基序与整合素结合，即发现整合素 $\alpha_V\beta_5$、$\alpha_V\beta_3$、$\alpha_5\beta_1$ 与 L1 相关。此外 L1 整合素的相互作用不仅发生在细胞表面　而且可通过 L1 外功能区的脱落，溶解性 L1 自分泌/旁分泌束缚 $\alpha_V\beta_5$ 整合素，L1 也可单独启动生长因子诱导的丝裂原激活蛋白激酶信号，并刺激细胞外调节激酶（Erk）1/2 介导的与细胞运动、侵袭和基质变化相关的转录表达的变化。

4. 选择素

选择素（selectin）是一类以糖基为其识别配体的黏附分子，属跨膜糖蛋白，主要有血小板选择素（platelet-selectin，P-选择素）、内皮细胞选择素（endothelium-selectin，E-选择素）及白细胞选择素（leukocyte-selectin，L-选择素）三种。正常情况下选择素处于惰性状态，不发生细胞黏附作用，当组织损伤时，细胞表面的选择素活化，介导细胞黏附到损伤部位。与肿瘤血运转移有关的 E-选择素主要在血管内皮细胞上表达，介导肿瘤细胞与血管内皮细胞之间的黏附，在肿瘤转移中具有重要地位。肿瘤细胞与 E-选择素的结合启动多个信号转导途径，进一步影响转移过程。

5. CD44

CD44 分子由单一基因所编码，但由于基因转录后的拼接和翻译后的修饰方式不同，其产物是一类具有高度异质性的单链膜表面糖蛋白，被称为 CD44 家族，分为标准型（standard isoform of CD44，CD44s）和变异型（splicing variant of CD44，CD44v）两种。CD44 分布极为广泛，在淋巴细胞和成纤维细胞表面均能检测到它的表达。转移性肿瘤细胞与血管内皮细胞结合后，还必须黏附并浸润到血管内皮下的基底膜和细胞外基质，CD44 在此过程中起重要作用。特定结构的 CD44 分子表达与肿瘤的恶性程度以及转移潜能之间具有显著相关性。CD44 基因敲除小鼠的体内实验证明，CD44 并不影响肿瘤的发生，但却影响肿瘤的转移。

综上所述，肿瘤的转移是由于肿瘤细胞之间同质黏附降低，肿瘤细胞与细胞外基质之间的正常连接被破坏，导致肿瘤细胞由原发部位释放，通过多种细胞黏附分子介导与肿瘤组织内皮细胞发生黏附及跨内皮细胞迁移。由于细胞黏附分子在肿瘤转移过程中具有重要的作用，因此，研究者们希望通过干扰细胞黏附分子的表达和调控来抑制肿瘤的转移。

二、细胞外基质降解酶在肿瘤侵袭和转移中的作用

从原位的增殖性肿瘤到侵袭转移癌的演变过程中，肿瘤细胞必须具备降解细胞外基质的能力。细胞外基质的降解主要依靠蛋白水解酶。蛋白水解酶包括：基质金属蛋白酶、丝氨酸蛋白酶、半胱氨酸蛋白酶、天冬氨酸蛋白酶；其中基质金属蛋白酶（matrix metalloproteinase，MMP）在细胞外基质（ECM）和基底膜（basement membrance，BM）的降解过程中发挥重要作用。

基质金属蛋白酶是一个大家族，目前已分离鉴别出 20 个成员，编号分别为 MMP1～20。它们是一组锌离子依赖性肽链内切酶，大小各异，底物不尽相同，能裂解维系蛋白质结构的肽链，主要参与结缔组织的降解。MMP 的家族成员具有相似的结构，一般由 5 个功能不同的结构域组成：①疏水信号肽序列；②前肽区，主要作用是保持酶原的稳定。当该区域被外源性酶切断后，MMP 酶原被激活；③催化活性区，有锌离子结合位点，对酶催化作用的发挥至关重要；④富含脯氨酸的铰链区；⑤C 端区，与酶的底物特异性有关。其中酶催化活性区和前肽区具有高度保守性。MMP 通过对细胞外基质中不同成分的降解，在肿瘤侵袭转移中起关键性作用。肿瘤细胞黏附到细胞外基质上是肿瘤浸润的必要步骤，黏附力较弱则移动性也弱，而黏附太紧密则不能分开和移动，故肿瘤细胞要成功地完成转移过程，必须具有黏附、脱黏附的能力。MMP 可以调节细胞黏附功能。

三、肿瘤血管生成在癌转移中的作用

肿瘤血管生成（angiogenesis）是指肿瘤细胞诱导的微血管生长以及肿瘤中血液循环建立的过程。肿瘤细胞进入血管通过血管系统启动癌细胞从原发病灶迁移到远端器官（血源性转移）。第一步不仅依靠肿瘤细胞内在的侵袭特性，而且还需要自身构成肿瘤基质部分的肿瘤血管网络，从以前存在的血管网络刺激肿瘤新生血管的生成，这使肿瘤细胞的转移成为一种危险因素。这一过程至少以两种途径进行：①肿瘤血管的增加增强了肿瘤细胞之间的接触范围及其潜在的逃逸途径，因此增加了进入血管内渗的可能性。②肿瘤血管与生理血管相比具有明显的表型，较大、曲折和渗漏导致基底膜易破碎及不完整的外膜和内膜，易使细胞进入。此外，肿瘤细胞血源性的蔓延也可形成淋巴转移。血管生成是通过分泌的血管生成因子和血管生成抑制因子的平衡而调节的，肿瘤基质对血管生成的调节是非常复杂的，不仅仅是简单地通过信号转化途径，以及肿瘤组织缺氧而导致肿瘤细胞血管生成因子的表达，如血管上皮生长因子（VEGF）和血小板源性生长因子（PDGF）。肿瘤基质的几种细胞类型对促成血管生成也起重要作用：①上皮细胞通过分泌血管生成因子，如血管生成素-2，从而影响内皮及其分化为成熟血管的活性状态；②浸润巨噬细胞和肥大细胞通过分泌另外的血管生成因子，包括 VEGF-A、FGF-2、TGF-β 和 IL-8，还有激活这些生长因子潜在形式的 MMP；③与癌症相关的纤维原细胞（CAF）通过分泌其他生长因子：细胞因子和趋化因子及通过调节 ECM 促成血管生成。因此，肿瘤细胞和肿瘤基质诱导的血管生成不仅启动原发肿瘤的生长，而且还促进肿瘤细胞血源性的扩散和转移生长。

四、肿瘤转移相关基因

近年来，肿瘤分子生物学的研究重点逐渐转入分离和克隆肿瘤转移基因（MG）与肿瘤转移抑制基因（MSG），并对这些基因的调控因子和转移过程中的作用机制进行深入研究，期望在基因水平揭示肿瘤转移的本质，为改进肿瘤的诊断方法和治疗手段提供依据。

(一) 肿瘤转移促进基因

1. *ras* 基因

ras 基因家族是较早发现的与肿瘤转移有关的癌基因。Thorgeirsson 首先证实，激活的 *c-H-ras* 基因转染 NIH3T3 细胞后，可使这些永生化的细胞获得转移表型和在裸鼠体内的成瘤性及高转移性。之后，Vousden 将 *c-H-ras* 基因转染高成瘤性、低转移性的小鼠乳腺癌细胞，结果该细胞转移能力明显增强。*H-ras* 转染非转移性淋巴瘤细胞 BW5147，也同样诱发了该细胞株的侵袭性和实验性转移潜能，且转移能力依赖于 *ras* 基因表达水平。W. Ebb 等对 *ras* 基因诱导转移的机制进行了研究，他们利用 *V 122-H-ras* 效应域突变剂破坏其激活下游靶区域的活性，无法诱导小鼠成纤维细胞 N IH3T 3 的肺转移。但有活化 *V 12-H-ras* 基因表达的成纤维细胞，则可通过 raf-MAPK-1/2 旁路引起肺转移。说明 *ras* 通过作用于信号转导途径发挥诱导转移功能。

2. *MTA1*

MTA1 基因是从人乳腺癌细胞系中分离出的鼠转移相关基因 *mta1* 的同源基因，Cui 等用荧光原位杂交技术（FISH）将该基因定位在 14q32.3。在消化道肿瘤的研究中，发现食管鳞状细胞癌、结肠癌、胃癌中 *MTA1* 均过度表达，且与淋巴结转移频率及肿瘤浸润深度明显相关，目前认为 *MTA1* 促进肿瘤转移的机制与信号转导和基因表达调控有关。Nawa 等推测 *MTA1* 基因编码的蛋白质含有脯氨酸富集区（SH3 结合基序）、锌指基序、亮氨酸拉链基序，而且在基因调节蛋白上经常发现 5 个拷贝的 SPXX 基序，认为 MTA1 蛋白在肿瘤细胞演变过程中起着细胞间信号转导的作用。其后 Mazumdar 等又进行了深入研究，发现 *MTA1* 基因编码的蛋白质定位于细胞核内，是 heregulin beta 1（HRG）诱导的组蛋白去乙酰化酶（HDAC）和核小体重组复合体的组分。MTA1 蛋白可直接与组蛋白去乙酰化酶和 ER-α 作用，在 ER（＋）乳腺癌细胞中 Hergulin/HER2 通路的激活能抑制雌激素受体元件（ERE）的启动转录作用，且扰乱细胞对雌二醇的反应，从而使肿瘤细胞更具有侵袭性和自主增殖能力。

3. *S100A4*

S100A4 又称 *CAPL*、*p9ka*、*mts1* 等，是 1989 年由 Sbralidze 从高转移性小鼠乳癌细胞中克隆出来的一种转移相关基因。1992 年克隆了人类 *S100A4* 基因。该基因位于染色体 1q21，编码 11 kDa 的蛋白质，为 S100 钙结合蛋白家族成员，S100A4 蛋白同应力纤维（stree fiber）、纤维型肌动蛋白和原肌球蛋白相连，调节细胞骨架动力和细胞的运动，具有调节癌细胞的增殖周期、细胞间黏附、侵袭和转移的功能。许多文献报道在乳腺癌、食管癌、胃癌、结直肠癌等中，该基因表达均增高，且与肿瘤的侵袭深度、转移密切相关，被认为是某些肿瘤独立的预后因素。S100A4 促进转移作用也可能与 p53 有关。Grigorian 等研究发现，Mts1 能结合到 p53C 端调控区域，其通过蛋白激酶 C 阻断 p53 全序列和 C 端多肽的磷酸化，干扰 p53 与 DNA 结合活性，进而影响 p53 的功能，从而促进肿瘤的转移。S100A4 与 E-cadherin 也存在密切关系，S100A4 的侵袭作用可能部分由于 E-cadherin 表达下降所致。目前认为 S100A4 的表达变化与多种因素有关。

4. Tiam-1

Tiam-1（T lymphoma invasion and metastasis 1）基因是 Hebets 等从 BW5147 小鼠 T 细胞淋巴瘤中分离出的促进转移相关基因。人的 *Tiam-1* 基因定位于 21p22.1，编码 177 kDa 的蛋白质，该基因在正常组织中不表达或低表达（大脑、睾丸组织除外），而在多种肿瘤（如神经母细胞瘤、黑色素瘤、乳腺癌、肺癌、卵巢癌等）中却大量表达，与肿瘤转移相关。Bourguignon 等对人 Tiam-1 蛋白的功能研究表明，Tiam-1 是 Rho GTP 酶（如 Racl）的鸟苷酸（GDP/GTP）互换因子，是小 GTP 结合蛋白中的 ras 超家族成员，它与细胞骨架蛋白（锚蛋白）在体内结合成复合体而发挥调节作用，进而控制细胞形态、黏附、运动、信号转导等，参与诱导肿瘤侵袭和转移。

（二）肿瘤转移抑制基因（MSG）

1. *nm23* 基因

1988 年 Steeg 从 K-1735 小鼠黑色素瘤高转移和低转移细胞株中，用差异筛选（differential screening）和减除杂交（substractive hybridization）技术，发现与癌细胞转移抑制基因，即命名为 *nm23*。高转移细胞株的 *nm23* mRNA 产生比低转移细胞株少 10 倍。Leone 将鼠 *nm23-1* cDNA 转染到具有高转移潜能的 K1735 黑色素瘤细胞株，发现 *nm23* 表达细胞的转移能力较对照组降低 39%，将该克隆接种到动物皮下后成瘤率也有所下降，在人类乳腺癌中，*nm23* 表达高于伴有淋巴结转移的侵袭性导管癌。Hennessy 用 Northern 杂交和免疫组化方法检测乳腺癌 *nm23* 的 mRNA 和蛋白质水平的表达，结果显示 *nm23* 表达与淋巴结转移呈负相关，与疾病生存期呈正相关，可作为乳腺癌转移与预后的参考指标。人类 *nm23* 基因有 2 种，*nm23-H1* 和 *nm23-H2*，其编码产物是一种二磷酸核苷酸激酶（NDPK），NDPK 参与 GDP 磷酸化成 GTP，激活 G 蛋白，进而介导细胞内信号传递，促进微管聚合，与 *mst-1* 基因产物作用相反。因此，*nm23* 与 *mst-1* 两者产物的比值影响细胞中微管的聚合状态，从而改变细胞的黏附能力和运动能力，介导肿瘤转移，两者比值降低则转移率上升。

2. *WDNM* 基因

此基因是最初从大鼠乳腺癌细胞株中利用减除杂交技术分离的一种转移负性调控基因。其转录产物比非转移性细胞株高 20 倍。正常细胞无 *WDNM* 基因表达。有人推测该基因编码的蛋白质可能属于肿瘤特异性抗原，其丢失可使癌细胞逃避体内免疫监视作用而促进转移。WDNM1 蛋白可能通过抑制由肿瘤相关蛋白酶产生的细胞外基质降解来发挥其转移抑制功能。

3. *ela* 基因

用 *H-ras* 癌基因和腺病毒 II 型 *ela* 基因（Ad2-ela）共转染鼠成纤维细胞，后者具有致瘤性而无转移性，单用 *ras* 转染可致细胞转移性，而 *ela* 单独对细胞无转化能力。可见 *ela* 基因具有一定的肿瘤转移抑制功能。Ad2-ela 共转染引起转移能力降低与 MHC-1 类抗原在细胞表面的产生增加有关。Ad2-ela 表达可上调 *MHC-1* 基因的转录，增强转移细胞对免疫细胞的敏感性，诱导宿主免疫细胞的识别和杀伤。*ela* 基因的转移抑制作用还可能与 IV 型胶原酶活性丧失有关，与 *nm23* 表达增加相伴。但 *ela* 基因及其产物的确切生理功能有待进一步探究。

肿瘤转移的基因调控是一个相当复杂的过程。高转移相关基因的过度表达或转移抑制基因的表达减少或缺失及几种基因之间的相互作用，均可导致肿瘤转移。转移的基因调控及针对转移的基因机制的实验治疗正逐步展开。这方面若取得长足的进步，将对改善肿瘤患者的预后具有深远影响。

第四节　表观遗传学修饰与肿瘤

肿瘤的形成受遗传学（genetic）修饰和表观遗传学（epigenetic）修饰的影响。长期以来人们一直认为遗传学修饰参与肿瘤的形成，近年来越来越多的证据表明，表观遗传修饰在肿瘤进展中同样具有非常重要的作用。表观遗传调控可以影响基因转录活性而不涉及 DNA 序列的改变

遗传学修饰是指基于基因序列改变所致基因表达水平变化，如基因突变、基因杂合丢失和微卫星不稳定等；而表观遗传学修饰则是指基于非基因序列改变所致基因表达水平变化，如 DNA 甲基化、组蛋白修饰和小 RNA 介导的沉默等，这三种特殊和互补的后生修饰（后天性）机制中任何一种遭到破坏，都将导致基因表达的异常，并引起癌症的发展或其他"后生的疾病"。尽管在这些精密的潜在机制中尚存在着许多未知之处，近年来我们还是通过实验胚胎学在这个领域得到了大量的证据，在这个领域的进一步研究将深化我们对肿瘤发生机制的理解，并有助于我们制订肿瘤的治疗与预防策略。

1. DNA 甲基化

关于表观遗传学研究较多的就是 DNA 甲基化（DNA methylation）。DNA 甲基化是指一个甲基基团通过共价基团与 CpG 二核苷酸中鸟苷酸的 $5'$ 端的胞嘧啶的 5-C（C^5）位点结合（CpG 二核苷酸的胞嘧啶 $5'$ 碳位共价键结合一个甲基基团）（图 9-2）。结构基因含有很多 CpG 结构，2CpG 和 2GpC 中两个胞嘧啶的 5 位碳原子通常被甲基化，且两个甲基基团在 DNA 双链大沟中呈特定三维结构。基因组中 $60\% \sim 90\%$ 的 CpG 都被甲基化，未甲基化的 CpG 成簇地组成 CpG 岛，位于结构基因启动子的核心序列和转录起始点。大量研究表明，DNA 甲基化在细胞分化过程中扮演着重要角色，DNA 甲基化能引起染色质结构、DNA 构象、DNA 稳定性及 DNA 与蛋白质相互作用方式的改变，从而控制基因表达。值得关注的是，DNA 甲基化异常与多种人类的肿瘤密切相关。DNA 甲基化由 DNA 甲基转移酶（DNMT）催化，在哺乳动物中，DNMT1——持续性 DNA 甲基转移酶，作用于仅有一条链甲基化的 DNA 双链，使其完全甲基化，可参与 DNA 复制双链中的新合成链的甲基化，DNM T1 可能直接与 HDAC（组蛋白去乙酰基转移酶）联合作用阻断转录；DNMT3A 和 3B——从头甲基转移酶，它们可甲基化 CpG，使其半甲基化，继而全甲基化。从头甲基转移酶可能参与细胞生长分化调控，其中 DNM T3B 在肿瘤基因甲基化中起重要作用。

在人类癌症中发现了两种形式的异常 DNA 甲基化：5-甲基-胞嘧啶全部缺失与基因启动子相关的超甲基化（特定的 CpG 岛）。事实上，在各种类型的癌症中均发现了广泛性的低甲基化和 CpG 岛的超甲基化。尽管存在着各种争论，但基因组的低甲基化与细胞原癌基因的激活有关（导致染色体的不稳定），与之相对应的是，基因启动子的超甲基化与基因的失活有关。

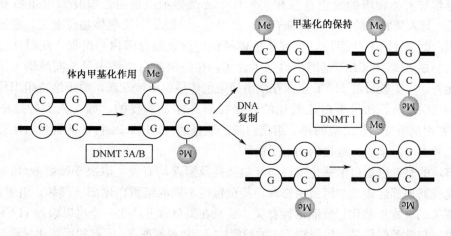

图 9-2　DNA 甲基化

　　因此，DNA 甲基化充当着"双刃剑"的角色，一方面，它通过原位的超甲基化沉默肿瘤抑制基因来促进肿瘤的发生；另一方面，它通过广泛性的低甲基化来触发细胞原癌基因的活化。

　　$p16^{INK4a}$ 是人类肿瘤研究中关于 DNA 甲基化被研究的最多的基因之一。$p16^{INK4a}$ 基因的蛋白质产物很早就被发现是一种肿瘤抑制物和细胞衰老的介导物，p16^{INK4a} 蛋白结合并抑制 CDK4/CDK6-细胞周期调节蛋白-D 激酶的活性，因此维持 pRB 处于它的非磷酸化和生长抑制状态，导致细胞周期停滞于 G_1 期。$p16^{INK4a}$ 启动子的后生性甲基化是可被检测到的多个人类癌症众多后生改变中的一种。此外，$p16^{INK4a}$ 通过启动子超甲基化被沉默具有高度的肿瘤特异性，在某些类型的肿瘤中是最早期的事件，因此，针对该基因制订肿瘤的预防策略具有积极作用。

　　DNA 去甲基化有两种方式：①被动途径，由于核因子 NF 结合甲基化的 DNA，使结合点附近的 DNA 不能被完全甲基化，从而阻断 DNM T1 的作用；②主动途径，是去甲基酶起作用，将甲基集团移去的过程。在 DNA 甲基化阻遏基因表达的过程中，甲基化 CpG 黏附蛋白起着重要作用。虽然甲基化 DNA 可直接作用于甲基化敏感转录因子 E2F、CREB、AP2、c-myc、NF-KB 使它们失去结合 DNA 的功能从而阻断转录，但是，甲基化 CpG 黏附分子可作用于甲基化非敏感转录因子（SP1、CTF、YY1），使它们失活，从而阻断转录。

2. 组蛋白修饰

　　翻译后的染色质蛋白（组蛋白）修饰（histone modification）是调控基因表达的重要表观遗传学机制，在以基因组 DNA 作为模板的细胞发育过程的调节中起着至关重要的作用。在哺乳动物基因组中，组蛋白则可以有很多修饰形式。一个核小体由两个 H2A，两个 H2B，两个 H3，两个 H4 组成的八聚体和缠绕在外面 147bp 的 DNA 组成。组成核小体的组蛋白核心部分状态大致是均一的，游离在外的 N 端则可以受到各种各样的修饰，包括组蛋白末端的乙酰化、甲基化、磷酸化、泛素化等，这些修饰都是组蛋白调节的主要内容，并共同构建"组蛋白密码"，从而扩展和调节基因编码，组蛋白调节具有多种生物学功能，具体包括基因转录，DNA 修复、重组和 DNA 复制，这些过程的异常与人类肿瘤的发生息息相关，几个证据提示组蛋白乙酰化和组蛋白复合体（HAT）与人类肿瘤的染色质修饰有关。已经显示 HAT 与染色体易位有关，通过特定的组蛋白改变 HAT 的活性，导致染色体易位，形成融合蛋白，出现"获得性功能"。这个机制在 *p300* 和 *CBP* 基因中得到了证实，在特定的白血病中可见同源染色体的易位。此外，一定数量的 HAT 蛋白在白血病染色体易位中被发现，对于维持组织稳态而言，加强对 HAT 和组蛋白乙酰化的紧密控制是非常重要的，如在人类肿瘤中筛查 *HAT* 基因的突变。相对应的是，组蛋白脱乙酰基酶（HDAC）负责从组蛋白尾移去乙酰基，并参与组蛋白乙酰化的动态平衡。

有趣的是，组蛋白甲基化也许与复制活性及阻遏均有关，取决于赖氨酸/精氨酸被甲基化或其他邻近残基的组蛋白修饰，甚至包括不同的组蛋白尾部。例如，组蛋白 H3 赖氨酸-4 的 3 次甲基化与复制活性有关，并且在染色体上产生一个可以添加 HAT 的结合位点。与此类似的是，组蛋白 H3 的赖氨酸-36 和赖氨酸-79 位点的甲基化显示与活性染色质和复制活性有关。与此相反的是，组蛋白 H3 的赖氨酸-9 位点的 3 次甲基化与沉默复制活性有关，具体是通过添加异染色质蛋白（HP1）和触发异染色质的形成有关，而组蛋白 H3 的赖氨酸-27 位点的甲基化与复制抑制和通过多分支复合物（PRC1）维持染色质沉默。与之在复制调控中的重要性一致之处在于，许多 HMT（histone methyltransferase）可导致细胞增殖和细胞异常，这与人类包括癌症在内的多种疾病均有关。

3. 微小 RNA 调控

微小 RNA（microRNA，miRNA）在表观遗传修饰中发挥极其重要的作用。这种后天形成的遗传类型小 RNA，指非编码 RNA 中除转录 RNA 以外的部分，包括微小 RNA、小干扰 RNA（small interfering RNA，siRNA）、核仁小分子 RNA 和核小 RNA。自 1993 年 Lee 等在秀丽新小杆线虫体内首次发现 miRNA 编码基因——*lin-4* 以来，人们相继在烟草、果蝇、线虫、斑马鱼、哺乳动物与人体内发现了大量 miRNA 编码基因。生物信息学预测，人类基因组内存在着约 1000 个 miRNA 基因，调控基因组内 1/3 左右基因的表达。研究发现，肿瘤细胞与正常组织来源的细胞之间 miRNA 表达谱具有明显差异，在许多人类肿瘤（肺癌、乳腺癌、脑癌、肝癌、结肠癌、白血病）中发现特异性 miRNA 表达谱系，此外，有些 miRNA 可充当癌基因或抑癌基因，超过 50％的 miRNA 基因位于肿瘤形成相关性脆弱基因位点，提示 miRNA 在肿瘤形成中扮演着重要角色。肿瘤中 miRNA 的过表达，如 *mir-17-92* 具有癌基因的功能，通过负向调节控制细胞分化或凋亡的肿瘤抑制基因和（或）癌基因从而启动肿瘤的发生；而 miRNA 的下调，如 *let-7* 具有抑癌基因的功能，可通过控制细胞分化或凋亡的癌基因抑制肿瘤。miRNA 表达谱分析将成为临床肿瘤诊断和预后的重要工具。

综上所述，表观遗传修饰主要包括 DNA 以及一些与 DNA 密切相关的蛋白质的化学修饰，另外某些非编码的 RNA 也在表观遗传修饰中起着重要的作用。因此表观遗传修饰能从多个水平上调控基因的表达，任何一方面的异常都将影响染色质结构和基因表达，同时也就意味着异常的表观遗传修饰会导致肿瘤发生。因此对表观遗传中各种因子与肿瘤发生、发展相关性的研究将有助于我们了解表观遗传机制，进而指导肿瘤的治疗和新药的研制（DNA 甲基转移酶抑制剂、组蛋白乙酰化酶抑制剂、靶向诱导 DNA 甲基化、小 RNA 的基因治疗等）表观遗传学蓝图已经展现在世人的面前，激励着人们去探索这片有着巨大潜力的前沿领域。

小 结

人类细胞中有几万个基因,这些基因在正常细胞中结构稳定、层次复杂、运行有序,对外界环境能做出迅速、准确应答,并且有自动纠错功能的信号网络。理论上,这样的完善网络不易发生紊乱,然而肿瘤细胞的出现正是这个完善网络发生错乱的结果。肿瘤生长表现为细胞不受控制地异常增生。科学家进一步发现,"失控"的肿瘤细胞,其实是因为癌基因(某些正常细胞生长发育不可缺少的功能性基因)与抑癌基因(一组监控细胞生长的功能性基因)的平衡失调和此消彼长而在人体组织和器官内"作乱"。因此可以说,20 世纪 80 年代初对癌基因和抑癌基因的发现,标志着肿瘤研究真正进入分子肿瘤学时代。

导致肿瘤细胞失控性增生的根本原因,正是细胞周期调控机制遭到破坏(包括驱动机制和监控机制的破坏)。监控机制的任何一个环节遭受破坏都将导致细胞基因组的不稳定性。基因组不稳定性在 DNA 水平上表现为基因突变、缺失、扩增、易位等现象;在染色体水平上表现为染色体畸形、异倍体和多倍体等现象。基因组不稳定使得突变基因的数量增加;同时,很大一部分的癌基因和抑癌基因又是细胞周期调控机制的组成部分。因此,在肿瘤发展过程中,行使监控职能的基因一旦"失察",将会使细胞周期调控机制进一步恶化,并导致细胞周期驱动机制的破坏,表现为强力"驱使"细胞出现癌变性生长。

侵袭和转移是恶性肿瘤的重要特征,也是患者死亡的主要原因。人类对肿瘤转移机制的研究已经有 100 多年历史,随着对肿瘤转移基因调控多阶段、多因素、多步骤理论的不断完善,对于这一复杂病理过程有了更深入的认识。发现并证实与肿瘤转移高度相关的基因成为当前肿瘤转移机制研究的热点,因为这些基因及其产物可能成为肿瘤抗转移治疗新的靶点或观察患者预后的重要指标。

表观遗传学是不涉及 DNA 序列改变的且在细胞分裂过程中可遗传的基因修饰作用,已发现的表观遗传变异有甲基化、组蛋白修饰、微小 RNA 调控等,而 DNA 甲基化作用是表观遗传现象形成的重要机制,将为肿瘤表观遗传学的研究揭开新的一页。

(邱小萍)

参 考 文 献

Arends J W. 2000. Molecular interactions in the vogelstein model of colorectal carcinoma. J Pathol, 190 (4): 412~416

Corson T W, Huang A, Tsao M S et al. 2005. KIF44 is a candidate oncogene in the 1q minimal region of genomic gain in multiple cancers. Oncogene, 1: 3

Crniccc I I, Strittmatter K, Cavallaro U, et al. 2004. Loss of neural cell adhesion molecuulee induces tumor metastasis by up-regulating lymphangiogenesis. Cancer Res, 64: 8630~8638

Croce C, Calin G A, Croce C M. 2006. MicroRNA signatures in human cancers, Nature Reviews Cancer, 6: 857~866

Guo W, Ciancotti F G. 2004. Intgrin signaling during tumour progression. Nat rev Mol Cell Biol. 5: 816~826

Kuroki T, Yendamuri S, Trapasso F etal. 2004. The tumor suppressor gene WWOX at FRAA16D is involved in pancreatic carcinogenesis. Clini Cancer Res, 10 (7): 2459~2465

Luedi P P, Hartemink A J, Jirtle R L et al. 2005. Genome-wide prediction of imprinted murine genes. Genome Research 15: 875~884

Marsit C J, McClean M D, Furniss C S et al. 2006. Epigenetic inactivation of the SFRP genes is associated with drinking, smoking and HPV in head and neck squamous cell carcinoma. International Journal of Cancer, 119: 1761~1766

Peters G, Gil J, Peters G. 2006. Regulation of the INK4b-ARF-INK4a tumour suppressor locus: all for one or one for all. Nature Reviews in Molecular and Cellular Biology 7: 667~677

Port D, Weremowicz S, Chin K et al. 2003. A neural survival factor is a candidate oncogene in breast cancer. Proc Natl Acad Sci USA, 10931~10936

Samowitz W S, Albertsen H, Herrick J et al. 2005. Evaluation of a large, population-based sample supports a CpG island methylator phenotype in colon cancer. Gastroenterology, 129: 837~845

Tarin D, Thompson E W, Newgreennnnn D F. 2005. The fallacy of epithelial mesenchhymal transition in neoplasia. Cancer Res. 65: 5996~6000

Chapter 9　Molecular　Oncology

Section 1　Oncogene

Oncogenes are genes whose normal activity promotes cell proliferation. Gain of function mutations in tumor cells creates forms that are excessively or inappropriately active. A single mutant allele may affect the phenotype of the cell. The non-mutant versions are properly called proto-oncogenes.

1.1　Virus Oncogene

Animal tumor viruses provided the first evidence of oncogenes. Tumor viruses fall into three broad classes: DNA viruses, Retroviruses and Acute transforming retroviruses.

a) DNA viruses normally infect cells lytically. They cause tumors by rare anomalous integrations into the DNA of non-permissive host cells (cells that do not support lytic infection). One way or another, integration of the viral genome implants the transcriptional activation or replication signals of the virus into the host genome and triggers cell proliferation. Some of the viral genes involved have been identified, such as those for the T-antigen of SV40 or E1A and E1B of adenoviruses. Unlike the classic retroviral oncogenes, these genes are virus-specific and do not have exact cellular counterparts.

b) Retroviruses have a genome of RNA. They replicate via a DNA intermediate, which is made using a viral reverse transcriptase. These viruses do not normally kill the host cell (HIV is an exception), and only rarely transform it. The genome of a typical retrovirus consists of three genes, *gag*, *pol* and *env*.

The RNA genome has terminal repeats (R), subterminal unique sequences (U5, U3) and three genes, *gag*, *pol* and *env*. A complicated scheme of splicing and post-translational processing leads to generate a variety of protein products. In an acute transforming retrovirus (bottom), one or more of the viral genes is replaced by a transduced cellular sequence, the oncogene. Initially it is translated into a fusion protein.

c) Acute transforming retroviruses are retrovirus particles which, unlike normal retroviruses, transform the host cell rapidly and with high efficiency. Their genomes include an additional gene, the oncogene. Usually the oncogene replaces one or more essential viral genes, so that these viruses are replication-defective. To propagate them, they are grown in cells which are simultaneously infected with a replication-competent helper virus that supplies the missing functions. Studies of acute transforming retroviruses have revealed more than 50 different oncogenes.

1.2　Cellular Oncogene

1.2.1　Cellular Oncogene and Proto-oncogene

A proto-oncogene is a normal gene that can become an oncogene due to mutations or increased expression. Proto-oncogenes code for proteins that help to regulate cell growth and differentiation. Proto-oncogenes are often involved in signal transduction and execution of mitogenic signals, usually through their protein products. Upon *activation*, a proto-oncogene (or its product) becomes a tumor inducing agent, an oncogene.

The first oncogene was discovered in 1970 and was termed *src* (pronounced *sarc* as in *sarcoma*). src was in fact first discovered as an oncogene in a chicken retrovirus. Experiments performed by Dr G. Steve Martin of the University of California, Berkeley demonstrated that the SRC was indeed the oncogene of the virus.

In 1976 Drs. J. Michael Bishop and Harold E. Varmus of the University of California, San Francisco demonstrated that oncogenes were defective proto-oncogenes, found in many organisms including humans. For this discovery Bishop and Varmus were awarded the Nobel Prize in 1989.

Oncogenes are mutated versions of genes involved in a variety of normal cellular functions.

It quickly became apparent that normal cells had counterparts of all the retroviral oncogenes and in fact that *v-onc* genes were transduced cellular genes. With a few exceptions, the *v-onc* gene products differ from their *c-onc* (proto-oncogene) counterparts by amino acid substitutions or truncations, which serve to activate the proto-oncogene.

Functional understanding of oncogenes began with the discovery in 1983 that the viral oncogene *v-sis* was derived from the normal cellular platelet-derived growth factor B (*PDGFB*) gene. Uncontrolled over-expression of a growth factor would be an obvious cause of cellular hyperproliferation. The roles of many cellular oncogenes (strictly speaking, proto-oncogenes) have now been elucidated. Gratifyingly, they turn out to control exactly the sort of cellular functions that would be predicted to be disturbed in cancer. Five broad classes can be distinguished: ① secreted growth factors (e.g. *SIS*); ② cell surface receptors (e.g. *ERBB*, *FMS*); ③ components of intracellular signal transduction systems (e.g. the *RAS* family, *ABL*); ④ DNA-binding nuclear proteins, including transcription factors (e.g. *MYC*, *JUN*); ⑤ components of the network of cyclins, cyclin-dependent kinases and kinase inhibitors that govern progress through the cell cycle (e.g. *MDM2*).

1. 2. 2 Classification and Function of Cellular Oncogene

There are several systems for classifying oncogenes, but there is not yet a widely accepted standard. They are sometimes grouped both spatially (moving from outside the cell inwards) and chronologically (parallelling the "normal" process of signal transduction). There are several categories that are commonly used.

1. 3 Activation Mechanism of Oncogene

Some of the best illustrations of molecular pathology in action are furnished by the various ways in which proto-oncogenes can become activated. Activation involves a gain of function. This can be quantitative (an increase in the production of an unaltered product) or qualitative (production of a subtly modified product as a result of a mutation, or production of a novel product from a chimeric gene created by a chromosomal rearrangement). These changes are dominant and normally affect only a single allele of the gene.

1. 3. 1 Gene Mutation

Activating mutations in oncogenes are somatic events. Constitutional mutations would probably be lethal. We have met one exception to this: specific activating point mutations in the *RET* oncogene cause multiple endocrine neoplasia or familial thyroid cancer, and sometimes these mutations are inherited. It is very unusual to be able to build a functioning organism out of cells containing an activated oncogene. These *RET* mutations must affect the behavior of only very specific cells in very special circumstances. Note however that nonactivating mutations in proto-oncogenes may be inherited constitutionally, if their effect is unrelated to cancer. For example, inherited mutations that inactivate the *KIT* oncogene produce piebaldism, while inherited loss-of-function mutations in *RET* predispose to Hirschsprung's disease.

The *H-RAS* gene is one of a family of *ras* genes that encode proteins involved in signal transduction from G-protein-coupled receptors. A signal from the receptor triggers binding of GTP to the RAS protein, and GTP-RAS transmits the signal onwards in the cell. RAS proteins have GTPase activity, and GTP-RAS is rapidly converted to the inactive GDP-RAS. Specific point mutations in *RAS* genes are frequently found in cells from a variety of tumors including colon, lung, breast and bladder cancers. These lead to amino acid substitutions that decrease the GTPase activity of the RAS protein. As a result, the GTP-RAS signal is inactivated more slowly, leading to excessive cellular response to the signal from the receptor.

1. 3. 2 Gene Insertion

Gene insertion, the incorporation of exogenous genetic material into a genome, can occur naturally or can be artificially induced.

In nature, mobile elements called insertion sequences exist. They encode only the information necessary for their insertion into DNA. Depending upon the insertion sequence, they can insert at specific regions or at random. As they do not carry other genes, they tend not to be used as tools of genetic research. However, since both their Insertion and exit from DNA can be disruptive, leading to the development of mutations, knowledge of their behavior is relevant to researchers. One of the disruptions, insertional duplication, is the duplication of a short region of the insertion sequence flanking the sequence itself. If the insertion sequence subsequently exits the genome, this duplicated area may be left behind, which can be disruptive if the host region of DNA codes for a protein. Another disruption to host DNA is the deletion of some of the host sequences upon integration of the insertion sequence.

1. 3. 3 Chromosomal Translocation

Tumor cells typically have grossly abnormal karyotypes, with multiple extra and missing chromosomes, many translocations and so on. Most of these changes are random, and reflect a general genomic instability which is a normal part of carcinogenesis. A huge research effort has been devoted to picking out tumor-specific changes superimposed on the background of random changes. Over 150 different tumor-specific breakpoints have now been recognized, and they reveal an important common mechanism in tumorigenesis.

The best-known tumor-specific rearrangement produces the Philadelphia (Ph[1]) chromosome, a very small acrocentric chromosome seen in 90% of patients with chronic myeloid leukemia. This chromosome turns out to be produced by a balanced reciprocal 9; 22 translocation. The breakpoint on chromosome 9 is within an intron of the *ABL* oncogene. The translocation joins most of the *ABL* genomic sequence onto a gene called *BCR* (breakpoint cluster region) on chromosome 22, creating a novel fusion gene. This chimeric gene is expressed to produce a tyrosine kinase related to the *ABL* product but with abnormal transforming properties.

Burkitt's lymphoma is a childhood tumor common in malarial regions of Central Africa and Papua New Guinea. Mosquitoes and Epstein-Barr virus are believed to play some part in the etiology, but activation of the *MYC* oncogene is a central event. A characteristic chromosomal translocation, t (8; 14) (q24; q32) is seen in 75%~85% of patients . The remainder have t (2; 8) - (p12; q24) or t (8; 22) (q24; q11). Each of these translocations puts the *MYC* oncogene close to an immunoglobulin locus, *IGH* at 14q32, *IGK* at 2p12 or *IGL* at 22q11. The Burkitt's lymphoma translocations do not create novel chimeric genes. Instead, they put the oncogene in an environment of chromatin that is actively transcribed in antibody-producing B-cells. Usually exon 1 (which is noncoding) of the *MYC* gene is not included in the translocated material. Deprived of its normal upstream controls, and placed in an active chromatin domain, *MYC* is expressed at an inappropriately high level.

Many other chromosomal rearrangements put one or another oncogene into the neighborhood of either an immunoglobulin (*IGG*) or a T-cell receptor (*TCR*) gene. Presumably the rearrangements arise by random malfunctioning of the recombinases that rearrange *IGG* or *TCR* genes during maturation of B and T cells, and are then selected for their growth advantage. Predictably, these rearrangements are characteristic of leukemias and lymphomas, but not solid tumors.

Many other rearrangements are known that produce chimeric genes. The products are normally transcription factors (or sometimes tyrosine kinases) which take their target specificity from one component gene, but couple it to an activation or ligand-binding domain from the other. This has been one of the most satisfying stories to emerge from cancer research, with several examples of clinical phenotypes being elegantly explained by a combination of cytogenetic and molecular genetic findings.

1. 3. 4 Gene Amplification

Many cancer cells contain multiple copies of structurally normal oncogenes. Breast cancers often amplify *ERBB2* and sometimes *MYC*; a related gene *NMYC* is usually amplified in late-stage neuroblastomas. Hundreds of extra copies may be present. They can exist as small separate chromosomes (*double minutes*) or as insertions within the normal chromosomes (*homogeneously staining regions*, *HSR*). The genetic events producing HSRs may be quite complex because they usually contain sequences derived from several different chromosomes. Similar gene amplifications are often seen in noncancer cells exposed to strong selective regimes—for example amplified dihydrofolate reductase genes in cells selected for resistance to methotrexate. In all cases the result is greatly to increase the level of gene expression.

1.4　Newly Discovered Oncogenes

1.4.1　*HOXA1* (*Hox* genes A1) Discovered in Breast Cancer

The *HOX* network contains 39 genes that act as transcriptional regulators and control crucial cellular functions during both embryonic development and adult life. Inside the network, this is achieved according to the rules of temporal and spatial co-linearity with $3'$ *HOX* genes acting on the anterior part of the body, central *HOX* genes on the thoracic part and lumbo-sacral *HOX* genes on the caudal region. *HOX* gene is expressed in normal breast tissue and in primary breast cancers by reverse-transcriptase-polymerase chain reaction (RT-PCR). 17 out of 39 *HOX* genes were expressed in the normal breast tissue. The expression of thoracic *HOX* genes tended to be similar in normal and neoplastic breast tissues suggesting that these genes are involved in breast organogenesis.

1.4.2　*CDC91 L1* (PIG-U) Discovered in Bladder Cancer

CDC91 L1 (*cell division cycle 91 -lik*, also called phosphatidylinositol glycan class U (PIG-U)) encodes for a transamidase complex unit in the glycosylphosphatidylinositol (GPI) anchoring pathway. *CDC91 L1* was amplified and overexpressed in about one-third of bladder cancer cell lines and primary tumors, as well as in oncogenic uroepithelial cells transformed with human papillomavirus (HPV) E7. Forced overexpression of *CDC91 L1* malignantly transformed NIH3T3 cells *in vitro* and *in vivo*. Overexpression of *CDC91 L1* also resulted in upregulation of the urokinase receptor (uPAR), a GPI-anchored protein, and in turn increased STAT-3 phosphorylation in bladder cancer cells.

1.4.3　Oncogenes Associated With Multiple Human Cancer

Kinesins comprise a superfamily of motor proteins that impact a wide array of cellular functions by coupling ATP hydrolysis to the regulated and targeted movement of specific intracellular cargo along microtubule filaments. Kinesins have been functionally linked to various biological phenomena, including, but not limited to, cargo-containing vesicle transport, mitotic spindle formation, chromosome segregation, midbody formation, and cytokinesis completion. The kinesin superfamily has been subdivided into 14 kinesin families and can be further defined by the position of the motor domain at the N terminus (N type), C terminus (C type), or internal region (I type). High KIF14 (Kinesin familily member 14, also known as KIF14), levels contribute to uncontrolled proliferation and cancer cell aneuploidy, perhaps by stimulating premature cytokinesis and/or evasion of the spindle checkpoint. Overexpression may also enable cells to bypass apoptosis because KIF14 knockdown increases apoptosis at multiple stages of the cell cycle.

Section 2　Tumor Suppressor Gene

2.1　Concept and Discovery of Tumor Suppressor Gene

Cell fusion experiments show that the transformed phenotype can often be corrected *in vitro* by fusion of the transformed cell with a normal cell. This provides evidence that tumorigenesis involves not only dominant activated oncogenes, but also recessive, loss-of-function mutations in other genes. These other genes are the tumor suppressor (TS) genes. Sometimes TS genes are called antioncogenes, but that is an unhelpful name because it wrongly implies that they are all specific antagonists or inhibitors of oncogenes. Some may be, but like oncogenes, TS genes can have a variety of functions.

TS genes have been discovered by three main routes:
- positional cloning of the genes causing rare familial cancers.

- defining chromosomal locations commonly deleted in tumor cells (by loss of heterozygosity analysis or comparative genomic hybridization).
- testing tumors for mutations in genes known to be involved in cell cycle regulation.

The rare eye tumor, retinoblastoma, has been the main test-bed for defining the concepts and methods of TS gene research.

2.2　Tumor Suppressor Genes

2.2.1　Rb

In humans, the protein is encoded by the *Rb* gene located on 13q14. 1-q14. 2. If both alleles of this gene are mutated early in life, the protein is inactivated and results in development of retinoblastoma cancer, hence the name *Rb*. It is not known why an eye cancer results from a mutation in a gene that is important all over the body.

Two forms of retinoblastoma were noticed: a bilateral, familial form and a unilateral, sporadic form. Sufferers of the former were 6 times more likely to develop other types of cancer later in life. This highlighted the fact that mutated *Rb* could be inherited and lent support to the two-hit hypothesis. This states that only one working allele of a tumor suppressor gene is necessary for its function (the mutated gene is recessive), and so both need to be mutated before the cancer phenotype will appear. In the familial form, a mutated allele is inherited along with a normal allele. In this case, should a cell sustain only one mutation in the other *Rb* B gene, all pRb in that cell would be ineffective at inhibiting cell cycle progression, allowing cells to divide uncontrollably and eventually become cancerous. Furthermore, as one allele is already mutated in all other somatic cells, the future incidence of cancers in these individuals is observed with linear kinetics. The working allele need not undergo a mutation *per se*, as loss of heterozygosity is frequently observed in such tumors.

However, in the sporadic form, both alleles would need to sustain a mutation before the cell can become cancerous. This explains why sufferers of sporadic retinoblastoma are not at increased risk of cancers later in life, as both alleles are functional in all their other cells. Future cancer incidence in sporadic *Rb* cases is observed with polynomial kinetics, not exactly quadratic as expected because the first mutation must arise through normal mechanisms, and then can be duplicated by LOH to result in a tumor progenitor.

pRb prevents the cell from replicating damaged DNA by preventing its progression along the cell cycle through G1 (first gap phase) into S (synthesis phase). pRb binds and inhibits transcription factors of the E2F family, which are composed of dimers of an E2F protein and a DP protein. The transcription activating complexes of E2 promoter-binding-protein-dimerization partners (E2F-DP) can push a cell into S phase. As long as E2F-DP is inactivated, the cell remains stalled in the G1 phase. When pRb is bound to E2F, the complex acts as a growth suppressor and prevents progression through the cell cycle. The pRb-E2F/DP complex also attracts a histone deacetylase (HDAC) protein to the chromatin, further suppressing DNA synthesis.

2.2.2　p53

In humans, p53 is encoded by the *TP53* gene located on the short arm of chromosome 17 (17p13. 1). p53 (also known as protein 53 or tumor protein 53), is a transcription factor. p53 is important in multicellular organisms, where it regulates the cell cycle and thus functions as a tumor suppressor that is involved in preventing cancer. As such, p53 has been described as "the guardian of the genome", "the guardian angel gene" and "the master watchman", referring to its role in conserving stability by preventing genome mutation.

Mutations that deactivate p53 in cancer usually occur in the DBD. Most of these mutations destroy the ability of the protein to bind to its target DNA sequences, and thus prevents transcriptional activation of these genes. As such, mutations in the DBD are recessive loss-of-function mutations. Molecules of p53 with mutations in the OD dimerise with wild-type p53, and prevent them from activating transcription.

Therefore OD mutations have a dominant negative effect on the function of p53.

Wild-type p53 is a labile protein, comprising folded and unstructured regions which function in a synergistic manner. p53 has many anti-cancer mechanisms:

- It can activate DNA repair proteins when DNA has sustained damage.
- It can also hold the cell cycle at the G_1/S regulation point on DNA damage recognition (if it holds the cell here for long enough, the DNA repair proteins will have time to fix the damage and the cell will be allowed to continue the cell cycle).
- It can initiate apoptosis, the programmed cell death, if the DNA damage proves to be irreparable.

2.2.3 *p16* Gene

p16 is a tumor suppressor gene. Mutations in *p16* increase the risk of developing a variety of cancers, notably melanoma. *p16* is an important gene in regulating the cell cycle.

Recent research on the *p16* gene indicates that its increased expression as organisms age reduces the proliferation of stem cells. This reduction in the division and production of stem cells protects against cancer while increasing the risks associated with cellular senescence.

The *p16* gene has an alternative reading frame (ARF) that encodes the p14 ARF protein. This is involved in stabilising p53 and is positively regulated by E2F.

p16 regulates the cell cycle by binding and deactivating various cyclin-CDK complexes. A study published in 2007 in the New England Journal of medicine established that there is a strong association between polymorphisms on chromosome 9p21.3 (SNP, rs1333049) and coronary artery disease. This region codes for the INK4 proteins p16^{INK4a} and p15^{INK4b}. The corresponding genes are CDKN2A and CDKN2B. The proteins may inhibit cell growth induced by Transforming Growth Factor-beta (TGF-β) (Figure 9-1).

Figure 9-1　Function of *p16* in the cell cycle (see page 291)

2.2.4　p15

Transcription of the p15^{INK4B} tumor suppressor gene, which encodes a cyclin-dependent kinase inhibitor, is positively regulated by transforming growth factor-β (TGF-β).

2.2.5　*p27* (and *p21*)

The *p27* ^{Kip1} gene has a DNA sequence similar to other members of the "Cip/Kip" family which include the *p21*$^{Cip1/Waf1}$ and *p57*^{Kip2} genes. In addition to this structural similarity the "Cip/Kip" proteins share the functional characteristic of being able to bind several different classes of Cyclin and Cdk molecules. For example, p27^{Kip1} binds to cyclin D either alone, or when complexed to its catalytic subunit Cdk4. In doing so p27^{Kip1} inhibits the catalytic activity of Cdk4, which means that it is prevents Cdk4 from adding phosphate residues to its principal substrate, the retinoblastoma (pRb) protein. Increased levels of the p27^{Kip1} protein typically cause cells to arrest in the G1 phase of the cell cycle. Likewise, p27^{Kip1} is able to bind other Cdk proteins when complexed to cyclin subunits such as Cyclin E/Cdk2 and Cyclin A/Cdk2.

2.2.6　APC

APC (adenomatosis polyposis coli) gene is classified as a tumor suppressor gene. The APC gene is located on the long (q) arm of chromosome 5 between positions 21 and 22, from base pair 112 118 468 to base pair 112 209 532. The protein made by the APC gene plays a critical role in several cellular processes that determine whether a cell may develop into a tumor. The APC protein helps control how often a cell divides, how it attaches to other cells within a tissue, or whether a cell moves within or away from a tissue. This protein also helps ensure that the chromosome number in cells produced through cell division is correct. The APC protein accomplishes these tasks mainly through association with other proteins, especially those that are involved in cell attachment and signaling. The activity of one protein in particular, B-catenin, is controlled by the APC protein (Wnt signaling pathway). Regulation of β-catenin prevents genes that stimulate cell division from being turned on too often and prevents cell overgrowth.

2.3 Newly Discovered Tumor Suppressor Genes

2.3.1 ST18 is a Newly Discovered TS Gene in Breast Cancer

ST18 (suppression of tumorigenicity 18,) gene is located within a frequent imbalanced region of chromosome 8q11. The ST18 gene encodes a zinc-finger DNA-binding protein with six fingers of the C2HC type (configuration Cys-X_5-Cys-X_{12}-His-X_4-Cys) and an SMC domain. ST18 has the potential to act as transcriptional regulator. ST18 is expressed in a number of normal tissues including mammary epithelial cells although the level of expression is quite low. In breast cancer cell lines and the majority of primary breast tumors, ST18 mRNA is significantly downregulated. A 160 bp region within the promoter of the ST18 gene is hypermethylated in about 80% of the breast cancer samples and in the majority of breast cancer cell lines. The strong correlation between ST18 promoter hypermethylation and loss of ST18 expression in tumor cells suggests that this epigenetic mechanism is responsible for tumor-specific downregulation. The ectopic ST18 expression in MCF 7 breast cancer cells strongly inhibits colony formation in soft agar and the formation of tumors in a xenograft mouse model.

2.3.2 *WWOX* is a Newly Discovered TS Gene in Pancrea Cancer

The ORF of *WWOX* is 1245 bp long, encoding a 414 amino acid protein. This gene is composed of nine exons. Loss of heterozygosity at the *WWOX* locus was observed in 4 cancer of pancreas primary tumors (27%). Methylation analysis showed that site-specific promoter hypermethylation was detected in two cell lines (22%) and treatment with the demethylating agent 5-aza-2'-deoxycytidine demonstrated an increase in the expression of *WWOX*. In addition, two primary tumor samples (13%) showed promoter hypermethylation including the position of site-specific methylation. Transcripts missing *WWOX* exons were detected in 4 cell lines (44%) and in two tumor samples (13%).

2.3.3 *BLU* is a Newly Discovered TS Gene in Nasopharyngeal Carcinoma (NPC)

BLU, also known as *ZMYND10* (zinc finger, MYND-type containing 10), spans 4.5 kb on 3p21.3. It encodes a 50 kDa protein containing a DUF500 protein binding motif on its N terminus and a MYND zinc finger DNA binding domain in its C terminus, which is commonly found in transcription repressors. *BLU* is identified as TS gene in NPC by examining both genetic and epigenetic changes of this gene in primary NPC tumors and cell lines. It was observed frequent alteration of mRNA expression of *BLU* in NPC. Absent or reduced mRNA levels of *BLU* were found in 78% of primary tumors and in all tumor cell lines examined, suggesting that the *BLU* function had been abrogated by transcriptional repression in most NPCs. It is suspected that *BLU/ZMYND10* has a function in cell cycle progression. *BLU/ZMYND10* is a stress-responsive gene regulated by E2F. It is commonly found to be downregulated in non-small cell lung cancer, esophagus squamous cell carcinoma and NPC. *BLU/ZMYND10* is inactivated or downregulated in 78% of primary NPC tumors and promoter hypermethylation was detected in up to 74% of the tumors. This suggested that *BLU/ZMYND10* plays a critical role in the tumorigenesis of NPC.

2.4 Tumor Suppressor Gene and Cell Cycle

Any cell at any time has three choices of behavior: it can remain static, it can divide or it can die (apoptosis). Some cells also have the option of differentiating. Cells select one of these options in response to internal and external signals. Oncogenes and tumor suppressor genes play key roles in generating and interpreting these signals.

Life would be very simple if the signal and response were connected by a single linear pathway, but this seems never to be the case. Rather, multiple branching, overlapping and partially redundant pathways control the behavior of the cell. Probably such complicated networks are necessary to confer stability and resilience on the extraordinarily complex machinery of a cell. Experimentally, unravelling the precise genetic circuitry of the controls is exceedingly difficult, partly because of their complexity and partly because it is difficult to distinguish direct from indirect effects in transfection or knockout experiments. DNA array technology is the best hope for dealing with this problem. Hybridizing poly $(A)^+$ RNA to arrayed probes representing hundreds or thousands of genes allows significant changes to be picked out from the web of interactions. Normal and cancer cells can be compared, or transfected cells and controls, to get an overall snapshot of the state of the web in a certain cellular condition.

2. 4. 1　Function of pRb, the *Rb* Gene Product

The *Rb* gene is widely expressed, encoding a 110 kDa nuclear protein, pRb. In normal cells pRb is inactivated by phosphorylation and activated by dephosphorylation. Active (dephosphorylated) pRb binds and inactivates the cellular transcription factor E2F1, function of which is required for cell cycle progression. The G_1-S checkpoint seems to be the most crucial in the cell cycle; 2~4 hours before a cell enters Sphase, pRb is phosphorylated. This releases the inhibition of E2F1 and allows the cell to proceed to S phase. Phosphorylation is governed by a cascade of cyclins, cyclin-dependent kinases and cyclin kinase inhibitors.

Rb gene mutations produce sporadic or inherited retinoblastoma, this being the classic example of Knudson's two-hit hypothesis. It is not clear why constitutional mutation of a gene so fundamental to cell cycle control should result specifically in retinoblastoma and a small number of other tumors, principally osteosarcomas. However, this is a common theme in molecular pathology: mutation of a gene produces a phenotypic effect in only a subset of the cells or tissues in which the gene is expressed and appears to have a function. The product of the *MDM2* oncogene (which is amplified in many sarcomas) binds and inhibits pRb, thus favoring cell cycle progression. Several viral oncoproteins (adenovirus E1A, SV40-T antigen, human papillomavirus E7 protein) also bind and sequester or degrade pRb.

2. 4. 2　Function of p53, the *TP53* Gene Product

p53 was first described in 1979 as a protein found in SV40-transformed cells, where it associated with the T-antigen. Later, the *TP53* gene which encodes p53 appeared as a dominant transforming gene in the 3T3 assay, and so was classed as an oncogene. Subsequently it transpired that while p53 from some tumor cells was oncogenic, p53 from normal cells positively suppressed tumorigenesis.

Loss or mutation of *TP53* is probably the commonest single genetic change in cancer. This reflects the central importance of p53, which has several functions in the cell. One is as a transcription factor. Tetramers of p53 bind DNA and can activate transcription of reporter genes placed downstream of a p53 binding site. However, p53 is believed to have a much broader role in the cell, which has been summarized as "the guardian of the genome". One of its guardian functions is to stop cells replicating damaged DNA. Normal cells with damaged DNA arrest at the G_1-S cell cycle checkpoint until the damage is repaired, but cells that lack p53 or contain a mutant form do not arrest at G_1. Replication of damaged DNA presumably leads to random genetic changes, some of which are oncogenic, similar to cells with a defective mismatch repair system.

Probably related to this, p53 has a crucial role in cell death. In response to oncogenic stimuli, cells undergo apoptosis (programmed cell death). Apoptosis has come to occupy a central place in our understanding of the cancer process. Tumor cells lacking p53 do not undergo apoptosis, and so escape the control. p53 may be knocked out by deletion, by mutation or by the action of an inhibitor such as the *MDM 2* gene product (which binds p53 and targets it for degradation; MDM2 also binds pRb, see above) or the E6 protein of papillomavirus.

Loss of heterozygosity assays confirmed the status of *TP53* as a tumor suppressor gene. *TP53* maps to 17p12, and this is one of the commonest regions of loss of heterozygosity in a wide range of tumors. Tumors that have not lost *TP53* very often have mutated versions of it. To complete the picture of *TP53* as a TS gene, constitutional mutations in *TP53* are found in families with the dominantly inherited Li-Fraumeni syndrome. Affected family members suffer multiple primary tumors, typically including soft tissue sarcomas, osteosarcomas, tumors of the breast, brain and adrenal cortex, and leukemia.

Section 3　Metastasis

Carcinomas are the most frequent type of human malignancies, and the vast majority of cancer deaths are caused by the formation of metastases rather than by the primary tumor itself. However, metastases are difficult to target by conventional cancer therapies, a curative regimen requires the detection of the disease before the primary tumor has spread. Yet, despite evolving clinical diagnostic tools, a large proportion of tumors have already metastasized by the time of diagnosis. The selective process of metastasis requires that cancer cells successfully complete several sequential, rate-limiting steps. They must detach from the primary tumor, invade the host stroma, intravasate into lymphatic or blood vessels spread to the capillary bed of distant organs, extravasate and proliferate in the receptive organ parenchyma. The succeeding metastatic cells have undergone changes in their proliferative, survival, migratory and invasive abilities and can be seen as winners of a "metastasis decathlon". It is now well established that tumor cell-autonomous changes are not sufficient to allow tumor progression and metastasis to occur. By analogy with the architecture of organs, tumors are not only composed of a "parenchyma" formed by the neoplastic cells, but also of a supportive "stroma", consisting of specific extracellular matrix (ECM) components, fibroblasts, adipocylcs, vascular cells, smooth muscle cells and cells of the haematopoietic system. During tumor progression and metastasis, an active crosstalk occurs between tumor cells and their stroma, mainly mediated by direct cell-cell contact or paracrine cytokine and growth factor signaling. reminiscent of the communication between epithelialing and mesenchymal cells during embryonic development. Such signaling may activate the tumor microenvironment at the primary and secondary tumor sites, thereby undergoing morphological changes (desmoplasia) and allowing or even supporting tumor outgrowth, invasion and metastasis. Hence, tumor cell-stroma interaction is an important new focus of research in the treatment of metastatic disease.

3.1　Loss of E-cadherin Function and Epithelial-Mesenchymal Transition

E-cadherin is a central player in the makeup of cell polarity and epithelial organization. With its extracelluar domains, this cadherin mediates, calcium-dependent homophilic cell-cell contact in adhesion junctions, while linking adhesion junctions to the actin cytoskeleton via a cytoplasmic cell adhesion complex consisting of α-catenin, β-catenin, γ-catenin, and p120 catenins. Thereby cell-cell adhesion can affect localization and function of cytoskeletal regulators and influence actin-guided cell motility. In most cancers of epithelial origin, E-cadherin-mediated cell-cell adhesion is most concomitantly with tumor progression and is correlated with advanced tumor grades and poor patient survival. Consistent with this observation, forced down regulation of E-cadherin function in a mouse model of carcinogenesis promotes tumor progression, invasion and metastasis. These and many more results from a variety of *in vivo* and *in vitro* experimental systems demonstrate that E-cadherin is an important molecule in tumor progression.

How does the loss of E-cadherin promote metastasis? In order to leave the coherent epithelial cell assembly of the primary tumor, malignant tumor cells have to acquire a fibroblastoid, migratory, and invasive phenotype. A similar process physiologically occurs during embryonic development, tissue remodeling, wound healing and inflammation and is termed epithelial-to-mesenchymal transition (EMT). A number of in vitro studies collectively support the idea that morphologic and molecular changes of tumor cells required for the first step of mechanisms mimic physiological EMT. In fact, full EMT of in vitro cultured tumor cells is thought to resemble, at least in part, the metastatic process in cancer patients. Nevertheless, this hypothesis is still under debate and evidence for carcinoma EMT *in vivo* has yet to be found . We support the view that EMT-like events occur in tumors and that studying these events may contribute to our understanding of metastasis formation.

Oncogenic events in tumor cells as well as growth factors secreted by tumor and stromal cells, including transforming growth factor-β (TGF-β) and fibroblast growth factor-2 (FGF-2), induce EMT. During EMT, cells progressively redistribute or downregulate their apical and basolateral epithelial-specific proteins, such as tight and adherens junction proteins (including E-cadherin), and re-express mesenchymal molecules, such as vimentin and N-cadherin. These changes lead to the abrogation of cell-cell contacts and the gain of cell motility necessary for invasion EMT is in contrast to the process of cell scattering, where different epithelial markers are reversibly downregulated and expression of mesenchymal proteins is not induced. Scattering can be mediated by a number of growth factors, including hepatocyte growth factor/scatter factor (HGF/SF), TGF-β, FGF-2, TGF-α and epidermal growth factor (EGF) via the PI3K or the Raf/MAPK signaling pathways. Loss of E-cadherin promotes, whereas maintenance of its expression inhibits, EMT and metastasis. E-cadherin is therefore considered a metastasis suppressor gene. Downregulation of E-cadherin mostly occurs at the transcriptional level. The E-cadherin promoter is frequently repressed by specific transcriptional repressors, including Snail 1 (previously Snail), Snail 2 (previously Slug), SIPI, δEF1, Twist and E12/E47, and by subsequent promoter hypermethylation. Consistent with this observation, some of these repressors have been found expressed specifically at the invasive front of human invasive hepatocellular and breast carcinoma. The expression of these repressors seems to be highly regulated by pathways, including canonical Wnt signaling, TGF-β, FGF, EGF, Stat3 and nuclear factor-κB (NF-κB) signaling. Notably, Snail 1 is a highly unstable protein. It is rapidly phosphorylated by glycogen synthase kinase 3β (GSK3β) and, subsequently-ubiquitylated and degraded by the proteasome pathway. As a result of transcriptional repression, the E-cadherin promoter is frequently found hyper-methylated in a large subset of cancer cases. E-cadherin can also be downregulated at the protein level. Receptor tyrosine kinases (RTKs), such as EGFR, c-Met, IGFIR, FGFR and the non-receptor tyrosine kinase c-Src can induce phosphorylation of E-cadherin and catenins, resulting in their ubiquitylation by the E3 ligase and subsequent endocytosis and degradation. Finally, secreted proteases such as matrix metalloprotease (MMP) -9, for example induced by TGF-β and I-IGF/SF, can cleave E-cadherin and disrupt cadherin-mediated cell-cell contacts.

Loss of E-cadherin during EMT disrupts adhesion junctions between neighboring cells and thereby supports detachment of malignant cells from the epithelial cell layer. Yet, migration and invasion of tumor cells is also promoted by the loss of interaction of E-cadherin with the cytoskeleton, subsequent changes in the activities of Rho family GTPases, most prominently Rac1, Cdc42 and RhoA, and the concomitant reorganization of the actin cytoskeleton. In epithelial cells, E-cadherin-mediated assembly of adherens junctions recruits and activates Rac1 and Cdc42. Thereby, E-cadherin-bound α-and p120-catenin directly interact with Rho family-specific guanine nucleotide exchange factors (GEFs) and with Rho family members. Moreover, activated Rac1 and Cdc42 promote and consolidate E-cadherin-mediated adhesion by sequestering their downstream effector IQGAP1, which in its free form inhibits the interaction of β-catenin with α-catenin/E-cadherin. After an early activation phase, where RhoA is required for E-cadherin clustering, RhoA gets gradually downregulated with a concomitant repression of cell migration. In contrast, forced expression of RhoA (and also RhoC) in tumor cells induces focal adhesions and stress fibers, promoting invasion and metastasis. Upon loss of E-cadherin function, RhoA, Rac1 and Cdc42 are released from adhesion junctions and promote cell migration and invasion. Consistent with the dual role of these molecules, the Rac1-specific GEF Tiam 1 localizes to adhesion junctions and promotes cell adhesion on substrates impeding cell migration (fibronectin, laminin), whereas it is found in lamellipodia when cells are cultured on cell motility-supporting collagen type 1. However, the mechanisms, by which ECM and thus integrin-mediated signaling affect the activity of RhoA, Rac1 and Cdc42, and with it E-cadherin function, remain to be elucidated.

Recent studies have shown that mesenchymal cadherins, in particular N-cadherin, enhance tumor cell motility and migration, thus exhibiting an opposite effect compared with E-cadherin. N-cadherin-induced tumor cell invasion can even overcome E-cadherin-mediated cell-cell adhesion. A novel concept based on the above observations is that a "cadherin switch" from epithelial to mesenchymal cadherins exists which supports the transition from a benign to an invasive, malignant tumor phenotype. Hence, in addition to the loss of E-cadherin, the gain of N-cadherin may critically contribute to tumor invasion and metastatic dissemination. In fact, N-cadherin has been show to induce tumor cell migration, most likely by stimulating FGF receptor-mediated signal transduction. Hence, changes in the expression of cadherins not only modulate cell-cell adhesion but also induce pro-metastatic signaling pathways.

3. 2　Integrins and Tumor Progression

In order to detach and migrate, tumor cells depend on changes in both cell-cell and cell-matrix interactions. It is assumed that strong cell-matrix adhesions need to be dissolved, whereas transient and weak adhesions are a prerequisite for migration. This complex situation has made it difficult to experinlentally determine the functional contribution of cell-matrix adhesion to tumor progression. With integrins being the prototype mediators of cell-matrix adhesion, they have been studied in great detail, also in tumor progression. Integrins are heterodimeric cell surface receptors consisting of two type-1 transmembrane subunits, α and β. They provide the essential link between the actin cytoskeleton and the ECM during cell migration. Different α-subunits and β-subunits form distinct integrin subtypes, which link ECM ligands, such as fibronectin, vitronectin, laminin and collagen, to the intracellular actin cytoskeleton. Importantly, binding to these ECM components activates integrins, which in turn induce intracellular signaling cascades modulating cell proliferation, survival, polarity, motility and differentiation. Thereby integrins mediate anchorage dependence, since they allow these processes only in appropriately adhering cells. However, during tumor progression, mutations in oncogenes and tumor-suppressor genes can enable nonadherent neoplastic cells to survive and proliferate in the absence of proper cell-matrix adhesion. Cancer cells that profit from integrin signals are being selected due to their survival and proliferation advantage. Many cancer types, including melanoma, glioblastoma, prostate, breast and ovarian cancer, exhibit increased expression of $\alpha_v\beta_3$ $\alpha_2\beta_1$ $\alpha_4\beta_1$ and $\alpha_6\beta_1$ integrins during tumor progression. Tumor cell migration and metastasis is supported by integrin-mediated focal adhesion and actin-omyosin-dependent contractility, as demonstrated for instance for integrins $\alpha_6\beta_4$ $\alpha_2\beta_1$ and $\alpha_v\beta_3$. Furthermore, activated integrins can recruit proteases, such as MMP 9, towards the site of attachment, and the subsequent ECM degradation supports migration and invasion of cells into the surrounding tissue. Unfortunately, various subsets of integrins seem to mediate different functions during progression of different cancer types, and more experimental work will be required to unravel their actual functional contribution.

Endothelial cell integrins have been studied in great detail during physiological and pathological angiogenesis and have been shown to play an important role in tumor angiogenesis, Distinct subtypes, including $\alpha_1\beta_1$ and $\alpha_2\beta_1$ integrins, support vascular endothelial growth factor (VEGF) -mediated angiogenesis in tumor xenotransplantion experiments. Endothelial $\alpha_5\beta_1$, $\alpha_v\beta_3$ and $\alpha_v\beta_5$ integrins are critically involved in both VEGF-mediated and FGF-mediated angiogenesis. Peptide antagonists against these integrins are able to interfere with tumor angiogenseis, angiogenesis in a number of experimental model systems, and initial clinical trials with some of these compounds are under way. Interestingly, experiments with mice lacking $\alpha_v\beta_3$ and $\alpha_v\beta_5$ integrins have demonstrated increased rather than reduced tumor growth, raising the possibility of a dual function of integrins at least in the angiogenic process. Thus, changes in integrin-mediated matrix adhesion on tumor cells, endothelial cells and certainly other cells of the tumor stroma can influence tumor cell migration, invasion and angiogenesis, all rate-limiting processes in tumor metastasis.

3. 3 Immunoglobulin Family Adhesion Molecules-NCAM and L1

Among many other cell adhesion molecules (CAMs), two have been specifically implicated in promoting tumor cell invasion and metastatic spread, NCAM (CD56) and L 1 (CD 171, L 1 CAM). Both structurally belong to the immunoglobulin (Ig) superfamily and are critical for central nervous system (CNS) development. They not only mediate static neuron-neuron adhesion but also are particularly involved in neurite outgrowth, axon guidance and neural cell migration. In addition to providing mechanical cell-cell and cell-ECM adhesion, they also activate signaling receptors and induce intracellular signaling cascades.

NCAM has been shown to play an important role in the progression to tumor malignancy. Both the shift of the adult NCAM 120 kDa isoform to the embryonic 140 kDa and 180 kDa isoforms and a general downregulation of NCAM expression are associated with poor prognosis in a few cancer types. In a transgenic mouse tumor model of pancreatic β cell carcinogenesis, loss of NCAM leads to the formation of lymph node metastasis mediated by the increased expression of the lymphangiogenic factors VEGF-C and D and with it up-regulated peri-tumoral lymphangiogenesis . Similar to its function in neurons, NCAM also modulates cell-matrix adhesion of tumor cells by binding to FGFR and inducing a variety of signaling pathways that lead toβ_1 integrin-mediated cell-matrix adhesion and neurite outgrowth.

While L1 exerts functions similar to NCAM in neurons, L1 is specifically expressed at the invasive front of gliomas, melanomas, lung, prostate, renal, ovarian and endometrial carcinoma, and its expression appears to correlate with metastasis. Similar to NCAM, L1 has been shown to enhance migration and invasion of tumor cells in vitro. L1 promotes cell-matrix adhesion and imi- gration of tumor cells on ECM proteins, such as fibronectin arid laminin, by associating with integrins through an integrin recognition motif. Namely, integrins $\alpha_v\beta_5$, $\alpha_v\beta_3$, and $\alpha_5\beta_1$ have been found to associate with L1. In these studies, integrin-mediated signaling events induced by associated L1 seemed to be responsible for promoted cell migration. Interestingly, L1 integrin interaction may not only occur at the cell surface but also by ectodomain shedding of L1 and subsequent autocrine/paracrine of L1 to $\alpha_v\beta_5$ integrin. Independent of integrin binding and activation, L1 has also been shown to promote growth factor-induced mitogen-activated protein kinase (MAPK) signaling and to subsequently stimulate extracellular regulated kinase (Erk) 1/2-mediated transcriptional changes in genes associated with cell motility and invasion and matrix remodeling . Such targets include the genes for cathepsin-L and-B, osteopontin (OPN), RhoC and CD44, all proteins implicated in tumor invasion and metastasis . Finally, L1 was recently identified as a direct target gene of activated Wnt signaling, together with the metalloprotease ADAM10. Co-induction of ADAM10 and L1 leads to proteolytic shedding of the L1 extracellular domain, which in turn induces tumor cell motility and enhanced tumorigenesis.

Together, these selected examples indicate that detachment and migration of tumor cells critically depends on altered cell-cell and cell-matrix adhesion. Yet, such alterations appear not to only affect cell adhesion but also a variety of signal transduction pathways, mediated by tyrosine kinase receptors or by integrins, which all seem to lead to increased tumor cell invasiveness and metastatic dissemination.

3. 4　Tumor Angiogenesis

The entry of tumor cells into blood vessels (intravasation) initiates dissemination of cancer cells from the primary tumor to distant organs via the blood vasculature (hematogenous metastasis). This first step not only depends on tumor-cell-intrinsic invasive properties but obviously also on the presence of a tumoral vascular net-work, which itself constitutes part of the tumor stroma. Stimulated sprouting of new tumor blood vessels from a preexisting vascular network (angiogenesis) represents a risk factor for the development of distant metastases in at least two ways. First, an increasing number of tumor blood vessels augment the contact area between tumor cells and their potential escape routes, thereby increasing the probability of intravasation. Second, tumor blood vessels display a distinct morphology in comparison to the physiological vascular bed, facilitating entry of cells because they are larger, tortuous and leaky due to a fragmented basement membrane and an incomplete pericyte lining . Furthermore, hematogenous spread of cancer cells can also occur as a result of lymphogenous metastasis. Since the lymph is drained into the venous system and lymph nodes display afferent and efferent blood vessels, tumor cells circulating in the lymphatic vasculature can successively enter the systemic circulation.

Angiogenesis is orchestrated by a fine-tuned balance between secreted pro- and anti-angiogenic factors . The contribution of the tumor stroma to the regulation of angiogenesis has also complicated the simplistic view that angiogenesis is mainly induced by transforming signaling pathways and by tumor hypoxia, resulting in the tumor cells' expression of angiogenic-factors, such as VEGF-A and PIGE. Rather, several cell types of the tumor stroma appear to contribute in a significant way ① Endothelial cells by secreting angiogenic factors, such as angiopoietin-2, which affect the activation status of the endothelium and its differentiation into mature vessels. ② Infiltrating macrophages and mast cells by secreting additional angiogenic factors, including VEGF-A, FGF-2, TGF-/3 and IL-8, and MMPs that activate latent forms of these growth factors. ③ Cancer-associated fibroblasts (CAFs) by secreting additional growth factors, cytokines and chemokines and by modulating. ④ The ECM and additional cells of the innate and adaptive immune system or of tissue homeostasis.

During tumor outgrowth, tumor cells and tumor stroma cells are stimulated for example by tumor hypoxia or lack of nutrition to produce angiogenic factors, such as VEGF-A and P1GF These factors, such as VEGF-A and P1GF These factors, together with various inflammatory chemokines and cytokines and other stimuli secreted by both neoplastic and tumor stroma.

Section 4   Epigenetic Modification and Tumor

4.1   DNA Methylation

The best-studied epigenetic mechanism is DNA methylation. The methylation of DNA refers to the co-valent addition of a methyl group to the 5-carbon (C^5) position of cytosine bases that are located 5′ to a guanosine base in a CpG dinucleotide. DNA methylation plays an important role in different cellular processes including gene expression, silencing of transposable elements, and defense against viral sequences. Importantly, aberrant DNA methylation is tightly connected to a wide variety of human cancer. DNA methylation is mediated by enzymes DNA methyltransferases (DNMTs) among which DNMT1 is the principal enzyme in mammals responsible for post-replicative methylation (known as maintenance of DNA methylation), and DNMT3A and 3B are responsible for methylation of new CpG sites (*de novo* methylation). The levels and patterns of DNA methylation undergo dramatic changes during embryonic development, starting with a wave of a profound demethylation during the cleavage stage and followed by a widespread *de novo* methylation after embryo implantation stage. Interestingly, in contrast to the maternal genome, which is only partially demethylated after fertilization; demethylation is a remarkably active and rapid process in the male genome resulting in an almost complete removal of methyl groups (within hours) after fertilization. Recent studies indicated that during early embryonic development the promoter methylation is accompanied by specific histone modifications that are typical of heterochromatin, and that abnormal epigenetic premarking during early embryonic development may predispose certain genes to cancer promoting DNA methylation events (Figure 9-2).

Figure 9-2   Schematic diagram of DNA methylation (see page 299)

4.2   Histone Modification

Post-translational marking of chromatin proteins (histones) is a major epigenetic mechanism of fundamental importance in the regulation of cellular processes that utilize genomic DNA as a template. Acetylation, methylation, phosphorylation and ubiquitination are major histone modifications, combination of which may constitute the "histone code" that extends and modulates the genetic code. Histone modifications play multifaceted roles for several cellular processes including gene transcription, DNA repair, recombination and DNA replication, and their deregulation is implicated in human malignancies. Several lines of evidence implicated histone acetyltransfases (HATs) responsible for this chromatin modification in human malignancies. HATs have been shown to be involved in chromosomal translocations in which resulting fusion protein exhibits a "gain-of-function" by altered HAT activity on specific histones. This is well exemplified by p300 and CBP genes that are found translocated on specific chromosomes in certain leukaemia. In addition, a number of HAT proteins have been found among leukemic translocations, underscoring the importance of a tight control of HATs and histone acetylation for tissue homeostasis. Furthermore, screening of human cancers identified mutations in HAT genes. For example, mutations in p300 and CBP have been found. These results thus suggest that HATs may act as tumor suppressor genes. Similarly, histone deacetylases (HDACs), enzymes responsible for the removal of acetyl groups from histone tails, participate in a dynamic equilibrium of histone acetylation. By analogy, deregulation of this process is believed to be implicated in human cancer, and several tumor suppressor genes such as Rb (retinoblastoma), APC (adenomatosis polyposis coli), and p53 may require HDAC activity for their function.

Histone methylation occurs on lysine and arginine residues and is carried out by a family of enzymes called histone methyltransferases (HMTs). Lysine residues can be mono-methylated, di-methylated, or tri-methylated, whereas, arginine can be either mono-methylated or di-methylated. Interestingly, histone methylation may be associated with either transcriptional activation or repression, depending on lysine/arginine modified by methylation and other histone modifications in the surrounding residues or even different histone tails. For example, tri-methylation of histone H3 lysine 4 is associated with transcriptional activation and creates a binding site for the chromodomain containing proteins which recruits HATs. Similarly, methylation of lysine 36 and lysine 79 of histone H3 appears to be associated with active chromatin and transcriptional activation. In contrast, tri-methylation of histone H3 lysine 9 contributes to transcriptional silencing by recruiting heterochromatin protein (HP1) and triggering formation of heterochromatin. Similarly, methylation of histone H3 lysine 27 is associated with transcriptional repression and maintenance of silent chromatin through recruitment of the polycomb complex (PRC1). Consistent with their important role in transcriptional control many HMTs are essential for cell proliferation and their deregulation is implicated in human diseases including cancer.

4.3　RNA-Associated Gene Silencing

This type of epigenetic inheritance involves RNA molecule (i. e. RNA interference or noncoding RNA) and is found to play an important role in the maintenance of the gene transcription state through multiple cell divisions. Accumulating evidence indicates that deregulation in microRNAs (miRNAs) are linked to several steps of cancer initiation and progression. miRNAs are relatively small noncoding RNAs (usually 20~22 nucleotides long) that are excised from longer (60~110 nucleotides) RNA precursor. miRNAs play essential roles in normal biological processes including development, proliferation, differentiation and cell death. Interestingly, miRNA appears to be able to act as either tumor suppressors or oncogenes by affecting distinct genes of gene families involved in critical biological processes such as proliferation and differentiation. Many miRNA genes are located in the genomic loci known as fragile sites and are therefore susceptible to either loss of amplification.

Several recent studies indicated that miRNA profiles differ significantly between cancer and normal tissues and also between different tumors. Interestingly, miRNA profiling revealed distinct patterns that may classify cancers according to the developmental lineage and differentiation status, lending miRNAs useful tools in cancer diagnostics and prognosis. Although we are just beginning to understand complexity of mechanisms regulated by miRNAs, future studies in this field are likely to provide important information to the overall knowledge in cancer biology.

Summary

There are ten thousands genes in human cells which compose stable structure and complicated layer and ordered running in normal cells. Otherwise they can fast and exactly response to surrounding and automatic correct signal network. Such consummate network is not easy to generate disorder, but appearance of tumor cells lead to the result of consummate network disorder. Tumorous growth can not be controlled and appear paraplasm. Scientists further discovered that loss of control tumor cells is disequilibrium of oncogenes (necessary functioning gene for some normal cells growth and development) and anti-oncogene (functionality genes of monitoring cell growth) in soma and organ. So discovery of oncogenes and anti-oncogene in 20 century 80 era indicated that tumor research has really entered molecular tumor era.

The basic reason which lead to out of control of tumor cells is mechanisms of cell cycle regulation destroyed (including driving and monitoring mechanism have been destroyed). Destroyed any component of monitoring mechanism will be lead to instability of cell genome. Genomic instability will appear gene mutation, deletion, amplification and translocation in DNA level, and appear chromosome abnormity, heteroploid and polyploidy in chromosome level. Chromosome abnormity made amount of mutant gene increased.

And these mutant genes are usually oncogenes and anti-oncogenes. In the same time most of oncogenes and anti-oncogenes are ingredient of mechanism of cell cycle regulation. So in development of tumor, monitoring genes are once to neglect theirs supervisory duties, mechanism of cell cycle regulation will further deteriorate and powerful "drive cells" will be cancerization growth.

Mankind has studied tumor metastasis for about one hundred years. With improving the theory of tumor metastasis gene regulation multiple stage, factor and step, we have further penetrated recognition for this complicated patho- process. We have found and confirmed that closely associated genes of tumor metastasis will become hot spot. These genes and their products become new target for tumor anti-metastasis therapy and important biomarkers of prognostic tool for detecting cancers. Epigenetic modification is a gene modification, which does not involve in change of DNA sequence and can inherit in cell division. There are DNA methylation, histone modification and microRNAs (miRNAs) regulation. The best-studied epigenetic mechanism is DNA methylation. It will disclose a new page for the study of epigenetics.

(Xiaoping Qiu)

第十章 分子生物学技术

毫无疑问，最近几十年中最引人瞩目的科学进展就是对生物大分子的分离、分析和操作。分子生物学和遗传学技术的发展与应用，使得人们对生物学的诸多研究领域，如生物化学、微生物学、生理学、药理学和医学等有了更加深入的了解。20世纪70年代早期限制性内切核酸酶的发现和克隆载体的发展使核酸分离和操作成为可能。基因探针标记和杂交技术的发展及改进为基因分析提供了有效的方法。这些技术以及DNA测序方法、蛋白质工程技术、PCR技术等，都是我们在分子水平上探寻生命现象及其本质强有力的研究手段。本章将介绍一些重要的分子生物学技术。

第一节 生物大分子的分离检测技术

21世纪注定是一个令生物学振奋的时代。各种分析DNA、RNA和蛋白质的新方法爆炸式地不断涌现，使得科学家们能以先前难以想象的方法研究细胞及各种生物大分子。如果我们想要完全探明细胞在对外界环境产生应答及细胞间相互作用时发生的所有事件，我们就必须先知道哪些基因是开启的，出现了哪些mRNA的转录以及哪些蛋白质在活动——它们在哪定位，它们之间相互作用的关系以及它们属于那些信号通路和调控网络。要回答这些问题，分离和检测这些大分子则是所有分析的第一步。在本节中，我们将简要回顾一些分离和检测生物大分子的主要方法，随后我们将介绍这些方法的最新进展及应用。

一、从组织和细胞中分离 DNA

真核DNA以核蛋白形式存在于细胞核中，制备DNA的原则是既要将DNA与蛋白质、脂类和糖类等分离，又要保持DNA分子的完整。蛋白酶K及链酶蛋白E的应用可以满足上述要求。蛋白酶K或链酶蛋白E均为广谱蛋白酶，其重要性能是能在SDS和EDTA（乙二胺四乙酸二钠）的存在下保持很高的活性。常用SDS、Triton X-100等去污剂溶解细胞膜。蛋白酶K或链酶蛋白E可将蛋白质降解成小肽或氨基酸，使DNA完整地分离出来；再用酚、氯仿抽提除去蛋白质，最后用无水乙醇沉淀DNA。CTAB（溴化十六烷三甲基铵）也可以沉淀基因组DNA，因其可以去除细菌和植物样品中的多糖而经常使用。在提取过程中染色体会发生机械断裂，产生大小不同的片段，因此分离基因组DNA应尽量采取温和条件，以保证得到较长的DNA。

从全血制备白细胞DNA，可用Triton X-100直接破裂血中红细胞和白细胞的细胞膜，释放出血红蛋白及细胞核。向核悬液中加SDS破裂核膜，并使DNA从核蛋白中解离，经蛋白酶K处理，酚、氯仿抽提等过程即可获得白细胞DNA。

二、RNA 制备

1. 总 RNA 和 mRNA 的制备

从纯实用角度来看，制备总 RNA 可以满足大部分研究的要求，如从 Northern 杂交检测 mRNA 种类，到 RT-PCR 进行 mRNA 起始材料定量等应用。对总 RNA 的偏好也反映出了纯化足量 mRNA（含量小于总 RNA 的 5%）的困难，纯化过程中易丢失某一特定种类 mRNA，以及定量出纯化到的微量 mRNA 的困难。有两种情况必须使用mRNA：构建 cDNA 文库和制备标记的 cDNA 用于基因阵列杂交。为了避免在 cDNA文库中得到大量的核糖体 RNA 克隆或者非特异标记的 cDNA，用 mRNA 进行这些实验十分关键。

从细胞和组织中提取总 RNA 有三种基本方法。大多数方法是靠离液剂（chaotrope），如胍盐或去污剂破碎细胞同时失活 RNA 酶。再用一种方法去除蛋白质、基因组 DNA 和其他细胞成分纯化得到 RNA。

（1）胍盐-氯化铯法

胍盐-氯化铯法（guanidium-cesium chloride method）用异硫氰酸胍裂解细胞，同时快速失活核酸酶。将裂解液经氯化铯或三氯乙酸铯超速离心得到纯化的 RNA。因为 RNA 的密度比 DNA 和大多数蛋白质大，经过 $12\sim24h > 32\,000 r \cdot min^{-1}$ 的离心，RNA会沉淀在底部。这是获得大量高质量 RNA 的经典方法。该方法无法制备小分子 RNA（如 5S RNA 和 tRNA）。

（2）单步或多步酸性胍盐-酚法

酸性胍盐-酚法（guanidium acid-phenol method）因其可以用 $2\sim4h$ 提取大量样品RNA，所以很大程度上替代了无法如此快速的超速离心法。各种大小的 RNA 分子都可以用此法得到，而且该技术可以方便的放大或缩小以处理不同大小的样品。单步法基于存在酚/氯仿有机相的情况下，RNA 分子溶解于含有 4mol 硫氰酸胍 pH4.0 的水相中。在如此低的 pH 下，DNA 分子留在有机相中，而蛋白质和细胞中其他大分子留在两相界面间。

（3）不利用酚的方法

酸性胍盐-酚法的一个缺点是要操作和处理危险的化学试剂——酚。因此基于离液盐，如胍盐存在下玻璃纤维滤膜结合核酸特性的无酚法（non-phenol-based method）得到了青睐。细胞先在胍盐缓冲液中裂解，裂解液经过有机溶剂，如乙醇或异丙醇稀释后通过玻璃纤维滤膜或树脂。洗去 DNA 和蛋白质，最后用水性缓冲液洗脱 RNA。这种技术得到的总 RNA 产量和酚法相同。因为不经过有机抽提，处理大量样品简单快速。这也是最快的 RNA 提取方法之一，通常在 1h 内完成。

从某些组织提取 RNA 需要改进方案以去除特定杂质或者在提取 RNA 前对组织进行预处理。纤维组织和富含蛋白质的组织，DNA 和核酸酶，这些都是提取总 RNA 所面对的独特的挑战。

mRNA 制备有两种基本策略：一步法直接从初始材料纯化 mRNA，两步法先分离总 RNA 再从中纯化 mRNA。通过一轮寡聚胸腺嘧啶核苷酸［oligo（dT）］-纤维素选择mRNA 除去 50%～70%核糖体 RNA。一轮选择可以满足大多数应用（如 Northern 分析和核酸酶保护分析）。用 50% 纯度的 mRNA 构建 cDNA 文库通常适用于鉴定大多数基因。但是要构建只含极少 rRNA 克隆的 cDNA 文库需要两轮 oligo（dT）选择去除大约 95% 的核糖体 RNA。但是每轮 oligo（dT）选择都会丢失 20%～50% 的 mRNA，所以多次选择会降低 mRNA 产量。使用标记的 cDNA 筛选基因阵列会被 rRNA 特异的探针强烈干扰，因此两轮 oligo（dT）选择是合适的。

2. RNA 制备方法优化

RNA 酶污染很普遍，因此配制溶液要特别注意。溶液应在一次性无 RNA 酶的塑料瓶或玻璃瓶中配制。玻璃器皿可经 180℃ 8h 或过夜烘烤除 RNA 酶或用商业 RNA 酶净化液处理。再者，容器可灌满 0.1%DEPC 在 37℃ 温浴 2h，用无菌水冲洗后 100℃ 加热 15min 或灭菌 15min 去除 RNA 酶。最常用的制备无 RNA 酶溶液的方法是焦炭酸二乙酯（DEPC）处理。DEPC 通过羧乙基化特定氨基酸侧链而失活 RNA 酶。DEPC 是潜在的致癌物，使用时要注意安全。多数方案指定在溶液中加入 DEPC 至 0.01%～0.1%，混匀后室温放置几小时至过夜。反应过程中会产生较大压力，长时间处理容器盖要松开。最后将溶液高压灭菌。该过程中残留的 DEPC 会水解，放出 CO_2 和乙醇。DEPC 在水中的半衰期为 30min，0.1%DEPC 溶液高压灭菌 15min 可以认为已不含 DEPC。应该注意的是，将 DEPC 浓度提高到 1% 可以抑制更多 RNA 酶，但也会抑制某些酶反应，所以通常并非越多越好。许多 RNA 研究用的试剂含有伯胺，如 Tris、MOPS、HEPES 和 PBS，这些不能用 DEPC 处理，其氨基会"吸住"DEPC 使其无法失活 RNA 酶。这些溶液应该用超纯试剂和 DEPC 处理过的水配制。RNA 酶存在于所有细胞和组织，因此取材生物死后它应该被迅速失活。样品应该立即冻存于液氮中或在离液溶液（如 GITC）中破碎。如果不冷冻，有些情况下即使有离液剂存在 RNA 酶也能恢复活性。快速完全裂解样品是纯化 RNA 的关键。过多的起始样品会使离液剂或其他 RNA 酶抑制剂无法发挥作用。

不管用什么纯化方法，提取过程最好不要被打断。如果无法避免，则应在 RNA 沉淀下来或有离液剂存在时停下。例如，用有机分离法，可以将样品在离液裂解液里匀浆后停止 RNA 抽提。样品可以在 -20℃ 或 -80℃ 保存几天不降解。

3. 纯化 RNA 的保存

RNA 酶或者 pH 高于 8 都可破坏 RNA。几周内保存 RNA 可用无 RNA 酶的 TE／EDTA 溶液分装，保存于 -20℃。长期应分装于 TE/EDTA、甲酰胺或以乙醇/盐沉淀混合物形式在 -80℃ 保存。

三、蛋白质的提取和分离

我们对蛋白质结构和功能的了解源自对许多单个蛋白质的研究。要细致研究一个蛋白质，研究者必须能够将其与其他蛋白质分离并拥有鉴定其性质的技术。这些必需的方法来自蛋白质化学——一门古老并在生物医学研究中保持重要地位的学科。分离蛋白质的方法利用了一种蛋白质和其他蛋白质不同的性质，包括大小、电荷和结合特性。

蛋白质的来源一般是组织或者微生物细胞。任何蛋白质提取的第一步都是破裂细胞，将它们的蛋白质释放到溶液中，这叫做粗提。如果需要，可以用不同等离心方法制备亚细胞成分或者分离特定的细胞器。制备好提取物或者细胞器后有各种方法来纯化其含有的一种或多种蛋白质。通常抽提物经过处理根据大小或者电荷将蛋白质分成不同的组分，这个过程叫分离。一般而言，大部分蛋白质可以溶于水、稀盐、稀碱或稀酸溶液，少数与脂类蛋白结合的蛋白质则溶于乙醇、丙酮、丁醇等有机溶剂。蛋白质在不同溶剂中溶解度的差别，主要取决于蛋白质中极性基团与非极性基团的比例，因此，采用不同溶剂和调整影响蛋白质溶解度的外界因素，如温度、pH、离子强度等，即可把所需的蛋白质和酶从细胞内复杂的组分中分离出来。

四、DNA 和 RNA 质量鉴定

最常见鉴定 DNA 和 RNA 质量的方法主要有紫外分光光度法（UV-spectrophoto-metric method）和电泳比较法。用紫外分光光度计测定 260nm 和 280nm 两个波长处的光吸收值，260nm 和 280nm 两处读数的比值（A260：A280）可反映核酸的纯度，用来作为最初的 DNA 和 RNA 质量估计。DNA 和 RNA 纯品的 A260：A280 分别为 1.8 和 2.0，如果样品中有蛋白质或酚的污染，则 A260：A280 将明显低于此值。

在 260nm 波长下，吸光度值为 1 时相当于双链 DNA 浓度 $50\mu g \cdot ml^{-1}$，单链 DNA 或 RNA 为 $40\mu g \cdot ml^{-1}$，相当于寡核苷酸 $20\mu g \cdot ml^{-1}$。

RNA 完整性最好用甲醛变性琼脂糖凝胶电泳鉴定。样品可以通过在凝胶上添加终浓度 $10mg \cdot ml^{-1}$ 溴化乙啶（EtBr）来观察。比较 28S rRNA 条带（多数哺乳动物细胞中大约位于 5 kb 处）和 18S rRNA 条带（多数哺乳动物细胞中大约位于 2 kb 处）。高质量的 RNA 中 28S 条带的亮度大概是 18S 条带的两倍（图 10-1）。

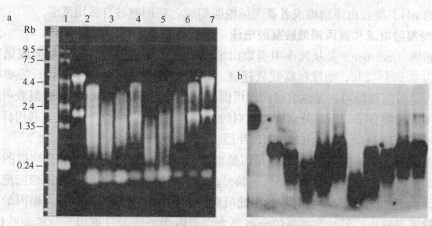

图 10-1　琼脂糖凝胶电泳检测 RNA 制备质量

a. 图片显示总 RNA 样品（5mg/泳道）从高质量完整的 RNA（2 泳道）到几乎完全降解的 RNA（7 泳道）；b. 同一张凝胶经过生物素标记的 GAPDH RNA 探针杂交用非同位素检测的自显影结果

最灵敏的 RNA 完整性的检测方法是用分析组织中低表达的高分子质量探针进行 Northern 分析。但是这种方法费时，多数情况下没有必要。

五、凝胶电泳技术

用来分离、鉴定、纯化 DNA 或者 RNA 以及蛋白质的凝胶电泳技术（gel electrophoresis）或许是分子生物学研究中应用最广泛的物理方法。这种技术不仅简单快速，而且能够解决其他方法，如密度梯度离心无法充分分离某些 DNA 片段的问题。此外，DNA 分子在凝胶中的定位可以直接通过低浓度的荧光嵌入染料中染色检测到，如溴化乙啶或者 SYBR Green。在紫外光下可以检测到凝胶中低至 20pg 的双链 DNA 条带。如果有需要，这些 DNA 片段可以从凝胶中回收以满足多种实验需求。

1. 原理

当一个带电荷的分子被置于电场中时，它将朝向与之带有相反电荷的电极泳动；核酸分子由于它们的糖-磷酸骨架中呈离子化状态的磷酸基团而常带有负电荷，因此在电场中将向正极迁移。然而蛋白质分子就没有这么简单，它们有的带正电荷，有的带有负电荷，更多的是中性的。因此必须通过 SDS 变性以使蛋白质分子完全带上负电荷。凝胶是由无数孔隙构成的复杂网络，核酸分子在其中移动的速度取决于其穿过这个网络的能力。对于在一定大小范围内的线状双链 DNA 片段，这将反映这个分子的大小（也就是 DNA 的长度）。我们不用考虑这些分子所带的电荷量，因为所有的核酸分子在单位长度所带的电荷量是一样的。松散的或者带有缺口的环状 DNA 分子要比带有相同净电荷的线状 DNA 分子迁移速度慢，而超螺旋 DNA 在电泳时的迁移率要比带有相同净电荷的环状 DNA 要快。混合组分在电泳时被分为不连续的"条带"；较小的组分在凝胶中具有更强的活动性。所呈现出的条带的模式被用于多种分析用途。

按常规制胶，胶被置于一块薄板中以成型，上面插入梳子以形成点样孔。制好的凝胶浸没在电泳缓冲液中，缓冲液为电泳提供离子以形成电流并且起到缓冲作用以保证相对恒定的 pH。凝胶由琼脂糖或者聚丙烯酰胺组成，它们有各自适用范围。

2. 琼脂糖凝胶电泳和聚丙烯酰胺凝胶电泳

琼脂糖（agarose）是从海藻中提取出来的一种多糖聚合物。琼脂糖凝胶通常采用的浓度为 0.5%～2%，浓度越高胶就越硬。琼脂糖凝胶非常容易制备：只需要在缓冲液中加入适量琼脂糖粉，加热溶解后即可倒胶，并且它是无毒的。琼脂糖凝胶可以分离很大范围的核酸分子，但是分离能力相对较低。通过调整琼脂糖的浓度，利用标准的电泳技术可以分离 200b～50 kb 的 DNA 片段。

聚丙烯酰胺（polyacrylamide）是丙烯酰胺分子的交联聚合体。聚合物链的长度取决于丙烯酰胺的浓度，通常为 3.5%～20%。聚丙烯酰胺凝胶的制备相比琼脂糖凝胶烦琐得多。由于 O_2 会抑制聚合过程，因此凝胶必须在两块玻璃板间制备。聚丙烯酰胺凝胶的分辨范围较小，但却有很高的分离能力。聚丙烯酰胺凝胶被用于分离 500 bp 以下的 DNA 片段，并且在适当条件下可以轻易区分长度相差一个碱基对的 DNA 片段。与琼脂糖凝胶不同的是聚丙烯酰胺凝胶还被广泛用于蛋白质的分离和鉴定（图 10-2）。

3. 脉冲场凝胶电泳

标准的琼脂糖凝胶可以分离范围在 200bp～50kb 的核酸分子。对于特别大的 DNA 分子需要一种特殊的凝胶电泳来分离。脉冲场凝胶电泳（pulsed-field gel electrophoresis）应运而生，被用于分析范围在 10kb～10Mb 的 DNA 分子。

在这种方法中，对凝胶施加交替的直角电场，由于电场方向、电流大小及作用时间都在交替地变换着，这就使得 DNA 分子能够随时地调整其泳动方向，以适应凝胶孔隙的无规则变化。与相对分子质量较小的 DNA 分子相比，相对分子质量较大的 DNA 分

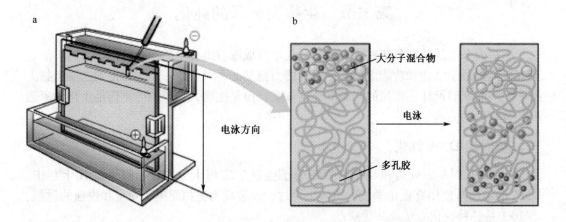

图 10-2　聚丙烯酰胺凝胶电泳

a. 电泳槽，在一块聚丙烯酰胺凝胶上可以同时进行几个样品的电泳，利用微量上样器将蛋白质溶液加到点样孔中，盖上电泳槽盖接通电源即可开始电泳，带有负电荷的 SDS-蛋白质复合体朝向凝胶的底部的正极迁移；b. 多孔隙的聚丙烯酰胺凝胶按照分子大小对蛋白质进行筛分，相对分子质量最小的泳动速度最快

子需要更多的次数来更换其构型和方位，以使其可以按新的方向泳动。因此在琼脂糖介质中的迁移速率也就显得更慢一些，从而达到了分离特别大的 DNA 分子的目的（图 10-3）。

　　对于从细菌到人类的所有生物，脉冲场凝胶电泳被用来解析它们的基因组结构并且用来克隆和分析大片段分子。

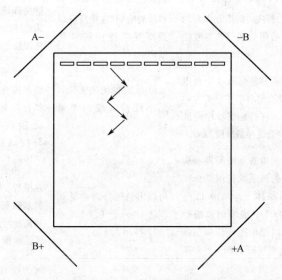

图 10-3　脉冲场凝胶电泳

图中方形代表一块琼脂糖凝胶，其中一系列水平的矩形小块表示点样孔，A 和 B 代表两组电极。当接通电极 A 时，DNA 分子向右下方的正极泳动，如图中第一个箭头表示。当电极 A 被关闭电极 B 立即被接通，DNA 分子又朝向左下方泳动。DNA 分子在电场方向不断改变的移动路径由箭头表示

第二节 生物大分子的纯化

生物大分子的研究包括对天然来源、化学合成或重组技术的产物的纯化。这可能是研究生物大分子结构和功能之前最严格、费力但又必不可少的一步。为得到这些生物大分子，科学家们设计一系列的实验室纯化方案，但是生物大分子的纯化仍是生化学家遇到的最麻烦的问题。

一、基因组 DNA 纯化

基因组 DNA 的分离和纯化对很多分子生物学应用十分关键。过去的几十年中，DNA 的纯化从使用有机化学试剂的多步骤过程改进成为更加简单、安全和快速的过程。表 10-1 是各种纯化方法及其优缺点。

表 10-1　纯化方法的优缺点

方法	原理	特点	限制
盐析和 DNA 沉淀	加入离液剂和 cosmotrope 利用溶解度不同分离细胞成分	产量较好，操作过程中剪切力小	加入 1% PVP(聚乙烯吡咯烷酮)可以减少酚污染；增加 CTAB 沉淀步骤可以去除多糖
用有机溶剂、离液剂抽提后沉淀 DNA	离液性胍盐裂解细胞，还原剂防止核酸氧化，酚将蛋白质和脂类溶解抽提到有机相使其与核酸分离	产量、纯度、速度好，适于处理少量样品	不经氯仿抽提会有酚残余，影响下游实验。盐浓度过高可能使有机相在水相上方。用 GTC/酚抽提液常会使 RNA 污染 DNA
玻璃奶/硅树脂策略	含盐情况下变性的核酸会结合到玻璃奶或者硅树脂上。用盐/乙醇洗去杂质和多余的离液剂后用盐或 TE 溶液洗脱核酸	快速简单。可直接在玻璃表面进行限制性内切核酸酶切/连接反应以提高复杂连接的转化效率	上样量太少、过度干燥和不合适的冲洗或洗脱缓冲液会影响产量。未除净的乙醇会造成样品溢出电泳凝胶孔或者无法被酶切
基于离子交换(AIX)的策略	每个碱基核酸带有 1 个负电荷，因此会结合带正电荷的纯化树脂。在低盐溶液中冲洗后，用高盐缓冲液洗脱 DNA	可以得到很纯的 DNA，但小规模实验的产量较低	一致性不好，回收率低，最后的洗脱步骤含有高盐溶液，需要脱盐。结合容量低，但随 pH 升高而增大，随 DNA 增大而减小。RNA 会与 DNA 竞争结合
基于羟基磷灰石(HA)的策略	核酸通过磷酸基团和羟基磷灰石中钙离子的作用结合到磷酸钙晶体上。从 $0.12 \sim 0.4\,mol \cdot L^{-1}$ 递增的磷酸根离子可以把单链核酸和双链 DNA 先后洗脱下来	可以出色地分离单链和双链 DNA 分子	质量和表现因厂商和批次而异。热洗脱步骤需要可靠的温度控制。羟基磷灰石机械稳定性不好。用到高盐溶液并稀释样品，需要附加沉淀步骤

为了将核酸浓缩并溶于合适的缓冲液，常使用乙醇（75%～80%）或异丙醇（终浓度 40%～50%）等溶剂在盐溶液中沉淀核酸。不考虑体积的话，使用乙醇较好，因为它会沉淀较少的盐并且沉淀物更容易干燥。聚乙二醇（PEG）选择性沉淀大分子 DNA，但它也更难干燥并会影响下游应用。三氯乙酸（TCA）可以沉淀更低分子质量的多聚物（小至 5kDa），但是沉淀后无法回收到具有功能活性的核酸。

二、质粒纯化

质粒（plasmid）提取是一项特殊的挑战，因为目的 DNA 必须与 DNA 杂质分开。分离方法利用了线形、闭环和超螺旋 DNA 的物理性质的差别。碱裂解、煮沸和其他变性方法的原理：闭环 DNA 会在冷却或中和时快速复性，而长的基因组 DNA 无法复性并和蛋白质、SDS、脂类"缠"在一起盐析出来。无论通过煮沸还是碱变性，复性步骤通常在低温进行以增强蛋白质和杂质核酸的沉淀或盐析。

另一种方法是用酶解、去污剂溶解和加热混合裂解细胞。裂解缓冲液通常含有溶菌酶、STE 和 Triton X-100 或 CTAB。细菌染色体 DNA 附着在细胞膜上沉淀出来。同样，这种方法得到的上清液中含有质粒和 RNA。特异性结合 DNA 片段的 PEG 被用来分离不同大小 DNA。6.5％的 PEG 溶液可以选择性沉淀澄清的细菌裂解液中的基因组 DNA。痕量的 PEG 可用氯仿抽提去除。

加入 EtBr 的氯化铯离心分离质粒 DNA 在大量制备 DNA 时特别有用。EtBr 和 DNA 的相互作用使核酸的密度降低。因为质粒小而且具有超螺旋结构，它比基因组 DNA 结合的 EtBr 少，提高了在密度梯度中的分离效率。

层析法（chromatography），如离子交换和凝胶过滤也可以在裂解后用来纯化质粒。使用层析法要先去除 RNA，因为 RNA 会在纯化过程中造成污染。RNA 酶 A 处理、RNA 特异沉淀、切面流超滤和硝酸纤维素膜结合等被用来脱盐、浓缩和一般的柱层析的样品制备。

加入只抑制细菌基因组复制的氯霉素或大观霉素（$300mg \cdot mL^{-1}$）可以极大提高低拷贝质粒的产量。但体外实验发现持续暴露于这些试剂会损伤 DNA。

使用 CsCl 密度梯度（CsCl gradient）分离 DNA 和杂质的方法十分有效但是费时。巨大的离心力使 Cs^+ 向离心管底迁移，直到离心力和扩散力达到平衡。在 CsCl 梯度中多糖会呈现随机缠绕的二级结构，DNA 为中等密度构象，RNA 因其二级结构呈现最大的密度。与核酸结合的染料可以改变其密度使其有效的与杂质分离。结合 EtBr 会降低 DNA 的密度。超螺旋 DNA 结合 EtBr 较线形 DNA 少，提高了利用它们密度的差别进行分离的效率。使用 EtBr 的 CsCl 离心法在质粒和柯斯质粒纯化中使用得十分普遍。

三螺旋树脂（triple helix resin）被用来纯化质粒和柯斯质粒是利用了它们在适宜的 pH、盐和温度条件下形成三螺旋而不是双螺旋构象。三螺旋亲和树脂是将一段合适的纯嘌呤序列插入质粒 DNA 中而将其互补序列交联到特定的层析树脂上得到的。三螺旋作用只在弱酸性 pH 下稳定，在碱性条件下会解离。弱酸性 pH 时的亲和力很高，这使得大量制备 DNA 时可以通过严格洗脱来去除基因组 DNA、RNA 和内毒素。

三、蛋白质纯化

1. 盐析与透析

各类蛋白质或酶初步分离后，包含目的蛋白的溶液常常需要进一步纯化处理。纯化处理的方法繁多各不相同。这是因各种蛋白质有不同的溶解性，以及 pH、温度、盐浓度和其他因素复杂作用的结果。蛋白质的溶解性一般在高盐浓度时会降低，最后析出，这个作用叫做"盐析"（salting out）。加入适量的盐可以选择性地沉淀一些蛋白质并使剩下的蛋白质留在溶液中。因此，调节混合蛋白质溶液中的中性盐浓度可使各种蛋白质分段沉淀。硫酸铵因其在水中的高溶解度常用来实现这一目的，是蛋白质盐析中常用的中性盐。

蛋白质在用盐析沉淀分离后，需要将蛋白质中的盐除去，常用的办法是透析（dialysis）。透析是一种利用蛋白质具有较大的相对分子质量从而将蛋白质和溶剂分离的方法。将部分纯化的抽提物放入用半透膜做成的袋子或管，当把它放入一个相对很大的有合适离子强度的缓冲液中，半透膜允许盐和缓冲液交换但是不允许蛋白质交换。因此透析袋将大的蛋白质留在袋内，同时让制备物中的其他溶质浓度改变直到它们和透析袋外的溶液达到平衡。例如，透析可以用来除去蛋白质制备物中的硫酸铵。

2. 层析

最有效的蛋白质分离纯化方法是利用蛋白质的电荷、大小、亲和性和其他性质的区别进行层析（chromatography）。层析又称为色谱技术。

图 10-4 为柱层析（column chromatography）模式，层析柱上固定着具有合适化学性质的多孔固体（固定相），缓冲溶液（流动相）从其中滤过。将含有蛋白质的溶液铺

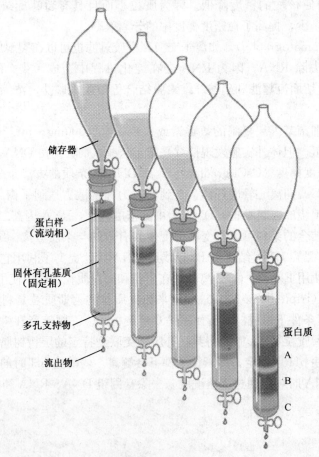

图 10-4　柱层析

标准的层析柱包括柱子内一个多孔的固体材料，通常为塑料或者玻璃。固体材料（基质）形成溶液（流动相）穿过的固定相。随着蛋白质在柱中的迁移，它们因和基质材料不同的相互作用而被不同程度地拖慢。整个蛋白质条带因此随着在柱中的移动而变宽。不同类型的蛋白质（如 A、B 和 C 显示为不同颜色）逐渐互相分离并在宽的总蛋白带中形成各自的带

在层析柱的顶端，它会在较大的流动相中作为一个不断扩展的条带滤过固体基质。不同蛋白质通过层析柱的快慢取决于其性质。在离子交换层析中（ion-exchange chromatography）（图 10-5 a），固体基质具有负电基团。流动相中，带有净正电荷的蛋白质比带

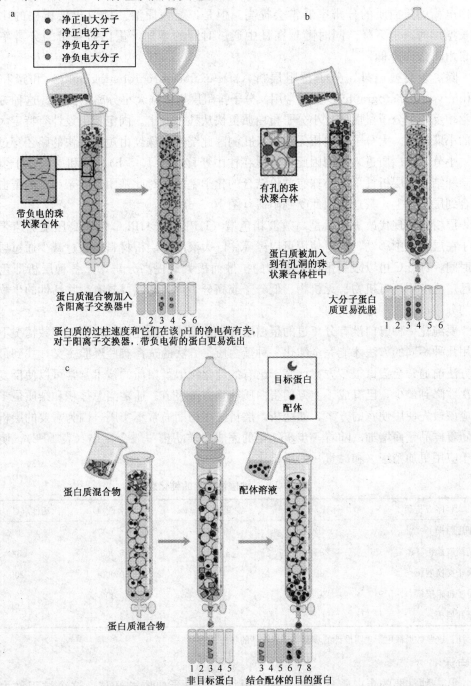

图 10-5　三种用于蛋白质纯化的层析方法

a. 离子交换层析利用给定 pH 下蛋白质净电荷的符号和数量分离蛋白质；b. 分子排阻层析，也叫做凝胶过滤，依据分子大小分离蛋白质；c. 亲和层析根据特异性结合分离蛋白质。洗去柱中未跟配体结合的蛋白质，结合的目的蛋白质可以用含有游离配体的溶液洗脱下来

有净负电荷的蛋白质移动慢得多，因为前者被与固定相的相互作用拖慢了很多。这两种蛋白质可以分成两个不同的条带。流动相（蛋白质溶液）中蛋白质条带的扩张是由不同性质蛋白质的分离和自由扩散造成的。随着层析柱长度的增加两类带有不同净电荷的蛋白质的分辨率通常会提高。但是，蛋白质溶液能流出层析柱的比例在柱长增加时通常下降。同时随着在柱中通过时间的增加，蛋白质条带的分辨率会因为自由扩散而降低。

除离子交换以外，分子排阻层析（size-exclusion chromatography）和亲和层析（affinity chromatography）也很常用。分子排阻层析依据大小分离蛋白质。这种方法与直觉相反，大分子蛋白质比小分子蛋白质更快从柱中流出。固定相由设计有特定大小孔洞的小珠构成。大分子蛋白质无法进入孔洞因此绕过小珠经由短并且快的路径穿过层析柱。小分子蛋白质进入孔洞因此更慢地在柱中迁移（图 10-5 b）。亲和层析基于蛋白质的亲和结合，层析柱里的小珠有共价结合的化学基团。对此基团具有亲和力的蛋白质会结合到层析柱上，其迁移因此受到阻碍（图 10-5 c）。

层析法经现代改进之后是高效液相色谱（HPLC）。HPLC 使用高压泵加速蛋白质分子在层析柱中移动，同时使用可以经受住巨大液压的层析材料。通过减少通过层析柱的时间，HPLC 可以限制蛋白质条带的扩散而极大地提高分辨率。各种层析即色谱技术已广泛应用于有机物、无机物、低分子或高分子化合物以及具有生物活性的生物大分子的分离、纯化和分析。

要纯化一种以前没有分离过的蛋白质通常依据成功的先例和常识。多数情况下，要使用几种不同的方法才能完全纯化一种蛋白质。方法的选择有些经验主义，找到最有效的方法前通常会尝试很多方案。基于已有的纯化相似蛋白质的技术常常可以使反复实验的次数降到最少。已有成千上万的蛋白质纯化方法报道。通常当总体积和杂质最多的时候应该最先使用便宜的方案，如盐析。层析法在初期通常不实用，因为需要的层析介质量随着样品量而增加。随着一步步纯化的完成，样品量一般会变小（表 10-2），使其适用于应用更加精细（和昂贵）的层析法。

表 10-2　一种假设的酶的纯化表

程序或步骤	体积/ml	总蛋白/mg	酶活性/U	比活性/(U·mg^{-1})
细胞粗提	1 400	10 000	100 000	10
硫酸铵盐析	280	3 000	96 000	32
离子交换层析	90	400	80 000	200
分子排阻层析	80	100	60 000	600
亲和层析	6	3	45 000	15 000

注：这些数据都用于表明设计的步骤实现后样品的状态。

3. 电泳

另一种重要的蛋白质分离技术基于电场中带电蛋白质的迁移，这个过程称为电泳（electrophoresis）。因为有其他更简单的方法，电泳通常不用于大量纯化蛋白质，而且电泳常不可逆地改变蛋白质结构，进而影响其功能。但是，电泳作为一种分析方法非常有用。其优点在于将蛋白质分离的同时将其纯化，使研究者可以快速鉴定混合物中不同蛋白质的数量或者特定蛋白制备品的纯度。而且，电泳可以确定一个蛋白质的关键性质，如等电点和大概的分子质量。

（1）聚丙烯酰胺凝胶电泳（PAEG）

蛋白电泳通常在交联成多聚体的聚丙烯酰胺凝胶上进行（图 10-6）。聚丙烯酰胺作为分子筛，大致按蛋白质的荷质比成比例的减慢其迁移速度。迁移也受蛋白质形状的影响。电泳时，推动大分子移动的力是电势 E。分子的电泳迁移率是分子的速度 V 对电势的比。电泳迁移率也等于分子的净电荷 Z 除以部分反应蛋白质形状的摩擦系数 f。因此：

$$\mu = \frac{V}{E} = \frac{Z}{f}$$

所以蛋白质在电泳凝胶中的迁移是其大小和形状的函数。

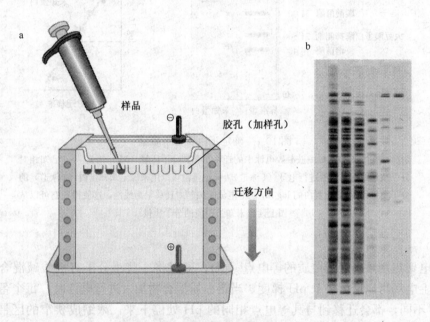

图 10-6　聚丙烯酰胺凝胶电泳。

a. 聚丙烯酰胺凝胶顶部的上样孔中加入了不同的样品。b. 凝胶上的每条带代表一个不同的蛋白质（或蛋白亚基）；小分子蛋白质在凝胶中移动比大分子蛋白质快因此出现在接近凝胶底部的地方

一种通常用来确定蛋白质纯度和分子质量的电泳方法用到了去污剂 SDS。SDS 与大部分蛋白质结合的数量大致和蛋白质分子质量成比例，大约每两个氨基酸残基结合一个 SDS 分子。SDS 的结合贡献了大量净负电荷，使蛋白质的本身电荷不再显著并使每个蛋白质都有相似的荷质比。另外，SDS 的结合改变了蛋白质的原始构型，大部分蛋白质都呈现相似的形状。因此 SDS 电泳几乎仅根据质量（分子质量）来分离蛋白质，小分子多肽迁移得更快。电泳过后的蛋白质可以通过添加只与蛋白质结合的物质，如考马斯蓝（coomassie blue）进行染色观察（图 10-6b）。因此，每当经过一个分离步骤时，凝胶上可见的蛋白质条带数目减少，研究者可以此监控蛋白质纯化程序。与凝胶中已知分子质量的蛋白质的位置相比较，可以很好地根据未鉴定蛋白质的位置确定其分子质量

（图 10-7）。如果蛋白质含有两个或更多不同的亚基，SDS 处理通常会使这些亚基分开，它们会各自显示一条单独的条带。

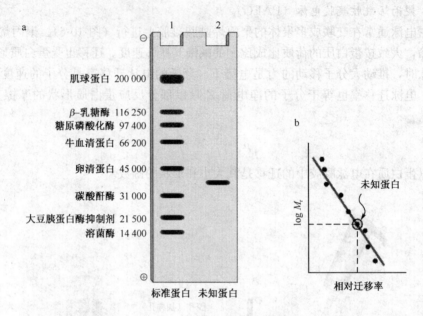

图 10-7　确定蛋白质分子质量

蛋白质在 SDS 聚丙烯酰胺凝胶电泳中的迁移率和其分子质量（M_r）相关。a. 已知相对
分子质量的标准蛋白进行电泳（1 泳道），标记蛋白可以用来确定未知蛋白（2 泳道）的
分子质量；b. 标记蛋白的 $\log M_r$ 比上电泳时的相对迁移量为线性，这使得我们可以从
图上读出未知蛋白质的分子质量

（2）等电聚焦电泳

等电聚焦用来确定蛋白质的等电点（pI）。通过将一些小分子有机酸碱混合并在凝胶上加上电场可以产生一个 pH 梯度。当蛋白质混合物加入进行电泳时，每个蛋白质因各自 pI 不同，都会迁移到与其等电点相同的 pH 处停下来，浓缩成狭窄的区带。这种分离纯化蛋白质的方法称聚丙烯酰胺凝胶等电聚焦电泳（IEF-PAGE）。最常用的载体两性电解质是"Ampholine"（图 10-8）。

（3）双向凝胶电泳

双向凝胶电泳（two-dimensional electrophoresis）又常被称为二维凝胶电泳。第一向是以蛋白质电荷差异为分离机制的等电聚焦电泳，不同等电点的蛋白质因此会在凝胶中分布于不同位置；第二向是以蛋白质分子质量差异为分离机制的 SDS-PAGE。结合等电聚焦和 SDS 电泳即成为双向电泳，它可以分辨复杂的蛋白质混合物。这种方法比两种电泳单独使用更灵敏。双向电泳可以分离相同分子质量不同 pI 或者具有相似 pI 但分子质量不同的蛋白质。双向凝胶电泳分离的结果不是条带，而是斑点。它是目前所有电泳技术中分辨率最高、信息最多的技术（图 10-9）。

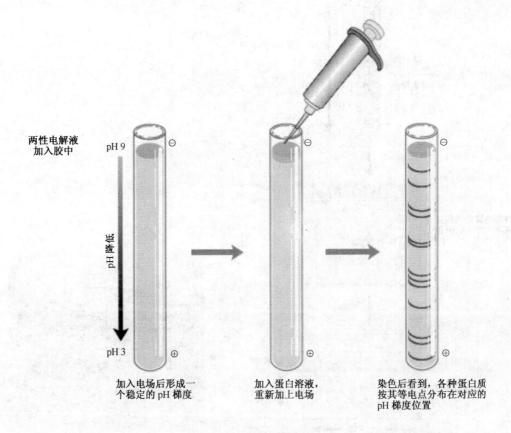

两性电解液
加入胶中

pH 9

pH 降低

pH 3

加入电场后形成一
个稳定的 pH 梯度

加入蛋白溶液，
重新加上电场

染色后看到，各种蛋白质
按其等电点分布在对应的
pH 梯度位置

图 10-8　等电聚焦

通过蛋白质等电点不同来分离蛋白质。稳定的 pH 梯度通过添加合适的两性电解质来实现

（4）未分离蛋白质的定量

为了纯化蛋白质，有必要找到一种检测和定量蛋白质的方法，从而可以在纯化过程的每个步骤中将目的蛋白质与其他众多蛋白质区分开。通常纯化必须在不了解其大小和物理性质的任何信息或者其出现在总蛋白提取物的哪个组分的情况下进行。对于酶蛋白，某种溶液或组织提取物中的量可以通过该酶产生的催化效果，即该酶出现时它的底物转化为反应产物速度的增加，来进行测量和鉴定。实现这一目的要知道：①该催化反应的总方程式，②确定底物消失或者产物出现的分析方法，③该酶是否需要辅因子，如金属离子或者辅酶，④酶活性对底物浓度的依赖，⑤最佳 pH，⑥酶保持稳定和高活性的温度范围。酶通常在其最佳 pH 和 25～38℃ 的适宜温度进行鉴定。而且，通常使用很高的底物浓度，这样通过实验测量的起始反应速率是和酶浓度成比例的。根据国际协议，酶活性的 1.0 个单位定义为 25℃ 时最佳测量条件下每分钟转化 1.0 mol 底物的酶量。酶活性是指溶液中酶的总单位数。比活性是每毫克总蛋白中酶的单位数（图 10-10）。

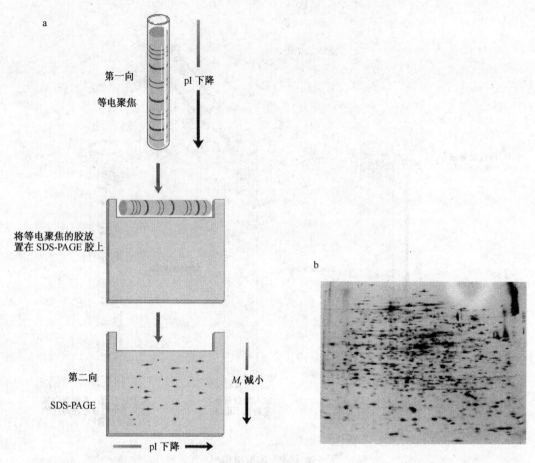

图 10-9　双向电泳

a. 蛋白质首先在圆柱形的凝胶中进行等电聚焦分离。然后凝胶条水平放置在平板形的第二块凝胶上，蛋白质通过 SDS-聚丙烯酰胺凝胶电泳分离。水平分离反应了 pI 的不同；垂直分离反映了分子质量不同。b. 使用双向电泳可分辨出大肠杆菌中超过 1000 种蛋白质

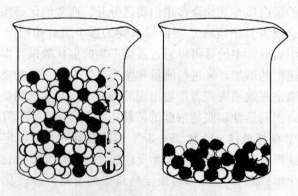

图 10-10　酶活性和比活性

这两个概念的区别用两大杯弹球来说明。两个杯子装有同样多的黑球和不同数目的白球。如果球代表蛋白质，黑球代表酶活性。第二个杯子有更高的比活性，因为这里黑球占总数的比例高很多

第三节　分子杂交技术

核酸分子杂交（molecular hybridization）是分子遗传学的基础研究工具，利用的是单链核酸分子间能退火形成双链分子（也就是杂交）的能力。能杂交形成双链分子的单链分子间必须有高度的碱基互补配对。通常核酸分子杂交分析是利用一段被标记的核酸分子作为探针（probe），在许多未经标记的核酸分子混合物中鉴别同源 DNA 或 RNA 分子。

当然，对基因表达和功能的研究不只停留在 DNA 和 RNA 水平，鉴别某种特定的蛋白质分子也是必需的。为了达到这个目的，我们用抗体（antibody）代替核酸探针以检测特定的蛋白质分子。

一、原理

核酸分子杂交是在两种单链核酸分子混合在一起的状态下进行的，包括已知序列的探针（如克隆的 DNA 片段或者化学合成的寡核苷酸）和通常包括含有靶分子的核酸混合物。如果初始探针或者靶分子是双链的，在杂交前必须变性为单链分子，一般采取热变性或碱变性的方法。当单链的探针与靶片段混合后，含有互补碱基序列的单链又重新聚集到一起。探针间和与探针互补的靶 DNA 分子间都能分别形成同质双链。然而，只有当单链的 DNA 探针和与之互补的靶 DNA 单链间退火形成异质双链时核酸分子杂交分析才是有效的（图 10-11）。杂交分析的原理实际上就是用已知序列的探针从成分

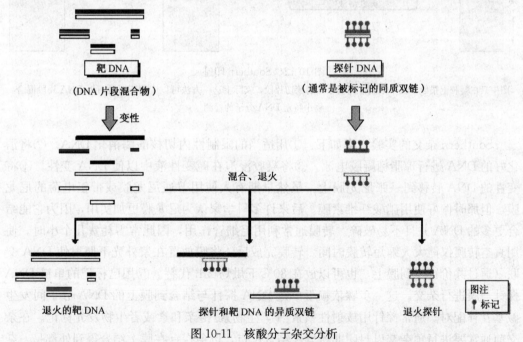

图 10-11　核酸分子杂交分析

图中的探针分子与目标 DNA 中的一段 DNA 片段中心区域高度同源。将变性过的探针和靶片段混合后能够退火形成三种形式的分子：靶 DNA 分子重新退火形成的同质双链（左下），探针与探针形成的同质双链（右下），DNA 探针和靶 DNA 间形成的异质双链（中间）。如果用适当的办法去除未与靶 DNA 结合的探针，形成的异质双链就能很容易地被检测出来

复杂的核酸混合体中寻找与之同源的靶核酸片段。探针必须经放射性标记、荧光标记或者其他方法标记以方便检测。

二、Southern 印迹

Southern 印迹（Southern blotting）是 DNA 与 DNA 杂交的技术。它是 20 世纪 70 年代由 Edward M. Southern 在爱丁堡大学发明的，并由此得名。简言之就是 DNA 分子经琼脂糖凝胶分离后被转移到膜上进行杂交。Southern 印迹被用于鉴别成分复杂的核酸混合物中具有特定序列的 DNA 分子。例如，Southern 印迹可以从基因组中定位某个特定的基因。所需的 DNA 样品量取决于探针的长短和特异性。短的探针具有的特异性更高。在最优条件下，可以检测到低至 0.1pg 的目的 DNA。图 10-12 显示了 Southern 印迹的基本步骤。

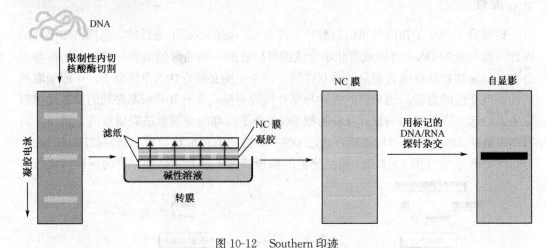

图 10-12　Southern 印迹

图中只在凝胶上描绘了三个限制性内切核酸酶酶切片段，实际上这一方法可以应用于对数百万 DNA 片段混合物中特定 DNA 分子的检测

Southern 杂交的实验方案如下：①用适当的限制性内切核酸酶消化 DNA。②将消化好的 DNA 进行琼脂糖凝胶电泳。③将凝胶浸泡在碱变性液中以使 DNA 变性。④将变性的 DNA 转移到一张尼龙膜上。虽然早期有人使用普通尼龙膜或带正电荷的尼龙膜，但那时偏好使用硝酸纤维素膜。后来许多科学家认为尼龙膜更加实用，因为它能结合更多的 DNA 并且不易破碎。转膜通常利用毛细管作用，因此常需耗费几个小时。使用真空转膜仪能大大缩短转膜时间。转膜完成后，将膜放置在紫外光下照射使 DNA 交联（通过共价键）到膜上，也可以放在 80℃下烘烤 2 h 代替。⑤用已标记的单链 DNA 探针与膜进行杂交。这一步骤依赖于单链 DNA 探针与结合到膜上的 DNA 分子间发生碱基互补配对。通常探针用放射性核素^{32}P，生物素/链亲和素或者生物发光标记。在杂交前通常要进行预杂交以封闭非特异性结合位点，以免探针在膜上结合得到处都是。杂交与预杂交使用相同的缓冲液，只是杂交时要在缓冲液中加入特异的探针。

如果使用放射性核素^{32}P 标记探针，可以通过放射自显影观察结果。生物素/链亲和素标记采用化学显色的方法检测结果，生物发光标记的探针在发光条件下观察结果。

三、Northern 印迹

Northern 印迹（Northern blotting）是 1977 年由 Alwine 等在斯坦福大学发明的一种 RNA 杂交技术。它的命名是源于 1975 年 Edward M. Southern 发明了 Southern 印迹。Northern 印迹的步骤和原理与 Southern 印迹基本一致，只不过将 DNA 换成 RNA 来处理（图 10-13）。

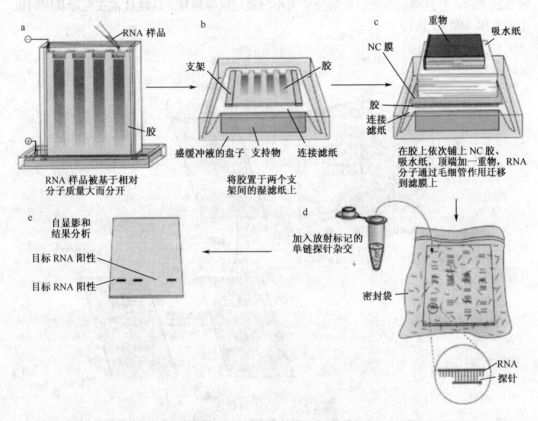

图 10-13　Northern 印迹

a. 从各类组织中抽提出来的 RNA 通过电泳被按照分子质量大小分开；b. 凝胶被置于毛细管转移装置上；c. 将硝酸纤维素膜置于胶上，其上再放置一叠吸水纸，在毛细管作用下核酸分子被流动的缓冲液带动，从凝胶中洗脱出来并聚集于硝酸纤维素膜；d. 滤膜与能和目的 mRNA 分子互补配对的放射性核素标记的单链 DNA 探针一同孵育；e. 洗脱掉未结合的探针后，经过放射自显影即可定位目标 mRNA

尽管现在处于实时荧光定量 PCR、核酸酶保护分析以及微阵列分析等高科技的时代，但 Northern 印迹仍然是检测和定量分析 mRNA 水平的金标准。它比用一些 PCR 方法来确定特定转录子大小和检测某种基因的不同转录子要可靠得多。Northern 印迹的应用除了在 mRNA 水平研究基因表达之外还包括检测 mRNA 转录子的大小，RNA 降解与 RNA 剪接，还能用来验证转基因或基因敲除的动物模型。

但是这项技术也存在着很多局限。首先，虽然我们可以通过信号的强度估计 mR-NA 的相对表达量，但是通常不能得知特异 mRNA 的绝对表达量。其次，这项技术的灵敏度较之核酸酶保护分析和 RT-PCR 低，并且需要相当大的 RNA 样品量。最后，

Northern 印迹的步骤较为烦琐并且费时。

四、Western 印迹

与 Southern 印迹和 Northern 印迹鉴别特异的 DNA 或者 RNA 类似，Western 印迹（Western blotting）技术利用蛋白质与各自抗体特异性结合的能力来鉴别和定位蛋白质，不论蛋白质是体内还是体外合成。Western 印迹能够从许多蛋白质的混合物中检测到目的蛋白，并且得到该蛋白质的分子大小与表达量等信息。这项技术在克隆基因时也十分有用（图 10-14）。

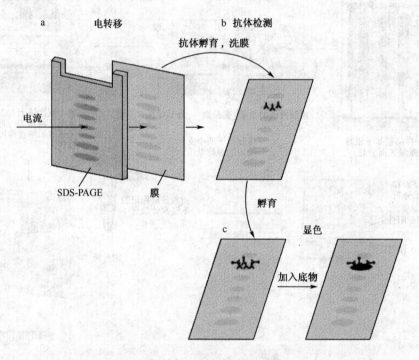

图 10-14　Western 印迹

a. 蛋白质的混合液经 SDS 凝胶电泳后被转移到膜上；b. 膜浸泡在含有能与目的蛋白特异结合的一抗（Ab1）的溶液中。充分的结合后，膜经过洗涤以除去未结合的一抗。c. 膜首先与含有能和一抗结合的二抗（Ab2）的溶液孵育以结合上二抗。二抗共价结合上碱性磷酸酶，催化后续的显色反应。最后，加入酶促反应底物，经固定的酶催化底物在原位形成一种深紫色的沉淀物，使含有目的蛋白的条带显现出来

为了检测特异蛋白，必须先得到该蛋白质的抗体。抗体可以是目的蛋白诱导产生的或者有时候可以购买到商业化的产品。由于硝化纤维素膜本身有许多可以非特异性结合蛋白的位点，会与抗体非特异结合，因此必须用非特异的蛋白质溶液封闭这些位点，如可以在杂交液中加入脱脂奶粉。将一抗加入牛奶溶液中它就能与目的蛋白结合，产生的抗体-蛋白质复合体可以通过标记的二抗检测到。通常碱性磷酸酶等报告酶被连接到二抗上，再加入化学发光剂 Lumiphos 或者 X-phos 以检测目的蛋白的条带。

第四节　DNA 和蛋白质序列测定

DNA 是细胞中遗传信息的主要携带者，分析了解 DNA 自然成为遗传研究的主要问题。我们的染色体，或者说基因组是由有序缠绕的 DNA 链组成。所有的生物，从细菌到人类，构成有机体的细胞中都含有 DNA，每个细胞中都含有整套的遗传信息。任何测序都是以得到整个基因组序列和鉴定一个完整的基因为目的的，但是最终目标是为了弄清楚某个基因是什么时候，在什么器官中怎么样开始表达的，也就是基因的表达谱式。一旦我们了解了一个基因在正常环境中的表达方式，我们就可以继续研究在另一个状态下基因表达的变化，如疾病。为了达到这个目标，必须了解并研究这个基因编码的某种蛋白质或几种蛋白质。

一、DNA 测序

DNA 测序（DNA sequencing）的魔力就在于它能将基因或基因组简化为有确定结构的化学分子。这是少有的能提供如此确定信息给生物学家的几种技术之一。目前在分子克隆实验室，测序的主要用途在对新克隆的 cDNA 进行定性研究，对某个克隆或突变的确认，检查某个新构建的突变的重现度，连接位点的精确度，以及 PCR 产物的保真度；在某些情况下，测序可以作为一个筛选工具来对一些特别感兴趣的基因的多态性和突变进行确认。

1. 原理

20 世纪 70 年代中期，科学家们就发明了快速高效的 DNA 测序方法。下面是两种几乎同一时间发布的方法：①链终止法（chain termination method），单链 DNA 分子的序列由 DNA 聚合酶合成的互补链确定，这些链会在特异的位置终止。②化学降解法（chemical degradation method），这个方法主要是用到一些化学试剂可以在特异的位置切割 DNA，DNA 的序列可用由此而产生的一组具有不同长度的 DNA 反应混合物来确定。开始两种方法都很受欢迎，但是近年来，特别是在基因组测序中，链终止法成为主要的测序方法。这一方面是因为化学降解法中的化学试剂是有毒性的，主要原因是链终止法更易自动化。在基因组计划中包括大量的测序反应，手工测序将花费很多时间。因此要在可接受的时间内完成测序计划，就必须有自动化测序（automated DNA sequencing）方法。

2. DNA 测序技术进展

传统的双脱氧链终止法用到了放射性核素标记，dNTP 混合液中有一部分是带放射性标记的，这样标记就能融入新的 DNA 链中。电泳后，将经过干燥处理的凝胶覆上 X 线片，再经过一定时间的曝光，X 线片显影得到一组不同条带组成的序列。由于 ^{32}P 发射的高能 β 射线常常会导致放射自显影图谱条带扩散、分辨率低，通常不用 ^{32}P 标记的核苷酸测序。^{35}S 或者 ^{33}P 标记被广泛使用。

大规模的基因组测序主要依赖于部分技术所带来的高效自动化。近年来主要的进步是荧光标记的 DNA 自动测序的发展。这种方法通常会对引物或者是 ddNTP 进行荧光标记。对 4 种碱基标记 4 种不同的荧光，因此和传统的 DNA 测序方法不同，4 个反应的产物可以在一个泳道中电泳。通过毛细管电泳能将新合成的不同长度的 DNA 带分开，见图 10-15。

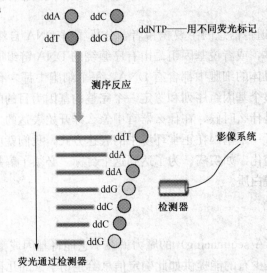

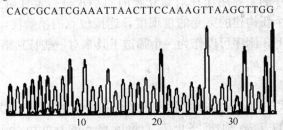

图 10-15　运用荧光标记的 ddNTP 的 DNA 自动测序

a. 每种 ddNTP 都标记一种不同的荧光基团，所以链终止反应能在一个管中
进行。在自动测序仪上，凝胶中的条带经过荧光检测器时，检测器就能分辨
出是哪种 ddNTP，然后将这些信号以图的形式输出。b.　一台自动测序仪的
输出文件。一个峰表示一个碱基

这种毛细管和人的头发差不多粗，2.5～7.5 cm 长，足够把大小相差 1 个碱基的条带分开。由于细小精密，灌胶和上样都是电脑控制的。短片段的迁移速度比较快，先跑出毛细管底部，长一点的片段由于受到凝胶的阻滞，需要更长的时间跑出毛细管底部。当这些 DNA 片段跑出毛细管时它们会通过激光检测器窗口，激光束会激发单链末端 ddNTP 上所带的荧光染料，使染料以特定的波长发光，或发出某种颜色的荧光。这些光信号被检测器检测到后将信息传给电脑。电脑将从检测器那里获得的信息以荧光吸收峰图的形式输出，这种荧光吸收峰图是光电探测器探测到的 4 种波长的信号图（图 10-15b）。由光电探测器检测到的真实的颜色接近绿、黄、橘黄、红色，计算机分别用其他的颜色代表 4 种波长的光信号以便我们更好地区分辨认。而且图中每个峰下面都标明对应的碱基的字母。因为连续的峰正好代表相差 1 个碱基的 DNA 片段，所以峰的顺序表明了 DNA 样品的碱基序列。

二、蛋白质测序

在几乎所有的生物学过程（催化作用、信号转导、维持结构等）蛋白质都发挥了至关重要的作用。尽管蛋白质测序比 DNA 测序复杂得多，蛋白质的氨基酸序列仍然能被直接测出。如果能够知道一个蛋白质的序列对鉴定这个蛋白质是非常有用的。从蛋白质的一级结构我们也能得到很多关于蛋白质功能和进化史方面的有价值的信息。此外，由于有大量的、完整或近乎完整的基因组序列资料，通过确定一小段蛋白质序列就足以鉴定出编码这个完整蛋白质的可读框。

1. 原理

用 Edman 自动降解法对蛋白质进行测序。先确定肽链的氨基酸组成。用 6molHCl在 110℃将肽链水解 24h 降解成氨基酸。水解产物可以通过带有磺化聚苯乙烯的离子交换柱分开。氨基酸是根据它的洗脱体积定性的，就是从柱子上洗下来氨基酸的缓冲液的体积，根据与水合茚三酮（ninhydrin）的反应定量。除了脯氨酸（因为带亚氨基）和茚三酮反应生成黄色物质，其他氨基酸生成深蓝色物质。与茚三酮反应后，溶液中氨基酸的浓度与溶液的吸光度成正比。这种方法能检测到低至 10 nmol 的氨基酸，这个量只是差不多拇指指纹里的量而已。要检测少到 ng 级别的氨基酸（10 pmol）可以用荧光胺代替茚三酮，它和 α 氨基反应能产生一种释放高强度荧光的物质。样品水解产物的层析色谱与标准的氨基酸混合物比较就能得出氨基酸的组成。接下来，通常利用 N 端残基和另一种化合物反应来鉴定。Frederick Sanger 最早用的是二硝基氟苯（FDNA）。因为丹磺酰氯（DNS）能形成荧光物质提高了检测的灵敏度而被广泛使用。它与不带电的氨基反应形成磺胺衍生物，这种物质在水解肽链的条件下不会降解。在 6 mol HCl 中水解 DNS-肽链能产生一个 DNS-氨基酸，这个化合物能够通过层析来鉴定。也因为丹磺酰氯能形成有荧光的磺胺类物质而成为一个重要的标记物。

尽管 DNS 鉴定末端残基的方法灵敏高效，但是却不能在一条链上重复使用，因为酸水解的过程中肽链已经降解了，也就没有序列的信息了。Pehr Edman 发明了一种标记末端氨基酸残基的方法，这种方法能将末端化合物切下来而不影响剩下肽链中其他氨基酸之间的肽键。Edman 降解法能连续不断地将肽链的末端氨基酸一个一个地释放出来。苯异硫氰酸酯（PITC）与多肽链的游离氨基反应生成 PTC-肽。然后在弱酸性条件下，形成末端氨基酸的环化物，留下少了一个氨基酸的肽链。这个环化物是 PTH-氨基酸，能通过层析法鉴定。Edman 过程在变短的肽链中可以重复进行，产生另一个 PTH-氨基酸，同样可以鉴定。若干次循环之后，就能知道整个肽链的序列（图 10-16）。

自动测序仪的发展大大缩短了蛋白质测序的时间。一个 Edman 降解的循环（从肽链上释放一个氨基酸并对它进行检测）只需要不到 1h。降解重复进行，一次可以检测 50 个左右的氨基酸残基。高压液相色谱提供了一种更灵敏的分辨氨基酸的方法（图 10-17）。气相测序仪可以在 pmol 水平上分析肽链或蛋白质。有了这么高的灵敏度，从 SDS-聚丙烯酰胺胶上的条带中洗下来的蛋白质样品都能进行分析。

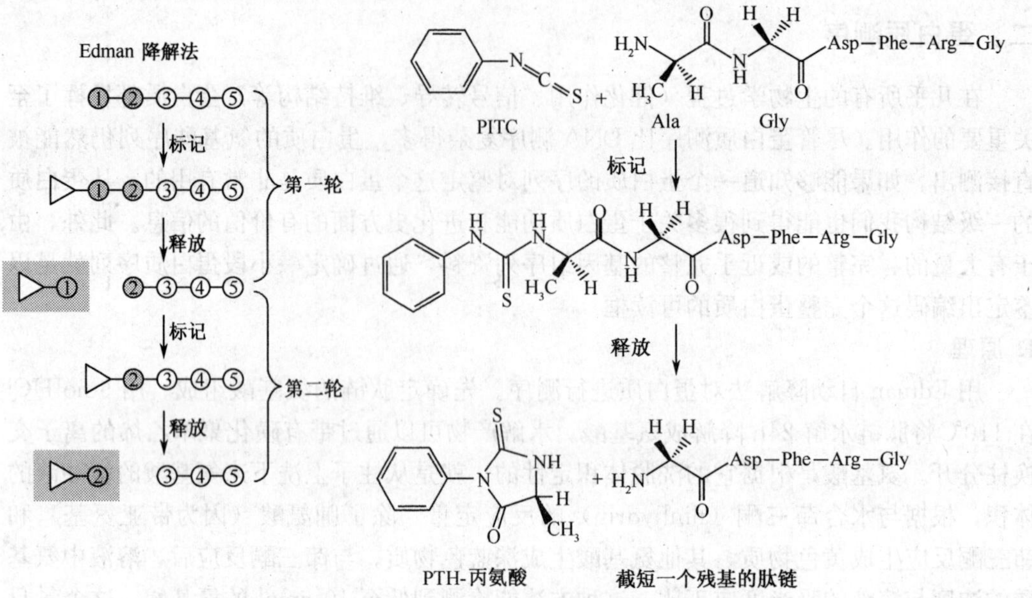

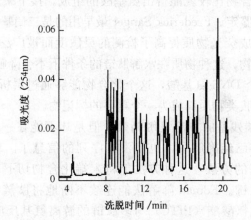

图 10-16　Edman 降解法（以一个五肽为例）(Berg and Tymoczko，2002)

不需要水解肽链就能释放标记的氨基酸末端残基（PTH-丙氨酸是第一轮被释放的）。因此，第二个循环就能检
测出剪短了的肽链末端氨基酸。再经过 3 个循环就能测出整个肽链的序列

图 10-17　分离 PTH-氨基酸（Berg and Tymoczko，2002）

通过高压液相色谱能快速分离 PTH-氨基酸。在这张高效液相色谱图
中，明显地表示了它的组分中含有数种 PTH-氨基酸，通过比较与已
知氨基酸洗脱的相对位置就能鉴定出未知氨基酸

2. 蛋白质测序的进展

使蛋白质测序发生革命性变化的技术是串联质谱（tandem mass spectrometry，MS/MS）。质谱是一种可以精确分析微量样品质量的方法。简单来说，它的原理是在仪器的真空环境中，物质的运行轨迹取决于其荷质比。对于较小的生物大分子，如肽链或质量比较小的蛋白质，分子质量的测定可以精确到 1Da。

通常先用胰蛋白酶将蛋白质样品降解成较小的肽段。然后用 HPLC 对混合肽段进行分离，用串联质谱分析。在串联质谱中，化合物通过连续的质谱以得到关于它们特性和结构的详细信息。在第一个质谱中，依据不同的荷质比化合物被分开，然后被降解的肽段在第二个质谱中分析。分析原始化合物和离子化的片段的质谱图就能辨认出目的蛋白及其结构。因为肽链离子的电荷状态（在离子化过程中肽链带上的正电荷或负电荷的电量）是已知的，所以几乎可以直接得出结果。在电荷状态已知的情况下，肽链的分子质量可以根据荷质比直接算出。将这个蛋白质的分子质量以及它的片段化离子的质谱图与蛋白质和多肽数据库比较就能得出它的序列和结构资料。这个技术意味着串联质谱和HPLC联用成为蛋白质组学研究的基础。

三、表达序列标签

当一些物种的全基因组测序工作完成时，研究人员都希望得到一个基本的资源即一个物种的所有基因的简要清单。一张基因清单，结合其物理性质和电子信息，就能研究复杂的生物体系里基因相互作用的方式。然而，很多对医药和农业有重要价值的物种还不具备基因组测序的条件，这时候 cDNA 提供了基因序列的初始资源。因此，可以预见在未来的一段时间转录产物的测序将依然会是一个人们感兴趣的重要领域。

1. 原理

1991 年，Venter 和他的同事们里程碑式的研究开创了一个高通量 cDNA 测序时代。基本的策略：随机挑选 cDNA 克隆，然后从插入片段的一端或两端测序。他们引入表达序列标签（expressed sequence tag，EST）来指代这种新的序列，它的特点就是短（一般400～600 bp），不太准确（错误率约有 2%）。主要的思想就是测 DNA 的一部分序列，这部分序列能代表在不同生物的细胞、组织、器官中表达的基因，然后用这些"标签"依据碱基匹配在部分染色体 DNA 中"钓"出基因。从基因组序列中找出基因要面对的困难在不同的生物中有所不同，与基因组大小、有无内含子、基因的编码序列中的间隔序列都有关系。EST 的应用加速了基因的发掘、辅助了整个基因组的注释，有助于基因结构分析，鉴定不同转录子的存在，指导单核苷酸多态性（SNP）的特性描述，并且有利于蛋白质组学的分析（图 10-18）。

2. EST 应用

EST 有多种用途。最早 EST 用于构建人类基因组的表达图谱，然后用来评估基因的覆盖率，还用来开发一些以基因为基础的标记图谱。伴随着全球测序工程中指数式增长的基因组数据，EST 数据库广泛用于预测基因结构，研究选择性剪切，区分在不同组织中表达的基因或者是特异表达的疾病基因，还被用于发现并描述潜在的单核苷酸多态性（SNP）位点。EST 数据的这些用途已经超出了它原本的在发现基因分析转录方面的应用。例如，EST 作为基因组的"路标"。正如开车的人需要一张地图来帮助找到目的地，科学工作者们在组成人类基因组的数十亿核苷酸中探索以寻找基因时，同样需要基因组的"地图"。一张地图要起到航标性的作用，它必须含有可靠的路标或者是类似路标的标志。已经被用来制作多张基因组图谱的序列标签位点（sequence tagged site，STS）作图法是目前最强大的作图技术。STS 是基因组（或染色体）中易辨认的单拷贝的短 DNA 序列。通常 EST 的 3′端都可以当作 STS，因为它们极有可能对某一特定物种是特异的，还提供了寻找特定表达基因的其他特征。又如，EST 作为发现基因的源泉，因为 EST 代表了基因组中大家感兴趣的那部分拷贝，而且都是表达的序列，

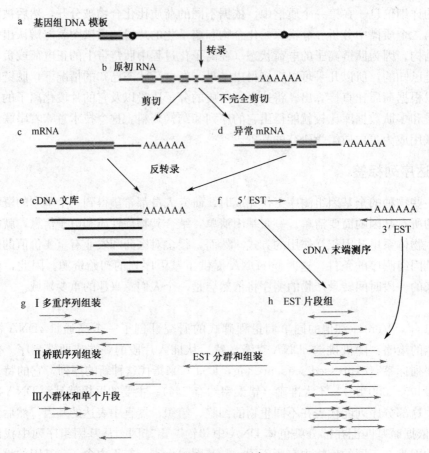

图 10-18　cDNA 克隆和 EST 测序（Rudd，2003）

基因组 DNA 模板（a）。基因转录后得到原初 RNA(b)。这个 RNA 能反映基因的结构，含有非编码区、内含子（方框之间的短线）和外显子（方框）序列。原初 RNA 完全剪切得到 mRNA（c），不完全剪切得到异常 mRNA 或叫做未完成 mRNA（d）。RNA 库反转录得到一个 cDNA 库（e）。将 3′端的 ploy(A) 尾巴作为一种筛选 mRNA 的标记，所以在 cDNA 文库中基因的 3′端比 5′端多。从 cDNA 的任一端测序得到 5′EST 和 3′EST(f)。这些序列被加入到这个生物原来的 EST 中（g），整理成合理的片段重叠群（h）。这种聚类和装配能产生大量代表单独 EST 的大的片段组。不同的片段组可能被拼接成一个假的聚类群。还有一些 EST 会聚到小的聚类群中或者只是单独的序列

所以多次在寻找与遗传疾病相关基因中表现出其强大的优势。EST 还有很多实用的优势，EST 测序迅速且成本低，每个 cDNA 只需一个测序反应，也不需要校正测序错误，因为错误不会妨碍对基因的鉴定。

第五节　聚合酶链反应

在所有的现代分子生物学技术中，聚合酶链反应（polymerase chain reaction，PCR）或许是最有用的技术之一。PCR 技术提供了一种在体外扩增 DNA 序列的手段。极其微量的目的基因经过 PCR 扩增后可以达到微克级别。PCR 非常灵敏，以至于可以

将单个细胞中提取出的 DNA 扩增到足够多的量以用于克隆或测序。因此，PCR 技术被广泛应用于临床诊断、遗传分析、基因工程及司法鉴定等领域。最值得一提的是，PCR 彻底革新并加速了整个重组 DNA 技术领域的发展。在这项技术发明之前，DNA 的克隆是通过培养细菌然后从中提取并纯化 DNA 实现的。PCR 使特异 DNA 序列能够快速的大量扩增，并且得到的 DNA 更加易于纯化并结构完好。

一、原理

PCR 是一种快速在体外从某种来源的 DNA 中扩增目标 DNA 的通用方法。通常，这种方法用于从各种不同序列的 DNA 混合物（如基因组 DNA 或者 cDNA 群体）中选择性扩增特异的 DNA 序列。为了达到选择性扩增的目的，我们必须对目的基因的 DNA 序列预先有所了解。有了这些序列信息，我们就能设计出两条寡核苷酸引物。引物（primer）具有靶序列特异性，长度一般为 15～25 个核苷酸。向变性后的 DNA 模板中添加引物后，引物能够在靶位点特异性地与互补的 DNA 序列结合。在合适的热稳定 DNA 聚合酶和 DNA 前体（4 种三磷酸脱氧核糖核苷酸：dATP、dCTP、dGTP 和 dTTP）的存在的条件下，以 2 条单链各自为模板起始合成 DNA 新链（图 10-19）。

PCR 之所以被称为链式反应是因为新合成的 DNA 链又能在后续的反应中作为模板继续合成新链。大约 25 个循环过后，PCR 产物中不仅包括起始的 DNA，更产生了约 10^5 个特异靶序列的拷贝，产物足以通过琼脂糖凝胶电泳观察到。反应由包含以下三步的连续循环构成：①变性（denaturation）：以人类基因组 DNA 为例通常是 93～95℃，以打开模板的双链结构。②退火（reannealing）：通常采用 50～70℃ 的某个温度，具体取决于我们所预期的二聚体产物 T_m 值（一般退火温度低于计算出的 T_m 值 5℃）。③延伸（extension）：DNA 的合成，通常在 70～75℃。

当 Kary Mullis 发明 PCR 技术的时候，他使用的是普通的 DNA 聚合酶。由于使 DNA 分开为 2 条单链所需的温度会破坏这种酶，因此它在每轮循环结束后都必须向反应管中加入新酶。幸运的是，后来发现了一些嗜热的微生物，它们的 DNA 聚合酶能够耐高温。例如，现在最常用的 TaqDNA 聚合酶是从一种嗜热细菌——嗜热水生菌（*Thermus aquaticus*）中提纯出来的，能够耐受 94℃ 的高温，而它最适合的工作温度是 80℃。

二、逆转录 PCR

逆转录 PCR（RT-PCR）是用来扩增 RNA 的 cDNA 拷贝的方法。RT-PCR 非常灵敏，被用于克隆 mRNA 的 5′ 端和 3′ 端，从小量的 mRNA 生成大量的 cDNA 文库等领域。另外，RT-PCR 也能用于鉴定转录序列的突变及多态性，当 mRNA 样本量有限或者目的 RNA 表达量很低时检测基因的表达强度。

RT-PCR 的第一步是在酶的作用下将 RNA 反转录成单链 cDNA 模板。一段寡聚脱氧核糖核苷酸引物结合到 mRNA 上后，在依赖 RNA 的 DNA 聚合酶的作用下延伸形成 cDNA 拷贝作为 PCR 的模板。根据实验的目的不同，用来合成 cDNA 第一链的引物也不同，可以是按照特异靶基因的序列设计的特异性引物，也可以是能与所有 mRNA 结合的通用引物。加入第二条引物和热稳定的 DNA 聚合酶后就开始了后续的 PCR 扩增

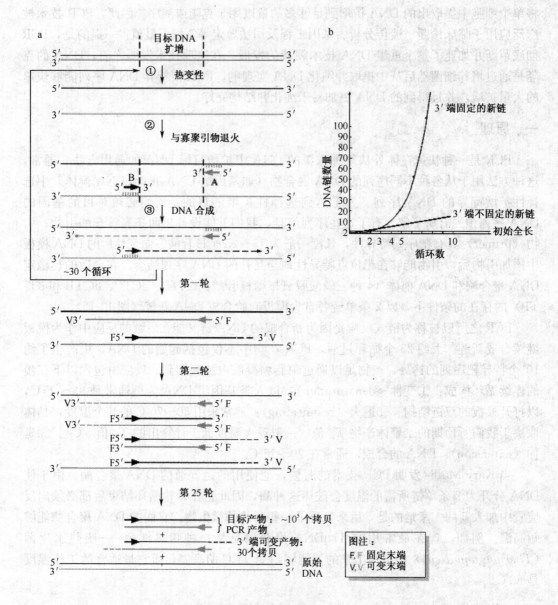

图 10-19 聚合酶链反应

a. 寡核苷酸引物 A 和 B 分别与模板 DNA 两条链上的某段序列互补，而这两段模板序列又分别位于待扩增
DNA 区段的两侧。退火的引物延伸合成新的 DNA 链；b. 几轮循环过后，具有期望长度的扩增产物——3′端
固定新链成为主要产物

阶段。在进行 RT-PCR 的过程中必须设立阳性对照和阴性对照（图 10-20）。

在可能的情况下，正义（sense）和反义（antisense）引物最好设计在靶 RNA 的不同外显子上以扩增 cDNA 产物。这样就能很容易地区分扩增产物是来自于 cDNA 还是污染的基因组 DNA。然而，对于不含内含子的基因来说很难区分是否污染了基因组序列。在这种情况下，RNA 用不含 RNA 酶的 DNA 酶处理并抽提是一种有效的解决方案。

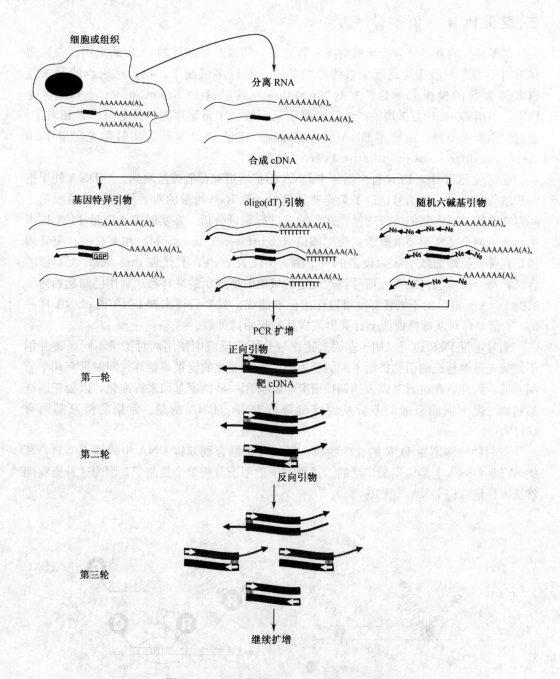

图 10-20　PCR 扩增 RNA 的不同方法示意图

cDNA 第一链的合成可以采用基因特异引物，oligo(dT) 或者随机六碱基序列其中的
一种作为引物。第二链的合成（第一个扩增循环）是由正义引物起始的。在正义
引物和反义引物的存在下，cDNA 继续得到扩增

三、定量 PCR

热循环仪的发展结合荧光检测技术催生了一项新的革新性的 PCR 的应用。在常规 PCR 中，我们只能检测到反应最终产物的量。在许多情况下，我们感兴趣的是未经 PCR 放大前的模板起始量。实时（real-time）或定量 PCR（quantitative PCR，Q-PCR）利用 DNA 指数扩增的线性阶段测定某个样品中特定序列的绝对量或者相对量。通过检测荧光染料，能描述 DNA 的合成过程。定量 PCR 又称为实时荧光定量 PCR（real-time fluorescent quantitative PCR，FQ-PCR）。

在定量 PCR 中，PCR 每个循环中 DNA 的扩增情况都能被监测到。当 DNA 处于指数扩增阶段时，荧光信号超过了荧光背景信号，PCR 产物量的对数值与起始模板量之间存在线性关系，我们可以选择在这个阶段进行定量分析。在实时荧光定量 PCR 技术中引入了两个非常重要的概念：荧光阈值（fluorescence threshold）和 CT 值。荧光阈值是在荧光扩增曲线上人为设定的一个值。C 代表 cycle，T 代表 threshold。CT 值的含义：每个反应管内的荧光信号到达设定的域值时所经历的循环数。利用已知起始拷贝数的 DNA 标准品的不同稀释度可以作出标准曲线，只要获得未知 DNA 或 cDNA 样品的 CT 值，即可从标准曲线上计算出该样品的起始拷贝数。

实时定量 PCR 也可以用于基因表达比较分析。反应中使用两对扩增不同区域并用不同荧光染料标记的引物扩增不同的基因。一些商业化的定量热循环仪同时具有几个监测通道。因此，在进行基因表达调控研究中都会用一些持家基因来标准化，以校正因样品初始浓度不同而造成的差异从而准确分析 DNA/cDNA 的量。常用的持家基因有 *GAPDH*，*β-actin*。

有两种检测定量 PCR 的化学物质。第一种是结合到双链 DNA 中的插入染料，其中 SYBR Green Ⅰ是应用最广泛的。这种方法适用于分析单个复制子，因为这种染料能够插入任何双链 DNA（图 10-21a）。

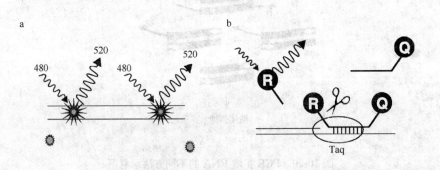

图 10-21　实时 PCR 中的化学物质

a. SYBR green Ⅰ（吸收波长为 480nm，激发波长为 520nm）与双链 DNA 结合时发出荧光；

b. Taqman 探针

第二种是利用与目标序列配对的引物或寡核苷酸，如 TaqMan 探针（图 10-21b）、分子信标探针和 Scorpion 引物。这段寡核苷酸标记有荧光基团和淬灭基团，本身不发出荧光信号，但当它们与模板退火（如分子信标）或者在延伸时荧光染料与淬灭剂分离（如 TaqMan 探针）而发出荧光。利用不同激发波长的荧光染料可以实现多元 PCR。

四、PCR 衍生技术

1. 随机扩增多态性 DNA

随机扩增多态性 DNA（randomly amplified polymorphic DNA，RAPD），常常被写作复数 RAPDs，发音"rapids"，部分原因是因为它是一种能够快速获取所要研究物种的大量基因信息的方法。RAPDs 技术的目的是检测 2 种物种间的亲缘关系。

RAPD 的原理：随机的短寡核苷酸序列能够与基因组中许多序列随机地互补结合。如果 2 段这样互补的序列恰好以正确的方向结合于 1 段基因组区域的 2 条链上，并且它们相隔的距离足够短，它们之间的 DNA 序列便能得到有效扩增。每段这样的扩增产物都是独立的，按照概率它们的长度也很可能是不同的。如果数量合适、大小不同的条带得到扩增，它们就能够通过凝胶电泳被分离开。不同的寡核苷酸序列能够扩增出一系列完全不同的位点。

在实际应用中，一般选择能够产生出 5~10 个 PCR 产物的引物长度。对于高等生物来说，引物的长度通常为 10 个碱基。PCR 产物通过凝胶电泳分离以检测相应条带的大小。用一组序列各不相同的引物重复进行以上程序就能得到某物种特有的条带组合，与其他物种的条带比较就能发现它们亲缘关系的远近，带型越接近说明它们的亲缘关系越近。虽然我们并不知道各个 PCR 产物条带来源于哪个基因，但这对分析亲缘关系来说并没有影响。因此判断的依据在于用一条（或者一组）引物能从某种物种中扩增出某种特有的条带而从其他物种中扩增出另外的条带，即使这些物种的亲缘关系十分接近。

2. 巢式 PCR

巢式 PCR（nested PCR）是 PCR 技术的一种变化形式，与一般 PCR 的区别在于用 2 对引物代替了 1 对引物对目标序列进行扩增。第一对 PCR 引物的扩增与标准的 PCR 十分相似。不过，第二对被称为巢式引物（因为它们位于第一轮扩增产物内）的引物结合到第一轮 PCR 产物片段上后将引发第二轮的扩增，只是产物比第一轮扩增的短。巢式 PCR 的优势在于，如果在第一轮扩增中出现了错误的扩增片段，在第二轮扩增中它再次被扩增的可能性非常小。因此，巢式 PCR 是一种特异性很强的 PCR 技术（图 10-22）。

3. 锚定 PCR

我们常常需要扩增临近已知 DNA 序列的未知 DNA，不论是在基因组水平还是 cDNA 水平。为了达到这个目的，一种被称为锚定 PCR（anchored PCR）的 PCR 方法应运而生。

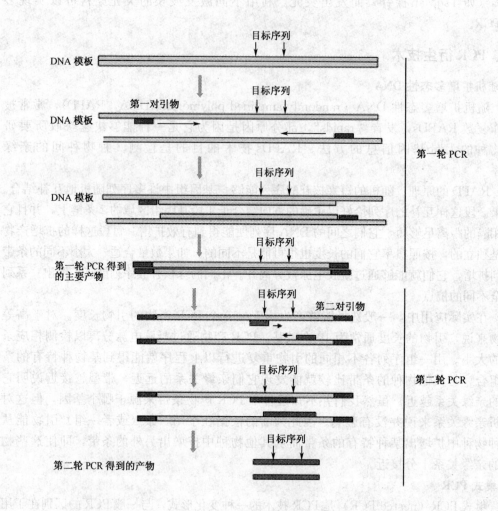

图 10-22 巢式 PCR 示意图

在第一轮 PCR 中，第一对引物结合到目标 DNA 模板两端。除了结合到特异的引物结合位点上
扩增目的序列之外，这对引物有可能结合到序列相近的其他位点上（图中没有表示出来），扩
增出错误的产物。在第二轮 PCR 中，第二对引物结合到在第一轮扩增出的特异序列的内侧进
行第二轮的 PCR 扩增

　　在锚定 PCR 的一对引物中，一条能特异的结合到靶序列上，另一条能够特异地结合到锚定序列上，成为锚定引物。锚定引物可以通过不同方式引入，如采用接头-引物的方法，或者通过修饰引物的 5′端来引入新的序列。图 10-23 为利用锚定 PCR 进行基因组步移的原理示意图。

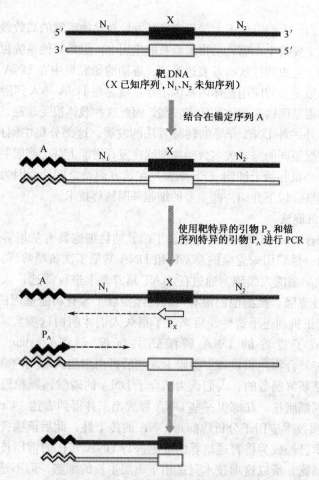

图 10-23 利用锚定 PCR 进行基因组步移

一段锚定序列（如一段双链寡聚核苷酸接头）首先被附加到 DNA 片段上，利用一条锚定
序列特异性引物和一条已知序列（X）特异性引物就能扩增得到包含部分已知序列 X 和与
X 临近的未知序列的片段。在这个例子中，为了描述得清晰只表示出了左边的锚定序列，
在引物 P_A 和 P_X 的作用下可以扩增出已知序列 X 旁的 N_1 序列

第六节 基 因 转 移

基因转移（gene transfer）就是将外源遗传信息以 DNA 的形式转化到细胞中。进行基因转移的原因有很多，但最重要的原因可能就是治疗疾病。有不同的转基因方法通过人工修饰过的载体（如病毒）将基因转入细胞中。转基因技术在基因操作技术中是一种重要的技术。

一、动物细胞的基因转移

动物细胞的基因转移已应用了 40 多年，目前该技术已应用于将 DNA 转入许多不同类型的培养细胞中，用来研究基因功能和调控或者生产大量的重组蛋白。动物细胞有利于生产重组动物蛋白，因为动物细胞可以进行细菌细胞和真菌所无法进行的真正的翻译后修饰。因此细胞培养物已用于商业化规模生产，如抗体、激素、生长因子、细胞因子和免疫用途的病毒衣壳蛋白。人们进行了大量研究来开发针对动物细胞的高效载体系统和转化方法。尽管这项研究主要集中在哺乳动物细胞系的应用上，但是其他系统也一样得到了广泛开发和应用，如用于昆虫的杆状病毒表达系统。近期的研究集中在 DNA 体内转入动物细胞方面。这一技术最重要的应用是体内基因治疗，就是将 DNA 导入到活体动物细胞中以治疗疾病。由于病毒基因输送载体的效率较高，因此这种载体很受欢迎，但是人们对安全性的考虑促进了另外一种 DNA 介导的转移方法的发展。这部分集中阐述体细胞的 DNA 转入。与植物的情况不同的是，大多数动物细胞在发育潜能上都受到限制，因此不能用于培育转基因动物。小鼠胚胎干细胞（ES 细胞）在这方面是个例外，因为它们取自生殖细胞形成之前的早期胚。以下介绍一些重要的细胞基因转移技术。

1. DNA/磷酸钙共沉淀法

Szbalska 和 Szybalski 1962 年最先报道了哺乳动物细胞具有从培养基里摄取外源提供的 DNA 的能力。他们用未克隆的总基因组 DNA 转染了次黄嘌呤-鸟嘌呤转磷酸核糖基酶（HPRT）缺陷型的人细胞。通过在 HAT 培养基上进行筛选，人们获得了少数几个 HPRT——阳性克隆，推测他们很有可能已经摄取了含有功能基因的 DNA 片段。当时摄取 DNA 的真正机制还不清楚。后来过了很久人们才明白这项实验中 DNA 转移的成功依赖于形成了良好的 DNA/磷酸钙共沉淀（DNA/calcium phosphate co-precipitate），这种复合物先积累到细胞膜上，然后再由细胞摄取到内部。在转染的时候这种共沉淀必须是新鲜制备的。人们认为结合了 DNA 的磷酸钙颗粒通过胞吞作用而由细胞摄取并转运到细胞核，在那里一些 DNA 释放出来并得到表达。Craham 和 Vander-Erb 1973 年将这项技术应用于分析腺病毒 DNA 的传染性，此后该项技术受到广泛的认同。现在这项技术已经成为将普通培养的细胞进行 DNA 转化的常用技术。但是，由于沉淀物必须包裹细胞，所以这项技术只适用于单层生长的细胞，而不适用于悬浮生长或成团生长的细胞。正如最开始描述的那样，细胞摄取 DNA 的比例不固定而且效率相当低（1%～2%），因此很大程度上限制了磷酸钙转染细胞的效率。摄取 DNA 的细胞中只有少量细胞可以稳定转化。改进的方法可以将一些细胞系的转染效率提高到 20%。使用另外一个缓冲液系统来对这项技术进行改进，可以使沉淀经过数小时缓慢形成，如果再使用高质量质粒 DNA，就可以使转化稳定且效率提高 100 倍。

2. 二乙氨乙基葡聚糖转染方法

磷酸钙方法适用于许多细胞类型，但是由于沉淀具有毒性，会对一些细胞系产生不利影响，因而难以转染。后来人们发明了另外一种化学转染方法来解决这个问题。这种方法使用了二乙氨乙基葡聚糖（DEAE-dextran），它是一种可溶性多聚阳离子糖类，可以促进 DNA 与胞吞作用之间的相互作用。这项技术最初是设计用于把病毒 DNA 转入细胞，但是后来成为了质粒 DNA 转移的一种方法。Lopata 等（1984）和 Sussman 及 Milman（1984）通过添加后处理步骤提高了最初方法的效率，如添加氯喹，可抑制体内囊泡的酸化作用。尽管对于许多细胞类型的瞬时转染都有很高的效率，但是 DEAE-dextran 仍不能用于形成稳定转化的细胞系。

3. 以磷脂作为输送基因的工具

不同于以上化学转染的另外一种方法是将 DNA 包装到一个磷脂（phospholipid）泡囊中，泡囊可以和靶细胞的细胞膜作用并促进 DNA 的摄取。Schaffner（1980）第一次用这种方法做了一次转化实验，他用含有质粒的细菌原生质体将 DNA 转入了哺乳动物细胞。简要地讲，用合适的质粒载体转化细菌细胞，再用氯霉素处理细胞以扩增质粒的拷贝数。用溶菌酶去掉细胞壁，再把得到的原生质体和哺乳动物细胞低速离心，使原生质体附着在动物细胞的单层膜上，最后用聚乙二醇诱导融合。Wiberg 等 1987 年使用了一个类似的策略，他们用红细胞的无血红蛋白空胞作为输送载体。这种方法获得的转化体数量非常多，但是这种方法工作量太大，因而难以成为转染的常规方法而广泛应用。但是，这种方法一个重要的优点是温和，大的 DNA 片段可以在不发生剪切的情况下进行转移。因此，用这种方法可以将带有酵母人工染色体（YAC）的去细胞壁的酵母（球粒体）和小鼠 ES 细胞融合，从而向小鼠细胞中引入 YAC，以形成 YAC 转基因鼠。

4. 脂质体转染

更普遍使用的是人工磷脂泡囊，它称为脂质体（liposome）。最初脂质体转染的方法所受到的限制是难于包装 DNA，而且转染效率并不高于磷酸钙法。但是，后来这一方法取得了突破，人们发现阳离子与中性脂混合物可以与 DNA 自发的形成稳定复合体，复合体可以与细胞膜进行高效作用，从而通过胞吞作用实现 DNA 摄取。这种低毒性的转染方法称为脂质体转染法，它是最容易进行的方法之一，适用于其他许多种方法难以转化的细胞类型，其中包括悬浮生长的细胞。这项技术适于瞬时和稳定转化，而且对于使用 YAC 和其他大 DNA 片段也足够温和。效率也远高于化学转染方法——高到一个培养皿中 90% 的细胞都能被转染。现在有大量不同的脂质混合物可供使用，它们对于不同细胞的效率不同。脂质体的一个独特的优点是它们可以转化活体细胞。将促进细胞融合的病毒蛋白、核定位信号以及识别特异细胞表面分子的各种分子偶联物与脂质体结合，可以提高转化效率，并且将转染定位到特异细胞类型上。脂质体用于基因治疗的进展已有系统的综述。

5. 电穿孔法

电穿孔（electroporation）是将细胞置于短暂的电脉冲中使细胞膜上形成瞬时的纳米大小的孔，DNA 通过这些孔进入细胞并转运到核。Wong 和 Neumann（1982）最先将这项技术应用于动物细胞，他们成功地将质粒 DNA 转入鼠成纤维细胞。电穿孔技术已用于许多其他类型的细胞。最关键的参数是电脉冲的强度和持续时间，这对于不同细胞必须通过实践来确定。但是，一旦把电穿孔的最佳参数确定下来，该方法就很容易进行，而且具有高重复性。这项技术需要高投入，因为需要一个专用的电容放电装置来精确控制脉冲长度和幅度。另外，和其他方法相比，这种方法需要的细胞量比较大，因为在大多数情况下，在电穿孔效率达到最高时细胞的死亡率会高达 50%。另外一种方法是用精确聚焦激光束来产生穿孔。尽管非常有效（可达 0.5% 的稳定转化），这项技术只适用于少量细胞，并没有得到广泛应用。

6. 直接转化方法

本部分讨论的最后一些方法包括直接将 DNA 转入细胞核的技术。其中一种方法是显微注射，它可以保证对靶细胞进行成功地操作，但是在一次实验中它只能用于少数细胞。这项技术已用于那些不适用其他转染方法的培养细胞，但是它的基本应用是将 DNA 和其他分子导入大的细胞，如卵母细胞、卵细胞和早期胚细胞。基因枪轰击是另一种直接转移的方法，它最初用于植物的转化。这项技术是用 DNA 包裹微小的金属颗粒，然后用强大的力量将粒子加速打入靶组织，这种强大的力量包括高压气体的释放和小水滴的放电等。在动物上，这项技术通常用于组织切片的多细胞转染，而不用于培养的细胞。这项技术也用于活体皮肤细胞的 DNA 转化。

二、转基因动物

转基因技术可以用来培育转基因动物，其中每个细胞都携带有新的遗传信息，和那些通过对基因组进行特意修饰而筛选到的突变体一样。完整动物是最终的实验系统，通过此系统可以研究基因的功能，特别是像发育这样的复杂的生物学过程。在大多数动物中，体细胞和生殖细胞（产生配子的细胞）在发育的早期阶段就已区分开来。因此，在动物中实现生殖细胞转化的唯一途径就是在生殖细胞形成之前就把 DNA 导入全能细胞。对小鼠生殖细胞系进行 DNA 导入是 20 世纪取得的最大成就之一，它为其他哺乳动物的基因转移铺平了道路。转基因哺乳动物不仅用来研究基因的功能和调控，也用于生产有价值的重组蛋白，如乳汁。目前也发展出数种方法，这些方法都需要把受精卵或早期胚胎从母体中移出，进行体外短暂培养后返回到代孕母体，在代孕母亲体内发育直至妊娠（图 10-24）。

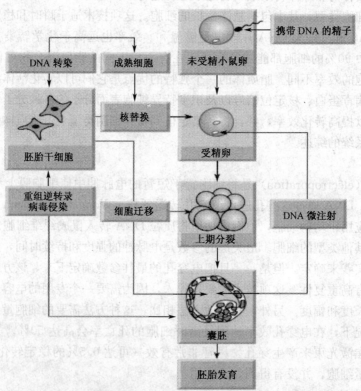

图 10-24　产生转基因哺乳动物的方法（Twyman and Old，2003）

在前面，我们讨论了转基因技术的早期发展，着重于可靠的转基因技术的发展及转移的基因在动植物中的表达。这一技术的标准运用是给基因组增加一些遗传信息，使得到的细胞和转基因生物产生新的功能。现在进行基因转移的过程已经变得很常规，于是转基因技术的重点开始转向对转移的基因进行更多的控制。新的技术同时在动物和植物中得到了应用。诱导表达系统的发展，使得对转入基因进行外部调控更为方便，而对位点特异性重组系统的探索，使得对靶基因进行精确修饰已非难事。在小鼠中，联合运用靶基因位点特异性重组和诱导转入基因的表达等技术，使转入的基因和内源基因在一定条件下可以选择性的开启或关闭。人们对基因沉默的其他方法也进行了研究，如反义RNA 的表达和 RNA 干扰。转基因动物和植物越来越成为基因组分析的工具。快速克隆插入突变的基因的标记系统的发展，以及诱捕载体的应用，这些实现了对基因表达和功能进行高通量分析。

第七节　基 因 敲 除

基因敲除（gene knock-out）是通过一定途径使机体特定的基因失活或缺失的技术。通常意义上的基因敲除，主要是应用 DNA 同源重组（homologous recombination）原理，用设计的一条 DNA 序列插入基因组代替原来的等位基因，从而达到基因敲除的目的。基因敲除有时也被称为基因打靶（gene targeting），它是一种简单的，可以在活的有机体中任意修饰遗传信息的一种技术。

1. 原理

基因敲除技术是利用同源重组原理而发挥其作用的。当一个设计者想用一段设计好的基因来替代一个等位基因而又不影响基因组的其他基因座的时候，就要用到同源重组。要进行同源重组的前提是，你必须知道你想改变的基因的 DNA 序列（图 10-25a）。只要掌握了这样的信息，你几乎可以利用一段构建好的基因代替任何一个你感兴趣的基因。

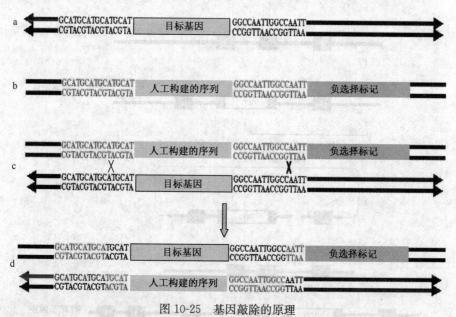

图 10-25　基因敲除的原理

a. 将被替换的靶基因。编码序列包含在上游侧翼序列和下游侧翼序列之中。指离靶基因的箭头代表着连续的染色体 DNA。b. 被用来替代野生型等位基因的体外构建好的序列。上下游序列分别和靶基因两侧序列相同。

c. 同源重组前的等位基因配对，体外构建好的序列在细胞核中与打靶位点处的 DNA 序列配对。d. 同源重组后的终产物。现在，染色体中包含部分侧翼序列以及代替野生型等位基因的全部的体外构建好的序列当你掌握了你的目的基因序列之后，下一步就是设计并构建一段你想插入到染色体中代替野生型等位基因的 DNA 序列。你可以在这段序列中添加任何一段 DNA 序列（有功能的和无功能的），不同的基因或者报告基因（抗生素抗性或者荧光蛋白），无论你想插入什么基因，这段序列必须包含与打靶位点序列相同的侧翼序列（图 10-25b）。这段体外构建好的包含有你感兴趣的打靶基因被转入细胞中，在减数分裂和有丝分裂中期同源染色体配对，并且在同源序列间发生同源重组（图 10-25c），同源重组可以发生在侧翼序列当中的任何地方，并且这个精确的位点是由细胞中的基因组决定的而不是设计者决定的。一旦细胞完成了这样的程序，就导致了一段新的 DNA 序列插入染色体。基因组的其他地方没有改变，只有打靶基因座被体外设计好的序列和其中的一段侧翼序列所代替（图 10-25d）。原来构建好的序列替代了我们感兴趣的靶基因，由于靶基因不能复制，它们在细胞分裂中将很快地消失，同时这段被插入了新的序列的染色体在细胞分裂中被精确的复制。

2. 打靶载体

打靶载体可以分成两大类：置换型载体和插入型载体。置换型载体通过与靶序列保持共线性而被线性化。通过在同源侧翼序列的 2 次重组交换，染色体序列被载体序列所代替。插入型载体在同源区域被线性化，同源重组产生了基因组序列一个副本。大多数无效突变的产生是通过置换型载体完成的（图 10-26）。

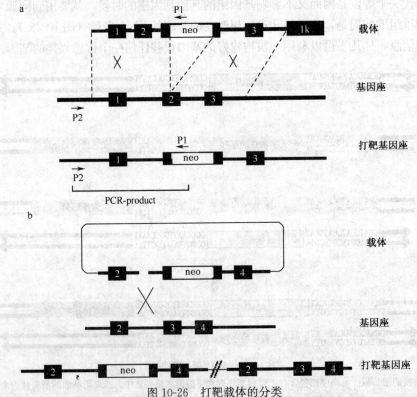

图 10-26　打靶载体的分类

　a. 置换型载体。与靶位载体同源同组后，打靶基因的第二个外显子由于插入了一个新霉素抗性标记基因而被破坏。b. 插入型载体。插入型载体在外显子 2 和外显子 3 之间的同源区被线性化，并且插入了新霉素抗性基因作为阳性选择的标记基因。同源重组使载体序列和部分基因组序列的副本整合到基因组中

3. 选择策略

同源重组在哺乳动物细胞中很少发生，因此需要一个强有力的选择策略去筛选发生了同源重组的细胞。一般情况下，被导入的基因由于插入了一个抗生素抗性基因而被破坏，如新霉素抗性基因。如果这个构建好的基因与内源基因发生同源重组，内源基因就遭到破坏，而抗性基因仍然保持自己的功能，这就使该细胞可以被含有新霉素抗性的培养基选择出来。然而，只要整合了新霉素抗性基因的细胞都可以表现出相应的抗性，而不管这种整合是否是通过同源重组整合进去的。为了能够选择出这些发生了同源重组的细胞，我们经常在构建的载体的末端接上胸腺激酶基因（*HSK-tk*），该基因来自单纯疱疹病毒（HSV）。发生 DNA 随机整合的细胞经常整合了整个载体序列，包括 *HSK-tk*，然而发生了同源重组的细胞中，由于同源重组只发生在同源序列之间，连接在载体末端的非同源基因 *HSV-tk* 没有整合进细胞 DNA。那些带有 *HSV-tk* 的细胞会被抗病毒药品 GANC 杀死，所以，发生同源重组的细胞因为拥有抗新霉素和抗 GANC 的唯一特性，可以被高效的选择出来（图 10-27）。

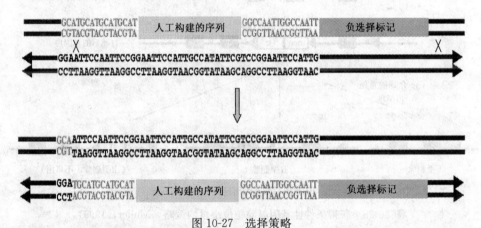

图 10-27　选择策略

打靶载体与基因组非同源序列排队，之后通过随机整合进基因组，负选择标记也跟着整合进基因组。
非同源重组的最终产物可以幸存于阳性选择。然而，GANC 能够杀死任何含有 *HSV-tk* 基因的细胞。
所以，我们通过抗性素抗性选择来筛选那些经过重组的细胞，通过 GANC 杀死那些没有进行过同源重
组的细胞，从而筛选出我们需要的经历过同源重组的细胞

4. 条件性基因打靶

传统的基因打靶使某个基因无效或者被修饰，并且让这种无效或者被修饰的效果在所有的组织中从开始发育到整个生命周期一直存在。最近，为了使靶基因敲除具有时间和组织特异性，一些方法已进一步发展了。当靶基因被完全失活会致死或者引起其他不利表型的产生，从而阻止我们对该基因的功能作进一步的分析时，条件性基因打靶的策略就显得非常有用。而且，如果一个假想的基因具有广泛的表达模式，组织特异性基因敲除的方式可能能够详细的说明该基因的表达产物在特定组织中所起到的生理作用，而不至于与该基因在有机体中其他组织中的作用相混淆。时间可控性的基因打靶技术可以区分长期缺失和突然缺失一个蛋白质的效果，也可以分析它们在不同发育阶段的功能。很多基因敲除小鼠会出现我们不期待的表型上的微小改变，这要归因于基因冗余或者是适应机制介导的发育的可塑性。基因在成熟个体的某个具体的时间被灭活，但可以保证

该基因在整个个体的发育过程中的功能仍是健全的，这就避免了机体对基因敲除后所产生的自身适应反应，因此，相比传统的基因敲除技术，通过这种时间可控的基因敲除技术可以得到我们比较期望得到的表型。

1）组织特异性基因敲除　第一代能够在组织特异性启动子的调控下表达 Cre 重组酶的转基因小鼠的产生，使得组织特异性基因敲除成为可能。这些小鼠与包含 Loxp 位点靶位小鼠杂交，使 Loxp 位点之间的基因组被删除或者被修饰。在 Cre 酶充分表达的所有细胞中，由于 Cre 介导的同源重组，双转基因小鼠将会进行一次基因修饰。我们可以通过在胚胎干细胞的同源重组中获得插入合适 Loxp 位点的小鼠，同时，转 Cre 基因的小鼠也可以通过传统的胚胎注射或者基因敲入来带入一个合适的启动子从而得到一个我们渴望得到的表达模式（图 10-28）。

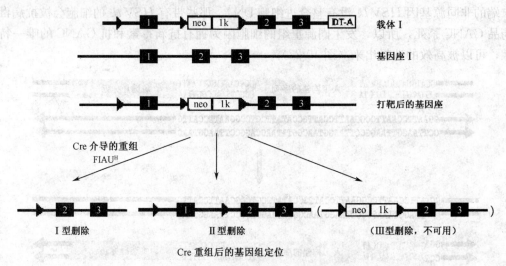

图 10-28　获得条件性基因敲除靶位小鼠的策略（Muller，1999）

第一步，通过同源重组在靶基因的沉默区引入三个 Loxp 位点。由于引入 Loxp 位点，经过 Cre 介导的同源重组的干细胞通过 FLAU 筛选，得到两个不同的基因组缺失产物（第一类缺失和第二类缺失，第三类缺失在 FLAU 选择中被排除）。这两种小鼠杂交得到双转基因小鼠，双转基因小鼠将会经历 Cre 介导的重组

2）可诱导的基因敲除　在很多情况下我们渴望能够从时间上控制基因的表达，比如我们可以通过诱导物来调控基因的表达。与组织特异性基因打靶相似，能够达到诱导性基因沉默的一个方法就是，在转录和转录后水平上控制基于 Cre 重组酶基因的表达系统。理论上，任何的诱导系统都应当满足以下几方面要求：①在缺失诱导物的时候没有其他副作用或者副作用很小。②在所有组织中都能够迅速高效的起作用。③诱导物对机体应该没有毒性或者其他的负面效应。用可诱导的启动子来控制 Cre 基因的表达，属于在转录水平上调控 Cre 酶活性的一些研究已经在进行（图 10-29a）。在经典的四环素诱导系统中（图 10-29b）应答基因在缺少配体或配体沉默失活的时候表达。在反向的四环素诱导系统中（图 10-29c），被修饰的转录激活物（rTA）只有在配体存在时才能与操纵子或者启动子结合以起始转录。因此，应答基因在该系统中一开始是沉默的，我们可以通过引入配体来启动应答基因的表达。无论是 tTA 还是 rTA 调控系统都向我们展示了可以在转基因小鼠中控制报告基因表达的功能。

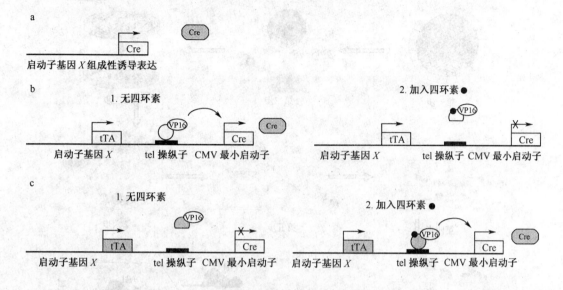

图 10-29 Cre 酶活性的转录调控（Muller，1999）

a. Cre 在转基因小鼠中的表达。b. Cre 在经典四环素诱导系统中表达。这个系统包括转录激活物 tTA，它会在缺少四环素的情况下同 tTA 依赖的转基因操纵子（tet operon）结合（用深色方框表示），从而激活 Cre 基因的转录。Cre 应答基因在缺失四环素的情况下表达，在引入四环素的情况下沉默。c. Cre 在反相四环素系统中的表达。在该系统下，经过修饰的转录激活物是只有在具有配体的情况下才与 DNA 结合，Cre 基因的表达才被开启

5. 基因敲除小鼠模型

基因敲除小鼠中某个基因的两个等位基因都可以被失活的等位基因替换。这是通过同源重组先得到两个等位基因中的一个基因被替换了小鼠，然后通过两代或者多代的选择育种，直到靶基因的两个等位基因被完全的灭活或者敲除。研究者可以通过观察完全缺失这个基因的基因敲除小鼠的表型来决定这个基因的功能。现在有很多获得基因敲除小鼠的方法，下面的这个方法是比较典型的（图 10-30）。

6. 基因敲除技术的应用与展望

基因敲除技术变革了小鼠遗传学研究领域，并且能够在体内分析基因的不同功能。目前，利用它既可以使遗传信息产生微小的突变，也可以使遗传信息发生染色体重排程度的变化。时空可控的组织特异性基因敲除技术也变得切实可行了。我们期待通过很多实验室的不懈努力，这些新方法在不久的将来能被很好地优化，以达到可以在任何组织任何时间随意可逆的调节基因的表达和失活的终极目标。尽管基因打靶方法已经很成熟，并且同源重组的效率已经在小鼠干细胞中得到优化，但是在大多数体细胞中，基因打靶的效率仍然很低。尽管如此，人们仍然对如何改进体细胞的遗传操纵技术有很大的兴趣，特别是在基因治疗的应用方面。此外，除了针对小鼠，在其他鼠类、果蝇，特别是在具有商业价值的家畜中进行基因打靶技术的研究，可能会发现新的有趣的应用。

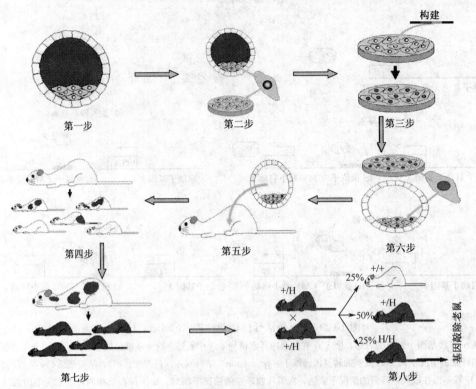

图 10-30　获得基因敲除小鼠的步骤

第一步，分离发育到囊胚期的胚胎。第二步，从这个胚胎中分离胚胎干细胞。并将这些细胞进行体外的组织培养。第三步，将体外构建好的打靶载体转入胚胎干细胞。neo/GANC 的选择培养基上筛选出经历过同源重组的细胞。第四步，将同源重组的细胞从培养皿中挑出并将其注射到一个来自白色小鼠的发育到囊胚期的胚胎。第五步，将该胚胎移植到假孕的白色小鼠的子宫内。第六步，产生出一系列小鼠。因为由重组的胚胎干细胞发育而来的皮肤细胞在小鼠皮肤上会产生灰色的斑点，很容易被发现。第七步，将这种灰白相间的小鼠与野生型纯白小鼠杂交。如果这个灰白相间小鼠的性腺是由重组的胚胎干细胞发育而来，那么所有的后代将会是灰色的。由于同源重组，灰色小鼠的每个细胞都是杂合的。第八步，将这些灰色的杂合小鼠杂交。鉴定出纯合子

第八节　DNA-蛋白质相互作用

DNA-蛋白质相互作用研究是分子生物学的重要内容。最典型的例子是 RNA 聚合酶和启动子（promoter）的相互作用。事实上，诸如此类的至关重要的相互作用在机体的生理生化过程中扮演着非常重要的角色，因此，建立一套精确定位和定量 DNA-蛋白质相互作用的实验技术就显得尤为重要了。在本节中，介绍 5 种目前主要的研究 DNA-蛋白质相互作用的方法。它们分别是滤膜结合（filter binding）实验、胶迁移率（gel mobility shift）实验、DNA 酶足迹（DNase footprinting）实验、硫酸二甲酯（dimethyl sulfate，DMS）足迹实验和染色质免疫共沉淀（chromatin inmunoprecipitation，ChIP）实验。

1. 滤膜结合实验

硝酸纤维素滤膜（NC 膜）长期以来用作溶液的过滤除菌。一次偶然的机会，当分子生物学家用它来过滤制备的 DNA 溶液的时候，发现滤出液中的 DNA 成分不见了！

通过后续的研究，发现了硝酸纤维素膜具有结合单链 DNA 的特性，但不能结合双链 DNA。此外，蛋白质也能很好结合到硝酸纤维素膜上。因此，如果双链 DNA 结合到蛋白质上，那么蛋白质-DNA 复合物也能够结合到 NC 膜上。基于上述原理，发明了用于研究 DNA-蛋白质相互作用的膜结合实验。（图 10-31）

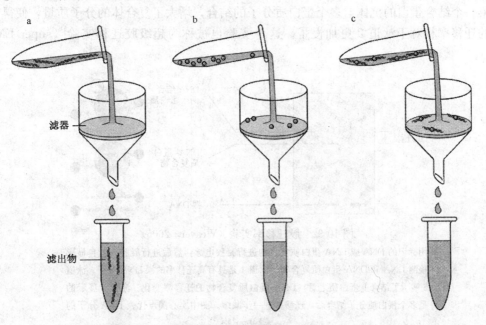

图 10-31　硝酸纤维素滤膜结合实验（Weaver，2005）

a. 双链 DNA。末端标记双链 DNA，然后滤膜过滤。滤膜上检测不到放射信号，说明双链 DNA 不能结合到膜上。b. 蛋白质。放射标记的蛋白质，滤膜过滤时能结合到膜上。c. 双链 DNA-蛋白质复合物。末端标记的双链 DNA 和未标记的蛋白质混合，通过 NC 膜，膜上能检测到放射信号，说明部分 DNA-蛋白质形成复合物，结合到膜上

　　如图 10-31a 所示，标记的双链 DNA 通过硝酸纤维素膜过滤。通过检测放射信号可以看到，放射标记的 DNA 全部能通过膜，进入到滤出液中。该实验表明双链 DNA 并不能"自己"单独结合到硝酸纤维素膜上。在图 10-31b 中，标记的蛋白质溶液通过滤膜的时候，全部被吸附到滤膜上。表明硝酸纤维素膜能很好地吸附蛋白质。图 10-31c 中，标记的 DNA 与相应的结合蛋白混合形成 DNA-蛋白质复合物。由于蛋白质能吸附到膜上，与之结合的 DNA 也被带到膜上，从而能在滤膜上检测到标记 DNA 的放射信号。滤膜结合实验就是通过这一原理来实现对 DNA-蛋白质相互作用的检测。

　　简单说来，基于双链 DNA 不能结合到硝酸纤维素膜这类的介质上，而蛋白质-DNA 复合物则可以结合这一原理，滤膜结合实验被用来测定 DNA-蛋白质的相互作用。研究者通过标记 DNA 双链的末端，然后与蛋白质混合孵育，硝酸纤维素膜过滤，并通过检测滤膜上的放射标记信号来检测 DNA-蛋白质的相互作用。

2. 胶迁移率实验

　　胶迁移率实验研究 DNA-蛋白质相互作用的方法是基于以下原理：小片段 DNA 分子在结合上蛋白质后电泳时在胶上的迁移速率将大大降低。因此，研究者可以将双链 DNA 进行标记后与蛋白质溶液进行混合孵育，然后进行电泳分析和标记信号的检测。

图 10-32 展示了三个样本的电泳迁移率。其中泳道 1 为裸 DNA 片段，可以看到这种小片段 DNA 在凝胶电泳中有很高的迁移率，迁移到了胶的底部。泳道 2 的 DNA 与蛋白质结合，迁移率大大降低。这一实验因此被称为凝胶电泳迁移率实验（EMSA）。泳道 3 中 DNA 片段结合了多个蛋白质分子，这些分子可以是其他 DNA 结合蛋白，也可以是第一个结合蛋白的抗体，多个蛋白质分子的结合，增大了复合体的分子质量，使得片段的迁移率相对于泳道 2 更加慢了。这个实验也被称为超级胶迁移实验（super EMSA）。

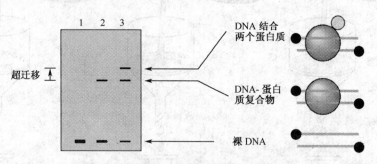

图 10-32　胶迁移率实验（Weaver，2005）
用标记的 DNA 或 DNA-蛋白质复合物进行凝胶电泳，然后进行凝胶放射自显影
检测 DNA 和 DNA-蛋白质复合物。泳道 1 是具有高迁移率的裸 DNA 分子。泳道
2 展示了结合上蛋白质之后，DNA-蛋白质复合物的迁移率变化。泳道 3 展示的
是多个蛋白质分子结合后，迁移率进一步减缓。图中实心圆点代表 DNA 分子的
末端标记物

归纳起来，基于小分子 DNA 片段在结合上蛋白质分子后在凝胶中的迁移率变小这一简单原理的，研究者可以通过胶迁移率实验研究 DNA 和蛋白质的相互作用。

3. DNA 酶足迹实验

足迹法是一类能准确定位 DNA 上与蛋白质相互作用位点的实验方法，其准确性可以达到一个碱基的精度。这类方法分许多种，其中 DNA 酶、硫酸二甲酯（DMS）和游离羟基足迹法是主要的三种。DNA 酶足迹法（图 10-33）的原理：蛋白质结合到 DNA 上，覆盖的 DNA 位点能保护 DNA 不被 DNA 酶降解，从而在 DNA 上留下"足迹"。实验首先要对 DNA 进行标记。两条链都能进行标记，但在一次实验中只标记其中的一条。然后蛋白质（图中深色所示）结合到 DNA 上。接下来，DNA-蛋白质复合物在低浓度的 DNA 酶 I 中处理，平均每个 DNA 分子上产生一次切割。接下来蛋白质从 DNA 上解离下来，分离的 DNA 片段进行高分辨率的聚丙烯酰胺凝胶电泳，通过设置未与蛋白质混合和不同蛋白质浓度的对照组，我们能看到连续条带中缺失的一段，这就是被保护的 DNA 片段，也就是 DNA 结合蛋白的"足迹"所在了。

简而言之，足迹实验就是研究 DNA 结合蛋白在 DNA 片段上的结合位点。通过将 DNA 结合蛋白与末端标记的 DNA 片段共孵育，用 DNA 酶消化复合物，消化后的 DNA 片段进行电泳分离，被 DNA 结合蛋白保护的 DNA 片段位点就能够在电泳图谱上以足迹的形式表现出来。

4. DMS 以及其他足迹实验

DNA 酶足迹实验提供了一种非常好的定位 DNA 分子上蛋白质结合位点的方法，

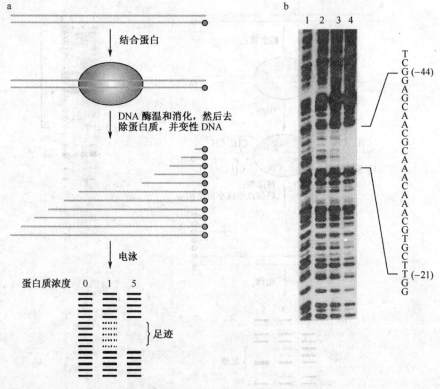

图 10-33 DNA 酶足迹实验（Weaver, 2005）

　a. 实验基本流程如下：对 DNA 的末端进行标记（灰色）后，与 DNA 结合蛋白混合孵育，孵育后
的混合物用低浓度的 DNA 酶 I 进行消化，从而在 DNA 分子上引入合适的切口。去除蛋白质，将
DNA 分子变性，获得末端标记的一系列 DNA 片段。最后，通过对这些系列片段进行电泳分离和
放射自显影进行检测分析。图示 3 条泳道展示了结合蛋白浓度依次为 0、1 和 5 个单位时的情形。

　b. 显示的是实际的实验结果。泳道 1～4 分别为加入了 0pmol、10pmol、18pmol 和 90pmol 结合蛋
白后的电泳结果。通过对 DNA 片段的测序，我们可以直接读出被保护区的碱基序列

但 DNA 酶作为一种大分子，在精细定位蛋白质结合位点时，难免会显得粗放。例如，某些 DNA 结合蛋白通常会改变 DNA 双螺旋分子的构相，而形成一些排斥 DNA 酶的结构，使得 DNA 酶不能接近等等。为了更精确的定位 DNA 上蛋白质的结合位点，发现更细致的相互作用，我们需要更小分子的足迹工具。DMS 足迹的出现，成为了完成这一细致工作的理想工具。图 10-34 示意了 DMS 足迹的基本过程。和 DNA 酶足迹法一样，首先，让末端带标记的 DNA 分子和 DNA 结合蛋白形成复合物。其次，在温和的条件下，DMS 将复合物中未被结合蛋白保护的 DNA 区域甲基化。理想的结果是：平均每个 DNA 分子在某个位置被带上一个甲基。再次，结合蛋白被移除。DNA 经哌啶处理，除去嘌呤上的甲基，产生去嘌呤位点，DNA 链在该位点被打断。最后，一系列的 DNA 分子片段通过电泳和放射自显影，形成连续的条带。而被结合蛋白保护的区域形成空白断区，与未加结合蛋白的对照组 DNA 标准泳道对比，就可以直接读出结合位点的碱基序列。除了 DMS 和 DNA 酶以外，在原理类似的情况下，一些金属螯合物也用来产生 DNA 足迹。

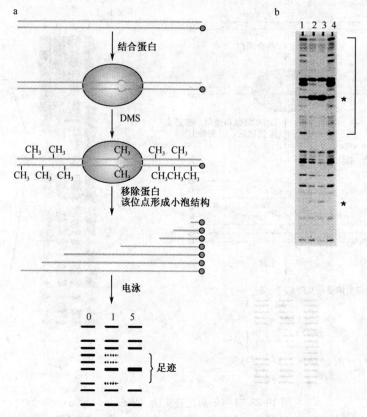

图 10-34　DMS 足迹实验（Weaver，2005）

a. 展示了 DMS 足迹实验的基本流程。与 DNA 酶足迹类似，DMS 足迹也是以与末端标记的 DNA 开始，然后与 DNA 结合蛋白结合。下一步，用 DMS 在 DNA 某些位置加上甲基（CH3）。接下来，用 Maxam-Gilbert 试剂去掉嘌呤上的甲基，在哌啶的处理下，DNA 链将在该区嘌呤位点上断裂，形成连续的只相差一个碱基的系列DNA 片段。这些片段通过电泳放射自显影，并与未加结合蛋白的 DNA 系列片段比对，就可以读出结合位点的碱基序列了。b. 实际的实验结果。泳道 1 和 4 没有加入结合蛋白，泳道 2 和 3 加入了一定浓度的结合蛋白。括号指示了一个典型的足迹区域。星号指示了特别敏感的小"泡区"

　　简而言之，除了 DNA 甲基化，DMS 足迹法采取了和 DNA 酶足迹法比较类似的原理。在 DMS 足迹实验中，甲基化试剂 DMS 取代了 DNA 酶扮演切割 DNA 分子的工具。DNA 分子在甲基化位点处被打断。未甲基化和超甲基化的位点在电泳图谱上展示出来，从而间接的反映被 DNA 结合蛋白保护的区域。游离羟基足迹法则使用含铜或铁离子的螯合物产生游离羟基来打断 DNA 链。

5. 染色质免疫共沉淀

　　染色质免疫共沉淀，能让研究者检测蛋白与活细胞内基因组的结合部位。例如，可以用于检测转录因子（transcription factor）及其他组分在特定时间是否和特定的启动子结合，检测某种调节蛋白是否和特定基因结合等实验。

　　概括地说，该技术的主要流程：向细胞中加入甲醛，使此时 DNA 上结合的蛋白质同 DNA 发生交联。随后将细胞裂解，DNA 断裂成 200～300 bp 的小片段。加入蛋白质抗体可以将与蛋白质结合的 DNA 片段从细胞内大量的 DNA 中分离出来，随后解除交联，除去蛋白质。为了确定 DNA 特定区域是否结合有蛋白质，利用设计好的引物进行 PCR，对特定区域（如启动子）进行扩增。如果该区域确实结合有蛋白质，DNA 片段将被分离出来，扩增也能进行。作为对照，以特定引物对另一段 DNA 区域（已知不能结合该蛋白质）进行 PCR，这样，将不能获得扩增产物（图 10-35）。

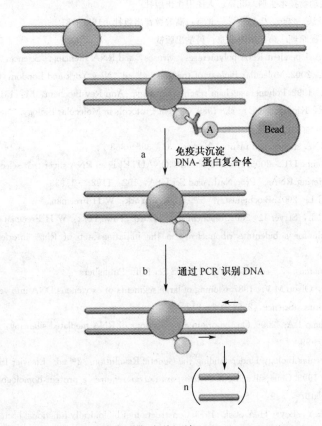

图 10-35　染色质免疫共沉淀（Weaver, 2005）

a. 免疫沉淀。目的蛋白结合到双链 DNA 特定位点上，并在甲醛处理下形成稳定交联。随后抗体结合到与目标蛋白相连的抗原表位，由于抗体的特异性，能准确的找到目标蛋白及与之结合的 DNA 形成的复合物。b. 目标 DNA 的识别。通过设计特异性引物对目标基因进行扩增，识别与目标蛋白特异性结合的 DNA 片段

小　结

1. DNA、RNA 和蛋白质的分离和纯化技术，如电泳、层析等，是分子生物学研究的重要基础。

2. DNA、RNA 和蛋白质的检测技术，如分子杂交、PCR 和序列测定，是基因克隆和表达调控研究的重要手段。

3. 转基因技术、基因敲除和 DNA-蛋白质相互作用是基因功能研究的重要方法。

（汪亚平　陈　庆）

参 考 文 献

阿里吉尔. 2002. 微注射和转基因实验指南. 张玉静译. 北京：科学出版社

劳为德. 2004. 转基因动物技术手册. 北京：人民卫生出版社

王镜岩，朱圣庚，徐长法. 2002. 生物化学. 北京：高等教育出版社. 168~181

吴乃虎. 1998. 基因工程原理. 第二版. 北京：科学出版社

Ahlquist P. 2002. RNA-dependent RNA polymerases, viruses, and RNA silencing. Science, 296：1270~1273

Alberts B, Johnson A. 2002. Molecular Biology of the Cell. 4th ed. New York and London：Garland Science

Arnheim N, Erlich H. 1992. Polymerase chain reaction strategy. Ann Rev Biochem, 61：131~156

Ausubel F M, Brent R, Kingston R E et al. 1989. Current Protocols in Molecular Biology. New York：Greene Publishing and Wiley Interscience

Baulcombe D. 2004. RNA silencing in plants. Nature. 31：356~363

Baumberger N, Baulcombe D C. 2005. Arabidopsis ARGONAUTE1 is an RNA slicer that selectively recruits microRNAs and short interfering RNAs. Proc Natl Acad Sci USA, 102：11928~11933

Berg J M, Tymoczko J L. 2002. Biochemistry. 5th ed. New York：W H Freeman

Berg J M, Tymoczko J L, Stryer L. 2002. Biochemistry. 5th ed. New York：W H Freeman and Co

Bernstein E. 2001. Role for a bidentate ribonuclease in the initiation step of RNA interference. Nature, 409：363~366

Brown T A. 2002. Genomes. 2nd ed. New York：BIOS Scientific Publishers

Burke D T, Carle G F, Olson M V. 1987. Cloning of large segments of exogenous DNA into yeast by means of artificial chromosome vectors. Science, 236：806~812

Cerutti H, Casas-Mollano J A. 2006. On the origin and functions of RNA-mediated silencing：from protists to man. Curr Genet, 50：81~99

Clark D P. 2005. Molecular Biology-Understanding the Genetic Revolution. 2nd ed. Elsevier Inc

Cogoni C, Macino G. 1999. Gene silencing in *Neurospora crassa* requires a protein homologous to RNA-dependent RNA polymerase. Nature, 399：166~169

Cohen S N, Chang A C Y, Boyer H W et al. 1973. Construction of biologically functional bacterial plasmids *in vitro*. Proc Natl Aca Sci USA, 70：3240~3244

Cooper G M, 2000. The cell-A Molecular Approach. 2nd ed. Sunderland (MA)：Sinauer Assoiates. Inc

Dymecki S M, Tomasiewicz H. 1998. Using Flp-recombinase to characterize expansion of Wnt1-expressing neural progenitors in the mouse. Dev Biol, 201：57~65

Dzitoyeva S, Nikola D, Hari M. 2003. Gamma-aminobutyric acid B receptor 1 mediates behavior-impairing actions of alcohol in Drosophila：adult RNA interference and pharmacological evidence. Proc Natl Acad Sci USA, 100：5485~5490

Edman P. 1949. A method for the determination of amino acid sequence in peptides. Archives of biochemistry, 22 (3)：475

Fire A, Xu S, Montgomery M K, Kostas S A, Driver S E, Mello C C. Potent and specific genetic interference by double-stranded RNA in *Caenorhabditis elegans*. Nature, 1998, 391：806~811

Fortunato A, Fraser A G. 2005. Uncover genetic interactions in *Caenorhabditis elegans* by RNA interference. Biosci Rep, 25：299~307

Gerstein A S. 2001. Molecular Biology Problem Solver：A Laboratory Guide. Wiley-Liss, Inc. 168～190, 198~219

Girard A, Sachidanandam R, Hannon G J et al. 2006. A germline-specific class of small RNAs binds mammalian Piwi proteins. Nature, 442: 199~202

Glick B R, Pasternak J J. 1994. Molecular Biotechnology: Principles and Applications of Recombinant DNA. Washington D C: ASM Press

Gregory R I. 2005. RISC couples microRNA biogenesis and posttranscriptional gene silencing. Cell, 123: 631~640

Griffiths A J F, Gelbart W M, Miller J H et al. Modern Genetic Analysis. 1999. New York: W H Freeman &Co

Guo S, Kemphues K J. 1995. par-1, a gene required for establishing polarity in C. elegans embryos, encodes a putative Ser/Thr kinase that is asymmetrically distributed. Cell, 81: 611~620

Hames B D, Glover D. 1995. DNA Cloning: A Practical Approach. Oxford, England: IRL Press

Hammond S M, Bernstein E, Beach D et al. 2000. An RNA-directed nuclease mediates post-transcriptional gene silencing in Drosophila cells. Nature, 404: 293~296

Hammond S M, Sabrina B, Caudy A A et al. 2001. Argonaute2, a link between genetic and biochemical analyses of RNAi. Science, 293: 1146~1150

Herzer S. DNA Purification. 2001. In: Gerstein A S. Molecular Biology Problem Solver: A Laboratory Guide. Hoboken: Wiley-Liss, Inc. 168~190

Holmquist G P, Ashley T. 2006. Chromosome organization and chromatin modification: influence on genome function and evolution. Cytogenet Genome Res, 114: 96~125

Jeremy W. Dale, Malcolm von Schantz, From Gene to Genomes: Concepts and Applications of DNA Technology. John Wiley & Sons Inc. 2002

Lee R C, Feinbaum R L, Ambros V. 1993. The C elegans heterochronic gene lin-4 encodes small RNAs with antisense complementarity to lin-14. Cell, 75: 843~854

Lippman Z, Martienssen R. 2004. The role of RNA interference in heterochromatic silencing. Nature, 431: 364~370

Liu Q, Rand T A, Kalidas S et al. 2003. R2D2, a bridge between the initiation and effector steps of the Drosophila. RNAi pathway Science, 301: 1921~1925

Lodish H, Berk A, Zipursky S L, et al. 1999. Molecular Cell Biology. New York: W H Freeman &Co

Luke A. 1977. DNA Sequencing. New York: BIOS Scientific Publishers

Macrae I J, Zhou K, Li F et al. 2006. Structural basis for double-stranded RNA processing by Dicer. Science, 311: 195~198

Makeyev E V, Bamford D H. 2002. Cellular RNA-dependent RNA polymerase involved in posttranscriptional gene silencing has two distinct activity modes. Mol Cell, 10: 1417~427

Martin L A, Smith T J, Obermoeller D et al. 2001. RNA Purification. In: Gerstein A S. Molecular Biology Problem Solver: A Laboratory Guide. Hoboken: Wiley-Liss Inc. 198~219

Matzke M A, Matzke A J M. 2004. Planting the Seeds of a New Paradigm. PLoS Biol, 2: e133 doi: 10.1371/journal. pbio. 0020133.

Matzke M, Matzke A J M, Kooter J M. 2001. RNA: guiding gene silencing. Science, 293: 1080~1083

McPherson M J, Hames B D, Taylor G. 1995. PCR: A Practical Approach. Oxford, England: IRL Press

Mello C C, Conte D. 2004. Revealing the world of RNA interference. Nature, 431: 338~342

Mello C C, Conte D J. 2004. Revealing the world of RNA interference. Nature, 431: 338~342

Morita T, Mochizuki Y, Aiba H. 2006. Translational repression is sufficient for gene silencing by bacterial small noncoding RNAs in the absence of mRNA destruction. Proc Natl Acad Sci USA, 103: 4858~4863

Muller U. 1999. Ten years of gene targeting: targeted mouse mutants, from vector design to phenotype analysis. Mech Dev, 82: 3~21

Nagy A. 2000. Cre recombinase: the universal reagent for genome tailoring. Genesis, 26: 99~109

Napoli C, Lemieux C, Jorgensen R. 1990. Introduction of a chalcone synthase gene into Petunia results in reversible co-suppression of homologous genes in trans. Plant Cell, 2: 279~289

Nathans D, Smith H O. 1975. Restriction endonucleases in the analysis and restructuring of DNA molecules. Ann Rev Biochem, 44: 273~293

Nelson D L, Cox M M. 2004. Principles of Biochemistry. 4th ed. W H Freeman & Co Ltd. 89~95, 318~319

Nicholson R H, Nicholson A W. 2002. Molecular characterization of a mouse cDNA encoding Dicer, a ribonuclease III ortholog involved in RNA interference. Mamm Genome, 13: 67~73

Palatnik J F, Allen E, Wu X et al. 2003. Control of leaf morphogenesis by microRNAs. Nature, 425: 257~263

Primrose S, Twyman R, Old B. 2001. Principles of Gene Manipulation , U. K. Blackwell Science

Rudd S. 2003. Expressed sequence tags: alternative or complement to whole genome sequences? Trends in Plant Science, 8 (7): 321~329

Saiki R K, Gelfand D H, Stoffel S et al. 1988. Primer-directed enzymatic amplification of DNA with a thermostable DNA polymerase. Science, 239: 487~491

Saito K, Nishida K M, Mori T et al. 2006. Specific association of Piwi with rasiRNAs derived from retrotransposon and heterochromatic regions in the Drosophila genome. Genes Dev, 20: 2214~2222

Sambrook J, David W. 2001. Russell, Molecular Cloning: A Laboratory Manual. 3rd ed. Cold Spring Harbor Laboratory Press

Sambrook J, Fritsch E F, Maniatis T. 1989. Molecular Cloning: A Laboratory Manual. 2nd ed. Plainview N Y: Cold Spring Harbor Laboratory Press

Sanger F. 1975. A rapid method for determining sequences in DNA by primed synthesis with DNA polymerase. Journal of molecular biology, 94: 444~448

Sanger F. 1981. Determination of nucleotide sequences in DNA. Science, 214: 1205~1210

Sanger F, Nicklen S, Coulson A R. 1977. DNA sequencing with chain-terminating inhibitors. Proc Natl Acad Sci USA, 74: 5463~5467

Scacheri P C, Crabtree J S, Novotny E A et al. 2001. Bidirectional transcriptional activity of PGK-neomycin and unexpected embryonic lethality in heterozygote chimeric knockout mice. Genesis, 30: 259~263

Schouten H J, Krens F A, Jacobsen E. 2006. Do cisgenic plants warrant less stringent oversight? Nat Biotechnol, 24: 753

Sedivy J M, Joyner A L. 1992. Gene Targeting Oxford University Press, Inc

Sharp P A. 2001. RNA interference 2001. Genes Dev, 15: 485~490

Sijen T, Fleenor J, Simmer F et al. 2001. On the role of RNA amplification in dsRNA-triggered gene silencing. Cell, 107: 465~476

Strachan T, Andrew P. 1999. Read, Human Molecular Genetics. 2nd ed. New York Garland Science

Strachan T, Read A P. 1999. Human Molecule Genetics. 2nd ed. New York: BIOS Scientific Publishers

Stram Y, Kuzntzova L. 2006. Inhibition of viruses by RNA interference. Virus Genes, 32: 299~306

Thompson S, Clarke A R, Pow A M et al. 1989. Germ line transmission and expression of a corrected HPRT gene produced by gene targeting in embryonic stem cells. Cell, 56: 313~321

Tijsterman M, Ketting R F, Plasterk R H. 2002. The genetics of RNA silencing. Ann Rev Genet, 36: 489~519

Tsai C S. 2007. BIOMACROMOLECULES: Introduction to Structure, Function and Informatics. Hoboken: John Wiley & Sons, Inc. 31

Uauy C, Distelfeld A, Fahima T et al. 2006. A NAC gene regulating senescence improves grain protein, zinc, and iron content in wheat. Science, 314: 1298~1301

Urbanek K B, Sawilska R D, Jedra M, et al. 2003. Detection of genetic modification in maize and maize products by ELISA test. Rocz Panstw Zakl Hig, 54: 345~353

Valasek M A. 2005. The power of real-time PCR. Adv Physiol Educ, 29: 151~159

Volpe T A, Kidner C, Hall I M et al. 2002. Regulation of heterochromatic silencing and histone H3 lysine-9 methylation by RNAi. Science, 297: 1833~1837

Wassenegger M, Heimes S, Riedel L et al. 2004. RNA-directed de novo methylation of genomic sequences in plants.

Cell，76：567～576

Watson J D，Gilman M，Witkowski J et al. 1992. Recombinant DNA. 2nd ed. New York：W H Freeman

Watson J D，Hopkins N H，Roberts J W et al. 1987. Molecular Biology of the Gene. 4th ed. Menlo Park，Benjamin Cummings

Weaver R F. 1999. Molecular Biology. New York：McGraw-Hill

Weaver R F. 2005. Molecular Biology. 3rd ed. McGraw-Hill，New York，126

Zamore P D. 2006. Essay：RNA interference：big applause for silencing in Stockholm. Cell，127：1083～1086

Chapter 10   Molecular Biology Techniques

There is no doubt that some of the most spectacular advances made in science over the past few decades have been in the isolation, analysis, and manipulation of biomacromolecules. This has led to a much greater understanding of mechanisms and processes across many fields of bioscience, such as biochemistry, microbiology, physiology, pharmacology, and the medical sciences to name a few. It has also led to the growth of the biotechnology industry, which seeks to develop and commercialize many of these important processes and methods. Much of these have come about because of the development of numerous molecular biology and genetic manipulation techniques. The discovery of restriction enzymes and the development of cloning vectors in the early 1970's opened the door to ways of isolating and manipulating nucleic acids that had never been thought possible. Gene probe labeling and hybridization were developed and refined to provide powerful methods of analysis. These, together with the development of DNA sequencing methods, protein engineering techniques and PCR, have all continued to contribute substantially to the understanding of biological processes at the molecular level. This chapter will introduce some important techniques in molecular biology.

Section 1   Separation and Detection of Biological Macromolecules

The 21st century promises to be a particularly exciting time for biology. New methods for analyzing DNA, RNA, and proteins are fueling an information explosion and allowing scientists to study cells and their macromolecules in previously unimagined ways. If we want to obtain a complete understanding of what takes place inside a cell as it responds to its environment and interacts with its neighbors, we need to know which genes are switched on, which mRNA transcripts are present, and which proteins are active—where they are located, with whom they partner, and to which pathways or networks they belong. Separation and detection of those macromolecules is the first step of all the analyses required to answer these questions. In this chapter, we briefly review some of the principal methods used to separation and detection of biological macromolecules. Then we also present the latest development and application of these techniques.

1. 1   Isolating DNA from Cells and Tissues

DNA of eukaryotes occurs as nucleoprotein in the nucleus. DNA preparation involves two principles: separating DNA from proteins, lipids and Carbohydrates etc. , and keeping the integrity of the DNA. The application of proteinase K and pronase E guarantees these two principles. Proteinase K and pronase E are broad-spectrum proteases, they can maintain high activity in the presence of SDS and the EDTA. Detergents are used to solubilize the cell membranes. Commonly used detergents include SDS, Triton X-100, and CTAB (hexadecyltrimethyl ammonium bromide). Proteinase K or pronase E is used to degrade protein into peptides and amino acids so that DNA can be separated completely; Phenol and chloroform are used to remove peptides and amino acids; anhydrous ethanol to precipitate DNA. CTAB to precipitate DNA, is also popular because it removes polysaccharides from bacterial and plant preparations. Extraction of genomic DNA should adopt mild conditions in order to keep the DNA intact or else the chromosome will be broken down into small fragments during the process.

When preparing DNA of leukocytes, destroy cell membranes of erythrocytes and leukocytes with Triton X-100 to liberate the hemoglobin and. Then add something like SDS to the nucleus solution and destroy the nuclear membrane to release DNA from nucleoprotein. At last, we can obtain DNA of leukocytes after digesting the protein with proteinase K, and removing peptides and amino acids with phenol and chloroform.

1.2 RNA Preparation

1.2.1 Total RNA and mRNA

From a purely application point of view, total RNA might suffice for most applications, and it is frequently the starting material for applications ranging from the detection of an mRNA species through Northern hybridization to quantification of mRNA through RT-PCR. The preference for total RNA reflects the challenge of purifying enough poly (A) RNA for the application (mRNA accounts for only less than 5% of cellular RNA), the potential loss of a particular message species during poly (A) purification, and the difficulty in quantifying small amounts of purified poly (A) RNA. Two situations where poly (A) RNA is essential are cDNA library construction and preparation of labeled cDNA for hybridization to gene arrays. To avoid generating cDNA libraries with large numbers of ribosomal clones and nonspecific labeled cDNA it is crucial to start with poly (A) RNA for these procedures.

There are three basic methods of isolating total RNA from cells and tissue samples. Most rely on a chaotropic agent such as guanidium or a detergent to break open the cells and simultaneously inactivate RNases. The lysate is then processed in one of several ways to purify the RNA away from protein, genomic DNA, and other cellular components.

1.2.1.1 Guanidium-cesium chloride method

This method employs guanidium isothiocyanate to lyse cells and simultaneously inactivate ribonucleases rapidly. The cellular RNA is purified from the lysate via ultracentrifugation through a cesium chloride or cesium trifluoroacetate cushion. Since RNA is denser than DNA and most proteins, it pellets at the bottom of the tube after $12\sim24$ hours of centrifugation at less about $32\ 000\text{r} \cdot \text{min}^{-1}$. This classic method yields the highest-quality RNA of any currently available technique. Small RNAs (e.g. 5S RNA and tRNAs) cannot be prepared by this method as they will not be recovered.

1.2.1.2 Single and multiple step guanidium acid-phenol method

The guanidium acid-phenol procedure has largely replaced the cesium cushion method because RNA can be isolated from a large number of samples in two to four hours (although it is somewhat cumbersome) without resorting to ultracentrifugation. RNA molecules of all sizes are purified, and the technique can be easily scaled up or down to process different sample sizes. The single step method is based on the propensity of RNA molecules to remain dissolved in the aqueous phase in a solution containing $4\text{mol} \cdot \text{L}^{-1}$ guanidium thiocyanate at pH 4.0 in the presence of a phenol/chloroform organic phase. At this low pH, DNA molecules remain in the organic phase, whereas proteins and other cellular macromolecules are retained at the interphase.

1.2.1.3 Non-phenol-based methods

One major drawback of using the guanidium acid-phenol method is the handling and disposal of phenol, a hazardous chemical. As a result phenol-free methods, based on the ability of glass fiber filters to bind nucleic acids in the presence of chaotropic salts like guanidium, have gained favor. As to the other methods, the cells are first lysed in a guanidium-based buffer. The lysate is then diluted with an organic solvent such as ethanol or isopropanol and applied to a glass fiber filter or resin. DNA and proteins are washed off, and the RNA is eluted at the end in an aqueous buffer. This technique yields total RNA of the same quality as the phenol-based methods. Since these are column-based protocols requiring no organic extractions, processing large sample numbers is fast and easy. This is also among the quickest methods for RNA isolation, usually completed in less than one hour.

RNA isolation from some tissues requires protocol modifications to eliminate specific contaminants, or tissue treatments prior to the RNA isolation protocol. Fibrous tissues and tissue rich in protein, DNA, and RNases present unique challenges for total RNA isolation.

For mRNA preparation, a one-step procedures purifies poly (A) RNA directly from the starting material. A two-step strategy first isolates total RNA, and then purifies poly (A) RNA from the product. One round of poly (A) RNA selection via oligo (dT) -cellulose typically removes $50\%\sim70\%$ of the ribosomal RNA. One round of selection is adequate for most applications (i.e. Northern analysis and ribonuclease protection assays). A cDNA library generated from poly (A) RNA that is 50% pure is usually sufficient to identify most genes, but to generate cDNA libraries with minimal rRNA clones, two rounds of oligo (dT) selection that will remove approximately 95% of the ribosomal RNA is necessary. 20%~50% of the poly (A) RNA can be lost during each round of oligo (dT) selection, so multiple rounds of selection will decrease mRNA yield. The use of labeled cDNA to screen gene arrays is severely compromised by the presence of rRNA-specific probes, so two rounds of poly (A) selection might be justified.

1.2.2　Maximizing the Yield and Quality of an RNA Preparation

RNase contamination is so prevalent, special attention must be given to the preparation of solutions. Solutions should be prepared in disposable, RNase-free plasticware or in RNase-free glassware prepared in the lab. Glassware can be made RNase-free by baking at 180℃ for 8 hours to overnight, or by treating with a commercial RNase decontaminating solution. Alternatively, RNase can be removed by filling containers with 0.1% DEPC, incubating at 37℃ for 2 hours, rinsing with sterile water and then either heating to 100℃ for 15 minutes, or autoclaving for 15 minutes to eliminate RNase. The most common method of preparing RNase-free solutions is diethylpyrocarbonate (DEPC) treatment. DEPC inactivates RNases by carboxyethylation of specific amino acid side chains in the protein. DEPC is a suspected carcinogen, and it should always be used with the proper precautions. Most protocols specify adding DEPC to solutions at a concentration 0.1%~0.01%, followed by mixing and room temperature incubation for several hours to overnight. The container lid should be loosened for the extended incubation because a considerable amount of pressure can form during the reaction. Finally, the solution is autoclaved; this inactivates the residual DEPC by hydrolysis, and releases CO_2 and ethanol as by-products. DEPC has a half-life of 30 minutes in water, and at a DEPC concentration of 0.1%, solutions autoclaved for 15 minutes can be assumed to be DEPC-free. Be aware that increasing the concentration of DEPC to 1% can inhibit more RNase but can also inhibit certain enzymatic reactions, so more is usually not better. It should be noted that many reagents commonly used in RNA studies contain primary amines, such as Tris, MOPS, HEPES, and PBS, and cannot be DEPC-treated because the amino group "sops up" the DEPC, making it unavailable to inactivate RNase. These solutions should be prepared with ultrapure reagents and DEPC-treated water. RNase is present in all cells and tissues; hence they must be immediately inactivated when the source organism dies. Samples should be immediately frozen in liquid nitrogen, or immediately disrupted in a chaotropic solution (i.e. gITC). In some cases RNase activity can eventually be restored even in the presence of a chaotrope if the extract is not frozen. Fast and complete lysis of any sample is arguably the most critical element of RNA purification. RNase inhibition provided by chaotropes and other reagents can be overwhelmed by adding too much starting material.

Ideally RNA should be purified without interruption, no matter which procedure is used. If a pause is unavoidable, stop when the RNA has precipitated or is in the presence of a chaotrope. For example, when using an organic isolation procedure, the RNA isolation can be stopped when the samples have been homogenized in a chaotropic lysis solution. They can be stored for a few days at −20℃ or −80℃ without degradation.

1.2.3　Storage of Purified RNA

RNase activity and pH up to 8 will destroy RNA. For short-term storage of a few weeks or less, store your RNA in RNase-free Tris-EDTA or 1mmol/L EDTA at −20℃ in aliquots. For long-term storage, RNA should be stored in aliquots at −80℃ in TE, 1mmol・L^{-1} EDTA, formamide, or as an ethanol/salt precipitation mixture.

1.3　Protein Separation and Preparation

Our understanding of protein structure and function has been derived from the study of many individual proteins. To study a protein in detail, the researcher must be able to separate it from other proteins and must have the techniques to determine its properties. The necessary methods come from protein chemistry, an old discipline that retains a central position in biochemical research. Methods for separating proteins take advantage of properties that vary from one protein to the next, including size, charge, and binding properties.

The source of a protein is generally tissue or microbial cells. The first step in any protein purification procedure is to break open these cells, releasing their proteins into a solution called a crude extract. If necessary, differential centrifugation can be used to prepare subcellular fractions or to isolate specific organelles. Once the extract or organelle preparation is ready, various methods are available for purifying one or more of the proteins it contains. Commonly, the extract is subjected to treatments that separate the proteins into different fractions based on a properties such as size or charge, a process referred to as fractionation. In general, most proteins can dissolve in water, sparse salt, sparse alkali or the sparse sour solution, and a few proteins that combined with lipids dissolve in organic solvents such as alcohol, acetone, butanol etc. The difference of protein solubility in different solvents depends on the proportion of polar group and non-polar group in proteins. So, by using different solvents and adjusting the factors, such as that the temperature, pH, ionic strength etc, that affect the protein solubility, we can separate the required proteins and enzymes from complicated component of cells.

1.4　Assessing Quality of DNA and RNA Preparation

The most common method to identify the quality of DNA and RNA are UV-spectrophotometric and electrophoresis comparation. The spectrophotometric values of 260nm and 280nm ware length were determined by a UV-spectrophotometer, then the purity of nucleic acid can be reflected by the ratio of A260 ∶ A280, which can be used to as an initial estimation of the quality of DNA and RNA. The ratio of A260 ∶ A280 of pure DNA and RNA are 1.8 and 2.0 respectively. If there are pollutions of proteins or phenols, the ratio of A260 ∶ A280 will obviously be lower than the normal value.

When the absorbency of a value equals one in the wavelength of 260nm, it corresponds to a concentration of $50\mu g \cdot ml^{-1}$ of a double-stranded DNA , $40\mu g \cdot ml^{-1}$ of single-stranded DNA or RNA and $20\mu g \cdot ml^{-1}$ of oligonucleotides.

The integrity of your RNA is best determined by electrophoresis on a formaldehyde agarose gel under denaturing conditions. The samples can be visualized by adding 10mg/ml of Ethidium Bromide (EtBr) (final concentration) to the sample before loading on the gel. Compare your prep's 28S rRNA band (located at approximately 5 Kb in most mammalian cells) to the 18S rRNA band (located at approximately 2 Kb in most mammalian cells). In high-quality RNA the 28S band should be approximately twice the intensity of the 18S band (Figure 10-1).

Figure 10-1　Assessing quality of RNA preparation via agarose gel electrophoresis (see page 321)

The most sensitive test of RNA integrity is Northern analysis using a high molecular weight probe expressed at low levels in the tissues being analyzed. However, this method of quality control is very time-consuming and is not necessary in most cases.

1.5　Gel Electrophoresis

Perhaps the most widely used physical method in all of molecular biology is gel electrophoresis, which separates, identifies and purifies fragments of DNA or RNA as well as proteins. This technique is simple, rapid to perform, and capable of resolving fragments of DNA that cannot be separated adequately by other procedures, such as density gradient centrifugation. Furthermore, the location of DNA within the gel can be determined directly by staining with low concentrations of fluorescent intercalating dyes, such as ethidium bromide or SYBR Green; bands containing as little as 20pg of double-stranded DNA can be detected by direct examination of the gel in UV. If necessary, these bands of DNA can be recovered from the gel and used for a variety of purposes.

1.5.1　Principle

When a charged molecular is placed in an electric field, it will migrate towards the electrode with the opposite charge; nucleic acid molecules, which have a consistent negative charge imparted by their phosphate backbone, will move towards the positive pole. However, proteins are not so convenient. Some have a positive charge, some have a negative charge, and most are neutral. They must be denatured by sodium dodecyl sulfate (SDS) to get an overall negative charge. In a gel, which contains of a complex network of pores, the rate at which a nucleic acid molecule moves will be determined by its ability to penetrate through this network. For linear fragments of double-stranded DNA within a certain size range, this will reflect the size of the molecule (i. e. the length of the DNA). We do not have to consider the amount of charge that the molecule carries, since all nucleic acids carry the same amount of charge per unit size. A circular DNA molecule that is relaxed or nicked migrates more slowly than does a linear molecule of equal mass, while supercoiled DNAs migrate more rapidly during electrophoresis than do less supercoiled or circular DNAs of equal mass. Mixture components separate into discrete "bands"; the smaller forms having more mobility through the gel. Patterns of such bands are then visualized for a variety of analytical purposes.

Most commonly, the gel is cast in the shape of a thin slab, with wells for loading the sample. The gel is immersed within an electrophoresis buffer that provides ions to carry a current and some type of buffer to maintain the pH at a relatively constant value. The gel itself is composed of either agarose or polyacrylamide, each of which has attributes suitable to particular tasks.

1.5.2 Agarose and Polyacrylamide gel Electrophoresis

Agarose is a polysaccharide extracted from seaweed. It is typically used at concentrations of 0.5%~ 2%. The higher the agarose concentration the "stiffer" the gel. Agarose gels are extremely easy to prepare: you simply mix agarose powder with buffer solution, melt it by heating, and pour the gel. It is also non-toxic. Agarose gels have a large range of separation, but relatively low resolving power. By varying the concentration of agarose, fragments of DNA from about $200 \sim 50\,000$ bp can be separated using standard electrophoretic techniques.

Polyacrylamide is a cross-linked polymer of acrylamide. The length of the polymer chains is dictated by the concentration of acrylamide used, which is typically between 3.5% and 20%. Polyacrylamide gels are significantly more annoying to prepare than agarose gels. Because oxygen inhibits the polymerization process, they must be poured between glass plates (or cylinders). Polyacrylamide gels have a rather small range of separation, but very high resolving power. In the case of DNA, polyacrylamide is used for separating fragments of less than about 500 bp. However, under appropriate conditions, fragments of DNA differing in length by a single base pair are easily resolved. In contrast to agarose, polyacrylamide gels are used extensively for separating and characterizing mixtures of proteins (Figure 10-2).

Figure 10-2 Polyacrylamide gel electrophoresis (see page 323)

1.5.3 Pulsed-Field gel Electrophoresis

Standard agarose gels can be used to separate nucleic acid that range from around 200bp to perhaps 50kb. Specialized gel electrophoresis are needed to separate DNA molecules that are extremely large. Pulsed-field gel electrophoresis (PFGE) is used for analysis of very large DNA molecules from 10kb~10 Mb.

In this method, alternating orthogonal electric fields are applied to a gel. Large DNA molecules become trapped in their reptation tubes every time the direction of the electric field is altered and can make no further progress through the gel until they have reoriented themselves along the new axis of the electric field. The larger the DNA molecule, the longer the time required for this realignment. Molecules of DNA whose reorientation times are less than the period of the electric pulse will therefore be fractionated according to size (Figure 10-3).

Figure 10-3 Pulsed-field gel electrophoresis (see page 323)

For all organisms, from bacteria to humans, PFGE is used to study genome organization and to clone and analyze large fragments.

Section 2　Purification of Biomacromolecules

Investigation of biomacromolecules involves their purification, either from biological sources or products derived from chemical synthesis or recombinant technology. It is perhaps the most demanding, laborious and yet essential step preceding structural and functional studies of biomacromolecules. Since laboratory scientists desiring specific biomacromolecules generally design purification processes, however, the purification of biomacromolecules is the most problematic process encountered by practicing biochemists.

2.1　Genomic DNA Purification

The isolation and purification of genomic DNA is crucial to many applications in molecular biology. Over the past decades, the purification of DNA has evolved from a multi-step process involving organic chemicals to a much simpler process that is safer and faster. The strengths and limitations of contemporary purification methods are listed in Table 10-1.

Table 10-1　Strengths and limitations of contemporary purification methods

Methods	Mechanisms	Features	Limitations
Salting out and DNA Precipitation	Based on the use of chaotropes and cosmotropes to separate cellular components based on solubility differences.	Yields are generally good. These procedures apply little mechanical stress, so shearing is generally not a problem.	Adding 1% polyvinylpyrrolidine to extraction buffer can absorb phenolic contaminants. Add a CTAB precipitation step to remove polysaccharides.
Extraction with Organic Solvents, Chaotropes, and DNA Precipitation	Chaotropic guanidinium salts lyse cells and denature proteins, and reducing agents prevent oxidative damage of nucleic acids. Phenol, which solubilizes and extracts proteins and lipids to the organic phase, sequestering them away from nucleic acids.	Yield, purity, and speed are good. Convenient for working with small numbers of samples.	Residual phenol can remain without chloroform steps and interfere with downstream applications. High salt concentrations can lead to phase inversion. When working with GTC/phenol-based extraction buffers, cross contamination of RNA with DNA is frequent.
Glass Milk/Silica Resin-Based Strategies	Nucleic acids bind to glass milk and silica resin under denaturing conditions in the presence of salts. Then wash away remaining impurities and excess chaotrope with salt/ethanol and nucleic acids are eluted from the glass in a salt or TE buffer.	Fast, simple, straightforward, and scalable. Allow restriction digestion/ligation reactions directly on the glass surface, improving transformation efficiency of complex ligation mixtures.	Underloading the sample, overdrying and improper wash or elution buffer will impair yields. Incomplete ethanol removal after wash steps will cause problems such as diffusion out of agarose gel wells ("unloadable" DNA/RNA) or undigestible DNA.
Anion Exchange (AIX) Based Strategies	Nucleic acids are very large anions with a charge of -1/base and -2/bp; hence they will bind to positively charged purification resins. After washes in low-salt buffers, the DNA is eluted in a high-salt buffer.	These methods can produce very pure DNA, but the yields in small-scale applications tend to be low.	Not the most robust method, and recoveries tend to be lower, and the final elution step involves high-salt buffers, which needs to be desalted. The binding capacities tend to be low ($0.25\sim2$ mg $\cdot$ ml^{-1} of resin), increase with pH, and decrease with increasing size of the DNA. RNA will compete with DNA for binding.

Methods	Mechanisms	Features	Limitations
Hydroxyapatite (HA) Based Strategies	Nucleic acids bind to crystalline calcium phosphate through the interaction of calcium ions on the hydroxyapatite and the phosphate groups of the nucleic acids. An increase in competing free phosphate ions from $0.12 \sim 0.4$ mol $\cdot$ L^{-1} will elute nucleic acids, with single-stranded nucleic acids eluting before double-stranded DNA.	Excellent separation of single-stranded from double-stranded DNA molecules	The quality and performance can vary from batch to batch and between manufacturers. Thermal elution procedures require reliable temperature control. Hydroxyapatite has poor mechanical stability. Hydroxyapatite procedures often employ high-salt buffers and lead to sample dilution, requiring an additional precipitation step.

To concentrate nucleic acids for resuspension in a more suitable buffer, solvents such as ethanol (75%~80%) or isopropanol (final concentration of 40%~50%) are commonly used in the presence of salt to precipitate nucleic acids. If volume is not an issue, ethanol is preferred because less salt will co-precipitate and the pellet is more easily dried. Polyethylene glycol (PEG) selectively precipitates high molecular weight DNA, but it is also more difficult to dry and can interfere with downstream applications. Trichloroacetic acid (TCA) precipitates even low MW polymers (down to 5kDa), but nucleic acids cannot be recovered in a functional form after precipitation.

2.2 Plasmid Purification

Plasmid purification holds a special challenge because the target DNA must be purified from DNA contaminants. Isolation strategies take advantage of the physical differences between linear, closed, and supercoiled DNA. Alkaline lysis, boiling. All other denaturing methods exploit the fact that closed DNA will renature quickly upon cooling or neutralizing, while the long genomic DNA molecules will not renature and remain "tangled" with proteins, SDS and lipids, which are salted out. Whether boiling or alkaline pH is the denaturing step, the renaturing step is usually performed in cold to enhance precipitation or salting-out of protein and contaminant nucleic acids.

One other method lyses cells by a combination of enzymatic breakdown, detergent solubilization, and heat. The lysis buffer usually contains lysozyme, STE, and Triton X-100 or CTAB. Bacterial chromosomal DNA remains attached to the membrane and precipitates out. Again, the aqueous supernatant generated by this method contains plasmid and RNA. Polyethylene glycol (PEG) has been used to separate DNA molecules by size, based on its size-specific binding to DNA fragments. A 6.5% PEG solution can be used to precipitate genomic DNA selectively from cleared bacterial lysates. Trace amounts of PEG may be removed by a chloroform extraction.

Isolation of plasmid DNA by cesium chloride centrifugation in the presence of ethidium bromide (EtBr) is especially useful for large-scale DNA preparations. The interaction of EtBr with DNA decreases the density of the nucleic acid; because of its supercoiled conformation and smaller size, plasmid incorporates less EtBr than genomic DNA, enhancing separation on a density gradient.

Chromatographic methods such as anion exchange and gel filtration may also be used to purify plasmids after lysis. For chromatography, RNA removal prior to separation is essential because the RNA will interfere with and contaminate the separation process. RNase A treatments, RNA specific precipitation, tangential flow filtration, and nitrocellulose filter binding have been employed to desalt, concentrate, and generally prepare samples for column purification.

The yield of low copy number plasmids can be improved dramatically by adding chloramphenicol or spectinomycin (300mg $\cdot$ ml^{-1}), which prevent replication of chromosomal but not plasmid DNA. However, extended exposure to such agents has been also shown to damage DNA *in vitro*.

The separation of DNA from contaminants based on density differences (isopycnic centrifugation) in CsCl gradients remains effective but time costly. High g forces cause the migration of dense Cs$^+$ ions to the bottom of the tube until centripetal force and force of diffusion have reached an equilibrium. Within a CsCl gradient, polysaccharides will assume a random coil secondary structure, DNA and RNA will have the highest density. Dyes that bind to nucleic acids and alter their density have been applied to enhance their separation from contaminants. The binding of EtBr decreases the apparent density of DNA. Super-coiled DNA binds less EtBr than linear DNA, enhancing their separation based on density differences. CsCl centrifugation is most commonly applied to purify plasmids and cosmids in combination with EtBr.

Triple helix resins have been used to purify plasmids and cosmids. This approach takes advantage of the adoption of a triple rather than a double helix conformation under the proper pH, salt, and tempera-ture conditions. Triple helix affinity resins are generated by insertion of a suitable homopurine sequence into the plasmid DNA and crosslinking the complement to a chromatographic resin of choice. The triple helix interaction is only stable at mild acidic pH; it dissociates under alkaline conditions. The interaction at mildly acid pH is very strong. This strong affinity allows for extensive washing that can improve the removal of genomic DNA, RNA, and endotoxin during large-scale DNA preparations.

2.3　Protein Purification

2.3.1　Salting Out and Dialysis

Once the extract or organelle preparation is ready, various methods are available for purifying one or more of the proteins it contains. Commonly, the extract is subjected to treatments that separate the pro-teins into different fractions based on a property such as size or charge, a process referred to as fractiona-tion. Early fractionation steps in purification utilize differences in protein solubility, which is a complex function of pH, temperature, salt concentration, and other factors. The solubility of proteins is generally lowered at high salt concentrations, an effect called "salting out". The addition of a salt in the right a-mount can selectively precipitate some proteins, while others remain in solution. Ammonium sulfate ($(NH_4)_2SO_4$) is often used for this purpose because of its high solubility in water.

A solution containing the protein of interest often must be further altered before subsequent purifica-tion steps are possible. Dialysis is a procedure that separates proteins from solvents by taking advantage of the proteins' larger size. The partially purified extract is placed in a bag or tube made of a semi-perme-able membrane. When this is suspended in a much larger volume of buffered solution of appropriate ionic strength, the membrane allows the exchange of salt and buffer but not proteins. Thus dialysis retains large proteins within the membranous bag or tube while allowing the concentration of other solutes in the protein preparation to change until they come into equilibrium with the solution outside the membrane. Dialysis might be used, for example, to remove ammonium sulfate from the protein preparation.

2.3.2　Chromatography

The most powerful methods for fractionating proteins make use of column chromatography, which takes advantage of differences in protein charge, size, binding affinity, and other properties.

Figure 10-4 shows a column chromatography model. A porous solid material with appropriate chemical properties (the stationary phase) is held in a column, and a buffered solution (the mobile phase) perco-lates through it. The protein-containing solution, layered on the top of the column, percolates through the solid matrix as an ever-expanding band within the larger mobile phase. Individual proteins migrate faster or more slowly through the column depending on their properties. In ion-exchange chromatography (Figure 10-5a), the solid matrix has negatively charged groups. In the mobile phase, proteins with a net positive charge migrate through the matrix more slowly than those with a net negative charge, because the migration of the former is retarded more by interaction with the stationary phase. The two types of protein can separate into two distinct bands. The expansion of the protein band in the mobile phase (the protein solution) is caused both by separation of proteins with different properties and by diffusive sprea-ding. As the length of the column increases, the resolution of two types of protein with different net charges generally improves. However, the rate at which the protein solution can flow through the column usually decreases with column length. And as the length of time spent on the column increases, the reso-lution can decline as a result of diffusive spreading within each protein band.

Figure 10-4　Column chromatography（see page 326）

Figure 10-5　Three chromatographic methods used in protein purification（see page 327）

There are two other variations of column chromatography in addition to ion exchange. One is the size-exclusion chromatography, which separates proteins according to size. In this method, large proteins emerge from the column sooner than small ones. The solid phase consists of beads with engineered pores or cavities of a particular size. Large proteins cannot enter the cavities, and so take a short path through the column, around the beads. Small proteins enter the cavities, and migrate through the column more slowly as a result (Figure 10-5b). The other is affinity chromatography, which is based on the binding affinity of a protein. The beads in the column have a covalently attached chemical group. A protein with affinity for this particular chemical group will bind to the beads in the column, and its migration will be retarded as a result (Figure 10-5c).

A modern refinement in chromatographic methods is HPLC, or high-performance liquid chromatography. HPLC makes use of high-pressure pumps that speed the movement of the protein molecules down the column, as well as higher-quality chromatographic materials that can withstand the crushing force of the pressurized flow. By reducing the transit time on the column, HPLC can limit diffusive spreading of protein bands and thus greatly improve resolution. The various laminar analysis, namely, chromatogram technology has been applied broadly in isolation, purification and analysis of the organic, inorganic, low molecular compounds or polymeric compounds and biomacromolecule with biological activity.

The approach to purify a protein that has not previously been isolated is guided both by established precedures and by common sense. In most cases, several different methods must be used sequentially to purify a protein completely. The choice of method is somewhat empirical, and many protocols may be tried before the most effective one is found. Trial and error can often be minimized by basing the procedure on purification techniques developed for similar proteins. Published purification protocols are available for many thousands of proteins. Common sense dictates that inexpensive procedures such as salting out be used first, when the total volume and the number of contaminants are greatest. Chromatographic methods are often impractical at early stages, because the amount of chromatographic medium needed increases with sample size. As each purification step is completed, the sample size generally becomes smaller (Table 10-2), making it feasible to use more sophisticated (and expensive) chromatographic procedures at later stages.

Table 10-2　Example of purifying a hypothetical enzyme

Procedure or step	Fraction volume /ml	Total protein /mg	Activity /units	Specific activity / (units · mg^{-1})
Crude cellular extract	1 400	10 000	100 000	10
Precipitation with ammonium sulfate	280	3 000	96 000	32
Ion-exchange chromatography	90	400	80 000	200
Size-exclusion chromatography	80	100	60 000	600
Affinity chromatography	6	3	45 000	15 000

Note: All data represent the status of the sample after the designated procedure has been carried out.

2.3.3　Electrophoresis

Another important technique for separation of proteins is based on the migration of charged proteins in an electric field, a process called electrophoresis. These procedures are not generally used to purify proteins in large amounts, because simpler alternatives are usually available and electrophoretic methods often adversely affect the structure and thus the function of proteins. Electrophoresis is, however, especially useful as an analytical method. Its advantage is that proteins can be visualized as well as separated, permitting a researcher to estimate quickly the number of different proteins in a mixture or the degree of purity of a particular protein preparation. Also, electrophoresis allows determination of crucial properties of a protein such as its isoelectric point and approximate molecular weight.

2.3.3.1　Polyacrylamide gel electrophoresis

Electrophoresis of proteins is generally carried out in gels made up of the cross-linked polymer polyacrylamide (Figure 10-6). The polyacrylamide gel acts as a molecular sieve, slowing the migration of proteins approximately in proportion to their charge-to-mass ratio. Migration may also be affected by protein shape. In electrophoresis, the force moving the macromolecule is the electrical potential, E. The electrophoretic mobility of the molecule is the ratio of the velocity of the particle molecule, V, to the electrical potential. Electrophoretic mobility is also equal to the net charge of the molecule, Z, divided by the frictional coefficient, f, which reflects in part a protein's shape. Thus:

$$\mu = \frac{V}{E} = \frac{Z}{f}$$

Figure 10-6　Electrophoresis（see page 329）

The migration of a protein in a gel during electrophoresis is therefore a function of its size and its shape.

An electrophoretic method commonly employed for estimation of purity and molecular weight makes use of the detergent sodium dodecyl sulfate (SDS). SDS binds to most proteins in amounts roughly proportional to the molecular weight of the protein, about one molecule of SDS for every two amino acid residues. The bound SDS contributes a large net negative charge, rendering the intrinsic charge of the protein insignificant and conferring on each protein a similar charge-to-mass ratio. In addition, the native conformation of a protein is altered when SDS is bound, and most proteins assume a similar shape. Electrophoresis in the presence of SDS therefore separates proteins almost exclusively on the basis of mass (molecular weight), with smaller polypeptides migrating more rapidly. After electrophoresis, the proteins are visualized by adding a dye such as coomassie blue, which binds to proteins but not to the gel itself (Figure 10-6b). Thus, a researcher can monitor the progress of a protein purification procedure as the number of protein bands visible on the gel decreases after each new fractionation step. When compared with the positions to which proteins of known molecular weight migrate in the gel, the position of an unidentified protein can provide an excellent measure of its molecular weight (Figure 10-7). If the protein has two or more different subunits, the subunits will generally be separated by the SDS treatment and a separate band will appear for each.

Figure 10-7　Estimating the molecular weight of a protein（see page 330）

2.3.3.2　Isoelectric focusing electrophoresis

Isoelectric focusing is a procedure used to determine the isoelectric point (pI) of a protein. A pH gradient is established by allowing a mixture of low molecular weight organic acids and bases to distribute themselves in an electric field generated across the gel. When a protein mixture is applied, each protein migrates until it reaches the pH that matches its pI. Proteins with different isoelectric points are thus distributed differently throughout the gel. This method is called isoelectric focusing polyacrylamide gel electrophoresis (IEF-PAGE). Ampholine is the usually used ampholytes (Figure 10-8).

Figure 10-8　Isoelectric focusing（see page 331）

2.3.3.3　Two-dimensional electrophoresis

Combining isoelectric focusing and SDS electrophoresis sequentially in a process called two-dimensional electrophoresis permits the resolution of complex mixtures of proteins. In the first dimension of two-dimensional electrophoresis, proteins are separated according to their electrical charge in an isoelectric focusing electrophoresis. Proteins of variant isoelectric points will distribute in different positions. In the second dimension, proteins are separated according to their molecular weight through a SDS-PAGE. This is a more sensitive analytical method than either electrophoretic method alone. Two-dimensional electrophoresis separates proteins of identical molecular weight that differ in pI, or proteins with similar pI values but different molecular weights. The separating results of two-dimensional electrophoresis are dots not bands. It is a method of the highest resolution power and richest information content among all the electrophoresis methods by now (Figure 10-9).

Figure 10-9 Two-dimensional electrophoresis (see page 332)

2. 3. 4 Quantification of Unseparated Proteins

To purify a protein, it is essential to have a way of detecting and quantifying that protein in the presence of many other proteins at each stage of the procedure. Often, purification must proceed in the absence of any information about the size and physical properties of the protein or about the fraction of the total protein mass it represents in the extract. For proteins that are enzymes, the amount in a given solution or tissue extract can be measured, or assayed, in terms of the catalytic effect the enzyme produces—that is, the increase in the rate at which its substrate is converted to reaction products when the enzyme is present. For this purpose one must know ① the overall equation of the reaction catalyzed, ② an analytical procedure for determining the disappearance of the substrate or the appearance of a reaction product, ③ whether the enzyme requires cofactors such as metal ions or coenzymes, ④ the dependence of the enzyme activity on substrate concentration, ⑤ the optimum pH, and ⑥ a temperature zone in which the enzyme is stable and has high activity. Enzymes are usually assayed at their optimum pH and at some convenient temperature within the range 25~38℃. Also, very high substrate concentrations are generally used so that the initial reaction rate, measured experimentally, is proportional to enzyme concentration. By international agreement, 1. 0 unit of enzyme activity is defined as the amount of enzyme causing transformation of 1. 0 mol of substrate per minute at 25℃ under optimal conditions of measurement. The term activity refers to the total units of enzyme in a solution. The specific activity is the number of enzyme units per milligram of total protein (Figure 10-10).

Figure 10-10 Activity versus specific activity (see page 332)

Section 3 Molecular Hybridization Technology

Nucleic acid hybridization is a fundamental tool in molecular genetics, this technology takes advantage of the ability of individual single-stranded nucleic acid molecules to form double-stranded molecules (that is, to hybridize to each other). For this to happen, the interacting single-stranded molecules must have a sufficiently high degree of base complementarity. Standard nucleic acid hybridization assays uses a labeled nucleic acid probe to identify related DNA or RNA molecules (that is, ones with a significantly high degree of sequence similarity) within a complex mixture of unlabeled nucleic acid molecules, the target nucleic acid.

Studies of gene expression and function require detection not only of DNA and RNA, but also specific proteins. For these studies, antibodies take the place of nucleic acid probes as reagents that can selectively react with unique protein molecules.

3. 1 Principle

Nucleic acid hybridization involves mixing single strands of two sources of nucleic acids, a probe which typically consists of a homogeneous population of identified molecules (e. g. cloned DNA or chemically synthesized oligonucleotides) and a target which typically consists of a complex, heterogeneous population of nucleic acid molecules. If either the probe or the target is initially double-stranded, the individual strands must be separated, generally by heating or by alkaline treatment. After mixing single strands of probe with single strands of target, strands with complementary base sequences can be allowed to reassociate. Complementary probe strands can reanneal to form homoduplexes, as can complementary target DNA strands. However, it is the annealing of a probe DNA strand and a complementary target DNA strand to form a labeled probe-target heteroduplex that defines the usefulness of a nucleic acid hybridization assay (Figure 10-11). The rationale of the hybridization assay is to use the identified probe to interrogate the target DNA by identifying fragments in the complex target which may be related in sequence to the probe. The probe must be labeled by radioactivity, fluorescence or some other way to enable its detection.

Figure 10-11   A nucleic acid hybridization assay (see page 333)

3.2   Southern Blotting

Southern blotting is a technique in which one DNA sample is hybridized to another DNA sample. It was named after Edward M. Southern who developed this procedure at Edinburgh University in the 1970s. To oversimplify, DNA molecules are transferred from an agarose gel onto a membrane. Southern blotting is designed to locate a particular sequence of DNA within a complex mixture. For example, Southern blotting could be used to locate a particular gene within an entire genome. The amount of DNA needed for this technique is dependent on the size and specific activity of the probe. Short probes tend to be more specific. Under optimal conditions, you can expect to detect 0.1 pg of the DNA for which you are probing.

This diagram (Figure 10-12) shows the basic steps involved in a Southern blot.

Figure 10-12   Southern blotting (see page 334)

The protocol of Southern blotting is as following: ① Digest the DNA with an appropriate restriction enzyme. ② Run the digest on an agarose gel. ③ Denature the DNA by soaking the gel in alkaline denaturing solution. ④ Transfer the denatured DNA to a nylon membrane. Traditionally, a nitrocellulose membrane is used, although nylon or a positively charged nylon membrane may be used. Many scientists feel nylon is better since it binds more and is less fragile. Transfer is usually done by capillary action, which takes several hours. To be faster, you may use a vacuum blot apparatus instead of capillary action. After you transfer your DNA to the membrane, treat it with UV light. This cross links (via covalent bonds) the DNA to the membrane. You can also bake the membrane at about 80℃ for a couple of hours. ⑤ Probe the membrane with labeled ssDNA probe. This is also known as hybridization. Whatever you call it, this process relies on the ssDNA hybridizing (annealing) to the DNA on the membrane due to the binding of complementary strands. Probing is often done with ^{32}P labeled ATP, biotin/streptavidin or a bioluminescent probe. A prehybridization step is required before hybridization to block non-specific sites, since you don't want your single-stranded probe binding just anywhere on the membrane. To hybridize, use the same buffer as for prehybridization, but add your specific probe.

If you used a radiolabeled ^{32}P probe, then you would visualize by autoradiograph. Biotin/streptavidin detection is done by colorimetric methods, and bioluminescent visualization uses luminesence.

3.3   Northern Blotting

Northern blotting is an RNA blotting technique which was developed in 1977 by Alwine et al. at Stanford University. It was named after the Southern blot technique, which blots for DNA invented by Edward M. Southern in 1975. The procedure for and theory behind Northern blotting is almost identical to that of Southern blotting, except you are working with RNA instead of DNA (Figure 10-13).

Figure 10-13   Northern blotting (see page 335)

Northern analysis despite its age in the high tech world of Real Time PCR, nuclease protection assays (PRAs) and microarrays, is still the gold-standard for the detection and quantitation of mRNA levels. The risk of artifacts is much smaller than with the PCR-based methods for determining the size of a specific transcript and for detecting the presence of different transcripts for a specific gene. Besides the study of gene expression at mRNA level, other applications of Northern blot include detection of mRNA transcript size, RNA degradation, RNA splicing and it's often used to confirm and check transgenic/knockout animals.

However, there are some limitations to this technique. First, although we can get a relative quantitative estimate of the strengths of the signals, it is generally not possible to obtain an absolute measurement of the amount of specific mRNA. Second, the technique is relatively less sensitive than nuclease protection assays and RT-PCR, and requires fairly large amounts of RNA. Third, the process of the technique is troublesome and time-consuming.

3. 4 Western Blotting

Much like Southern and Northern blot identify specific DNA or RNA, Western blotting (Figure 10-14) is a technique used to identify and locate proteins based on their ability to bind to specific antibodies, no matter whether the protein has been synthesized *in vivo* or *in vitro*. Western blot ting analysis can detect the protein of interest from a mixture of a great number of proteins, and give you the information about the size and expression amount of the protein. This technique is also very useful in the cloning of genes.

Figure 10-14 Western blotting (see page 336)

To detect a specific protein, an antibody to that protein must be available. An antibody can either be produced for the protein of interest or sometimes purchased commercially. The nitrocellulose membrane itself has many non-specific sites that can bind proteins, including antibodies. These sites must be blocked with a non-specific protein solution such as re-hydrated powdered milk. The primary antibody is added in the milk solution and binds to the protein of interest. The antibody protein complex is detected using a secondary antibody that has a label attached to it. Often a reporter enzyme such as alkaline phosphatase is linked to the secondary antibody, and the addition of lumiphos or X-phos to the blot allows detection of the protein band.

Section 4　Determination of DNA and Protein Sequences

Since it was discovered that DNA is the material in the cell that carries our genetic information, understanding DNA has become a primary focus of genetic research. Our genome consists of neatly wound strands of DNA. All living organisms, from bacteria to human beings, contain DNA in each of their cells. Each cell contains the entire genetic code for that organism. Although important goals of any sequencing project is to obtain a genomic sequence and identify a complete set of genes, the ultimate goal is to understand when, where, and how a gene is turned on, a process commonly referred to as gene expression. Once we begin to understand where and how a gene is expressed under normal circumstances, we can then study what happens in an altered state, such as in disease. To accomplish the latter goal, however, researchers must identify and study the protein (s), coded for by gene (s).

4. 1 DNA Sequencing

The power of DNA sequencing is its ability to reduce genes and genomes to chemical entities of defined structures. Few other techniques provide biologists with such certainty and comfort. In molecular cloning laboratories, DNA sequencing today is used chiefly to characterize newly cloned cDNA; to confirm the identity of a clone or mutation; to check the fidelity of a newly created mutation, ligation junction, or product of a polymerase chain reaction; and, in some cases, as a screening tool to identify polymorphisms and mutations in genes of particular interest.

4. 1. 1 Principle

Rapid and efficient methods for DNA sequencing were first developed in the mid-1970s. Two different procedures were published at almost the same time: ① The chain termination method, in which the sequence of a single-stranded DNA molecule is determined by enzymatic synthesis of complementary polynucleotide chains, these chains terminating at specific nucleotide positions; ② The chemical degradation method, in which the sequence of a double-stranded DNA molecule is determined by treatment with chemicals that cut the molecule at specific nucleotide positions. Both methods were equally popular to begin with but the chain termination procedure has gained ascendancy in recent years, particularly for genome sequencing. This is partly because the chemicals used in the chemical degradation method are toxic, and because it has been easier to automate chain termination sequencing. The genome project involves a huge number of individual sequencing experiments, or it would take many more years to perform all these by hand. Automated sequencing techniques are therefore essential if the project is to be completed in a reasonable time-span.

4.1.2 Advances in DNA Sequencing

Traditional dideoxy sequencing methods have employed radioisotope labeling: the dNTP mix contains a proportion of radiolabeled nucleotides which are incorporated within the growing DNA chains. Following electrophoresis, the gel is dried and an autoradiographic film is placed in contact with the dried gel. After a suitable exposure time, the film is developed, giving a characteristic pattern of dark bands. ^{32}P-labeled nucleotides are not very suitable for this purpose: the high energy β-radiation causes considerable scattering of the signal, leading to diffuse bands. Instead, ^{35}S-takeled or ^{33}P-labeled nucleotides had been used.

Large-scale DNA sequencing efforts are dependent on improving efficiency by partial automation of the technologies involved. One major improvement in recent years has been the development of automated procedures for fluorescent DNA sequencing. These procedures generally use primers or dideoxynucleotides to which are attached fluorophores (Figure 10-15a). The use of different fluorophores in the four base-specific reactions means that, unlike conventional DNA sequencing, all four reactions can be loaded into a single lane. The newly synthesized DNA strands, each labeled with one of four dyes, are now sorted by length using capillary electrophoresis.

Figure 10-15 Automated DNA sequencing with fluorescently labeled dideoxynucleotides (see page 338)

This tube is not much thicker than a human hair and is $1\sim3$ feet long, sufficient to separate strands that differ in length by only one base. Because of the small dimensions involved, preparation of the capillary and loading of the sample are computer controlled. Shorter DNA strands migrate through the gel material more quickly, and come out the bottom of the capillary first, while longer strands become tangled in the gel material and take longer to emerge out the bottom.

As the strands emerge out the bottom of the capillary they pass through a laser beam that excites the fluorescent dye attached to the dideoxynucleotide at the end of each strand. This causes the dye to fluoresce, or glow, at a specific wavelength, or color. This color is then detected by a photocell, which feeds the information to the computer.

The computer displays the information received from the photocell as an electrophogram, which is a tracing of signal received by the photodetector in each of the four wavelengths (Figure 10-15b). Although the real colors seen by the photodetector are close to green, yellow, orange and red, the computer assigns four colors to each of the four tracings to make it easier to tell them apart. It also prints the letter of the appropriate base below each of the signal peaks. Because successive peaks correspond to DNA segments differing in length by one nucleotide, the sequence of peaks reveals the sequence of bases in the original DNA sample.

4.2 Protein Sequencing

Proteins play crucial roles in nearly all biological processes—in catalysis, signal transmission, and structural support. Although more complex than the sequencing of nucleic acids, protein molecules can also be sequenced: that is, the liner order of amino acids in a protein chain can be directly determined. The ability to determine a protein's sequence is very valuable for protein identification. A wealth of information about a protein's function and evolutionary history can often be obtained from the primary structure. Furthermore, because of the vast resource of complete or nearly complete genome sequences, the determination of even a small stretch of protein sequence is often sufficient to identify the gene which encodes that protein by finding a matching open-reading frame.

4.2.1 Principle

Protein is sequenced by automated Edman degradation. The first step is to determine the amino acid composition of the peptide. The peptide is hydrolyzed into its constituent amino acids by heating it in

6mol $\cdot$ L^{-1} HCl at 110°C for 24 hours. Amino acids in hydrolysates can be separated by ion-exchange chromatography on columns of sulfonated polystyrene. The identity of the amino acid is revealed by its elution volume, which is the volume of buffer used to remove the amino acid from the column, and quantified by reaction with ninhydrin. Amino acids treated with ninhydrin give an intense blue color, except for proline, which gives a yellow color because it contains a secondary amino group. The concentration of an amino acid in a solution, after heating with ninhydrin, is proportional to the optical absorbance of the solution. This technique can detect a microgram (10nmol) of an amino acid, which is about the amount present in a thumbprint. As little as a nanogram (10pmol) of an amino acid can be detected by replacing ninhydrin with fluorescamine, which reacts with the α-amino group to form a highly fluorescent product. A comparison of the chromatographic patterns of the sample hydrolysate with that of a standard mixture of amino acids would show the amino acid composition.

The next step is often to identify the N terminal amino acid by labeling it with a compound that forms a stable covalent bond. Fluorodinitrobenzene (FDNB) was first used for this purpose by Frederick Sanger. Dabsyl chloride is now commonly used because it forms fluorescent derivatives that can be detected with high sensitivity. It reacts with an uncharged α-NH2 group to form a sulfonamide derivative that is stable under conditions that hydrolyze peptide bonds. Hydrolysis of a sample dabsyl-peptide in 6mol $\cdot$ L^{-1} HCl would yield a dabsyl-amino acid, which could be identified as dabsyl-alanine by its chromatographic properties. Dabsyl chloride, too, is a valuable labeling reagent because it forms fluorescent sulfonamides.

Although the dabsyl method for determining the amino-terminal residue is sensitive and powerful, it cannot be used repeatedly on the same peptide, because the peptide is totally degraded in the acid-hydrolysis step and thus all sequence information is lost. Pehr Edman invented a method for labeling the amino-terminal residue and cleaving it from the peptide without disrupting the peptide bonds between the other amino acid residues. The Edman degradation sequentially removes one residue at a time from the amino end of a peptide (Figure 10-16). Phenyl isothiocyanate reacts with the uncharged terminal amino group of the peptide to form a phenylthiocarbamoyl derivative. Then, under mildly acidic conditions, a cyclic derivative of the terminal amino acid is liberated, which leaves an intact peptide shortened by one amino acid. The cyclic compound is a phenylthiohydantoin (PTH) -amino acid, which can be identified by chromatographic procedures. The Edman procedure can then be repeated on the shortened peptide, yielding another PTH-amino acid, which can again be identified by chromatography. More rounds of the Edman degradation will reveal the complete sequence of the original peptide.

Figure 10-16　The Edman degradation (see page 340)

The development of automated sequencers has markedly decreased the time required to determine protein sequences. One cycle of the Edman degradation—the cleavage of an amino acid from a peptide and its identification—is carried out in less than 1 hour. By repeated degradations, the amino acid sequence of some 50 residues in a protein can be determined. High-pressure liquid chromatography (HPLC) provides a sensitive means of distinguishing the various amino acids (Figure 10-17). Gas-phase sequenators can analyze picomole quantities of peptides and proteins. This high sensitivity makes it feasible to analyze the sequence of a protein sample eluted from a single band of an SDS-polyacrylamide gel.

Figure 10-17　Separation of PTH-Amino Acids (see page 340)

4.2.2　Advances in Protein Sequencing

One method that revolutionized protein sequencing is tandem mass spectrometry (MS/MS). Mass spectrometry is a method in which the mass of very small samples of a material can be determined with great accuracy. Very briefly, the principle is that material travels through the instrument (in a vacuum) in a manner that is sensitive to its mass/charge ratio. For small biological macromolecules such as peptides and small proteins, the mass of a molecule can be determined with the accuracy of single Dalton.

Protein samples are generally first exposed to the enzyme trypsin, which digests the proteins into their component peptides. These peptides are then separated by HPLC and analyzed by MS/MS. In MS/MS, compounds are sent through successive mass spectrometers in order to obtain detailed information about their identity and structure. This involves isolating compounds with specific mass-to-charge ratios in the first mass spectrometer, and then breaking these compounds apart and analyzing the fragments in a second mass spectrometer. The identity and structure of the original compounds can then be worked out from the mass spectra produced by both the original compounds and the fragment ions. This is all fairly straight-forward as long as the charge states of the peptide ions (the number of positive or negative charges added to the peptides during the ionisation process) are known. This is because if the charge state is known then the mass of the peptides can be calculated from their mass-to-charge ratios. The identity and structure of the peptide can then be worked out by comparing its mass and the mass spectra produced by its fragment ions with protein and peptide databases. This ability means that MS/MS, in conjunction with high performance liquid chromatography (HPLC), has become a cornerstone of proteomic studies.

4.3 ESTs (Expressed Sequence Tags)

At a time when the genomes of many species have been sequenced completely, a fundamental resource expected by many researchers is a simple list of all of an organism's genes. A gene list, together with associated physical reagents and electronic information, allows one to begin to investigate the ways in which many genes interact in the complex system of the organism. However, many species of medical and agricultural importance have not yet been prioritized for genomic sequencing, and expressed cDNAs have provided the primary source of gene sequences. Furthermore, when the genomic sequence of an organism becomes available, a collection of cDNA sequences provides the best tool for identifying genes within the DNA sequence. Thus, we can anticipate that the sequencing of transcribed products will remain a significant area of interest well into the future.

4.3.1 Principle

The era of high-throughput cDNA sequencing was initiated in 1991 by a landmark study from Venter and his colleagues. The basic strategy involves selecting cDNA clones at random and performing a single sequencing read from one or both ends of their inserts (Figure 10-18). They introduced the term EST to refer to this new class of sequence, which is characterized by being short (typically about 400~600 bases) and relatively inaccurate (around 2% error). The idea is to sequence bits of DNA that represent genes expressed in certain cells, tissues, or organs from different organisms and use these "tags" to fish a gene out of a portion of chromosomal DNA by matching base pairs. The challenge associated with identifying genes from genomic sequences varies among organisms and is dependent upon genome size as well as the presence or absence of introns, the intervening DNA sequences interrupting the protein coding sequence of a gene.

Figure 10-18 Summary of cDNA cloning and expressed sequence tag (EST) sequencing (see page 342)

ESTs accelerate gene discovery, complement genome annotation, aid gene structure identification, establish the viability of alternative transcripts, guide single nucleotide polymorphism (SNP) characterization and facilitate proteome analysis.

4.3.2 EST Applications

ESTs are versatile and have multiple uses. ESTs were first used to construct expression maps of the human genome, then to assess the gene coverage and to develop and map gene-based site markers. With the exponential rise in genomic data in the global sequencing projects, databases of ESTs are widely used in gene structure prediction, to investigate alternative splicing, to discriminate between genes exhibiting tissue or disease specific expression and for the discovery and characterization of candidate SNPs. The usefulness of EST data has extended well beyond its original application in gene finding and in transcriptome analysis. For example, ESTs as genome landmarks, just as a person driving a car may need a map to find a destination, scientists searching for genes also need genome maps to help them to navigate through the billions of nucleotides that make up the human genome. To make a map navigational sense, it must include reliable landmarks or "markers". Currently, the most powerful mapping technique, and one that has been used to generate many genome maps, relies on sequence tagged site (STS) mapping.

An STS is a short DNA sequence that is easily recognizable and occurs only once in a genome (or chromosome). The 3′ ESTs serve as a common source of STSs because of their likelihood of being unique to a particular species and provide the additional feature of pointing directly to an expressed gene. Further, ESTs as gene discovery resources, Because ESTs represent a copy of just the interesting part of a genome, that which is expressed, they have proven themselves again and again as powerful tools in hunt for genes involved in hereditary diseases. ESTs also have a number of practical advantages in that their sequences can be generated rapidly and inexpensively, only one sequencing experiment is needed per each cDNA generated, and they do not have to be checked for sequencing errors because mistakes do not prevent identification of the gene.

Section 5   Polymerase Chain Reaction

Of all technical advances in modern molecular biology, the polymerase chain reaction (PCR) is one of the most useful. The PCR provides a means of amplifying DNA sequence. Starting with incredibly tiny amount of any particular DNA molecule, the PCR can be used to generate microgram quantities of DNA. PCR is sufficiently sensitive that it can amplify the DNA from a single cell into amounts sufficient for cloning or sequencing. Consequently, PCR is used in clinical diagnosis, genetic analysis, genetic engineering and forensic analysis. In particular, PCR has revolutionized and speeded up the whole area of recombinant DNA technology. Previously, cloned DNA was made by growing up bacterial cultures and extracting and purifying the DNA. PCR allows the rapid generation of large amounts of specific DNA sequences that are easier to purify and less damaged.

5.1   Principle

PCR is a rapid and versatile *in vitro* method for amplifying defined target DNA sequences present within a source of DNA. Usually, the method is designed to permit selective amplification of a specific target DNA sequence (s) within a heterogeneous collection of DNA sequences (e. g. total genomic DNA or a complex cDNA population). To permit such selective amplification, DNA sequence information from the target sequences is required. This information is used to design two oligonucleotide primers (amplimers) which are specific for the target sequence and often about 15~25 nucleotides long. After primers are added to denatured template DNA, they bind specifically to complementary DNA sequences at the target site. In the presence of a suitably thermostable DNA polymerase and DNA precursors (the four deoxynucleoside triphosphates, dATP, dCTP, dGTP and dTTP), they initiate the synthesis of new DNA strands that are complementary to the individual DNA strands of the target DNA segment, and overlap each other (Figure 10-19).

Figure 10-19   Polymerase chain reaction (see page 344)

The PCR is a chain reaction because newly synthesized DNA strands will act as templates for further DNA synthesis in subsequent cycles. After about 25 cycles of DNA synthesis, the products of the PCR will include, in addition to the starting DNA, about 2^{25} copies of the specific target sequence, an amount which is easily visualized as a discrete band of a specific size when submitted to agarose gel electrophoresis. A thermostable DNA polymerase is used because the reaction involves sequential cycles composed of three steps: ① Denaturation, typically at about 93~95℃ for human genomic DNA. ② Re-annealing at temperatures usually from about 50~70℃ depending on the Tm of the expected duplex (the annealing temperature is typically about 5℃ below the calculated Tm). ③ DNA synthesis, typically at about 70~75℃.

When Kary Mullis invented PCR in 1987, he used normal DNA polymerase. Since the temperature needed to separate DNA into single strands destroys this enzyme, he had to add a fresh dose of polymerase to each tube every cycle. Luckily, suitably thermostable DNA polymerases have been obtained from microorganisms whose natural habitat is hot springs. For example, the widely used Taq DNA polymerase is obtained from *Thermus aquaticus* and is thermostable up to 94℃, with an optimum working temperature of 80℃.

5.2　Reverse-Transcription PCR

Reverse transcription PCR (RT-PCR) is a method used to amplify cDNA copied of RNA. Sensitive and versatile, RT-PCR is used to retrieve and clone the 5′ and 3′ termini of mRNAs and to generate large cDNA libraries from very small amounts of mRNA. In addition, RT-PCR can be easily adapted to identify mutations and polymorphisms in transcribed sequences and to measure the strength of gene expression when the amounts of available mRNA are limited and /or when the RNA of interest is expressed at very low levels.

The first step of RT-PCR is the enzymatic conversion of RNA to a single-stranded cDNA template. An oligodeoxynucleotide primer is hybridized to the mRNA and is then extended by an RNA-dependent DNA polymerase to create a cDNA copy that can be amplified by PCR. Depending on the purpose of the experiment, the primer for first-strand cDNA synthesis can be specifically designed to hybridize to a particular target gene or it can bind generally to all mRNAs (Figure 10-20).

Figure 10-20　Schematic representation of various methods for amplification of RNA by PCR（see page 345）

A second primer and a thermostable DNA polymerase are then added for the subsequent PCR-driven amplification step. Positive and negative controls should always be included when setting up RT-PCRs.

Wherever possible, oligonucleotides that bind to sequences located in different exons of the target RNA should be used as sense and antisense primers for amplification of cDNA product. In this way, amplification products derived from cDNA and contaminating genomic DNA can be easily distinguished. However, transcripts of intronless genes cannot be differentiated unambiguously from contaminating genomic sequences. In these circumstances, treating the RNA preparation with RNase-free DNase may be helpful.

5.3　Quantitative PCR

With the development of thermal cyclers incorporating fluorescent detection, PCR has a new, innovative application. In routine PCR, the critical result is the final quantity of amplicon generated after the process. In most circumstances, what we are interested in is the initial amount of the templates before PCR amplification. Real-time or Quantitative PCR (Q-PCR) and RT-PCR use the linearity of DNA amplification to determine absolute or relative amounts of a known sequence in a sample. By using a fluorescent reporter in the reaction, it is possible to measure DNA generation.

In quantitative PCR, DNA amplification is monitored at each cycle of PCR. The amount of fluorescence increases above the background, when the DNA is in the log linear phase of amplification. The point at which the fluorescence becomes measurable is called cycle threshold (CT) or crossing point, C represents cycle, T represents threshold. The cycle at which the fluorescence from a sample crosses the threshold is called the cycle threshold. By using multiple dilutions of a known amount of standard DNA, a standard curve can be generated of log concentration against CT. The amount of DNA or cDNA in an unknown sample can then be calculated from its CT value.

Real time PCR also lends itself to relative studies. A reaction may be performed using primers unique to each region to be amplified and tagged with different fluorescent dyes. Several commercially available quantitative thermal cyclers include multiple detection channels. In this multiplex system, the amount of target DNA/cDNA can be compared to the amount of a housekeeping sequence (e.g. GAPDH or β-actin).

Two types of detection chemistries are used for quantitative PCR. The first uses an intercalating dye that incorporates into double-stranded DNA. Of these fluorescent dyes, SYBR ® Green I dye is the most common one used (Figure 10-21a). This detection method is suitable when a single amplicon is being studied, as the dye will intercalate into any double-stranded DNA generated.

Figure 10-21　Real-time PCR chemistries（see page 346）

The second detection method uses a primer or oligonucleotide specific to the target of interest, as in TaqMan ® probes (Figure 10-21b), Molecular Beacons ™, or Scorpion primers. The oligonucleotide is labeled with a fluorescent dye and quencher. The oligonucleotide itself has no significant fluorescence, but fluoresces either when annealed to the template (as in molecular beacons) or when the dye is clipped from the oligo during extension (as in TaqMan probes). Multiplex PCR is possible by using dyes with different fluorescent emissions for each primer.

5.4 PCR Ramifications

5.4.1 Randomly Amplified Polymorphic DNA (RAPD)

Randomly Amplified Polymorphic DNA, or RAPD, is usually found in the plural as RAPDs and is pronounced "rapids", partly because it is a quick way to get a lot of information about the genes of an organism under investigation. The purpose of RAPDs is to test how closely related two organisms are.

The principle of RAPD is, short oligonucleotides of random sequence are, complementary, just by chance, to numerous sequences within the genome. If two complementary sequences are present on opposite strands of a genomic region in the correct orientation and within a close enough distance from each other, the DNA between them can become amplified by PCR. Each amplified fragment will be independent of all others and, by chance, will likely be of different length as well; if few enough bands are amplified, all will be resolvable from each other by gel electrophoresis. Different oligonucleotides will amplify completely different sets of loci.

In practice, the length of the primers is chosen to give five to ten PCR products. For higher organisms, primers of around 10 bases are typical. The bands from PCR are separated by gel electrophoresis to measure their sizes. The procedure is repeated several times with primers of different sequence. The result is a diagnostic pattern of bands that will vary in different organisms, depending on how closely they are related. Although we do not know in which particular genes the PCR bands originated, this does not matter in measuring relatedness. Diagnosis therefore relies on having a primer (or set of primers) that reliably give a band of a particular size with the target organism and give different bands with other organism, even those closely related.

5.4.2 Nested PCR

Nested PCR is a variation of the polymerase chain reaction (PCR), in that two pairs (instead of one pair) of PCR primers are used to amplify a fragment. The first pair of PCR primers amplifies a fragment similar to a standard PCR. However, a second pair of primers called nested primers (as they are nested within the first fragment) bind inside the first PCR product fragment to allow amplification of a second PCR product which is shorter than the first one (Figure 10-22). The advantage of nested PCR is that if the wrong PCR fragment was amplified, the probability is quite low that the region would be amplified a second time by the second set of primers. Thus, Nested PCR is a very specific PCR amplification.

Figure 10-22 Schematic representation of nested PCR (see page 348)

5.4.3 Anchored PCR

It is often desirable to be able to amplify previously uncharacterized DNA sequences that neighbor a known DNA sequence, either at the genomic or cDNA level. To do this a form of anchored PCR is used (Figure 10-23).

Figure 10-23 Genome walking by anchored PCR (see page 349)

One of the primers is specific for the target sequence and the second primer is specific for a common sequence that can be introduced in different ways, such as by using a linker-primer method, or by using primers that are modified at the 5' end so as to introduce a novel sequence.

Section 6   Gene Transfer

Gene transfer is a technique that unrelated genetic information in the form of DNA is inserted into cells. There are different reasons to do gene transfer. Perhaps foremost among these reasons is the treatment of diseases using gene transfer to supply patients with therapeutic genes. There are also different ways to transfer genes. Some of these methods involve the use of a vector such as a virus that has been specifically modified so it can take the gene along with it when it enters the cell. It is a very important technology in gene manipulation.

6.1 Gene Transfer to Animal Cells

Gene transfer to animal cells has been practiced now for over 40 years. Techniques are available for the introduction of DNA into many different cell types in culture, either to study gene function and regulation or to produce large amounts of recombinant protein. Animal cells are advantageous for the production of recombinant animal proteins because they perform authentic post-translational modifications that are not carried out by bacterial cells and fungi. Cell cultures have therefore been used on a commercial scale to synthesize products such as antibodies, hormones, growth factors, cytokines and viral coat proteins for immunization. There has been intense research into the development of efficient vector systems and transformation methods for animal cells. Although this research has focused mainly on the use of mammalian cell lines, other systems have also become popular, such as the baculovirus expression system, which is used in insects. More recently, research has focused on the introduction of DNA into animal cells *in vivo*. The most important application of this technology is *in vivo* gene therapy, i. e. the introduction of DNA into the cells of live animals in order to treat disease. Viral gene-delivery vectors are favored for therapeutic applications because of their efficiency, but safety concerns have prompted research into alternative DNA-mediated transfer procedures. This section focuses on the introduction of DNA into somatic cells. Unlike the situation in plants, most animal cells are restricted in terms of their developmental potential and cannot be used to generate transgenic animals. Mouse embryonic stem cells (ES cells) are exceptional in this respect because they are derived from the early embryo prior to the formation of the germ line. Gene transfer to ES cells is therefore discussed in the next, along with the introduction of DNA into oocytes, eggs and cells of the early embryo. Following are some important gene-transfer techniques to animal cells.

6.1.1 DNA/Calcium Phosphate Coprecipitate Method

The ability of mammalian cells to take up exogenously supplied DNA from their culture medium was first reported by Szybalska and Szybalski (1962). They used total uncloned genomic DNA to transfect human cells deficient for the enzyme hypoxanthineguanine phosphoribosyl transferase (HPRT). Rare HPRT-positive cells, which had presumably taken up fragments of DNA containing the functional gene, were identified by selection on HAT medium. At that time, the actual mechanism of DNA uptake was not understood. Much later, it was appreciated that successful DNA transfer in such experiments was dependent on the formation of a fine DNA/calcium phosphate coprecipitate, which first settles on to the cells and is then internalized. The precipitate must be formed freshly at the time of transfection. It is thought that small granules of calcium phosphate associated with DNA are taken up by endocytosis and transported to the nucleus, where some DNA escapes and can be expressed. The technique became generally accepted after its application, by Graham and Van der Erb (1973), to the analysis of the infectivity of adenoviral DNA. It is now established as a general method for the introduction of DNA into a wide range of cell types in culture. However, since the precipitate must coat the cells, this method is suitable only for cells growing in monolayers, not those growing in suspension or as clumps. As originally described, calcium phosphate transfection was limited by the variable and rather low proportion of cells that took up DNA (1%~2%). Only a small number of these would be stably transformed. Improvements to the method have increased the transfection frequency to 20% for some cell lines. A variant of the technique, using a different buffer system, allows precipitate to form slowly over a number of hours, and this can increase stable transformation efficiency by up to 100-fold when using high-quality plasmid DNA.

6.1.2 Diethylaminoethyl Dextran Transfection Method

The calcium phosphate method is applicable to many cell types, but some cell lines are adversely affected by the coprecipitate due to its toxicity and are hence difficult to transfect. Alternative chemical transfection methods have been developed to address this problem. One such method utilizes diethylaminoethyl dextran (DEAE-dextran), a soluble polycationic carbohydrate that promotes interactions between DNA and the endocytotic machinery of the cell. This technique was initially devised to introduce viral DNA into cells, but was later adapted as a method for plasmid DNA transfer. The efficiency of the original procedure was improved by Lopata et al. (1984) and Sussman and Milman (1984) by adding after-treatments, such as exposure to chloroquine, to inhibit the acidification of endosomal vesicles. Although efficient for the transient transfection of many cell types, DEAE-dextran cannot be used to generate stably transformed cell lines.

6.1.3　Phospholipids as Gene-Delivery Vehicles

An alternative to chemical transfection procedures is to package DNA inside a phosopholipid vesicle, which interacts with the target cell membrane and facilitates DNA uptake. The first example of this approach was provided by Schaffner (1980), who used bacterial protoplasts containing plasmids to transfer DNA into mammalian cells. Briefly, bacterial cells were transformed with a suitable plasmid vector and then treated with chloramphenicol to amplify the plasmid copy number. Lysozyme was used to remove the cell walls, and the resulting protoplasts were gently centrifuged on to a monolayer of mammalian cells and induced to fuse with them, using polyethylene glycol. A similar strategy was employed by Wiberg et al. (1987), who used the haemoglobin-free ghosts of erythrocytes as delivery vehicles. The procedures are very efficient in terms of the number of transformants obtained, but they are also labour-intensive and have not been widely adopted as a general transfection method. However, an important advantage is that they are gentle, allowing the transfer of large DNA fragments without shearing. Yeast cells with the cell wall removed (sphaeroplasts) have therefore been used to introduce yeast artificial chromosome (YAC) DNA into mouse ES cells by this method, for the production of YAC transgenic mice.

6.1.4　Liposomes as Gene-Delivery Vehicles

Liposomes, artificial phospholipid, have been more widely used. Initial liposome-based procedures were hampered by the difficulty encountered in encapsulating the DNA, and provided a transfection efficiency no better than the calcium phosphate method. However, a breakthrough came with the discovery that cationic/neutral lipid mixtures can spontaneously form stable complexes with DNA that interact productively with the cell membrane, resulting in DNA uptake by endocytosis. This low-toxicity transfection method, commonly known as lipofection, is one of the simplest to perform and is applicable to many cell types that are difficult to transfect by other means, including cells growing in suspension. This technique is suitable for transient and stable transformation, and is sufficiently gentle to be used with YACs and other large DNA fragments. The efficiency is also much higher than that of chemical transfection methods—up to 90% of cells in a culture dish can be transfected. A large number of different lipid mixtures are available, varying in efficiency for different cell lines. A unique benefit of liposome gene-delivery vehicles is their ability to transform cells in live animals. Transfection efficiency has been improved and targeting to specific cell types has been achieved by combining liposomes with viral proteins that promote cell fusion, nuclear targeting signals and various molecular conjugates that recognize specific cell surface molecules. The development of liposomes for gene therapy has been comprehensively reviewed.

6.1.5　Electroporation

Electroporation involves the generation of transient, nanometre-sized pores in the cell membrane, by exposing cells to a brief pulse of electricity. DNA enters the cell through these pores and is transported to the nucleus. This technique was first applied to animal cells by Wong and Neumann (1982), who successfully introduced plasmid DNA into mouse fibroblasts. The electroporation technique has been adapted to many other cell types. The most critical parameters are the intensity and duration of the electric pulse, and these must be determined empirically for different cell types. However, once optimal electroporation parameters have been established, the method is simple to carry out and highly reproducible. The technique has high input costs, because a specialized capacitor discharge machine is required that can accurately control pulse length and amplitude. Additionally, larger numbers of cells may be required than for other methods because, in many cases, the most efficient electroporation occurs when there is up to 50% cell death. In an alternative method, pores are created using a finely focused laser beam. Although very efficient (up to 0.5% stable transformation), this technique is applicable to only small numbers of cells and has not gained widespread use.

6.1.6 Direct Transfer Methods

A final group of methods considered in this section encompasses those in which the DNA is transferred directly into the cell nucleus. One such procedure is microinjection, a technique that is guaranteed to generate successful hits on target cells but that can only be applied to a few cells in any one experiment. This technique has been applied to cultured cells that are recalcitrant to other transfection methods, but its principal use is to introduce DNA and other molecules into large cells, such as oocytes, eggs and the cells of early embryos. Particle bombardment is another direct delivery method, initially developed for the transformation of plants. This involves coating small metal particles with DNA and then accelerating them into target tissues using a powerful force, such as a blast of high-pressure gas or an electric discharge through a water droplet. In animals, this technique is most often used to transfect multiple cells in tissue slices rather than cultured cells. It has also been used to transfer DNA into skin cells *in vivo*.

6.2 Transgenic Animals

Gene-transfer techniques can be used to produce transgenic animals, in which every cell carries new genetic information, as well as designer mutants with specific preselected modifications to the genome. The whole animal is the ultimate assay system, which is used to investigate gene function, particularly for complex biological processes, such as development. In most animals, the somatic cells and germ cells (the cells that give rise to gametes) separate at an early developmental stage. Therefore, the only way to achieve germ-line transformation in animals is to introduce DNA into totipotent cells prior to the developmental stage at which the germ line forms. The ability to introduce DNA into the germ line of mice is one of the greatest achievements of the twentieth century and has paved the way for the transformation of other mammals. Genetically modified mammals have been used not only to study gene function and regulation, but also as bioreactors producing valuable recombinant proteins, e.g. in their milk. Several methods for germ-line transformation have been developed, all of which require the removal of fertilized eggs or early embryos from donor mothers, brief culture *in vitro* and then their return to foster-mothers, where development continues to term. These methods are discussed below and summarized in Figure 10-24.

Figure 10-24　DNA transfer in mice（see page 352）

In the previous, we discussed the early development of transgenic technology, which focused on the evolution of reliable techniques for gene transfer and transgene expression in animals and plants. The standard use of this technology was to add genetic information to the genome, generating a gain of function in the resulting cells and transgenic organisms. Now that such gene-transfer processes are routine, the focus of transgenic technology has shifted to the provision of more control over the behaviour of transgenes. New techniques have evolved in parallel in animals and plants. Advances have come about through the development of inducible expression systems that facilitate external regulation of transgenes and the exploitation of site-specific recombination systems to make precise modifications in target genomes. In mice, the combination of gene targeting, site-specific recombination and inducible transgene expression makes it possible to selectively switch on and off both transgenes and endogenous genes in a conditional manner. Other routes to gene silencing have also been explored, such as the expression of antisense RNA and the recently described phenomenon of RNA interference. Transgenic animals and plants are also being used increasingly as tools for the analysis of genomes. The development of tagging systems for the rapid cloning of genes disrupted by insertional mutagenesis and the use of entrapment vectors allow the high-throughput analysis of gene expression and function.

Section 7   Gene Knock-Out

Gene knock-out is a technique using certain approach to make a specific gene of a organism deactivated or deleted. In the usual sense of gene knock-out, the introduction of a null mutation in a gene by a designed alteration in a cloned DNA sequence that is then introduced into the genome by homologous recombination and replacement of the normal allele. Gene knock-out sometimes be called gene targeting, which is quite simple, a willful modification of genetic information in a living organism.

7. 1   Principle

The technique of gene knock-out takes advantage of the phenomenon known as homologous recombination. When an investigator wants to replace one allele with an engineered construct but not affect any other locus in the genome, then the method of choice is homologous recombination. To perform homologous recombination, you must know the DNA sequence of the gene you want to replace (Figure 10-25a). With this information, it is possible to replace any gene with a DNA construct of your choosing.

Figure 10-25   The principle of gene knock-out (see page 353)

The next step is to design and fabricate the DNA construct you want to insert into the chromosome in place of the wild-type allele. This construct may contain any DNA sequence of your choosing, this means you can insert different alleles (both functional and non-functional ones), different genes or reporter genes (e. g. antibiotic resistance or green fluorescent protein). Regardless of what you want to insert, you must include some flanking DNA that is identical in sequence to the targeted locus (Figure 10-25b).

The engineered construct is added to cells that contain the targeted gene of interest. By mechanisms that are poorly understood but are similar to what occur during meiosis and mitosis when homolgous chromosomes align along the metaphase plane, the engineered construct finds the targeted gene and recombination takes place within the homolgous sequences (Figure 10-25c). The recombination may take place anywhere within the flanking DNA sequences and the exact location is determined by the cells and not the investigators.

Once the cells have performed their part of the procedure, the end result is a new piece of DNA inserted into the chromosome. The rest of the genome is unaltered but the single targeted locus has been replaced with the engineered construct and some of its flanking DNA (Figure 10-25d). The original engineered construct has taken up the targeted gene of interest but since it cannot replicate in a nucleus, it is lost quickly in dividing cells, in while the modified chromosome replicates faithfully, including its new insert.

7. 2   Targeting Vector

Targeting vectors can be classified as either replacement or insertion vectors (Figure 10-26). A replacement vector (Figure 10-26a) is linearized in such a way that the vector sequences remain colinear with the target sequences. Chromosomal sequences are replaced by vector sequences by a double crossover event involving the flanking homologous regions. An insertion vector (Figure 10-26b) is linearized within the region of homology and homologous recombination will lead to a duplication of genomic sequences. The vast majority of null mutants generated to date have employed replacement type vectors.

Figure 10-26   Classification of targeting vectors (see page 354)

7. 3   Selection Strategies

Homologous recombination is a rare event in mammalian cells, and thus a powerful selection strategy is required to detect those cells in which it has occurred. Most commonly, the introduced gene construct has its sequence disrupted by an inserted antibiotic resistance gene such as that for neomycin resistance. If this construct undergoes homologous recombination with the endogenous copy of the gene, the endogenous gene is disrupted but the antibiotic resistance gene remains functional, allowing cells that have incorporated the gene to be selected in culture for resistance to the neomycin. However, antibiotic resistance on its own shows only that the cells have taken up and integrated the neomycin-resistance gene. To be able to select for those cells in which homologous recombination has occurred, the ends of the construct usually carry the thymidine kinase gene from the herpes simplex virus (HSV-tk) (Figure 10-27). Cells that incorporate DNA randomly usually retain the entire DNA construct including HSV-tk, whereas homologous recombination between the construct and cellular DNA, the desired result, involves the exchange of homologous DNA sequences so that the nonhomologous HSV-tk genes at the ends of the construct are eliminated. Cells carrying HSV-tk are killed by the antiviral drug ganciclovir, and so cells with homologous recombination have the unique feature of being resistant to both neomycin and ganciclovir, allowing them to be selected efficiently when these drugs are added to the cultures.

Figure 10-27 selection strategies (see page 355)

7.4 Conditional Gene Targeting

Conventional gene targeting leads to inactivation or modification of a gene in all tissues of the body from the onset of development throughout the whole lifespan. More recently, methods have been developed aimed at controlling gene targeting in a time-dependent and tissue-dependent manner. These so called conditional gene targeting approaches are particularly useful in cases where complete gene inactivation leads to a lethal or otherwise adverse phenotype that prevents a more detailed analysis. Moreover, if a given gene has a widespread pattern of expression, tissue-specific gene inactivation may define physiological roles of the gene product in a certain tissue, without compromising other functions in the organism. Control of gene targeting in a time dependent manner allows the differentiation between effects of chronic versus acute depletion of a protein and also the analysis of functions at different time points in development. Many knock-out mice revealed an unexpectedly minor phenotype that was attributed either to gene redundancy, or to adaptive mechanisms mediating developmental plasticity. Gene inactivation at a specific time point in the adult, leaving gene function intact throughout ontogeny, should prevent these adaptive responses and therefore phenotypes are expected to be more severe in conditional, as opposed to conventional, knock-out mice.

7.4.1 Tissue-Specific Knock-out

Tissue-specific gene inactivation may be achieved by first generating transgenic mice that express Cre recombinase under the control of a tissue-specific promoter. These mice may then be crossed to target mice that contain *loxP* sites flanking the genomic region to be deleted or modified. Double transgenic mice will carry a modified gene copy, due to Cre-mediated recombination, in all cells in which Cre was sufficiently expressed. Target mice harbouring appropriately engineered *loxP* sites can be obtained by homologous recombination in ES cells (Figure 10-28), whereas Cre-transgenic mice are either generated by conventional oocyte injection, or alternatively, by "knock-in" gene targeting behind a suitable promoter exhibiting the desired expression pattern. The expression level of Cre recombinase will determine the efficiency of gene modification, while temporo-spatial control is mainly dependent on the type of promoter driving Cre expression (Figure 10-29a).

Figure 10-28 Generation of target mice for conditional knock-out strategies (see page 356)

Figure 10-29 Transcriptional regulation of Cre activity (see page 357)

7.4.2 Inducible Gene Knock-out

For many applications it would be desirable to control the function of a gene product in a time-dependent manner, regulated for example by the administration of an inducer. One way to achieve this kind of inducible gene silencing is, in analogy to tissue-specific gene targeting, to use a Cre recombinase based system that may be controlled either at the transcriptional or at the post-transcriptional level. Ideally, any inducible system should fulfil the following requirements: ① no (or only minor) background activity in the absence of the inducer; ② fast and efficient response in all tissue; and ③ no toxic, or otherwise adverse, effects should be caused by the inducer.

Several approaches have been pursued to put Cre expression under the control of inducible promoters, which belong to the transcriptional regulation of Cre activity (Figure 10-29). In the classical tet system (Figure 10-29b) the responder gene is expressed in the absence of the ligand and silenced upon introduction of the ligand. In the reverse tet system (Figure 10-29c), a modified transactivator (rTA) is used that binds to the operator/promoter only in the presence of ligand. Therefore, responder gene expression is normally silent, but can be switched on by ligand administration. Both tTA- and rTA-regulated systems have been shown to control the expression of reporter genes in transgenic mice.

A knock-out mouse has had both alleles of a particular gene replaced with an inactive allele. This is usually accomplished by using homologous recombination to replace one allele followed by two or more generations of selective breeding until a breeding pair are isolated that have both alleles of the targeted gene inactivated or knocked out. Knock-out mice allow investigators determine the role of a particular gene by observing the phenotype of individuals that lack the gene completely.

There are several variations to the procedure of producing knock-out mice; the following is a typical example (Figure 10-30).

Figure 10-30　The procedure of producing knock-out mice (see page 358)

7.5　The Application and the Future of Gene Knock-Out

Gene knock-out has revolutionized the field of mouse genetics and allowed the analysis of diverse aspects of gene function *in vivo*. It is now possible to engineer specific genetic alterations ranging from subtle mutations to chromosomal rearrangements and more recently, even tissue-specific gene knock-out with temporo-spatial control has become feasible. With continuous efforts from many laboratories these new approaches are expected to be further optimized in the near future, with the ultimate goal of reversibly switching on and off any gene in a given tissue at a chosen time-point. Although gene targeting methods have been refined, and the efficiency of homologous recombination has been optimized in murine ES cells, targeting of most somatic cells is still much less efficient. There is, however, considerable interest to improve the genetic manipulation of somatic cells, especially with regard to applications in somatic gene therapy. Moreover, work directs at establishing gene targeting in species other than mice such as rats, *Drosophila* and in particular in commercial livestock, may lead to interesting new applications.

Section 8　Assaying DNA-Protein Interactions

One of the recurring themes of molecular biology is the study of DNA-protein interactions. We have already discussed RNA polymerase-promoter interactions, and we will encounter many more examples. Therefore, we need methods to quantify these interactions and to determine exactly what part of the DNA interacts with a given protein. We will consider here two methods for detecting protein-DNA binding and three examples of method to show DNA bases interact with a protein, they are filter binding, gel mobility shift, DNase footprinting, DMS footprinting and chromatin immunoprecipitation.

8.1　Filter Binding

Nitrocellulose membrane filters have been used for decades to filter-sterilize solutions. Part of the folklore of molecular biology is that someone discovered by accident that DNA can bind to such nitrocellulose filters because they lost their DNA preparation that way. Weather this story is true or not is not important. What is important is that nitrocellulose filter can indeed bind DNA, but only under certain conditions. Single-stranded DNA binds readily to nitrocellulose, but double-stranded by itself does not. On the other hand, protein does bind, and if a protein is bound to double-stranded DNA the DNA-protein complex will bind. This is the basis of the assay portrayed in Figure 10-31.

Figure 10-31 Nitrocellulose filter binding assay (see page 359)

In Figure 10-31a, labeled double-strand DNA is poured though a nitrocellulose filter. The amount of label in the filtrate (the material that passes though the filter) and in the filter-bound material is measured, which shows that all the labeled material has passed though the filter into the filtrate. This confirms that double-stranded DNA does not bind to nitrocellulose. In Figure 10-31b, a solution of a labeled protein is filtered, showing that all the protein is bound to the filter. This demonstrates that proteins bind by themselves to the filter. In Figure 10-31c, double-stranded DNA is again labeled, but this time it is mixed with a protein to which it binds. Because the protein binds to the filter, the protein-DNA complex will also bind, and the radioactivity is found bound to the filter, rather than in the filtrated. Thus, filter binding is a direct measure of DNA-protein interaction.

In summary, filter binding as a means of measuring DNA-protein interaction is based on the fact that double-stranded DNA will not bind by itself to a nitrocellulose filter or similar medium, but a protein-DNA complex will. Thus, one can label a double-stranded DNA, mix it with a protein, and assay protein-DNA binding by measuring the amount of label retained by the filter.

8. 2 Gel Mobility Shift

Another method for detecting DNA-protein interaction relies on the fact that a small DNA has a much higher mobility in gel electrophoresis than the same DNA does when it bound to a protein. Thus, one can label a short, double-stranded DNA fragment, then mix it with a protein, and electrophorese the complex. Then one subjects the gel to autoradiography to detect the labeled species. Figure 10-32 shows the electrophoretic mobilities of three different species. Lane 1 contains naked DNA, which has a very highmobility because of its small size. Recall from earlier chapter that DNA electropherograms are conventionally depicted with there origins at the top, so high - mobility DNAs are found near the bottom, as shown here. Lane 2 contains the same DNA bound to protein, and its mobility is greatly reduced. This is the origin of the name for this technique: gel mobility shift assay or electrophoretic mobility shift assay (EMSA). Lane 3 depicts the behavior of the same DNA bound to two proteins. The mobility is reduced still further because of the greater mass of protein clinging to the DNA. This is called a supper shift. The protein could be another DNA-binding protein, or a second protein that binds to the first. It can even be an antibody that specifically binds to the first protein.

Figure 10-32   Gel mobility shift assay (see page 360)

In summary, a gel mobility shift assay detects interaction between a protein and DNA by the reduction of the electrophoretic mobility of a small DNA that occurs on binding to a protein.

8. 3 DNase Footprinting

Footprinting is a method for detecting protein-DNA interactions that can tell where the target site lies on the DNA and even which bases are involved in protein binding. Several methods are available, but three are very popular: DNase footprinting, dimethylsulfate (DMS), and hydroxyl radical footprinting. DNase footprinting (Figure 10-33) relies on the fact that a protein, by binding to DNA, covers the binding site and so protects it from attack by DNase. In this sense, it leaves its "footprint" on the DNA. The first step in a footprinting experiment is to end-label the DNA. Either strand can be labeled, but only one strand per experiment. Next, the protein (yellow in the figure) is bound to the DNA. Then the DNA-protein complex is treated with DNase 1 under wild conditions (very little DNase), so that an average of only one occurs per DNA molecule. Next, the protein is removed from the DNA, the DNA strands are separated, and the resulting fragments are electrophoresed on a high-resolution polyacrylamide gel alongside size markers (not shown). Of course, fragments will arise from the other end of the DNA as well, but they will not be detected because they are unlabeled. A control with DNA alone (no protein) is always included, and more than one protein concentration is usually used so the gradual disappearance of the bands in the footprinting region reveals that protection of the DNA depends on the concentration of added protein. The footprint represent the region of DNA protected by the protein, and therefore tells where the protein binds.

Figure 10-33 DNase footprinting (see page 361)

In summary, footprinting is means of finding the target DNA sequence, or binding site, of a DNA-binding protein. DNase footprinting is performed by binding the protein to its end-labeled DNA target, then attacking the DNA-protein complex with DNase. When the resulting fragments are electrophpresed, the protein binding site shows up as a gap, or "footprint" in the pattern, in which the protein protects the DNA from degradation.

8.4   DMS Footprinting and Other Footprinting Methods

DNase footprinting gives a good idea of the location the binding site for the protein, but DNase is a macromolecule and is therefore a rather blunt instrument for probing the fine detail of the binding site. That is, gaps may occur in the interaction between protein and DNA that DNase would not fit into and therefore would not detect. Moreover, DNA-binding proteins frequently perturb the DNA within the binding region, distorting the double helix. These perturbations are interesting, but are not generally detected by DNase footprinting because the protein keeps the DNase away. More detailed footprinting requires a smaller molecule that can fit into the nooks and crannies of the DNA-protein complex and reveal more of the subtleties of the interactions. A favorite tool for this job is the methylating agent dimethyl-sulfate (DMS).

In summary, DMS footprinting follows as a similar principal, except that the DNA methylating agent DMS, instead of DNase, is used to attack the DNA-protein complex. The DNA is then broken at the methylated sites. Unmethylated (or hypermethylated) sites show up on electrophoresis of the labeled DNA fragments and demonstrate where the protein bound to the DNA. Hydroxyl radical footprinting uses copper-containing or iron-containing organometallic complexes to generate hydroxyl radicals that break DNA strands (Figure 10-34).

Figure 10-34 DMS Footprinting (see page 362)

In addition to DNase and DMS, other regents are commonly used to footprint protein-DNA complexes by breaking DNA except where it is protected by bound protein. For example, organometallic complexes containing copper or iron act by generating hydroxyl radical that attack and break DNA strands.

8.5   Chromatin Immunoprecipitation

This technique, often just called ChIP, enables an investigator to identify where a given protein is bound in the genome of a living cell. Thus, for example, it is possible to determine whether components of the transcriptional machinery are bound to a given promoter at a given time. It is also possible to determine whether a specific regulatory protein is bound at a given gene, and so on.

In outline, the technique is performed as follow: formaldehyde is added to cells, cross-linking to the DNA, all proteins are bound to DNA at that moment. The cells are then lysed and the DNA is broken into small fragments (200~300bp each). Using an antibody specific for the protein of interest, the fragments of DNA attached to that protein can be separated from the majority of the DNA in the cell. The cross-linking is then reversed and the protein removed. To determine whether a particular region of DNA is bound by the protein, PCR is performed using primers designed to amplify that particular region (a promoter, for example). If the protein had indeed been bound there, DNA will be present and get amplified. As a control, PCR primers targeting another region of DNA (one to which the protein is known not to bind) are used; in that case, no DNA should be amplified (Figure 10-35).

Figure 10-35 chromatin immunoprecipitation (see page 363)

Summary

a) The separation and purification techniques of DNA, RNA and protein such as electrophoresis and chromatography etc. are significant foundations of molecular biology research.

b) The detection techniques of DNA, RNA and protein such as molecular hybridization, PCR and sequencing etc. are important means of gene cloning and expression regulation research.

c) Transgenic technique, gene knock-out and DNA-protein interaction are important methods of gene function research.

(Yaping Wang　Qing Chen)

c. The deduced amino acid sequences of DNA, RNA, and protein, such as reproduction, production. It is important to put important areas of gene, binding site, pattern of translocation sequence.

e. Transgenic technique, gene knock out, and DNA protein interaction, indiscriminate method of gene function research.

Jun Chen

第十一章　生物信息学基本原理和应用

　　21 世纪是生命科学的世纪，其里程碑就是历时 13 年、耗资数十亿的人类基因组计划的完成。因为该计划将为最终揭示人体构造之谜奠定坚实的数据基础；人类基因组计划以及各种模式生物基因组计划是我们了解人类基因组功能的重要部分。所有这些工作都涉及大量数据的处理工作，而且数据量也正在以科学史上前所未有的速度增长着。这些情况表明，仅靠传统的研究手段是不够的，生物学已不再是仅仅基于实验观察的科学，理论和计算将发挥越来越巨大的作用，通过大型计算机设备对海量数据进行处理就成为了必要。在这样一个背景下，生物信息学（bioinformatics）作为一门新的学科领域应运而生。

　　生物信息学可以分为两个基本大类：生物信息管理学和计算生物学。前者指开发各种计算工具和方法，尽可能充分使用这些生物医学数据，包括获得、描述、存储、分析和可视化这些数据。后者指开发数据分析的理论方法以及进行数据建模，并利用计算机模拟技术进行生物学、行为学以及社会体系的研究，揭示这些复杂数据所包含的生物学奥秘。本章旨在介绍生物信息学的一般概念、原理和方法，以及在生物学中的基本应用，最后介绍一些常用的数据库。

第一节　生物信息学概述

一、生物信息学概念及诞生

　　什么是生物信息学？生物信息学是 20 世纪 80 年代末随着人类基因组计划的启动而兴起的一门新的交叉学科，最初常被称为基因组信息学。"bioinformatics"这一名词是从何而来的呢？林华安博士（Dr. Hwa A. Lim）认识到将计算机科学与生物学结合起来的重要意义，开始留意为这一领域构思一个合适的名称。起初，考虑到与将要支持他主办一系列生物信息学会议的佛罗里达州立大学超级计算机计算研究所的关系，他使用的是"CompBio"；不久，他便进一步把它更改为"bio-informatics（或 bio/informatics）"。但由于当时的电子邮件系统与今日不同，该名称中的"-"或"/"符号经常会引起许多系统问题，于是林博士将其去除，"bioinformatics"就正式启用，一门崭新的、如火如荼的、拥有巨大发展潜力的生物信息学悄然而坚定地诞生和发展起来了。

　　生物信息学就其诞生而言，是一门历史悠久的学科，因为早在计算机初创期的 1956 年就已经在美国田纳西州的 Gatlinburg 召开过首次"生物学中的信息理论讨论会"；而就其发展而言，却是一门相当年轻的学科，因为继 20 余年的沉默之后，伴随着八九十年代计算机技术的迅猛发展，它才得以大发展。它的诞生和发展是应时所需，是历史的必然，并快速渗透到生物科学的每一个角落，以至于人们在意识到它的存在之前就已经离不开它了！生物信息学已经成为生物医学、农学、遗传学、细胞生物学等学科发展的强大推动力量，也是药物设计、环境监测的重要组成部分。它既是当今生命科学和自然科学的重大前沿领域之一，也是 21 世纪自然科学的核心领域之一。

二、生物信息学发展与成熟

国际上一直非常重视生物信息学的发展，各种专业研究机构和公司如雨后春笋般涌现出来，生物科技公司和制药工业内部的生物信息学部门的数量也与日俱增。从数据库的角度来讲，早在 20 世纪 60 年代，美国就建立了手工搜集数据的蛋白质数据库。美国洛斯阿拉莫斯国家实验室 1979 年就已经建立起 GenBank 数据库，欧洲分子生物学实验室 1982 年就已经提供核酸序列数据库 EMBL 的服务，日本也于 1984 年着手建立国家级的核酸序列数据库 DDBJ，并于 1987 年开始提供服务。美国于 1988 年在国会的支持下成立了美国国家生物技术信息中心（NCBI），其目的是进行计算分子生物学的基础研究，构建和发布分子生物学数据库；欧洲于 1993 年 3 月就着手建立欧洲生物信息学研究所（EBI），日本也于 1995 年 4 月组建了自己的信息生物学中心（CIB）。目前，绝大部分的核酸和蛋白质数据库由美国、欧洲和日本的 3 家数据库系统产生；他们共同组成了 DDBJ/EMBL/GenBank 国际核酸序列数据库，每天交换数据，同步更新。其他一些国家，如德国、法国、意大利、瑞士、澳大利亚、丹麦和以色列等，在分享网络共享资源的同时，也分别建有自己的生物信息学机构、二级或更高级的具有各自特色的专业数据库以及自己的分析技术，服务于本国生物（医学）研究和开发，有些服务也对全世界开放。

我国对生物信息学领域也越来越重视，在一些著名院士和教授的带领下，在各自领域取得了一定成绩，有的在国际上还占有一席之地，但从全国总体上来看与国际水平差距很大。

生物信息学不仅仅是一门学科，它更是一种重要的研究开发工具。早在 1962 年，Zuckerkandl 和 Pauling 就将序列变异分析与其演化关系联系起来，从而开辟了分子演化的崭新研究领域；1964 年，Davies 开创了蛋白质结构预测的研究；1970 年，Needleman 和 Wunsch 发表了广受重视的两序列比较算法；1974 年，Ratner 首先运用理论方法对分子遗传调控系统进行处理分析；1975 年，Pipas 和 McMahon 首先提出运用计算机技术预测 RNA 二级结构；1976 年之后生物学数据分析技术大量涌现，1980 年 Science 第 209 卷就曾发表关于计算分子生物学的综述。正如我们现在所看到的那样，在 20 世纪八九十年代，生物学数据分析技术在国外更是获得了突飞猛进的发展。人类基因组计划及模式生物的基因组测序工作的完成表明，只有经过生物信息学手段的分析处理，我们才能获得对基因组的正确理解，因此可以说生物信息学兴盛于人类基因组计划，人类基因组计划为生物信息学创造了施展身手的巨大空间。

生物信息学发展初期，由于没有专业的期刊，起初的专业文献都散落在各种其他领域的期刊中，1970 年，出现了 *Computer Methods and Programs in Biomedicine* 这本相关期刊；1985 年 4 月，就有了第一种生物信息学专业期刊——*Computer Application in the Biosciences*；现在，我们可以看到的专业期刊已经很多了，包括书面期刊和网上期刊两种，如 *Bioinformatics*（原 *Computer Applications in the Biosciences*）、*Acta Biotheoretica*、*Bio Informatics Technology & Systems*、*Bioinform Newsletter*、*Briefings in Bioinformatics* 和 *Journal of Computational Biology* 等。

三、生物信息学的研究范围

生物信息学从广义来说是用数理和信息科学的观点、理论和方法去研究生命现象、组织和分析呈指数增长的生物学数据的一门学科。首先是研究遗传物质的载体 DNA 及其编码的大分子蛋白质，以计算机为其主要工具，发展各种软件，对日渐增长的浩如烟海的 DNA 和蛋白质的序列和结构进行收集、整理、存储、发布、提取、加工、分析和研究，目的在于通过这样的分析逐步认识生命的起源、进化、遗传和发育的本质，破译隐藏在 DNA 序列中的遗传语言，揭示人体生理和病理过程的分子基础，为人类疾病的预测、诊断、预防和治疗提供最合理和有效的方法或途径。其研究重点主要体现在基因组学和蛋白质组学两方面，具体说，是从核酸和蛋白质序列出发，分析序列中所蕴藏的结构与功能的生物信息。目前基因组学的研究出现了几个重心转移：①将已知基因的序列与功能联系在一起的功能基因组学研究。②从作图为基础的基因分离转向以序列为基础的基因分离。③从研究疾病的起因转向探索发病机制。④从疾病诊断转向疾病易感性研究。生物芯片的应用将为上述研究提供最基本和必要的信息及依据，已成为基因组信息学研究的主要技术支撑。生物信息学的发展为生命科学研究的进一步突破及药物研制过程革命性变革提供了契机。就人类基因组来说，得到序列仅仅是第一步，后一步的工作是所谓后基因组时代（post-genome era）的任务，即收集、整理、检索和分析序列中表达的蛋白质结构与功能的信息，找出规律。生物信息学将在其中扮演至关重要的角色。大致可分三类：①数据库的建立与优化。国际上著名的公共数据库有 GeneBank、SWISS-PORT、PIR、PDB，另外一些公司还有内部数据库。②培养生物信息学专业人员。③数据库的理论研究、软件的研制、序列的排列比较（alignment）、对新序列的识别与预测等。

第二节　生物信息学在生物医学中的应用

一、新基因的发现及功能预测

发现新基因是当今基因组研究的重中之重。而使用生物信息学的方法更是发现新基因的重要手段。新基因的发现不仅对于人类认识世界，了解其他物种有着深远的影响，也可以为治疗某些疑难杂症带来新的契机。由于发现新基因过程中会产生大量的生物信息数据，在这种情况下，就要求建立庞大数据信息的数据库以能够支持快速、大批量数据的查询、更新与分析，以此来阐明和理解这些数据所包含的生物学意义。

1. 从基因组序列预测新基因

这种方法实质上是把基因组中编码蛋白质和非编码蛋白质的区域区分开，将这些序列与已知基因数据库进行比较，进而发现新的基因。

2. 通过多序列比对从基因组 DNA 序列中预测新基因

在生物学的研究中，一个常用的方法，就是通过比较分析以获取有用的信息和知识。对于一个新发现的序列，我们无法得知用什么序列同它进行比对，而数据库相似性搜索使我们能够从数据库中存在的数十万个序列中挑选出与新发现的序列有关联的可能序列。

3. 电子克隆发现新基因

在脊椎动物中，大量的基因，如人类和老鼠的基因都已经被测定并存入基因库中。这也促使了表达序列标签（EST）工程的产生。EST 是对应于某一种 mRNA 的一个 cDNA 克隆的一段序列。主要用途是在数据库搜索中，用 EST 片段进行 cDNA 克隆以分离出感兴趣的基因，包括其他模型生物中的同源基因。关于 EST 技术应用的首次报道是 Adams 等从三种人脑组织 cDNA 文库随机挑取 609 个克隆进行测序，得到一组人脑组织的 EST，生物信息学分析结果表明其中 36 个代表已知基因，337 个代表未知基因。另外，以网络数据库中已有的疾病生物学信息为基础，建立高通量 EST 分析平台可以寻找肿瘤差异表达基因。

二、蛋白质结构、功能的预测

随着功能基因组及蛋白质组学研究技术的发展，产生了许多蛋白质相关数据库及分析软件，如 ExPASy。ExPASy 工具包中的 AAComp Ident 程序能根据氨基酸组成辨识蛋白质，提供氨基酸组成、蛋白质的名称、等电点和相对分子质量以及它们的估算误差，所属物种或物种种类或全部标准蛋白的氨基酸组成等信息。ExPASy 工具包中的 AACompSim 可用于发现蛋白质间较弱的相似关系。数据库的相似性搜索是确定蛋白质功能最可靠的方法，一个显著的匹配应至少有 25％的相同序列和超过 80 个氨基酸的区段。目前，已有 BLASTP 等多种数据库搜索工具，BLASTP 搜索速度快，能快速发现匹配良好的序列。当 BLASTP 不能发现显著匹配时，可利用搜索速度较慢但灵敏度高的工具软件，如 FASTA 进行搜索，如果 FASTA 也无法得到相应结果，则可选用根据 Smith-Waterman 算法设计的搜索程序，如 BLITZ。

三、基因调控网络的研究

在生命科学研究中，人们发现对细胞某种生物学行为的影响涉及多种基因的表达调控与网络调节问题。随着分子生物学及高通量分析技术的发展，产生了大量细胞生物学行为调控的基因表达数据并形成了不同的数据库，相关的分析手段也随之诞生。目前常用的利用生物信息学研究基因调控网络的方法：①进行多物种直系同源基因调控区的比较分析，发现保守的序列，进而确认新的顺式作用元件，并与已知的顺式作用元件比较、检验其结果，此类方法被广泛应用并取得了较好的结果。例如，Stavrinides 等对人、小鼠、大鼠和狗进行了多物种全基因组联配分析，结果中包括了大部分已知的转录因子结合位点和尚未确定的保守基序；②基于已有调控元件知识，收集实验研究确认的顺式作用元件和反式作用因子，总结其特征，依据结构特征类似则功能类似的原则来发现新的调控元件。这类方法发展出预测已知转录因子可能结合位点的主流数据库和预测方法。

四、生物信息学在临床疾病诊治中的作用

一旦明确了人类全部基因在染色体上的位置及其序列特征，包括单核苷酸多态性（single nucleotide polyrnorphism，SNP）及其表达规律和产物 RNA、蛋白质的特征以

后，人们就可以有效了解各种疾病发生的分子机制，进而发展适宜的诊断和治疗手段；或对疾病的潜在基因进行预测和优先排序。通过构建肿瘤高表达 cDNA 基因文库，用生物信息学软件分析差异表达的基因来揭示肿瘤发生的分子水平变化，为寻找新的治疗靶基因提供线索。现在普遍认为 SNP 研究是人类基因组计划走向应用的重要步骤。这主要是因为 SNP 将提供一个强有力的工具，用于高危群体的发现、疾病相关基因的鉴定等方面的研究。

五、生物信息学在药物开发中的应用

传统药物开发方法耗时长、成本高。生物信息学方法为药物研制提供了更多的、潜在的靶标，使现代药物开发模式发生了巨大的变化，药物开发的过程也明显缩短。生物信息学可用于药物靶标基因的发现和验证。有许多数据库可用来获得在正常或疾病状态下不同组织基因表达的差异，进而得到候选基因作为药物靶标，特异性地针对某一种疾病。另外，还可根据蛋白质功能区和三维结构的预测来对药物靶标进行鉴定，以便早期了解所研究蛋白质的属性，预测它是否适用于药物作用。生物信息学利用功能基因组学、蛋白质组学等学科所提供的丰富的数据资源以及开发出来的一些算法软件，可快速实现对靶标的识别。

总之，生物信息学作为一门综合系统科学，可发挥其独特的桥梁作用和整合作用。它以数学和统计学的方法，挖掘数据和模式识别，或利用临床数据库、基因型-生物表现型关系数据库，研究生物医学和进行基因体功能分析，使人们能够从各生物学科众多分散的观测资料中，获得对生物学系统和生物学过程运作机制的理解，最终达到自由应用于实践的目的。应用生物信息学研究方法分析生物数据，提出与疾病发生、发展相关的基因或基因群，再进行实验验证，是一条高效的研究途径。生物信息学已广泛地渗透到医学的各个研究领域中，在疾病相关基因的发现、疾病临床诊断、疾病的个体化治疗、新的药物分子靶点的发现、创新药物设计以及基因芯片的设计与数据处理等医学应用研究方面将发挥重要作用。

第三节　常用生物信息学数据库介绍

生物信息学是一门研究生物系统中信息现象的学科。但目前的生物信息学基本上只是分子生物学与信息技术（尤其是因特网技术）的结合体，因为现在的生物技术的迅速发展不可否认是以计算机为主的信息技术直接参与的结果。生物信息学的研究材料和结果就是各种各样的生物学数据，其研究工具是连接网络的各种计算机，研究方法包括对生物学数据的搜索（收集和筛选）、处理（编辑、整理、管理和显示）及利用（计算、模拟）。

随着生物信息学的发展，生物信息学数据库的数量在不断的递增，内部结构也日趋复杂，其功能也在不断的细化。数据库种类繁多，分类方法也各有不同。依据数据类型可分为核酸序列数据库、蛋白质序列数据库、三维分子结构数据库。按数据层次可分为基本数据库、复合数据库、二次数据库。按其来源分为原始数据库、衍生数据库、集成数据库和知识数据库。这些数据库已经成为分子生物学、生物医药工业、农业科学和环

境科学等学科的必不可少的重要成分。本节主要介绍几个常用的生物信息学数据库。

一、核酸数据库

为了更高效的提取信息，欧美国家及日本等相继成立了生物信息资源和研究中心。美国的国家生物技术信息中心（NCBI）成立于 1988 年 11 月 4 日，1992 年 10 月开始负责维护 GenBank 核酸序列数据库，同时也维护一系列其他的相关数据库，如在线孟德尔遗传数据库（OMIM）、人类基因组基因图谱数据库等。其中 GenBank 与欧洲生物信息研究所（EBI）的欧洲分子生物学实验室（EMBL）核酸序列数据库、日本国立遗传学研究所（National Institute of Genetics）的日本 DNA 数据库（DDBJ）一起构成了国际核酸序列数据库合作计划。

（一）GenBank

GenBank 包含了所有已知的核酸序列和蛋白质序列，以及与它们相关的文献著作和生物学注释。它是由 NCBI 建立和维护的。GenBank 每天都会与 EMBL 数据库、DDBJ 数据库交换数据，使这三个数据库的数据同步。GenBank 的数据可以从 NCBI 的 FrP 服务器上免费下载完整的库，或下载积累的新数据。NCBI 还提供广泛的数据查询、序列相似性搜索以及其他分析服务，用户可以从 NCBI 的主页上找到这些服务。

GenBank 里的所有数据记录被划分在若干个文件里，如细菌类、病毒类、灵长类、啮齿类，以及 EST 数据、基因组测序数据、大规模基因组序列数据等 16 类，其中 EST 数据等又被各自分成若干个文件。

（二）EMBL 核酸序列数据库

EMBL 核酸序列数据库由 EBI 维护的核酸序列数据构成，由于与 GenBank 和 DDBJ 的数据合作交换，它也是一个全面的核酸序列数据库。该数据库由 Oracal 数据库系统管理维护，查询、检索可以通过因特网上的序列提取系统（SRS）任务完成。向 EMBL 核酸序列数据库提交序列可以通过基于 Web 的 WEBIN 工具，也可以用 Sequin 软件来完成。

（三）DDBJ 数据库

DDBJ 数据库创建于 1984 年，由日本国立遗传学研究所遗传信息中心维护。它首先反映日本所产生的 DNA 数据，同时与 GenBank、EMBL 合作，互通有无，同步更新，每年四版。DDBJ 也是一个全面的核酸序列数据库。可以使用其主页上提供的 SRS 工具进行数据检索和分析。可以用 Sequin 软件向该数据库提交序列。

（四）GDB

人类基因组数据库（GDB）是人类基因图谱和疾病的数据库。GDB 的目标是构建关于人类基因组图谱和测序。目前 GDB 中有人类基因组区域〔包括基因、克隆、扩增引物 PCR 标记、断点、细胞遗传标记（cytogenetic marker）、易碎位点（fragile）、EST 序列、综合区域（syndromic ons）、重叠片段（contig）和重复序列〕；人类基因组图谱（包括细胞遗传图谱、连接图谱、放射性杂交图谱、库重叠图谱和综合图谱等）；人类基因组内的变异（包括突变和多态性，等位基因频率数据）。GDB 数据库以对象模型来保存数据，提供基于 Web 的数据对象检索服务，用户可以搜索各种类型的对象，并以图形方式看基因组图谱。

二、蛋白质序列数据库

随着人类基因组计划的不断深入以及测序技术的不断进步，蛋白质序列信息也成指数级增长，蛋白质序列数据库就是主要以蛋白质的一级结构作为数据源，并辅以序列来源、序列发布时间、序列参考文献、序列特征等内容加以注释，最终形成数据文件，存放于数据库。目前规模较大的综合型蛋白质序列数据库有 PIR、SWISS-PROT、PROS-ITE 等。

（一）PIR 和 PSD

PIR 是蛋白质信息资源（protein information resource）的缩写。这是一个国际蛋白质序列数据库，它包含所有序列已知的自然界中野生型蛋白质的信息。此库的主要目的是提供按同源性和分类学组织的综合的、非冗余的数据库，其中包括来自几十个完整基因组的蛋白质序列。所有序列数据都经过整理，超过 99% 的序列按蛋白质家族分类。国际蛋白质序列数据库（PSD）是由美国华盛顿的全国生物医学研究基金会（NBRF）所支持的，慕尼黑蛋白质序列信息中心（MIPS）和日本国际蛋白质序列数据库（JIP-ID）共同维护的国际上最大的公共蛋白质序列数据库。PSD 的注释中还包括对许多序列、结构、基因组和文献数据库的交叉索引，以及数据库内部条目之间的索引。每季度都放行一次完整的数据库，每周可以得到更新部分。

（二）SWISSS-PROT

SWISS-PROT 是经过注释的蛋白质序列数据库，由欧洲生物信息研究所（EBI）维护。数据库由蛋白质序列条目构成，每个条目包含蛋白质序列、引用文献信息、分类信息、注释等，注释中包括蛋白质的功能、转录后修饰、特殊位点和区域、二级结构、四级结构、与其他序列的相似性、序列残缺与疾病的关系、序列变异等信息。SWISS-PROT 中尽可能减少了冗余序列，并与其他 30 多个数据库建立了交叉引用，其中包括核酸序列库、蛋白质序列库和蛋白质结构库等。利用序列提取系统（SRS）可以方便地检索 SWISS-PROT 和其他 EBI 的数据。SWISS-PROT 只接受直接测序获得的蛋白质序列，序列提交可以在其 Web 页面上完成。北京大学生物信息中心有 SWISS-PROT 镜像，可以通过检索工具 SRS 查询。

（三）PROSITE

PROSITE 由专家根据生物知识审编的、SWISS-PROT 蛋白质序列中有生物意义的位点、模式和轮廓的数据库。涉及的序列模式包括酶的催化位点、配体结合位点、与金属离子结合的残基、二硫键的半胱氨酸、与小分子或其他蛋白质结合的区域；除了序列模式之外，PROSITE 还包括由多序列比对构建的 profile，能更敏感的发现序列与 profile 的相似性。PROSITE 的主页上提供各种相关检索服务。

三、蛋白质结构数据库

将通过实验研究，如基于 X 射线和磁共振（NMR）分析所获得的关于蛋白质、酶、病毒、糖类和核酸的晶体结构数据收集起来，就形成了生物大分子的结构数据库。虽然其中序列的数量远比不上蛋白质序列数据库，但其数据量也显然在呈指数增长。

（一）PDB

蛋白质数据库（protein data bank，PDB）由美国 Brook haven 国家实验室建立。PDB 收集的数据来源于 X 射线晶体衍射和磁共振实验测定的生物大分子三维结构数据，经过整理和确认后存档而成，是国际上唯一的生物大分子结构数据档案库。RCSB 的主服务器（research collaboratory for structure bioinfomatics 负责管理）和世界各地的镜像服务器提供数据库的检索和下载服务。

（二）SCOP

蛋白质结构分类（structural classification of protein，SCOP）数据库详细描述了已知蛋白质结构之间的关系。分类基于若干层次：家族，描述相近的进化关系；超家族，描述远源的进化关系；折叠子，描述空间几何结构的关系；折叠类，所有折叠子被归于全 α、全 β、α/β、α+β 多结构域等几大类。SCOP 还提供了一个非冗余的 ASTRAIL 序列库，这个库通常被用来评估各种序列的比对算法。此外，SCOP 还提供一个 PDB-ISL 中介序列库，通过与这个库中序列的两两比对，可以找到与未知结构序列远缘的已知结构序列。

（三）CATH

CATH 数据库是一个新的对蛋白质结构域进行等级分类的数据库，它通过半自动的方法对 Brook haven 蛋白质数据库中的单一或者多结构域蛋白结构进行等级分类，非蛋白质结构、模型以及纯 alphac 结构都没归在 CATH 中，而且收集的蛋白质晶体结构或者磁共振结构的分辨率要求＜0.3mm。分类按照 4 个水平：簇［class（C）］、构件［architecture（A）］、拓扑结构［topolo（T）]和同源超家族［homologous supefamily（H）］。

（四）FSSP

结构相似蛋白家族（families of structurally similar protein）基于 PDB 数据库中现有蛋白质三维结构，用自动结构对比程序 Dali 逐一比较而形成的折叠单元和家族分类库。它以 PDB 非冗余数据库作为数据源，进行彻底、全面的三级结构比较，而且数据库的升级以及维护都是 DALL 搜索引擎支持的。此库在 PDB 库每次新版后自动更新。

（五）MMDB

蛋白质模型数据库（molecular modeling database，MMDB），由 NCBI 的 MMDB

组维护。这是 Entrez 检索工具所使用的三维结构数据库，它以 ASN.1 格式反映 PDB 库中的结构和序列数据，引文连接到 MEDLINE。MMDB 有一个配套的三维结构显示程序 Cn3D。

综上所述，世界著名的一级核酸数据库有 GenBank、EMBL 和 DDBJ 等；蛋白质序列数据库有 SWISS-PROT、PIR、UNIPROT 等；蛋白质 X 射线晶体三维结构数据库有 PDB 等。国际上二级生物学数据库非常多，针对不同的研究内容和需要而各具特色，如 GDB、TRANSFAC、SCOP 等。

生物信息数据库的发展是十分惊人的，但也存在诸多问题。大多数数据库对于数据的创新、精确性和准确性没有权威评价，数据过多、重复、分类较粗等。因此需要生物信息学专家们在数据库结构设计、数据处理、数据提取、数据的重新组合、专一性等几方面进行更进一步的完善。我国的生物信息学数据库也蓬勃发展起来。北京大学于1997 年 3 月成立了生物信息学中心，华大基因研究中心是我国目前测序能力最强的单位，中山大学生物信息中心与法国巴斯德研究所合作于 1999 年 9 月开通了"法国巴斯德亚洲研究网"，中国科学院上海生命科学研究院也于 2000 年 3 月成立了生物信息学中心，分别维护着国内两个专业水平较高的生物信息学网站。但是，我国尚未形成比较完整有效地生物信息数据库系统，现有的数据库的质量也有待提高，服务有待改善。随着我国经济的快速增长，在科研方面的投入也会大大提高，这些情况将进一步得到改善和提高。相信在不久的将来，更加完备，整合充分的生物信息学数据库将为我国生物学、医学等提供更加便捷的服务，从而有力推进我国生命科学的进一步发展。这一美好未来值得期待。

小　结

生物信息学是现代生命科学与计算机科学、数学、物理科学、化学等领域相互交叉而形成的一门新兴学科。它通过对分子生物学实验数据的获取、加工、存储、检索与分析，进而达到揭示这些数据所蕴涵的生物学意义的目的。随着人类基因组计划的不断推进，生物信息学已经成为当今生命科学和自然科学的核心领域和最具活力的前沿领域之一。人类对基因的认识，从以往对单个基因的了解，上升到在整个基因组水平上考察基因的组织结构和信息结构，考察基因之间在位置、结构和功能上的相互关系。这就要求生物信息学在一些基本的思路上要做本质的观念转变。和其他学科一样，生物信息学也需要理论和实践的紧密结合。在实际运用中掌握生物信息学的相关技术，从而为生物学研究服务。生物信息学基于分子生物学与多种学科交叉而成的新学科，但并非各种学科的简单堆砌，相互之间还需要更加紧密的联系。

本章首先结合人类基因组计划介绍生物信息学的历史发展和概况，然后介绍了生物信息学的主要研究内容，包括生物信息数据库和生物信息学主要问题、生物序列的相似性比较及其数据库搜索、基因预测、基因组进化和分子进化、蛋白质结构预测。此外，本章还介绍了生物信息学在生物学方面的应用，最后介绍了一些常用的生物信息学数据库。在处理大规模数据方面，需要行之有效的一般性方法；而对于大规模数据内在的生成机制也还需要进一步探索和明确，这使得充分发挥生物信息学的作用还需要科学家们做进一步的努力。要想得到真正的解决，最终不能从计算机科学得到，而需要从生物学自身，从数学上的新思路来获得本质性的动力。

毫无疑问，正如 Dulbecco 1986 年所说："人类的 DNA 序列是人类的真谛，这个世界上发生的一切事情，都与这一序列息息相关。"但要完全破译这一序列以及相关的内容，我们还有相当长的路要走。通过本章的学习，可以为生物学研究提供一个全新的思维视角，从而达到加速生物学发展的目的。

（左泽华）

参 考 文 献

杨咏梅. 2007. 生物信息学在医学研究中的应用进展. 医学综述, 13 (22)：1681～1683

张晓东. 2006. 生物信息学数据库研究进展. 生物信息学, 4 (3)：143～145

Eleanor J, Whitfield et al. 2006. Bioinformatics database infrastructure for biotechnology research. Journal of Biotechnology, 124：629～639

http：//en. wikipedia. org/

Kanehisal M, Bork P. 2003. Bioinformatics in the post-sequence era. Nature Genetics, 33：305～310

Rhee S Y. 2006. Bioinformatics and Its Applications in Plant Biology. Annu Rev Plant Biol, 57：335～360

Saeys Y. 2007. A review of feature selection techniques in bioinformatics. Bioinformatics, 23 (19) 2507～2517

Chapter 11   The Application and the Basic Principles of Bioinformatics

The twenty-first century is the century of life sciences and the milestone is well-known Human Genome Project, which spends 13 years and at a cost of billions of dollars. Because it will ultimately reveal the structure of the human body and establish a solid data foundation; The Human Genome Project and various model organisms genome project is an important part of our understanding of the function on the human genome. All of these are related to the large amount of data processing, and the amount of data is also sharply increased at a unprecedented speed in the history of science. These factors indicate that only through traditional means of research is of no avail, biology is no longer the only based on the scientific test observation, theory and computing will play an increasingly significant role, and through large-scale computer equipment for mass data processing has become a necessary. Bioinformatics as a new subject areas came into being in such a context.

Bioinformatics can be divided into two basic categories: biological information management and computational biology. The first category, that is, the development of a variety of tools and methods of calculation, as possible as they can to full use of these bio-medical data, including access, description, storage, analysis and visualization of data. The latter category refers to the study of the development and application of data analysis and theoretical modeling data, as well as the use of computer simulation technology, biology, behavior and social system of research, which reveals the complexity of the data contained in the mysteries of biology. This chapter aims to introduce the general concepts, principles and methods of bioinformatics, as well as the application of the basic biology; some of common database will be introduced at last.

Section 1   Outline of Bioinformatics

1. 1   The Conception of Bioinformatics and it's Emerged

What is bioinformatics: Bioinformatics is a new cross-disciplinary emerged with the Human Genome Project at the late of 80's and often referred to genome informatics at first. How is the term of bioinformatics come from? In the late of 1980s, Dr. Lin Huaan was aware of the importance of combing the computer science with the biology, and then, he began to pay attention to this area and got the idea of a proper name. At first, because of having to take the supercomputer calculation institute into his account for this institute will support him to host a series of bioinformatics conference, he used the "CompBio"; in the near future, he further change it as the "bio-informatics (or bio/informatics)". But the e-mail system at that time is different from today's, with the name of _ or / symbols would often lead to many system problems, so it were removed by Dr. Yu, "bioinformatics" started to be used then, from then on, a new, full swing, and tremendous potential for development of bioinformatics quietly and raised up and development.

As far as the birth of Bioinformatics be concerned, it is a rather old-fashioned discipline, because the first "information theory biology seminars" had been hold at Gatlinburg of Tennessee during start-up phase of computer in 1956; but as for its development was concerned, it was a relatively young discipline yet, because after more than 20 years of stasis, it was able to acquire their own development only accompanied by 80~90s rapid development of computer technology.

The birth and development of bioinformatics is meet the times' needs and it is a historical necessity, and quickly penetrates into every aspect of the biological sciences, as people aware of its existence before they cannot do without it! Bioinformatics has become a powerful driving force of disciplines development on bio-medicine, agronomy, genetics, cell biology and other disciplines as well as an important part of drug design, environmental monitoring. It is one of the major fronts of the contemporary life sciences and natural sciences and will be one of the core areas of natural sciences in the 21st century as well.

1. 2　Bioinformatics Development and Maturity

The international community has attached great importance to the development of bioinformatics, a variety of professional research institutions and companies have sprung up out of, the number of bioinformatics departments in biotechnology companies and pharmaceutical industry is also increasing. From the database perspective, back in the 60s, the United States had been established a protein database base on manual collection. Los Alamos National Laboratory has been established the GenBank database in 1979, the European Molecular Biology Laboratory has been provided on the EMBL Nucleotide Sequence Database of service in 1982, and Japan established a national database of nucleic acid sequence (DDBJ) in 1984, which provided services in 1987.

The United States established the National Center for Biotechnology Information (NCBI) in support of the Congress in 1988, the purpose was to calculate the basis of molecular biology research, build and distribute molecular biology databases; Europe was embarking on the establishment of a European Bioinformatics Institute (EBI) in March 1993, Japan, in April 1995, set up their own information Biology Center (CIB). At present, most of the nucleic acid and protein database by the United States, Europe and Japan have three database systems; they jointly formed the DDBJ/ EMBL/GenBank International Nucleotide Sequence Database, they exchanged data daily, synchronization update. Some other countries such as Germany, France, Italy, Switzerland, Australia, Denmark and Israel, in the sharing of network resources, meanwhile, respectively, have their own bioinformatics institutions, secondary or higher level specialized databases with their own characteristics, as well as their own analysis technology, and service in their own biological (medical) research and development, some services are also open in the world.

Domestic bioinformatics is also increasing emphasis on the field, under the leadership of some well-known academicians and professors, some achieved certain performance record in their respective fields, and some also has a place in the international arena, but from the overall point of view, the gap of the national and international level is tremendous.

Bioinformatics is not only a scientific disciplines, it is an important tool for research and development. As early as in 1962, Zuckerkand and Pauling correlate sequence variation analysis and its evolution, which opened up a brand-new areas of research on molecular evolution; Davies created a study of protein structure prediction in 1964; Needleman and Wunsch published important two sequence comparison algorithm in 1970; the first time, Ratner used the theoretical methods of molecular genetics regulation and control system for processing and analysis in 1974; in 1975, Pipas and McMahon were first proposed using computer technology to predict RNA secondary structure; With a large number of biology data analysis techniques emerged in 1976, Science have been published a review on the calculating molecular biology in its 209 volumes of 1980; as we now see, biological data analysis techniques have a rapid development in foreign countries during 80~90s.

Human Genome Project and the model organism genome sequencing of the work completed shows that the only means of bioinformatics analysis, we can obtain a correct understanding of the genome, it can be said bioinformatics flourished from the Human Genome Project, because the Human Genome Project created a huge space for Bioinformatics at first time to display its virtuosity.

At the initial development stage of Bioinformatics, for the absence of the specialized journals, the professional literature initially scattered in various journals in other fields. By 1970, a relevant journal, *Computer Methods and Programs in Biomedicine* was emerged. The first bioinformatics professional journals, *Computer Application in the Biosciences*, appeared till April of 1985; now, we can see a lot of professional journals, including the paper periodicals and online journals, such as the *Bioinformatics* (formerly *Computer Applications in the Biosciences*), *Acta Biotheoretica*, *Bio Informatics Technology & Systems*, *Bioinform Newsletter*, *Briefings in Bioinformatics* and *Journal of Computational Biology* and etc.

1. 3　The Scope of Bioinformatics Research

From a broad point of view, Bioinformatics is discipline to study the phenomenon of life, organize and analyze the exponential increasing biological data by using the opinion, theory and methods of mathematical theory and information science.

First of all is to study the carrier of genetic material DNA and its encoded macromolecular protein, make the computer as its main tool for the development of software; collecting, collating, storage, release, extracting, processing, analysis and research on the tremendous amount of vast of DNA and protein sequence and structure, in order to deciphering the hidden in the DNA sequence of the genetic language by the analysis methods to realize the origin of life, evolution, genetics and the nature of development. Reveal the human physiological and pathological process of the molecular basis and offer the most reasonable and effective methods or pathways for human disease prediction, diagnosis, prevention and treatment.

Their research mainly focus on two aspects including genomics and proteomics, specifically, from the nucleic acid and protein sequence, analyze the biological information in the sequence which expressed structure and function.

At present, genomics research appeared several centrobaric transferring: firstly, functional genomics research with making the known gene's sequence and function correlated. Secondly, make a map-based gene isolation shift to sequence-based gene isolation. Thirdly, the research transferred from study the origin of disease to exploring the causes of disease pathogenesis. The fourth, disease diagnosis shifted to the study of disease susceptibility. The application of Biochip will provide the most basic and necessary information and the basis, it will become a major technical support of genomic informatics research. The development of bioinformatics provided an opportunity for further breakthroughs in life sciences and revolutionary change on drug development process. As the human genome, got the sequence is only the first step, the next work is fulfilling the task of so-called post-genome era, that is, the collection, collation, retrieval and analysis of expressed protein structure and function information of sequence, to find out the laws. There are three categories: ① Database establishment and optimization. Internationally renowned public database: GenBank, SWISS-PORT, PIR, PDB and some other companies have their internal database. ② Cultivating bioinformatics professionals. ③ Theoretical studies on the database software development and alignment), the new sequence of identification and prediction.

Section 2 Bioinformatics Applications in Biomedicine

2.1 The New Gene Discovery and Function Prediction

Discovery of new genes is the top priority of genome research, and the use of bioinformatics methods is an important means of discovery of the new genes. The discovery of new genes not only has far-reaching implications for human understanding of the world and aware of other species, but also brings about new opportunities for the treatment of certain Difficult and Complicated diseases. As results of bioinformatics have developed very rapidly, a làrge amount of data will generate at any time. Under such circumstances, it calls for the establishment of a huge database of information to be able to support the rapid, high-volume data queries and updates and analysis in order to clarify and understand the large amount of data contained in biological significance.

2.1.1 Predict New Genes From the Genome Sequence

This approach is essentially distinguish encoding proteins regions from non-coding proteins in the genome, and compares these sequences with the known gene database, you can discover new genes.

2.1.1.1 Predict new genes through multiple sequence alignment from the genomic DNA sequence

A common method in biology research is through a comparative analysis of access to useful information and knowledge. For a newly discovered sequence, we do not know what sequence to compare with, the database similarity search allows us to select sequence associated interested ones from the hundreds of thousands database.

2. 1. 1. 2　New genes found by Electronic cloning

In vertebrates, a large number of genes, such as human and mouse genes have been determined and deposited in the gene bank. It has stimulated expressed sequence tags (EST) project generated.

Expressed Sequence Tags is a section of sequence of cDNA cloning which correspond to a certain mRNA. The main use is in the database search, using EST fragments clone cDNA to isolate the interest gene, including other homologene of the models biology. The first reported on the EST technology is ADAMS randomly picked 609 clones sequenced from three kinds of human brain cDNA library, they got a series of ESTs of human brain tissue, bioinformatics analysis showed that 36 of them represent known genes, 337 of them on behalf of unknown genes. In addition, on the basis of biological information of existing disease at web database, the establishment of high-throughput EST analysis platform can find tumor differentially expressed genes.

2. 2　Protein Structural and Functional Prediction

With functional genomics and proteomics research technology development, a number of protein-related databases and analysis software, such as ExPASy, are generated. AA Comp Ident program of ExPASy toolkit can identify protein based on amino acid composition; provide amino acid composition, the name, isoelectric point and molecular weight of protein as well as their estimates of error, their species or species category or amino acid composition of all standard protein, etc. AACompSim in ExPASy toolkit can be used to found a weak similar relationship between proteins. Database similarity search is the most reliable method to determine the function of proteins; a remarkable match should be at least 25% of the same sequence of and exceed 80 amino acids sections. At present, there are various BLASTP database search tool, the BLASTP search is fast, it can quickly find matching sequences. When BLASTP cannot found the significant match, it can use some slow but highly sensitive search software, such as the FASTA to search, if the FASTA cannot get results, we can choose search procedures designed according to Smith-Waterman algorithm, such as BLITZ.

2. 3　The Studies of Gene Regulatory Networks

In life science research, it was found influence on cells' certain biological behavior involves in a variety of gene expression control and network regulates issues. With the development of molecular biology and high-throughput analysis technology, resulting in a large number of gene expression data in cell biology behavior regulations and formed a different database, the relevant analytical tools also emerged.

At present, there are two ways to study gene regulatory networks by the use of common bioinformatics: comparative analysis of multi-species orthologous gene regulation and found conservative sequence, then confirm the new cis-acting element, by comparing with the test results of the known cis-element, such methods have been widely used and have achieved good results.

For example, Stavrinides had a multi-species genome-wide matched analysis in human, mice, rats and dogs, the results was including most of the known transcription factor binding sites and conservative motif which has not yet been determined; Based on existing regulatory elements knowledge, collect experimental studies to confirm the cis-and trans-acting element and sum up its characteristics, according structural characteristics similar to the function of similar principles to discover new regulatory elements. Such methods can develop to predict the known transcription factor, mainstream databases and forecasting methods of possible binding sites as well.

2. 4　The Role of Bioinformatics Diseases in Clinical Diagnosis and Treatment

Once identify all human genes' location in the chromosome and sequence features, including single nucleotide polymorphisms (SNP) and its rules of expression and the future of the product of RNA, proteins characteristic, people will be able to effectively understand the molecular mechanism of a variety of diseases, then develop appropriate means of diagnosis and treatment; predicted and preferential ordered the potential disease gene; summarized different bioinformatics strategy; by constructing overexpressing tumor cDNA library, and then through bioinformatics software analyze the differentially expressed genes to reveal the tumor changes at molecular level, provide clues for find a new target of gene therapy. It is now widely recognized that SNP research is an important step of the Human Genome Project making for the application. This is mainly because the SNPs will provide a powerful tool for the discovery of high-risk groups, disease-related genes identified, and pharmacological research as well.

2. 5 Bioinformatics in Drug Development Applications

Traditional medicines to find new drugs are time-consuming and costly. Bioinformatics methods will provide more potential targets for drug development. Bioinformatics can be used for drug target gene discovery and validation. There are many databases can be used to obtain the differences of gene expression in different organizations which in the normal and disease state.

By searching these databases, candidate gene can be obtained as a drug target, specifically aim at a certain disease. In addition, according to protein functional areas and three-dimensional structure prediction to identify drug targets in order to understand the properties of proteins at early stage and predict whether it should be applied in drugs. By using functional genomics, proteomics and other subjects provided rich information resources, as well as a number of algorithms software; Bioinformatics can quickly realize the target identification. Thus, modern drug development model is undergoing dramatic changes, shortening the process of looking for drugs.

In short, bioinformatics, as an integrated system of science, can play its unique role as a bridge and integration. Using mathematics and computing methods, ongoing the data mining and pattern recognition, or using clinical databases, genotype -performance-based relational database of biological and genetic structure of three-dimensional modeling of biomedical research and analysis of gene function, allowing people understand the biological systems and the operating mechanism of the biological process from the multi-disciplines biological scattered observational data, and ultimately to achieve the purpose of freedom applies to practice. Application of bioinformatics research methods analysis of biological data, with the disease occurrence and development of related genes or gene cluster, further experimental verification, is a highly efficient research pathway. Bioinformatics has been widely infiltrated into the whole medical research field, in the disease-related gene discovery, disease diagnosis and treatment of individual diseases, the new molecular targets for drug discovery, drug design and innovation in the design of gene chips and data-processing applications such as medical research will play an important role.

Section 3 Introduction of the Commonly Used Bioinformatics Database

Bioinformatics is ultimately a study of the phenomenon of biological systems of information subjects. However, bioinformatics is basically a molecular biology and information technology (especially Internet technology) combination, as the current rapid development of biotechnology cannot be denied as a result of computer-based information technology directly involved in the outcome. Bioinformatics research materials and resulted in a wide range of biological data, its research tools are a variety of online computers, research methods, including the search for biological data (the collection and screening), processing (editing, arrangement, management and display) and utilization (computing, simulation).

With the development of bioinformatics, bioinformatics databases constantly increasing number of internal structure of the growing complexity of its functions are also constantly refining. A wide range of databases, classification methods are different. Based on the data can be divided into sub-types: nucleic acid sequence database, protein sequence database, three-dimensional molecular structure of the database. According to the data can be divided into sub-levels: basic databases, complex databases, and second database. According to their sources can be divided into original database, derivative database, integrated databases and knowledge databases. These databases were essential component of disciplines such as molecular biology, bio-pharmaceutical industry, agricultural science and environmental science et al. This section mainly introduces a number of commonly used bioinformatics databases.

3. 1 Nucleic Acid Database

In order to extract information more efficient, Europe American, Japan and other countries have set up a bio-information resources and research center. United States National Center for Biotechnology Information (NCBI), it was established in November 4, 1988, which was starting to be responsible for the maintenance of GenBank nucleic acid sequence database in October. Meanwhile, it maintains a range of other relevant databases, such as online database of Mendelian Inheritance in Man (OMIM), the human genome database. GenBank with the European Bioinformatics Institute (EBI) and the European Molecular Biology Laboratory (EMBL) nucleic acid databases, Japan's National Institute of Genetics (National Institute of Genetics) Japanese DNA database (DDBJ) together constitute the International Nucleotide Sequence Database Partnership Program.

3. 1. 1   GenBank

GenBank database contains all known nucleic acid sequence and protein sequence, and with them the relevant literature and biology notes. It is from the establishment and maintenance of NCBI. Genbank exchange data every day with the European Molecular Biology Laboratory (EMBL) database and Japan's DNA database (DDBJ), so that these three data synchronization. Genbank data can free download complete library available from the NCBI server FrP, or download the accumulation of new data. NCBI also provides a wide range of data query, sequence similarity search and other analytical services; users can find these services from the NCBI's home page.

All the data records in Genbank is divided in several documents, such as bacteria, viruses, class, primates, rodents, as well as EST data, genome sequencing data, a large-scale genomic sequence data and other 16 categories, including EST data which was divided into a number of their documents.

3. 1. 2   EMBL Nucleotide Sequence Database

EMBL Nucleotide Sequence Database maintained by the EBI nucleotide sequence data structure, it is a comprehensive database of nucleic acid sequence with the Genbank and DDBJ data exchanged. The database was managed and maintained by the Oracal database system, query retrieval via the Internet on the serial extraction system (SRS) to complete mission. Nucleotide Sequence Database sequence can be submitted to the EMBL through the Web-based WEBIN tool and can also be used by the sequin software.

3. 1. 3   DDBJ Database

DDBJ database was founded in 1984 by Japanese National Institute of Genetics information Center for the maintenance. First of all, it reflects Japan's DNA data generated at the same time with the Genbank, EMBL cooperation and exchange of needed goods, synchronous updating, the annual Fourth Edition. DDBJ is also a comprehensive database of nucleic acid sequence. Can use its home page of the SRS provides tools for data retrieval and analysis. Sequin software can be used submitting sequences to the database.

3. 1. 4   GDB

Human Genome Database (GDB) is the human genome and disease databases. GDB's goal is to build on the human genome mapping and sequencing. At present, GDB contains: the human genome (including genes, cloning, amplimers PCR markers, breakpoint, cytogenetic markers, fragile sites, EST sequences, the syndromic onsites, contigs and repeats; human genome (including cytogenetic mapping, connectivity map, radioactive hybrid map, content contigs map and integrated map, etc.); variation within the human genome (including mutation and polymorphism, allele frequency data). GDB database save data by object model, Web-based data object retrieval service; users can search all types of objects and got the graphical genome map.

3. 2   Protein Sequence Database

With the constant deepening of Human Genome Project plan as well as the sequencing of the continuous advancement of technology, protein sequence information into exponential growth, protein sequence database is mainly in these sequences is the primary structure of protein as a data source, supplemented by sequence sources, Sequence release time, reference sequence, sequence characteristics of the content of the Notes, and ultimately the formation of data files, stored in a database. At present, large-scale integrated-type protein sequence database: PIR, SWISS-PROT, PROSITE such.

3.2.1 PIR and PSD

The Protein Information Resource (PIR) is an integrated public resource of protein informatics that supports genomic and proteomic research and scientific discovery. PIR maintains the Protein Sequence Database (PSD), an annotated protein database containing over 283 000 sequences covering the entire taxonomic range. Family classification is used for sensitive identification, consistent annotation, and detection of annotation errors. The superfamily curation defines signature domain architecture and categorizes memberships to improve automated classification. To increase the amount of experimental annotation, the PIR has developed a bibliography system for literature searching, mapping, and user submission, and has conducted retrospective attribution of citations for experimental features. PIR also maintains NREF, a non-redundant reference database, and iProClass, an integrated database of protein family, function, and structure information. PIR-NREF provides a timely and comprehensive collection of protein sequences, currently consisting of more than 1 000 000 entries from PIR-PSD, SWISS-PROT, TrEMBL, RefSeq, GenPept, and PDB. The PIR web site (http://pir.georgetown.edu) connects data analysis tools to underlying databases for information retrieval and knowledge discovery, with functionalities for interactive queries, combinations of sequence and text searches, and sorting and visual exploration of search results. The FTP site provides free download for PSD and NREF biweekly releases and auxiliary databases and files.

3.2.2 SWISSS-PROT

SWISS-PROT data manually read the Notes database of protein sequences, maintained by the European Bioinformatics Institute (EBI) maintenance. Database is constituted by the protein sequence entries, each of which contains the protein sequence, references information, classified information, notes and so on. The Notes contain the function of proteins, post-transcriptional modification, special sites and regions, secondary structure, quaternary structure, similarity as other sequence, sequence defects and disease relations, information of sequence variation. SWISS-PROT reduces redundant sequence and establishes a cross-reference data with other more than 30 data including nucleic acid sequence database, protein sequence databases and protein structure database. The sequence extraction system (SRS) can easily search SWISS-PROT and other EBI data. SWISS-PROT only accepts the protein directly sequenced and the sequence is completed on its Web pages. Peking University Center for Biotechnology Information has SWISS-PROT mirror, you can query through the search tools SRS.

3.2.3 PROSITE

PROSITE is SWISS-PROT protein sequence in accordance with biological knowledge edited by experts and has the biological significance of sites, models and contour database, involving sequential patterns, the enzyme catalytic sites, ligand binding sites, metal ion. Binding residues, the cytokine disulfide bonds, regions combining with small molecules or other protein. Besides sequence model, PROSITE also includes profile built by a multiple sequence alignment, which can find a more sensitive sequence similar as itself. PROSITE's home page provides a variety of relevant search services.

3.3 Protein Structure Database

The collection of experimental study such as proteins, enzymes, viruses, carbohydrates and nucleic acid crystal structure which are obtained by based on x-rays and magnetic resonance (NMR) analysis obtained forms the database of structural and biological macromolecules. Although the number of sequence in such databases is far less than that of protein sequence databases, it grows exponentially.

3.3.1 PDB

Protein Data Bank (PDB) is established by Brook Haven National Laboratory of U.S. Data collected by PDB come from the experimental determination on three-dimensional structure of biological macromolecules determined through x-ray crystallography and nuclear magnetic resonance (NMR). After the collation and confirmation, data are the only international data archive of structure of biological macromolecules. RCSB (Research Collaboratory for Structure Bioinfomatics responsible for the management) and the mirror servers provide the retrieve and download services.

3.3.2 SCOP

Protein Structure Classification (SCOP) database describes in details the relationship between known protein structures. Classification bases on a number of levels: family, description of the similar evolutionary relationship; superfamily, description of the evolution of the relationship between distal; fold, description of the relationship between the geometry of space; folding type, all fold to be attributed to the whole a, full-p, a /, a + f3 and structure of the domain, and several other major categories. SCOP also provides a non-redundant sequence database ASTRAIL which is often used to assess a variety of sequence alignment algorithm. In addition, SCOP also provides a PDB-ISL intermediate sequence library. Through contrasting the library sequences than 22 rights, it is possible to find the known structure of sequences far from the known structure Sobrinus sequence.

3.3.3 CATH

CATH database is a new one classifying the domain of the protein structure. Through the semi-automatic method, CATH classifies single or multi-domain protein structure in Brookhaven database, while non-protein structure, model or structure of pure alpha is excluded, and the resolution requirement of collected protein crystal structure or NMR structure is less than 0.3mm. There are four classifying levels: cluster (class (C)), Component (architecture (A)), topology (topolo (T)) and homologous superfamily (homologous superfamily (H)).

3.3.4 FSSP

Basing on existing three-dimensional structure of proteins in PDB data, structurally similar protein family (families of structurally similar proteins) use an automatic structure comparison program to create Dali folding unit and family classification of the Treasury. The non-redundant PDB database is its data sources, so it conducts a thorough and comprehensive three-tier structure, and database upgrades and maintenance are Dall search engine support. This library in the PDB database automatically updated after each new version.

3.3.5 MMDB

Protein Model Database (Molecular Modeling Database) is maintained by NCBI's MMDB group, which are three-dimensional structure of the database used by Entrs. It reflects PDB database's structure and sequence data by ASN. 1 format and connects to Citation MEDLINE. MMDB has a complementary three-dimensional structure display program Cn3D.

To sum up, the famous nucleic acid database in the world contain GenBank database, EMBL and DDBJ database libraries of nucleic acid, etc.; protein sequence database SWISS-PROT, PIR, UNIPROT, etc. the protein X-ray crystal three-dimensional structure database contain PDB, et. Secondary biology international database is very large and with different research needs of the content and features, such as the human genome database GDB, the transcription factor binding site database TRANSFAC, protein structure classification database SCOP family, and so on.

The development of databases of biological information is alarming, but there also exist many problems. Most of the databases lack the authoritative evaluation on innovation, precision and accuracy, and the data is duplicate and coarse classification. Therefore it demands the bioinformatics experts to be more specific and improving at the database structure design, processing, extraction, and re-combination. Bioinformatics in China also flourishes. Peking University established a Bioinformatics Center in March 1997, Genomics Research Center is Chinese current sequencing capacity of the strongest units, Guangzhou Sun Yat-sen University Center for Biotechnology Information in cooperation with the French Pasteur Institute in September 1999 opened a French Pasteur Research Network in Asia - Chinese Academy of Sciences, Shanghai Institutes for Biological Sciences set up a bioinformatics center in March 2000. Each of them respectively has been maintaining two domestic higher level professional Bioinformatics Web site. However, our country has not yet formed a relatively completive effectively complete biological information database system, and the existing database needs to be improved the quality and service. As Chinese rapid economic growth, investment in scientific research will be greatly improved, further enhancing current condition. In the near future, the more complete, fully integrated bioinformatics database for Chinese biology and medicine will provide more efficient services; thereby effectively promote the further development of life sciences. This is worth looking forward to a bright future.

Summary

Bioinformatics is a new branch of science constituted by cross-cutting of modern life sciences and areas such as computer sciences, mathematics, chemistry, etc. Through the acquisition, processing, storage, retrieval and analysis of the experimental data on molecular biology, bioinformatics can reveal the biological significance implied by these data. Bioinformatics has become a core territory and one of the most viable frontier sciences along with the continuous advancement of Human Genome Project. Realization of genes, from the previous understanding of each single, has now risen to the level of whole genome to study gene information structure and to inspect the interrelation of gene location, structure and function. This requires bioinformatics, like other disciplines, to closely integrate its theory with practice. Only by linking technology and application can bioinformatics service the biological research efficiently. Bioinformatics based on closely correlation, but not simple pile of molecular biology and a variety of other subjects.

This chapter first introduces Human Genome Project and the history, development and overview of bioinformatics. Then it introduces the main researches of bioinformatics, including bioinformatics databases, bio-sequences similarity comparison, gene and protein structure prediction, and genome and molecular revolution. Besides, this chapter also introduced the applications of bioinformatics in biology and some commonly used bioinformatics databases. To deal with large-scale data, it needs some effective general methods; while to further explore and identify the internal generate mechanisms of these large-scale data and to bring into full play of bioinformatics, scientists need make additional endeavor. In that case, the final solution can not be gained from computer science, but from getting essential power of biology itself and new mathematical ideas.

There is no doubt that, as Dulbecco said in 1986, the human DNA sequence is the essence of humanity, everything happened in the world closely related with this sequence. However, we still have a long way to go to decipher the full sequence, as well as its coherent content. This chapter provides a firenew perspective of biology research which may help speed up the development of biology.

(Zehua Zuo)

索　引